Ship Stability for Masters and Mates

To my wife Hilary and our family

Ship Stability for Masters and Mates

C. B. Barrass and D. R. Derrett

ELSEVIER

AMSTERDAM • BOSTON • HEIDELBERG • LONDON • NEW YORK • OXFORD
PARIS • SAN DIEGO • SAN FRANCISCO • SINGAPORE • SYDNEY • TOKYO
Butterworth-Heinemann is an imprint of Elsevier

BH

Butterworth-Heinemann is an imprint of Elsevier
The Boulevard, Langford Lane, Kidlington, Oxford, OX5 1GB
225 Wyman Street, Waltham, MA 02451, USA

First published by Stanford Maritime Ltd 1964
Third edition (metric) 1972
Reprinted 1973, 1975, 1977, 1979, 1982
Fourth edition 1984
Reprinted 1985
Fourth edition by Butterworth-Heinemann Ltd 1990
Reprinted 1990 (twice), 1991, 1993, 1997, 1998, 1999
Fifth edition 1999
Reprinted 2000 (twice), 2001, 2002, 2003, 2004
Sixth edition 2006
Seventh edition 2012

Notices

Knowledge and best practice in this field are constantly changing. As new research and experience broaden our understanding, changes in research methods, professional practices, or medical treatment may become necessary.

Practitioners and researchers must always rely on their own experience and knowledge in evaluating and using any information, methods, compounds, or experiments described herein. In using such information or methods they should be mindful of their own safety and the safety of others, including parties for whom they have a professional responsibility.

To the fullest extent of the law, neither the Publisher nor the authors, contributors, or editors, assume any liability for any injury and/or damage to persons or property as a matter of products liability, negligence or otherwise, or from any use or operation of any methods, products, instructions, or ideas contained in the material herein.

British Library Cataloguing in Publication Data
A catalogue record for this book is available from the British Library

Library of Congress Cataloguing in Publication Data
A catalog record for this book is available from the Library of Congress

ISBN: 978-0-08-097093-6

Printed and bound in the UK
12 13 14 15 10 9 8 7 6 5 4 3 2 1

Contents

PART II Endnotes

PART III Appendices

Acknowledgments

I gladly acknowledge with grateful thanks, the help and time afforded to me by the following personnel in the Maritime Industry:

Captain D.R. Derrett, author of *Ship Stability for Masters and Mates*, 3rd edition (metric) 1972, published by Stanford Maritime Ltd.
Captain Sergio Battera, Vice-Chief (Retired) Pilot, Co-operation of Venice Port and Estuary Authority.
Nick Savvides, Editor of *The Naval Architect*, Royal Institute of Naval Architects, in London.
Alan Sleigh, Head of Testing Services, Scottish Qualification Authority (SQA), in Dalkeith, Midlothian.
Colin Thomas, Chief Examiner, Department for Transport/Maritime and Coastguard Agency (DfT/MCA), in Southampton.
Darren Dodd, Managing Director of Saab Tank Control (UK) in Wokingham.
Christopher Murman, Naval Architect, Floating Point Designz, in Sydney, Australia.
Farhan Saeed, Lecturer in Maritime Studies, at Liverpool John Moores University.
Dr Peter Wright, Senior Lecturer in School of Marine Science and Technology, at Newcastle upon Tyne University.
Captain Chris Heeks, Head Teacher of Maritime Studies Institute, Sydney, Australia.

Preface

This book is written specifically to meet the needs of students studying for their Transport Certificates of Competency for Deck Officers and Engineering Officers and STCW equivalent International qualifications.

It is primarily for Chief Mates and Officers on Watch (Officers in Charge) on board merchant ships. The book will also prove to be extremely useful to other maritime personnel shown listed in the Introduction.

Note

Throughout this book, when dealing with transverse stability, BM, GM, and KM will be mentioned and used in calculations. When dealing with trim, i.e. longitudinal stability, then BM_L, GM_L, and KM_L will be used to denote the longitudinal considerations.

Therefore, there will be no suffix 'T' for transverse stability calculations but there will be a suffix 'L' for the longitudinal stability text and diagrams.

C.B. Barrass

Introduction

In 2006, I updated the 1999 edition of *Ship Stability for Masters and Mates*. At the request of the publishers, I have now refreshed, updated, and hopefully improved on this 2006 edition.

I have introduced further topics in keeping with certain developments within the Maritime Shipping Industry during 2006–2012. At the same time, I have attempted to improve the standard of content and presentation as and when required.

Changes to the previous edition include the following:

- Nine chapters of the 2006 book being deleted and replaced with nine new chapters in this book
- Details of FLNG 'Facility' design of 2011
- Website information for maritime establishments in the UK
- Turning circle diameters in shallow waters
- Triple 'E' container ships of 2011
- Qflex and Qmax LNG vessels of 2007–2012
- Ship squat improvements in research over the period 2006–2012
- Latest examination papers for Chief Mate and Officer On Watch (Officer In Charge) courses
- Thirty possible reasons why the *Herald of Free Enterprise* capsized at Zeebrugge
- Equations and graphs for position of LCB about amidships
- Recent worldwide ship groundings over the period 2006–2012
- General particulars of vessels delivered to shipowners 2006–2012
- Statical stability curves in waves having peaks at amidships or at end drafts
- Hydrostatic information for static vessels having trim instead of being on even keel
- Dividing the trim chapter in the 2006 book into three chapters, to improve presentation
- Separating cross curves and hydrostatic curves data into two chapters
- Combining two chapters on stability relationship with large suspended weights
- Ship stability data sheets as per MCG/SQA organizations
- Revamped pages on nomenclature of ship terms.

My main aims and objectives for this seventh edition of the book are:

1. To help Masters, Mates, and Engineering Officers prepare for their MCA/SQA written and oral examinations.
2. To provide knowledge at a basic level for those whose responsibilities include the day-to-day safe operation of ships.
3. To give Maritime students and Marine Officers an awareness of ship stability problems and to suggest solutions to these problems that are easily understandable.
4. To act as a good quick reference source for those officers who obtained their Certificates of Competency a few months/years prior to joining their ship, port, or dry dock.
5. To assist students of Naval Architecture/Ship Technology in studies on ONC, HNC, and HND and their initial years on undergraduate degree courses.
6. To advise drydock personnel, Ship Designers, DfT Ship Surveyors, Port Authorities, Marine Consultants, Nautical Study Lecturers, Marine Superintendents, etc. in their deliberations regarding ship stability.

Maritime Courts are continually dealing with cases involving ships that have gone aground, collided, or capsized. If this book helps to prevent further incidents of this sort, then the efforts of the late Captain D.R. Derrett and myself will have been worthwhile.

Finally, it only remains for me to wish each student every success in the exams and to wish those working within the Shipping Industry continued success in your chosen career. I hope you find this book to be a useful addition to your bookshelf.

C.B. Barrass

17 July 2012

PART I

Linking Ship Stability and Ship Motions

CHAPTER 1

Group Weights, Water Draft, Air Draft, and Density

Group Weights in a Ship

The first estimate that the Naval Architect makes for a new ship is to estimate the lightweight.

Lightweight: This is the weight of the ship itself when completely *empty*, with boilers topped up to working level. It is made up of steel weight, wood and outfit weight, and the machinery weight. This lightweight is evaluated by conducting an inclining experiment normally just prior to delivery of the new vessel. Over the years, this value will change.

Deadweight: This is the weight that a ship *carries*. It can be made up of oil fuel, fresh water, stores, lubricating oil, water ballast, crew and effects, cargo, and passengers. This deadweight will vary, depending on how much the ship is loaded between light ballast and fully loaded departure conditions.

Displacement: This is the weight of the volume of water that the ship displaces:

$$\text{Displacement} = \text{Lightweight} + \text{Deadweight}$$

Hence

$$W = Lwt + Dwt$$

Water draft: This is the vertical distance from the waterline down to the keel. If it is to the top of the keel, then it is draft molded. If it is to the bottom of the keel, then it is draft extreme. Draft molded is used mainly by naval architects. Draft extreme is used mainly by masters, mates, port authorities, and dry-dock personnel.

Air draft: This is the quoted vertical distance from the waterline to the highest point on the ship when at *zero forward speed*. It indicates the ability of a ship to pass under a bridge spanning a waterway that forms part of the intended route.

Figures in May 2011 stipulate that, for the Panama Canal, this air draft is to be no greater than 57.91 m. For the St Lawrence Seaway the maximum air draft is to be 35.5 m. For the Suez Canal, the maximum air draft is to be 68 m.

Ship Stability for Masters and Mates. http://dx.doi.org/10.1016/B978-0-08-097093-6.00001-3

In order to go beneath bridges over rivers, some vessels have telescopic masts. Others have hinged masts to lessen the chances of contact with the bridge under which they are sailing.

Occasionally, the master or mate needs to calculate the maximum cargo to discharge or the minimum ballast to load to safely pass under a bridge. This will involve moment of weight estimates relating to the final end drafts.

What must be remembered is that the vessel is at a forward speed. Therefore, allowances have to be made for the squat components of mean bodily sinkage and trim ratio forward and aft (see Chapter 43).

Effect of Change of Density when the Displacement is Constant

When a ship moves from water of one density to water of another density, without there being a change in mass, the draft will change. This will happen because the ship must displace the same mass of water in each case. Since the density of the water has changed, the volume of water displaced must also change. This can be seen from the formula:

$$\text{Mass} = \text{Volume} \times \text{Density}$$

If the density of the water increases, then the volume of water displaced must decrease to keep the mass of water displaced constant, and vice versa.

The Effect on Box-Shaped Vessels

$$\text{New mass of water displaced} = \text{Old mass of water displaced}$$

$$\therefore \text{New volume} \times \text{New density} = \text{Old volume} \times \text{Old density}$$

$$\frac{\text{New volume}}{\text{Old volume}} = \frac{\text{Old density}}{\text{New density}}$$

$$\textit{But}\ \text{Volume} = \text{L} \times \text{B} \times \text{Draft}$$

$$\therefore \frac{\text{L} \times \text{B} \times \text{New draft}}{\text{L} \times \text{B} \times \text{Old draft}} = \frac{\text{Old density}}{\text{New density}}$$

or

$$\frac{\text{New draft}}{\text{Old draft}} = \frac{\text{Old density}}{\text{New density}}$$

Example 1

A box-shaped vessel floats at a mean draft of 2.1 meters, in dock water of density 1020 kg per cubic meter. Find the mean draft for the same mass displacement in salt water of density 1025 kg per cubic meter.

$$\frac{\text{New draft}}{\text{Old draft}} = \frac{\text{Old density}}{\text{New density}}$$

$$\text{New draft} = \frac{\text{Old density}}{\text{New density}} \times \text{Old draft}$$

$$= \frac{1020}{1025} \times 2.1 \text{ m}$$

$$= 2.09 \text{ m}$$

Ans. New draft = 2.09 m.

Example 2

A box-shaped vessel floats upright on an even keel as shown in fresh water of density 1000 kg per cubic meter, and the center of buoyancy is 0.50 m above the keel. Find the height of the center of buoyancy above the keel when the vessel is floating in salt water of density 1025 kg per cubic meter.

Note. The center of buoyancy is the geometric center of the underwater volume and for a box-shaped vessel must be at half draft, i.e. KB = ½ draft.

In Fresh Water

$$\text{KB} = 0.5 \text{ m}, \text{ and since KB} = \frac{1}{2}\text{ draft, then draft} = 1 \text{ m}$$

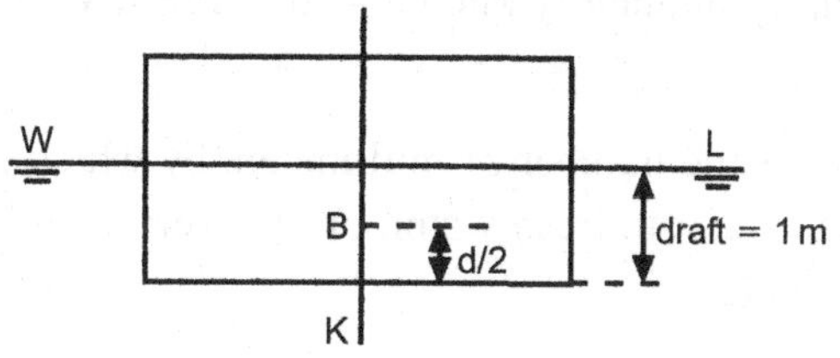

Figure 1.1

In Salt Water

$$\frac{\text{New draft}}{\text{Old draft}} = \frac{\text{Old density}}{\text{New density}}$$

$$\text{New draft} = \text{Old draft} \times \frac{\text{Old density}}{\text{New density}}$$

$$= 1 \times \frac{1000}{1025}$$

$$\text{New draft} = 0.976 \text{ m}$$

$$\text{New KB} = \frac{1}{2} \text{ new draft}$$

Ans. New KB = 0.488 m, say 0.49 m.

The Effect on Ship-Shaped Vessels

It has already been shown that when the density of the water in which a vessel floats is changed the draft will change, but the mass of water in kg or tonnes displaced will be unchanged, i.e.

$$\text{New displacement} = \text{Old displacement}$$

or

$$\text{New volume} \times \text{New density} = \text{Old volume} \times \text{Old density}$$

$$\therefore \frac{\text{New volume}}{\text{Old volume}} = \frac{\text{Old density}}{\text{New density}}$$

With ship shapes this formula should not be simplified further as it was in the case of a box shape because the underwater volume is not rectangular. To find the change in draft of a ship shape due to change of density a quantity known as the 'Fresh Water Allowance' must be known.

The *Fresh Water Allowance* is the number of millimeters by which the mean draft changes when a ship passes from salt water to fresh water, or vice versa, whilst floating at the loaded draft. It is found by the formula:

$$\text{FWA (in mm)} = \frac{\text{Displacement (in tonnes)}}{4 \times \text{TPC}}$$

The proof of this formula is as follows:

$$\text{To show that FWA (in mm)} = \frac{\text{Displacement (in tonnes)}}{4 \times \text{TPC}}$$

Consider the ship shown in Figure 1.2 to be floating at the load Summer draft in salt water at the waterline, WL. Let V be the volume of salt water displaced at this draft.

Now let W_1L_1 be the waterline for the ship when displacing the same mass of fresh water. Also, let 'v' be the extra volume of water displaced in fresh water.

The total volume of fresh water displaced is then V + v.

$$\begin{aligned}
\text{Mass} &= \text{Volume} \times \text{Density} \\
\therefore \text{Mass of SW displaced} &= 1025\ \text{V} \\
\text{and Mass of FW displaced} &= 1000(\text{V} + \text{v}) \\
\text{but Mass of FW displaced} &= \text{Mass of SW displaced} \\
\therefore 1000(\text{V} + \text{v}) &= 1025\ \text{V} \\
1000\ \text{V} + 1000\ \text{v} &= 1025\ \text{V} \\
1000\ \text{v} &= 25\ \text{V} \\
\text{v} &= \text{V}/40
\end{aligned}$$

Now let w be the mass of salt water in volume v, in tonnes, and let W be the mass of salt water in volume V, in tonnes:

$$\therefore \text{w} = \text{W}/40$$

$$\text{but w} = \frac{\text{FWA}}{10} \times \text{TPC}$$

$$\frac{\text{FWA}}{10} \times \text{TPC} = \text{W}/40$$

or

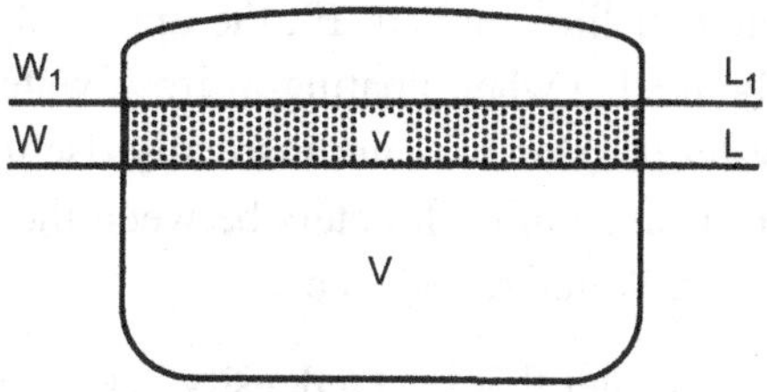

Figure 1.2

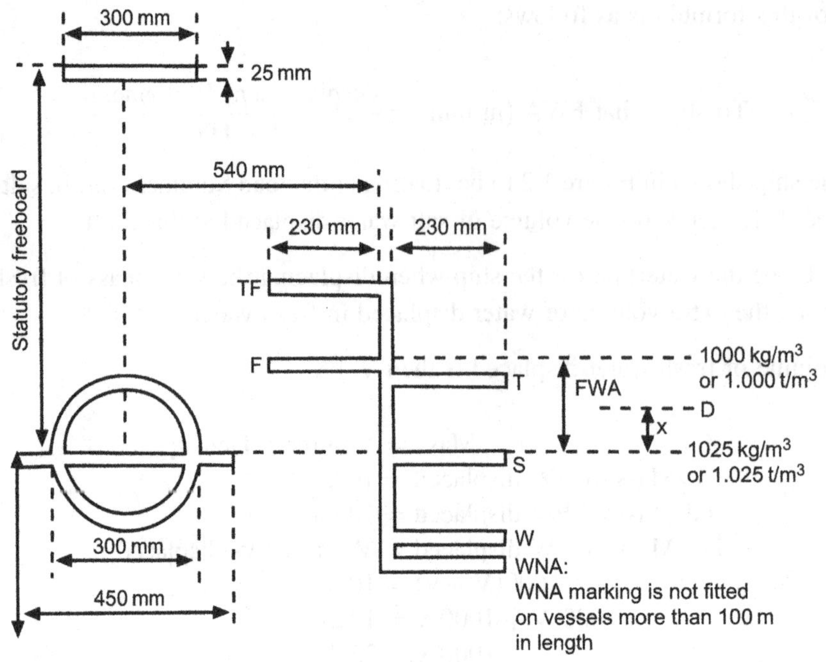

Figure 1.3

$$\text{FWA} = \frac{\text{W}}{4 \times \text{TPC}} \text{ mm}$$

where

$$\text{W} = \text{Loaded salt water displacement in tonnes}$$

Figure 1.3 shows a ship's load line marks. The center of the disk is at a distance below the deck line equal to the ship's statutory freeboard. Then 540 mm forward of the disk is a vertical line 25 mm thick, with horizontal lines measuring 230 × 25 mm on each side of it. The upper edge of the one marked 'S' is in line with the horizontal line through the disk and indicates the draft to which the ship may be loaded when floating in salt water in a Summer Zone. Above this line and pointing aft is another line marked 'F', the upper edge of which indicates the draft to which the ship may be loaded when floating in fresh water in a Summer Zone. If loaded to this draft in fresh water the ship will automatically rise to 'S' when she passes into salt water. The perpendicular distance in millimeters between the upper edges of these two lines is therefore the ship's Fresh Water Allowance.

When the ship is loading in dock water that is of a density between these two limits, 'S' may be submerged such a distance that she will automatically rise to 'S' when the open sea and salt water is reached. The distance by which 'S' can be submerged, called the *Dock Water Allowance*, is found in practice by simple proportion as follows:

$$\text{Let } x = \text{The Dock Water Allowance}$$
$$\text{Let } \rho_{DW} = \text{Density of the dock water}$$

Then

$$\frac{x \text{ mm}}{\text{FWA mm}} = \frac{1025 - \rho_{DW}}{1025 - 1000}$$

or

$$\text{Dock Water Allowance} = \frac{\text{FWA}(1025 - \rho_{DW})}{25}$$

Example 3

A ship is loading in dock water of density 1010 kg per cubic meter. FWA = 150 mm. Find the change in draft on entering salt water.

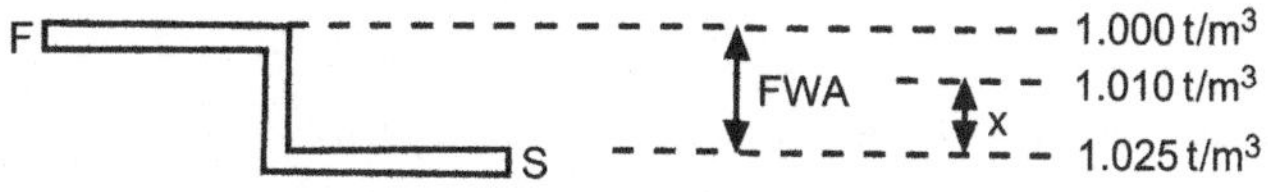

Figure 1.4

$$\text{Let } x = \text{The change in draft in millimeters}$$

$$\text{Then } \frac{x}{\text{FWA}} = \frac{1025 - 1010}{25}$$

$$x = 150 \times \frac{15}{25}$$

$$x = 90 \text{ mm}$$

Ans. Draft will decrease by 90 mm, i.e. 9 cm.

Example 4

A ship is loading in a Summer Zone in dock water of density 1005 kg per cubic meter. FWA = 62.5 mm, TPC = 15 tonnes. The lower edge of the Summer load line is in the waterline to port and is 5 cm above the waterline to starboard. Find how much more cargo may be loaded if the ship is to be at the correct load draft in salt water.

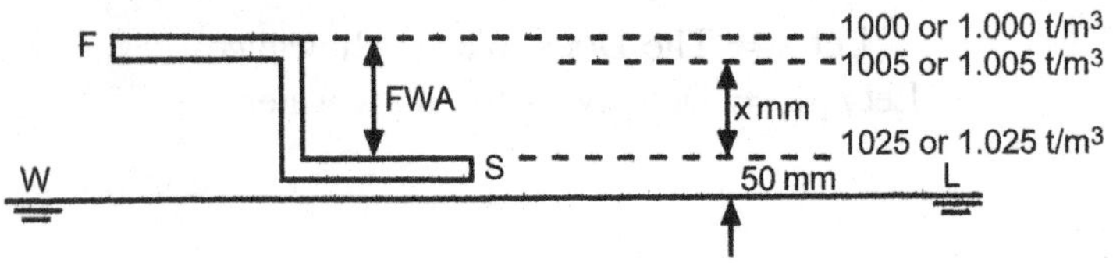

Figure 1.5

Note. This ship is obviously listed to port and if brought upright the lower edge of the 'S' load line on each side would be 25 mm above the waterline. Also, it is the upper edge of the line that indicates the 'S' load draft and, since the line is 25 mm thick, the ship's draft must be increased by 50 mm to bring her to the 'S' load line in dock water. In addition 'S' may be submerged by x mm.

$$\frac{x}{FWA} = \frac{1025 - \rho_{DW}}{25}$$

$$x = 62.5 \times \frac{20}{25}$$

$$x = 50 \text{ mm}$$

$$\therefore \text{Total increase in draft required} = 100 \text{ mm or } 10 \text{ cm}$$

and

$$\text{Cargo to load} = \text{Increase in draft} \times \text{TPC}$$
$$= 10 \times 15$$

Ans. Cargo to load = 150 tonnes.

Effect of Density on Displacement when the Draft is Constant

Should the density of the water in which a ship floats be changed without the ship altering her draft, then the mass of water displaced must have changed. The change in the mass of water displaced may have been brought about by bunkers and stores being loaded or consumed during a sea passage, or by cargo being loaded or discharged. In all cases:

$$\text{New volume of water displaced} = \text{Old volume of water displaced}$$

or

$$\frac{\text{New displacement}}{\text{New density}} = \frac{\text{Old displacement}}{\text{Old density}}$$

or

$$\frac{\text{New displacement}}{\text{Old displacement}} = \frac{\text{New density}}{\text{Old density}}$$

Example 1

A ship displaces 7000 tonnes whilst floating in fresh water. Find the displacement of the ship when floating at the same draft in water of density 1015 kg per cubic meter, i.e. 1.015 t/m^3.

$$\frac{\text{New displacement}}{\text{Old displacement}} = \frac{\text{New density}}{\text{Old density}}$$

$$\text{New displacement} = \text{Old displacement} \times \frac{\text{New density}}{\text{Old density}}$$

$$= 7000 \times \frac{1015}{1000}$$

Ans. New displacement = 7105 tonnes.

Example 2

A ship of 6400 tonnes displacement is floating in salt water. The ship has to proceed to a berth where the density of the water is 1008 kg per cubic meter. Find how much cargo must be discharged if she is to remain at the salt water draft.

$$\frac{\text{New displacement}}{\text{Old displacement}} = \frac{\text{New density}}{\text{Old density}}$$

or

$$\text{New displacement} = \text{Old displacement} \times \frac{\text{New density}}{\text{Old density}}$$

$$= 6400 \times \frac{1008}{1025}$$

$$\text{New displacement} = 6294 \text{ tonnes}$$

$$\text{Old displacement} = 6400.0 \text{ tonnes}$$

Ans. Cargo to discharge = 6400 − 6294 = 106 tonnes.

Example 3

A ship 120 m × 17 m × 10 m has a block coefficient 0.800 and is floating at the load Summer draft of 7.2 meters in fresh water. Find how much more cargo can be loaded to remain at the same draft in salt water.

$$\text{Old displacement} = L \times B \times \text{Draft} \times C_b \times \text{Density}$$

$$= 120 \times 17 \times 7.2 \times 0.800 \times 1.000 \text{ tonnes}$$

$$\text{Old displacement} = 11{,}750 \text{ tonnes}$$

$$\frac{\text{New displacement}}{\text{Old displacement}} = \frac{\text{New density}}{\text{Old density}}$$

$$\text{New displacement} = \text{Old displacement} \times \frac{\text{New density}}{\text{Old density}}$$

$$= 11{,}750 \times \frac{1025}{1000}$$

$$\text{New displacement} = 12{,}044 \text{ tonnes}$$

$$\text{Old displacement} = 11{,}750 \text{ tonnes}$$

Ans. Cargo to load = 12,044 − 11,750 = 294 tonnes.

Note. This problem should not be attempted as one involving TPC and FWA.

Exercise 1

Density and Draft

1. A ship displaces 7500 cubic meters of water of density 1000 kg per cubic meter. Find the displacement in tonnes when the ship is floating at the same draft in water of density 1015 kg per cubic meter.
2. When floating in fresh water at a draft of 6.5 m a ship displaces 4288 tonnes. Find the displacement when the ship is floating at the same draft in water of density 1015 kg per cubic meter.
3. A box-shaped vessel 24 m × 6 m × 3 m displaces 150 tonnes of water. Find the draft when the vessel is floating in salt water.
4. A box-shaped vessel draws 7.5 m in dock water of density 1006 kg per cubic meter. Find the draft in salt water of density 1025 kg per cubic meter.
5. The KB of a rectangular block that is floating in fresh water is 50 cm. Find the KB in salt water.

6. A ship is lying at the mouth of a river in water of density 1024 kg per cubic meter and the displacement is 12,000 tonnes. The ship is to proceed up river and to berth in dock water of density 1008 kg per cubic meter with the same draft as at present. Find how much cargo must be discharged.
7. A ship arrives at the mouth of a river in water of density 1016 kg per cubic meter with a freeboard of 'S' m. She then discharges 150 tonnes of cargo, and proceeds up river to a second port, consuming 14 tonnes of bunkers. When she arrives at the second port the freeboard is again 'S' m, the density of the water being 1004 kg per cubic meter. Find the ship's displacement on arrival at the second port.
8. A ship loads in fresh water to her salt water marks and proceeds along a river to a second port consuming 20 tonnes of bunkers. At the second port, where the density is 1016 kg per cubic meter, after 120 tonnes of cargo have been loaded, the ship is again at the load salt water marks. Find the ship's load displacement in salt water.

The TPC and FWA, etc.

9. A ship's draft is 6.40 meters forward and 6.60 meters aft. FWA = 180 mm. Density of the dock water is 1010 kg per cubic meter. If the load mean draft in salt water is 6.7 meters, find the final drafts F and A in dock water if this ship is to be loaded down to her marks and trimmed 0.15 meters by the stern (center of flotation is amidships).
10. A ship floating in dock water of density 1005 kg per cubic meter has the lower edge of her Summer load line in the waterline to starboard and 50 mm above the waterline to port. FWA = 175 mm and TPC = 12 tonnes. Find the amount of cargo that can yet be loaded in order to bring the ship to the load draft in salt water.
11. A ship is floating at 8 meters mean draft in dock water of relative density 1.01. TPC = 15 tonnes and FWA = 150 mm. The maximum permissible draft in salt water is 8.1 m. Find the amount of cargo yet to load.
12. A ship's light displacement is 3450 tonnes and she has on board 800 tonnes of bunkers. She loads 7250 tonnes of cargo, 250 tonnes of bunkers, and 125 tonnes of fresh water. The ship is then found to be 75 mm from the load draft. TPC = 12 tonnes. Find the ship's deadweight and load displacement.
13. A ship has a load displacement of 5400 tonnes. TPC = 30 tonnes. If she loads to the Summer load line in dock water of density 1010 kg per cubic meter, find the change in draft on entering salt water of density 1025 kg per cubic meter.
14. A ship's FWA is 160 mm, and she is floating in dock water of density 1012 kg per cubic meter. Find the change in draft when she passes from dock water to salt water.

■

CHAPTER 2

Transverse Statical Stability

Introduction

1. The center of gravity of a body 'G' is the point through which the force of gravity is considered to act vertically downwards with a force equal to the weight of the body. KG is VCG of the ship.
2. The center of buoyancy 'B' is the point through which the force of buoyancy is considered to act vertically upwards with a force equal to the weight of water displaced. It is the center of gravity of the underwater volume. KB is VCB of the ship.
3. To float at rest in still water, a vessel must displace her own weight of water, and the center of gravity must be in the same vertical line as the center of buoyancy.
4. KM = KB + BM. Also KM = KG + GM.

See Figure 2.1(a).

Definitions

1. *Heel.* A ship is said to be heeled when she is inclined by an external force, for example when the ship is inclined by the action of the waves or wind.
2. *List.* A ship is said to be listed when she is inclined by forces within the ship, for example when the ship is inclined by shifting a weight transversely within the ship. This is a fixed angle of heel.

The Metacenter

Consider a ship floating upright in still water as shown in Figure 2.1(a). The centers of gravity and buoyancy are at G and B respectively. Figure 2.1(c) shows the righting couple. GZ is the righting lever.

Now let the ship be inclined by an external force to a small angle (θ) as shown in Figure 2.1(b). Since there has been no change in the distribution of weights, the center of gravity will remain at G and the weight of the ship (W) can be considered to act vertically downwards through this point.

When heeled, the wedge of buoyancy WOW_1 is brought out of the water and an equal wedge LOL_1 becomes immersed. In this way a wedge of buoyancy having its center of gravity at g is

Ship Stability for Masters and Mates. http://dx.doi.org/10.1016/B978-0-08-097093-6.00002-5

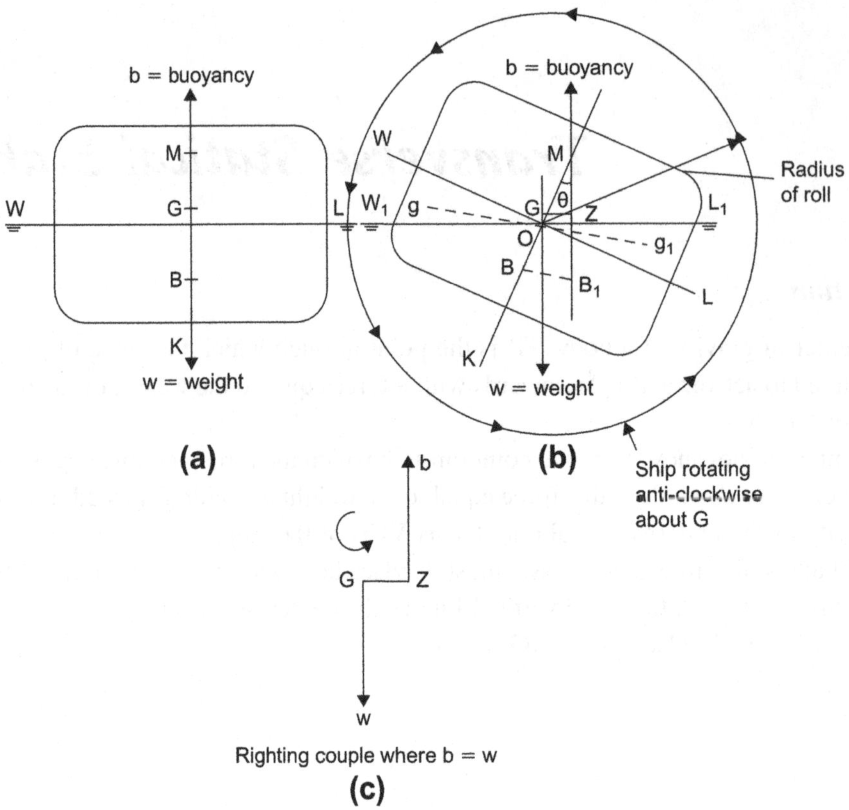

Figure 2.1:
Stable Equilibrium.

transferred to a position with its center of gravity at g_1. The center of buoyancy, being the center of gravity of the underwater volume, must shift from B to the new position B_1, such that BB_1 is parallel to gg_1, and $BB_1 = (v \times gg_1)/V$, where v is the volume of the transferred wedge and V is the ship's volume of displacement.

The verticals through the centers of buoyancy at two consecutive angles of heel intersect at a point called the *metacenter.* For angles of heel up to about 15° the vertical through the center of buoyancy may be considered to cut the centerline at a fixed point called the initial metacenter (M in Figure 2.1(b)). The height of the initial metacenter above the keel (KM) depends upon a ship's underwater form. Figure 2.2 shows a typical curve of KMs for a ship plotted against draft.

The vertical distance between G and M is referred to as the *metacentric height.* If G is below M the ship is said to have positive metacentric height, and if G is above M the metacentric height is said to be negative.

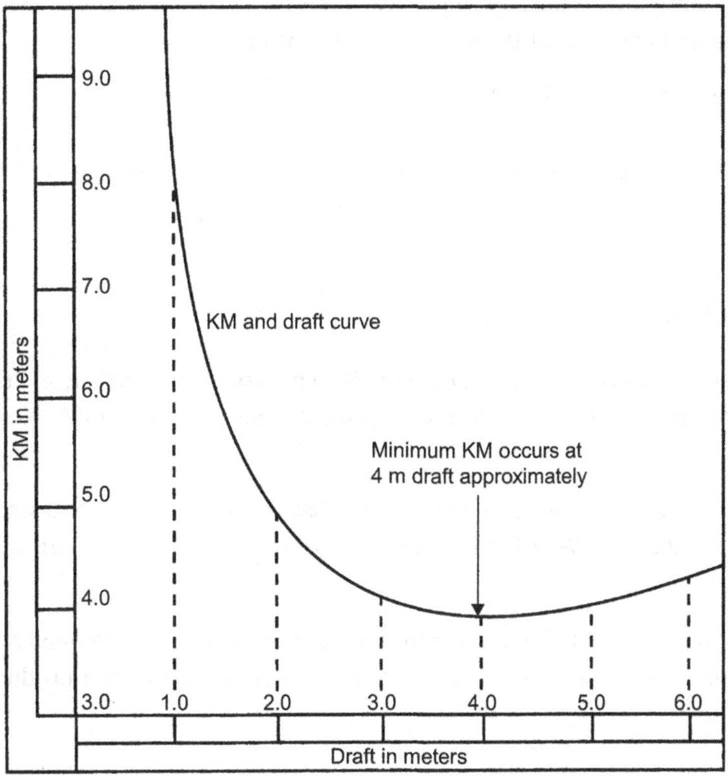

Figure 2.2

Equilibrium

Stable Equilibrium

A ship is said to be in stable equilibrium if, when inclined, she tends to return to the initial position. For this to occur the center of gravity must be below the metacenter; that is, the ship must have positive initial metacentric height. Figure 2.1(a) shows a ship in the upright position having a positive GM. Figure 2.1(b) shows the same ship inclined to a small angle. The position of G remains unaffected by the heel and the force of gravity is considered to act vertically downwards through this point. The center of buoyancy moves out to the low side from B to B_1 to take up the new center of gravity of the underwater volume, and the force of buoyancy is considered to act vertically upwards through B_1 and the metacenter M. If moments are taken about G there is a moment to return the ship to the upright. This moment is referred to as the *Moment of Statical Stability* and is equal to the product of the force 'W' and the length of the lever GZ, i.e.

$$\textit{Moment of Statical Stability} = W \times GZ \ \textit{tonnes meters}$$

The lever GZ is referred to as the *righting lever* and is the perpendicular distance between the center of gravity and the vertical through the center of buoyancy.

At a small angle of heel (less than 15°):

$$GZ = GM \times \sin\theta \text{ and Moment of Statical Stability} = W \times GM \times \sin\theta$$

Unstable Equilibrium

When a ship that is inclined to a small angle tends to heel over still further, she is said to be in unstable equilibrium. For this to occur the ship must have a negative GM. Note how G is above M.

Figure 2.3(a) shows a ship in unstable equilibrium that has been inclined to a small angle. The moment of statical stability, W × GZ, is clearly a capsizing moment that will tend to heel the ship still further.

Note. A ship having a very small negative initial metacentric height GM need not necessarily capsize. This point will be examined and explained later. This situation produces an angle of loll.

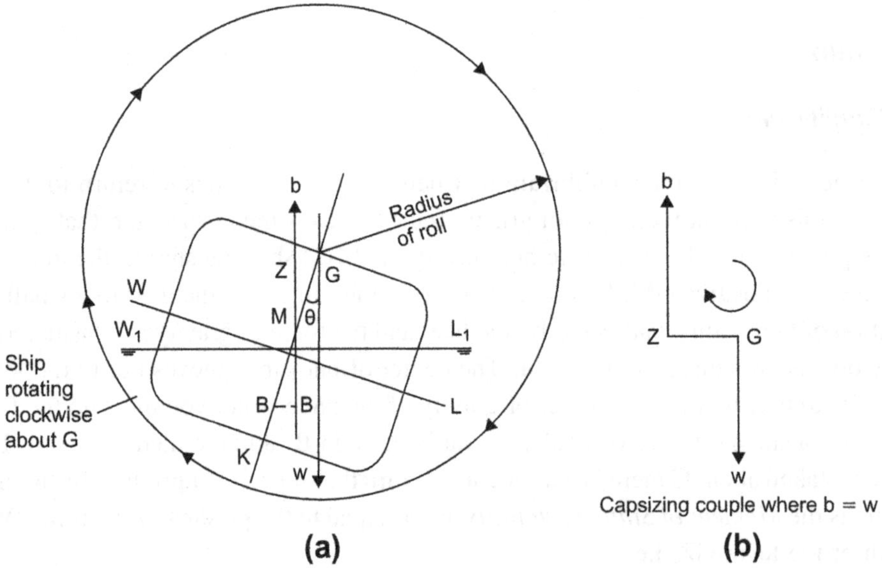

Figure 2.3:
Unstable Equilibrium.

Neutral Equilibrium

When G coincides with M as shown in Figure 2.4(a), the ship is said to be in neutral equilibrium, and if inclined to a small angle she will tend to remain at that angle of heel until another external force is applied. The ship has zero GM. Note that KG = KM.

$$\text{Moment of Statical Stability} = W \times GZ, \text{ but in this case } GZ = 0$$
$$\therefore \text{Moment of Statical Stability} = 0; \text{ see Figure 2.4(b)}$$

Therefore there is no moment to bring the ship back to the upright or to heel her over still further. The ship will move vertically up and down in the water at the fixed angle of heel until further external or internal forces are applied.

Correcting Unstable and Neutral Equilibrium

When a ship in unstable or neutral equilibrium is to be made stable, the effective center of gravity of the ship should be lowered. To do this one or more of the following methods may be employed:

1. Weights already in the ship may be lowered.
2. Weights may be loaded below the center of gravity of the ship.
3. Weights may be discharged from positions above the center of gravity.
4. Free surfaces within the ship may be removed.

The explanation of this last method will be found in Chapter 3.

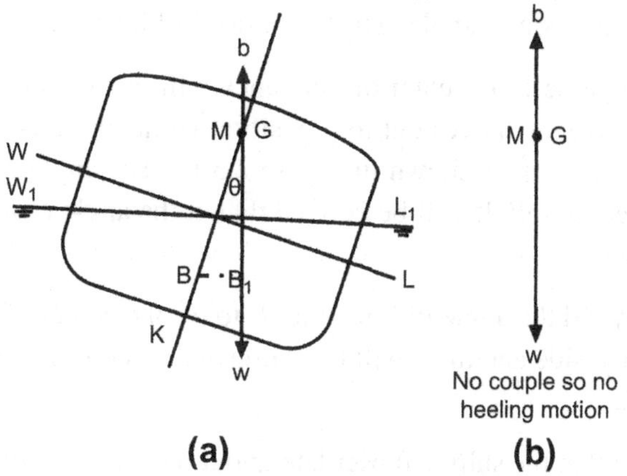

Figure 2.4:
Neutral Equilibrium.

Stiff and Tender Ships

The *time period* of a ship is the time taken by the ship to roll from one side to the other and back again to the initial position.

When a ship has a comparatively large GM, for example 2–3 m, the righting moments at small angles of heel will also be comparatively large. It will thus require larger moments to incline the ship. When inclined she will tend to return more quickly to the initial position. The result is that the ship will have a comparatively short time period, and will roll quickly – and perhaps violently – from side to side. A ship in this condition is said to be 'stiff', and such a condition is not desirable. The time period could be as low as 8 seconds. The effective center of gravity of the ship should be raised within that ship (see Figure 47.2).

When the GM is comparatively small, for example 0.16–0.20 m, the righting moments at small angles of heel will also be small. The ship will thus be much easier to incline and will not tend to return so quickly to the initial position. The time period will be comparatively long, for example 25–35 seconds, and a ship in this condition is said to be 'tender'. As before, this condition is not desirable and steps should be taken to increase the GM by lowering the effective center of gravity of the ship (see Figure 47.2).

The officer responsible for loading a ship should aim at a happy medium between these two conditions whereby the ship is neither too stiff nor too tender. A time period of 15–25 seconds would generally be acceptable for those on board a ship at sea (see Figure 47.2).

Negative GM and Angle of Loll

It has been shown previously that a ship having a negative initial metacentric height will be unstable when inclined to a small angle. This is shown in Figure 2.5(a).

As the angle of heel increases, the center of buoyancy will move out still further to the low side. If the center of buoyancy moves out to a position vertically under G, the capsizing moment will have disappeared, as shown in Figure 2.5(b). The angle of heel at which this occurs is called the *angle of loll*. It will be noticed that at the angle of loll, the GZ is zero. G remains on the centerline.

If the ship is heeled beyond the angle of loll from θ_1 to θ_2, the center of buoyancy will move out still further to the low side and there will be a moment to return her to the angle of loll, as shown in Figure 2.5(c).

From this it can be seen that the ship will oscillate about the angle of loll instead of about the vertical. If the center of buoyancy does not move out far enough to get vertically under G, the ship will capsize.

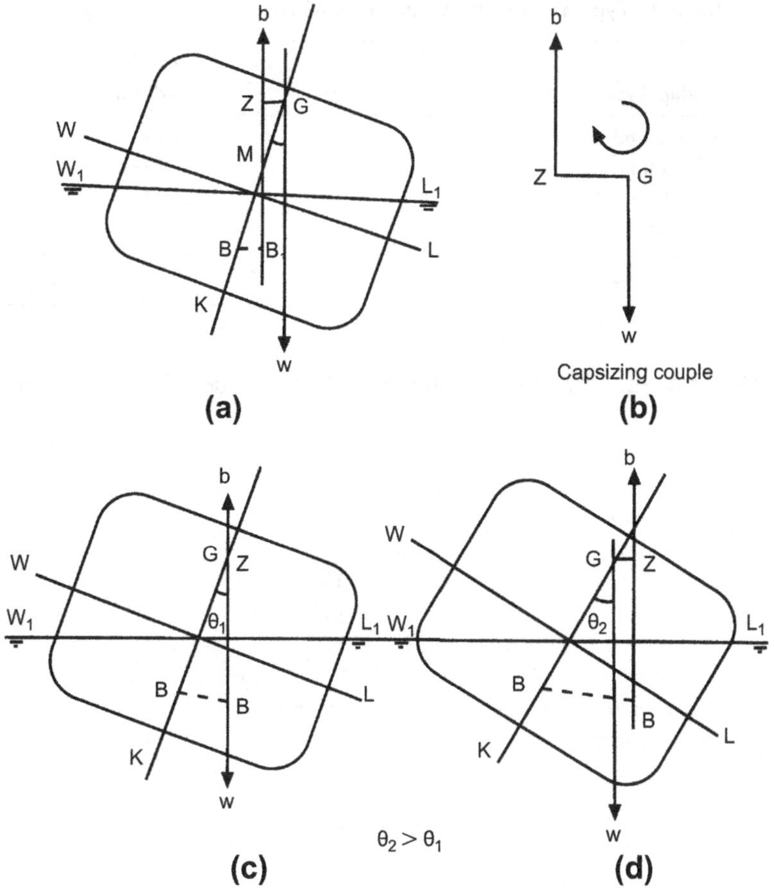

Figure 2.5

The angle of loll will be to port or starboard and back to port depending on external forces such as wind and waves. One minute it may flop over to 3°P and then suddenly flop over to 3°S.

There is always the danger that G will rise above M and create a situation of unstable equilibrium. This will cause capsizing of the ship.

The GM Value

GM is crucial to ship stability. Table 2.1 shows *typical* working values of GM for several ship types all at *fully loaded* drafts.

At drafts below the fully loaded draft, due to KM tending to be larger in value it will be found that corresponding GM values will be *higher* than those listed in Table 2.1. For all conditions

Table 2.1: Typical working values of GM for different ship types.

Ship Type	GM at Fully Loaded Condition (m)
General cargo ships	0.30–0.50
Oil tankers	0.50–2.00
Double-hull supertankers	2.00–5.00
Container ships	1.50–2.50
Ro-ro vessels	1.50 approximately
Bulk ore carriers	2–3

of loading the Department for Transport (DfT) stipulate that the GM must never be less than 0.15 m.

Exercise 2

1. Define the terms 'heel', 'list', 'initial metacenter', and 'initial metacentric height'.
2. Sketch transverse sections through a ship, showing the positions of the center of gravity, center of buoyancy, and initial metacenter, when the ship is in: (a) stable equilibrium; (b) unstable equilibrium; and (c) neutral equilibrium.
3. Explain what is meant by a ship being (a) tender and (b) stiff.
4. With the aid of suitable sketches, explain what is meant by 'angle of loll'.
5. A ship of 10,000 t displacement has an initial metacentric height of 1.5 m. What is the moment of statical stability when the ship is heeled 10°?

CHAPTER 3

Effect of Decreasing Free Surface on Stability

When a tank is completely filled with a liquid, the liquid cannot move within the tank when the ship heels. For this reason, as far as stability is concerned, the liquid may be considered as a static weight having its center of gravity at the center of gravity of the liquid within the tank.

Figure 3.1(a) shows a ship with a double-bottom tank filled with a liquid having its center of gravity at g. The effect when the ship is heeled to a small angle θ is shown in Figure 3.1(b). No weights have been moved within the ship, and therefore the position of G is not affected. The center of buoyancy will move out to the low side indicated by BB_1.

$$\text{Moment of statical stability} = W \times GZ$$
$$= W \times GM \times \sin\theta$$

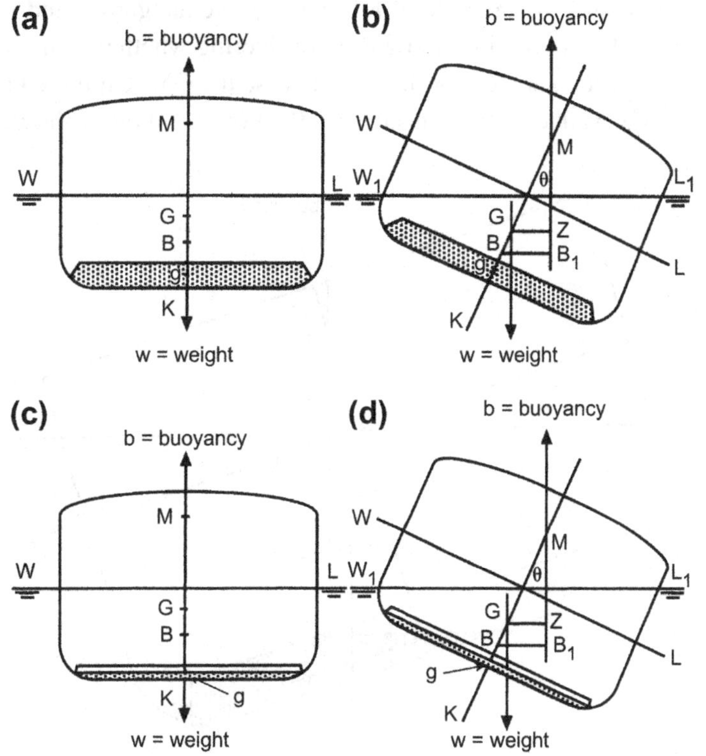

Figure 3.1

Ship Stability for Masters and Mates. http://dx.doi.org/10.1016/B978-0-08-097093-6.00003-7

Now consider the same ship floating at the same draft and having the same KG, but increase the depth of the tank so that the liquid now only partially fills it, as shown in Figure 3.1(c) and (d).

When the ship heels, as shown in Figure 3.2, the liquid flows to the low side of the tank such that its center of gravity shifts from g to g_1. This will cause the ship's center of gravity to shift from G to G_1, parallel to gg_1.

$$\begin{aligned}\text{Moment of statical stability} &= W \times G_1Z_1\\ &= W \times G_vZ_v\\ &= W \times G_vM \times \sin\theta\end{aligned}$$

where G_v is the virtual position of G.

This indicates that the effect of the free surface is to reduce the effective metacentric height from GM to G_vM. GG_v is therefore the virtual loss of GM due to the free surface. Any loss in GM is a loss in stability.

If a free surface is to be created in a ship with a small initial metacentric height, the virtual loss of GM due to the free surface may result in a negative metacentric height. This would cause the ship to take up an angle of loll that may be dangerous and in any case is undesirable. This should be borne in mind when considering whether or not to run water ballast into tanks to correct an angle of loll, or to increase the GM. Until the tank is full there will be a virtual loss of GM due to the free surface effect of the liquid. It should also be noted

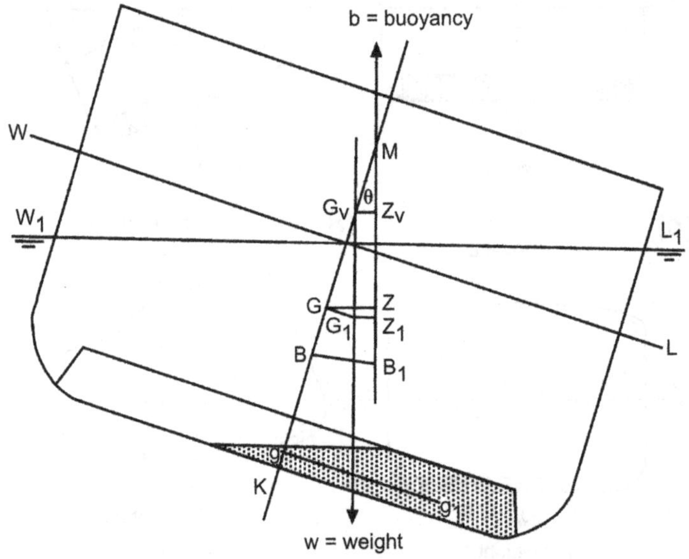

Figure 3.2

from Figure 3.2 that even though the distance GG_1 is fairly small it produces a relatively large virtual loss in GM (GG_v).

Correcting an Angle of Loll

If a ship takes up an angle of loll due to a very small negative GM it should be corrected as soon as possible. GM may be, for example, −0.05 to −0.10 m, well below the DfT minimum stipulation of 0.15 m.

First make sure that the heel is due to a negative GM and not due to uneven distribution of the weights on board. For example, when bunkers are burned from one side of a divided double-bottom tank it will obviously cause G to move to G_1, away from the center of gravity of the burned bunkers, and will result in the ship listing, as shown in Figure 3.3.

Having satisfied oneself that the weights within the ship are uniformly distributed, one can assume that the list is probably due to a very small negative GM. To correct this it will be necessary to lower the position of the effective center of gravity sufficiently to bring it below the initial metacenter. Any slack tanks should be topped up to eliminate the virtual rise of G due to free surface effect. If there are any weights that can be lowered within the ship, they should be lowered. For example, derricks may be lowered if any are topped; oil in deep tanks may be transferred to double-bottom tanks, etc.

Assume now that all the above action possible has been taken and that the ship is still at an angle of loll. Assume also that there are double-bottom tanks that are empty. Should they be filled, and if so, which ones first?

Before answering these questions consider the effect on stability during the filling operation. Free surfaces will be created as soon as liquid enters an empty tank. This will give a virtual

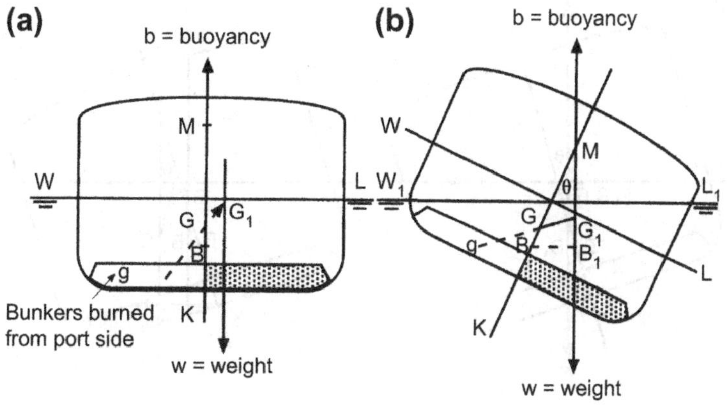

Figure 3.3

rise of G, which in turn will lead to an increased negative GM and an increased angle of loll. Therefore, if it is decided that it is safe to use the tanks, those which have the smallest area can be filled first so that the increase in list is cut to a minimum. Tanks should be filled one at a time.

Next, assume that it is decided to start by filling a tank that is divided at the center line. Which side is to be filled first? If the high side is filled first the ship will start to right herself but will then roll suddenly over to take up a larger angle of loll on the other side, or perhaps even capsize. Now consider filling the low side first. Weight will be added low down in the vessel and G will thus be lowered, but the added weight will also cause G to move out of the center line to the low side, increasing the list. Free surface is also being created and this will give a virtual rise in G, thus causing a loss in GM, which will increase the list still further.

Figure 3.4(a) shows a ship at an angle of loll with the double-bottom tanks empty and in Figure 3.4(b) some water has been run into the low side. The shift of the center of gravity from G to G_v is the virtual rise of G due to the free surface, and the shift from G_v to G_1 is due to the weight of the added water.

It can be seen from the figure that the net result is a moment to list the ship over still further, but the increase in list is a gradual and controlled increase. When more water is now run into the tank the center of gravity of the ship will gradually move downwards and the list will start to decrease. As the list decreases, water may be run into the other side of the tank. The water will then be running in much more quickly, causing G to move downwards more quickly. The ship cannot roll suddenly over to the other side as there is more water in the low side than in the high side. If sufficient weight of water is loaded to bring G on the center line below M, the ship should complete the operation upright.

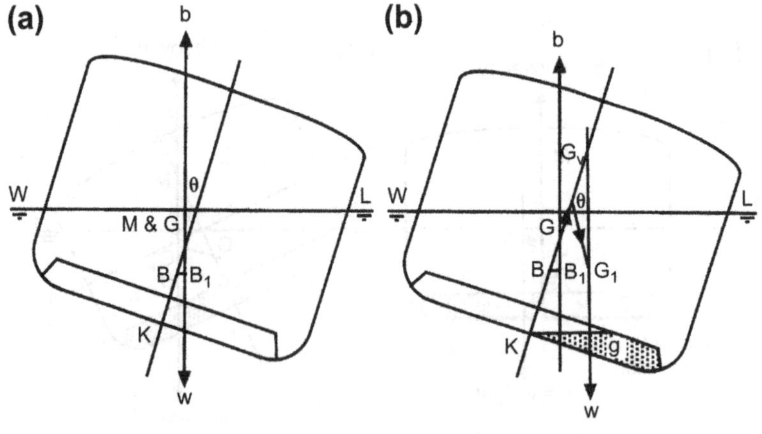

Figure 3.4

To summarize:

1. Check that the list is due to a very small negative GM, for example −0.05 to −0.10 m.
2. Top up any slack tanks and lower weights within the ship if possible.
3. If the ship is still listed and it is decided to fill double-bottom tanks, start by filling the low side of a tank that is adequately subdivided.
4. The list is bound to be increased in the initial stages.
5. Never start by filling tanks on the high side first.
6. Always *calculate the effects first* before *authorizing action* to be taken to ballast any tanks.

Exercise 3

1. With the aid of suitable sketches, show the effect of slack tanks on a ship's stability.
2. A ship leaves port upright with a full cargo of timber, and with timber on deck. During the voyage, bunkers, stores, and fresh water are consumed evenly from each side. If the ship arrives at her destination with a list, explain the probable cause of the list and how this should be remedied.
3. A ship loaded with timber, and with timber on deck, berths with an angle of loll away from the quay. From which side should the timber on deck be discharged first and why?

CHAPTER 4

TPC and Displacement Curves

Introduction

The TPC is the mass that must be loaded or discharged to change the ship's mean draft by 1 cm. When the ship is floating in salt water it is found by using the formula:

$$TPC_{SW} = \frac{WPA}{100} \times \rho_{SW} = \frac{WPA}{97.56}$$

where

ρ_{SW} = water density of 1.025 t/m^3
WPA = the area of the waterplane in square meters

The area of the waterplane of a box-shaped vessel is the same for all drafts if the trim is constant, and so the TPC will also be the same for all drafts.

In the case of a ship the area of the waterplane is not constant for all drafts, and therefore the TPC will reduce at lower drafts, as shown in Figure 4.1. The TPCs are calculated for a range of drafts extending beyond the light and loaded drafts, and these are then tabulated or plotted on a graph. From the table or graph the TPC at intermediate drafts may be found.

TPC Curves

When constructing a TPC curve the TPCs are plotted against the corresponding drafts. It is usually more convenient to plot the drafts on the vertical axis and the TPCs on the horizontal axis.

Example

(a) Construct a graph from the following information:

Mean draft (m)	3.0	3.5	4.0	4.5
TPC (tonnes)	8.0	8.5	9.2	10.0

(b) From this graph find the TPCs at drafts of 3.2, 3.7, and 4.3 m.

Ship Stability for Masters and Mates. http://dx.doi.org/10.1016/B978-0-08-097093-6.00004-9

(c) If the ship is floating at a mean draft of 4 m and then loads 50 tonnes of cargo, 10 tonnes of fresh water, and 25 tonnes of bunkers, whilst 45 tonnes of ballast are discharged, find the final mean draft.

(a) For the graph see Figure 4.1

(b) TPC at 3.2 m draft = 8.17 tonnes
TPC at 3.7 m draft = 8.77 tonnes
TPC at 4.3 m draft = 9.68 tonnes

(c) TPC at 4 m draft = 9.2 tonnes

Loaded cargo	50 tonnes
Fresh water	10 tonnes
Bunkers	25 tonnes
Total	85 tonnes
Discharged ballast	45 tonnes
Net loaded	40 tonnes

$$\text{Increase in draft} = \frac{w}{TPC} = \frac{40}{9.2} = 4.35\ \text{cm}$$

Increase in draft = 0.044 m

Original draft = 4.000 m

New mean draft = 4.044m

Note. If the net weight loaded or discharged is very large, there is likely to be a considerable difference between the TPCs at the original and the new drafts, in which case the procedure to find the change in draft is as follows.

First find an approximate new draft using the TPC at the original draft, then find the TPC at the approximate new draft. Using the mean of these two TPCs find the actual increase or decrease in draft.

■

Displacement Curves

A displacement curve is one from which the displacement of the ship at any particular draft can be found, and vice versa. The draft scale is plotted on the vertical axis and the scale of

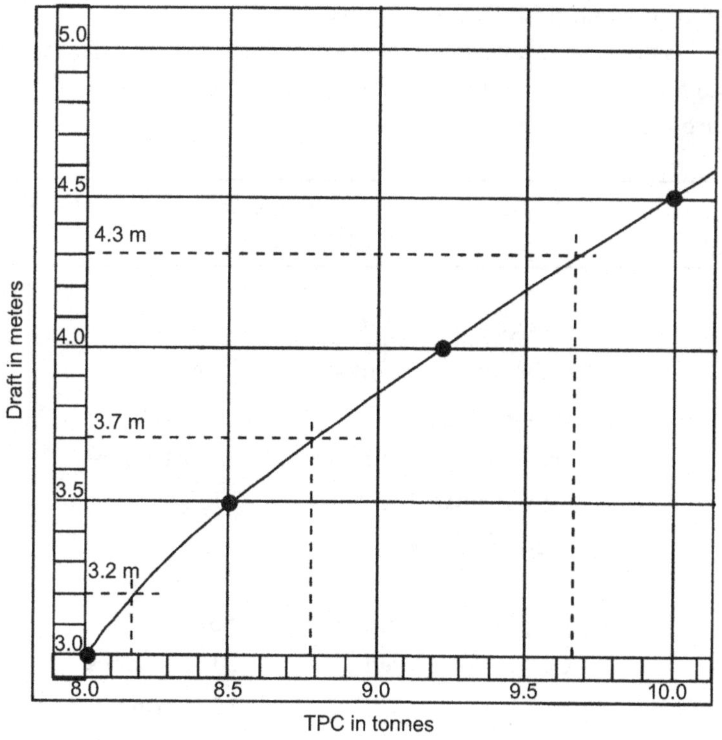

Figure 4.1

displacements on a horizontal axis. As a general rule the largest possible scale should be used to ensure reasonable accuracy. When the graph paper is oblong in shape, the length of the paper should be used for the displacement scale and the breadth for the drafts. It is quite unnecessary in most cases to start the scale from zero as the information will only be required for drafts between the light and load displacements (known as the boot-topping area).

Example

(a) Construct a displacement curve from the following data:

Draft (m)	3	3.5	4	4.5	5.0	5.5
Displacement (tonnes)	2700	3260	3800	4450	5180	6060

(b) If this ship's light draft is 3 m and the load draft is 5.5 m, find the deadweight.

(c) Find the ship's draft when there are 500 tonnes of bunkers, 50 tonnes of fresh water, and stores on board.

(d) When at 5.13 m mean draft the ship discharges 2100 tonnes of cargo and loads 250 tonnes of bunkers. Find the new mean draft.

(e) Find the approximate TPC at 4.4 m mean draft.

(f) If the ship is floating at an existing mean draft of 5.2 m and the required load mean draft is 5.5 m, find how much more cargo may be loaded.

(a) See Figure 4.2 for the graph.

(b)

Load	Draft 5.5 m	Displacement	6060 tonnes
Light	Draft 3.0 m	Displacement	2700 tonnes

Deadweight = 3360 tonnes

(c) Light displacement	2700 tonnes
Bunkers	+500 tonnes
Fresh water and stores	+50 tonnes
New displacement	3250 tonnes

∴ Draft = 3.48 m

(d) Displacement at 5.13 m	5380 tonnes
Cargo discharged	−2100 tonnes
	3280 tonnes
Bunkers loaded	+250 tonnes
New displacement	3530 tonnes

∴ New draft = 3.775 m

(e) At 4.5 m draft the displacement is	4450 tonnes
At 4.3 m draft the displacement is	−4175 tonnes
Difference to change the draft 0.2 m	275 tonnes
Difference to change the draft 1 cm	$\frac{275}{20}$ tonnes

∴ TPC = 13.75 tonnes

(f) Load draft 5.5 m	Displacement	6060 tonnes
Present draft 5.2 m	Displacement	−5525 tonnes
	Difference	535 tonnes

∴ Load = 535 tonnes

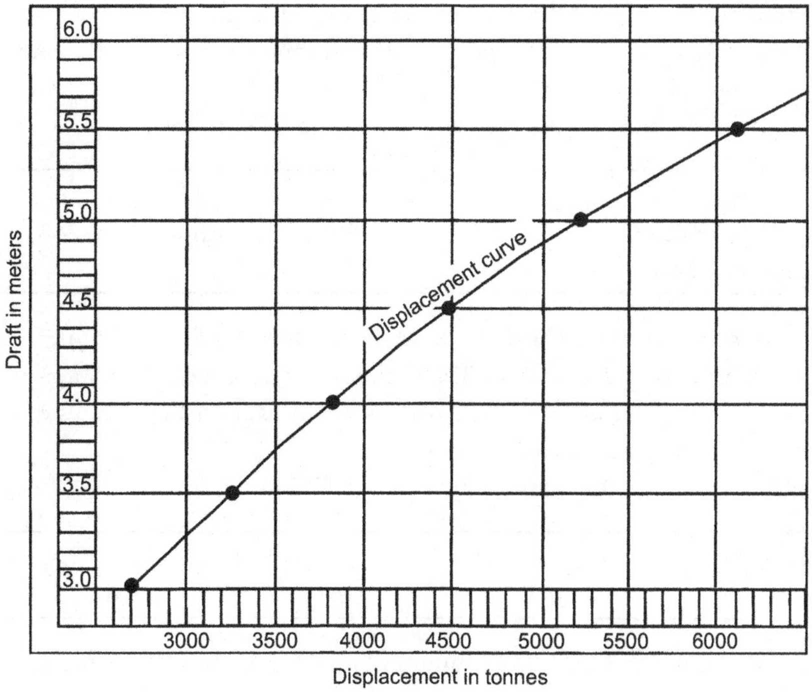

Figure 4.2

In fresh water, the TPC is calculated as follows:

$$TPC_{FW} = \frac{WPA}{100}$$

$$\text{or} \quad TPC_{FW} = TC_{SW} \times \frac{1.000}{1.025}$$

Exercise 4

TPC Curves

1. (a) Construct a TPC curve from the following data:

Mean draft (m)	1	2	3	4	5
TPC (tonnes)	3.10	4.32	5.05	5.50	5.73

(b) From this curve find the TPC at drafts of 1.5 and 2.1 m.

(c) If this ship floats at 2.2 m mean draft and then discharges 45 tonnes of ballast, find the new mean draft.

2. (a) From the following information construct a TPC curve:

Mean draft (m)	1	2	3	4	5
Area of waterplane (square meters)	336	567	680	743	777

(b) From this curve find the TPCs at mean drafts of 2.5 and 4.5 m.

(c) If, while floating at a draft of 3.8 m, the ship discharges 380 tonnes of cargo and loads 375 tonnes of bunkers, 5 tonnes of stores, and 125 tonnes of fresh water, find the new mean draft.

3. From the following information construct a TPC curve:

Mean draft (m)	1	3	5	7
TPC (tonnes)	4.7	10.7	13.6	15.5

Then find the new mean draft if 42 tonnes of cargo is loaded whilst the ship is floating at 4.5 m mean draft.

Displacement Curves

4. (a) From the following information construct a displacement curve:

Displacement (tonnes)	376	736	1352	2050	3140	4450
Mean draft (m)	1	2	3	4	5	6

(b) From this curve find the displacement at a draft of 2.3 m.

(c) If this ship floats at 2.3 m mean draft and then loads 850 tonnes of cargo and discharges 200 tonnes of cargo, find the new mean draft.

(d) Find the approximate TPC at 2.5 m mean draft.

5. The following information is taken from a ship's displacement scale:

Displacement (tonnes)	335	1022	1949	2929	3852	4841
Mean draft (m)	1	1.5	2	2.5	3	3.5

(a) Construct the displacement curve for this ship and from it find the draft when the displacement is 2650 tonnes.

(b) If this ship arrived in port with a mean draft of 3.5 m, discharged her cargo, loaded 200 tonnes of bunkers, and completed with a mean draft of 2 m, find how much cargo she discharged.

(c) Assuming that the ship's light draft is 1 m, find the deadweight when the ship is floating in salt water at a mean draft of 1.75 m.

6. (a) From the following information construct a displacement curve:

Displacement (tonnes)	320	880	1420	2070	2800	3680
Draft (m)	1	1.5	2	2.5	3	3.5

(b) If this ship's draft light is 1.1 m and the load draft 3.5 m, find the deadweight.
(c) If the vessel had on board 300 tonnes of cargo, 200 tonnes of ballast, and 60 tonnes of fresh water and stores, what would be the salt water mean draft?

7. (a) Construct a displacement curve from the following data:

Draft (m)	1	2	3	4	5	6
Displacement (tonnes)	335	767	1270	1800	2400	3100

(b) The ship commenced loading at 3 m mean draft and, when work ceased for the day, the mean draft was 4.2 m. During the day 85 tonnes of salt water ballast had been pumped out. Find how much cargo had been loaded.
(c) If the ship's light draft was 2 m find the mean draft after she had taken in 870 tonnes of water ballast and 500 tonnes of bunkers.
(d) Find the TPC at 3 m mean draft.

8. (a) From the following information construct a displacement curve:

Draft (m)	1	2	3	4	5	6
Displacement (tonnes)	300	1400	3200	5050	7000	9000

(b) If the ship is floating at a mean draft of 3.2 m, and then loads 1800 tonnes of cargo and 200 tonnes of bunkers, and also pumps out 450 tonnes of water ballast, find the new displacement and final mean draft.
(c) At a certain draft the ship discharged 1700 tonnes of cargo and loaded 400 tonnes of bunkers. The mean draft was then found to be 4.5 m. Find the original mean draft.

■

CHAPTER 5

Form Coefficients

The Coefficient of Fineness of the Waterplane Area (C_w)

The coefficient of fineness of the waterplane area is the ratio of the area of the waterplane to the area of a rectangle having the same length and maximum breadth.

In Figure 5.1 the area of the ship's waterplane is shown shaded and ABCD is a rectangle having the same length and maximum breadth:

$$\text{Coefficent of fineness } (C_W) = \frac{\text{Area of waterplane}}{\text{Area of rectangle ABCD}}$$

$$= \frac{\text{Area of waterplane}}{L \times B}$$

$$\therefore \text{Area of the waterplane} = L \times B \times C_W$$

Example 1

Find the area of the waterplane of a ship 36 meters long, 6 meters beam, which has a coefficient of fineness of 0.800.

$$\text{Area of waterplane} = L \times B \times C_W$$

$$= 36 \times 6 \times 0.800$$

Ans. Area of waterplane = 173 square meters.

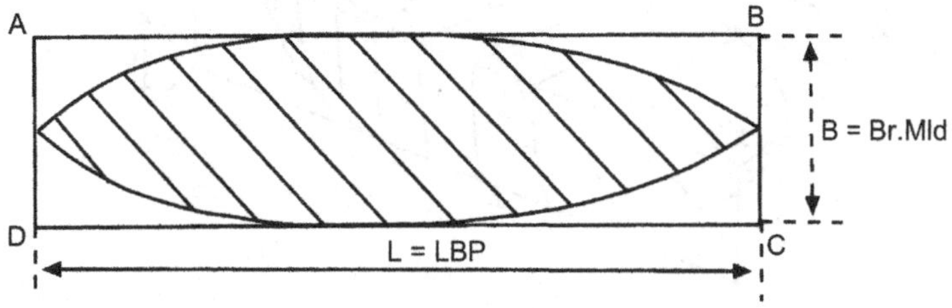

Figure 5.1

Ship Stability for Masters and Mates. http://dx.doi.org/10.1016/B978-0-08-097093-6.00005-0

Example 2

A ship 128 meters long has a maximum beam of 20 meters at the waterline, and coefficient of fineness of 0.850. Calculate the TPC at this draft in salt water.

$$\text{Area of waterplane} = L \times B \times C_w$$
$$= 128 \times 20 \times 0.850$$
$$= 2176 \text{ square meters}$$
$$TPC_{SW} = \frac{WPA}{97.56}$$
$$= \frac{2176}{97.56}$$

Ans. TPC $_{SW}$ = 22.30 tonnes.

The Block Coefficient of Fineness of Displacement (C_b)

The block coefficient of a ship at any particular draft is the ratio of the volume of displacement at that draft to the volume of a rectangular block having the same overall length, breadth, and depth.

In Figure 5.2 the shaded portion represents the volume of the ship's displacement at the draft concerned, enclosed in a rectangular block having the same overall length, breadth, and depth.

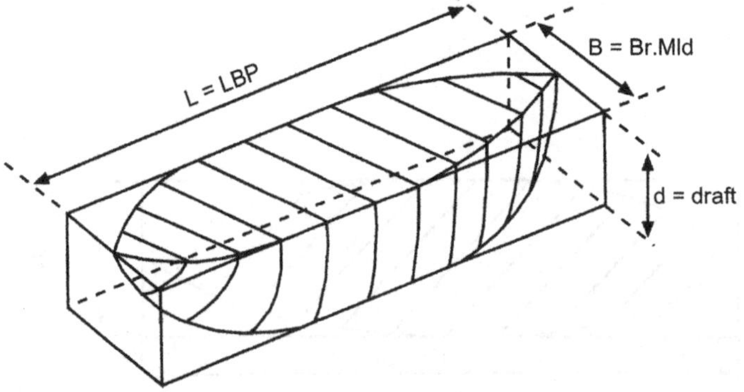

Figure 5.2

$$\text{Block coefficient } (C_b) = \frac{\text{Volume of displacement}}{\text{Volume of the block}}$$

$$= \frac{\text{Volume of displacement}}{L \times B \times \text{draft}}$$

$$\therefore \text{Volume of displacement} = L \times B \times \text{draft} \times C_b$$

Example 3

A ship 64 meters long, 10 meters maximum beam, has a light draft of 1.5 meters and a load draft of 4 meters. The block coefficient of fineness is 0.600 at the light draft and 0.750 at the load draft. Find the deadweight.

$$\begin{aligned}
\text{Light displacement} &= (L \times B \times \text{draft} \times C_b) \text{ cubic meters} \\
&= 64 \times 10 \times 1.5 \times 0.600 \\
&= 576 \text{ cubic meters} \\
\text{Load displacement} &= (L \times B \times \text{draft} \times C_b) \text{ cubic meters} \\
&= 64 \times 10 \times 4 \times 0.750 \\
&= 1920 \text{ cubic meters} \\
\text{Deadweight} &= \text{Load displacement} - \text{Light displacement} \\
&= (1920 - 576) \text{ cubic meters} \\
\text{Deadweight} &= 1344 \text{ cubic meters} \\
&= 1344 \times 1.025 \text{ tonnes}
\end{aligned}$$

Ans. Deadweight = 1378 tonnes.

The Midships Coefficient (C_m)

The midships coefficient to any draft is the ratio of the transverse area of the midships section (A_m) to a rectangle having the same breadth and depths.

In Figure 5.3 the shaded portion represents the area of the midships section to the waterline WL, enclosed in a rectangle having the same breadth and depth.

$$\text{Midships coefficient } (C_m) = \frac{\text{Midships area } (A_m)}{\text{Area of rectangle}}$$

$$= \frac{\text{Midships area } (A_m)}{B \times d}$$

or

$$\text{Midships area } (A_m) = B \times d \times C_m$$

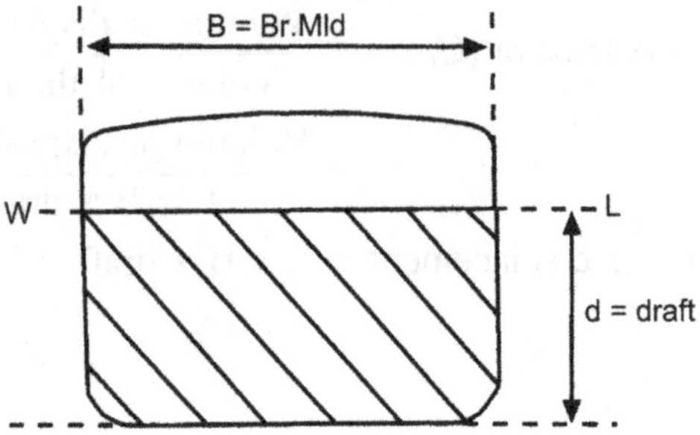

Figure 5.3

The Prismatic Coefficient (C_p)

The prismatic coefficient of a ship at any draft is the ratio of the volume of displacement at that draft to the volume of a prism having the same length as the ship and the same cross-sectional area as the ship's midships area. The prismatic coefficient is used mostly by ship-model researchers.

In Figure 5.4 the shaded portion represents the volume of the ship's displacement at the draft concerned, enclosed in a prism having the same length as the ship and a cross-sectional area equal to the ship's midships area (A_m).

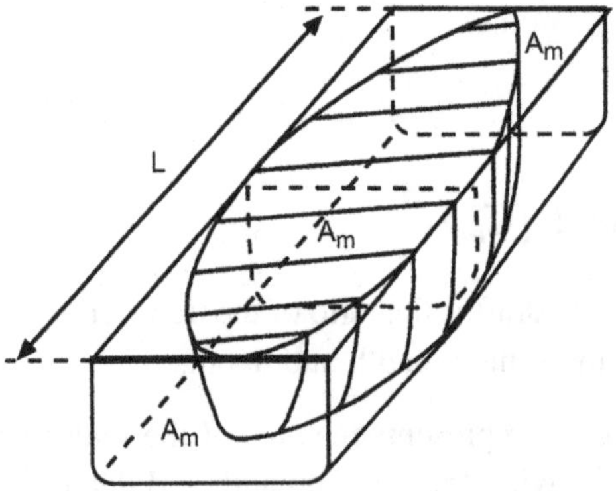

Figure 5.4

$$\text{Prismatic coefficient } (C_p) = \frac{\text{Volume of ship}}{\text{Volume of prism}}$$

$$= \frac{\text{Volume of ship}}{L \times A_m}$$

or

$$\text{Volume of ship} = L \times A_m \times C_p$$

Note

$$C_m \times C_p = \frac{A_m}{B \times d} \times \frac{\text{Volume of ship}}{L \times A_m}$$

$$= \frac{\text{Volume of ship}}{L \times B \times d}$$

$$= C_b$$

$$\therefore C_m \times C_p = C_b \qquad \text{or}$$

$$C_p = \frac{C_b}{C_m}$$

Note that C_p is always slightly higher than C_b at each waterline.

Having described exactly what C_w, C_b, C_m, and C_p are, it would be useful to know what their values would be for several ship types. For most merchant vessels, the coefficients are usually rounded off to *three* decimal places. However, for very large vessels such as supertankers or ULCCs, they are rounded off to *four* decimal places.

First of all it must be remembered that all of these form coefficients will never be more than unity. To be so is not physically possible.

For the C_b values at *fully loaded drafts* Table 5.1 gives good typical values.

To estimate a value for C_w for these ship types at their *fully loaded* drafts, it is useful to use the following rule-of-thumb approximation:

$$C_w = \left(\frac{2}{3} \times C_b\right) + \frac{1}{3} \quad \underline{\text{@ Draft Mld only!}}$$

Table 5.1: Typical C_b values at fully loaded drafts.

Ship Type	Typical C_b Fully Loaded	Ship Type	Typical C_b Fully Loaded
ULCC	0.850	General cargo ship	0.700
Supertanker	0.825	Passenger liner	0.575–0.625
Oil tanker	0.800	Container ship	0.575
Bulk carrier	0.775–0.825	Coastal tug	0.500

Medium-form ships (C_b approx. 0.700), full-form ships ($C_b > 0.700$), fine-form ships ($C_b < 0.700$).

Hence, for the oil tanker, C_w would be 0.867, for the general cargo ship C_w would be 0.800, and for the tug C_w would be 0.667 in fully loaded conditions.

For merchant ships, the midships coefficient or midship area coefficient is 0.980–0.990 at fully loaded draft. It depends on the rise of floor and the bilge radius. Rise of floor is almost obsolete nowadays.

As shown before,

$$C_p = \frac{C_b}{C_m}$$

Hence for a bulk carrier, if C_b is 0.750 with a C_m of 0.985, the C_p will be:

$$C_p = \frac{0.750}{0.985} = 0.761 \ \underline{@ \text{ Draft Mld}}$$

C_p is used mainly by researchers at ship-model tanks carrying out tests to obtain the least resistance for particular hull forms of prototypes.

C_b and C_w change as the drafts move from fully loaded to light-ballast to lightship conditions. Figure 5.5 shows the curves at drafts below the fully loaded draft for a general cargo ship of 135.5 m LBP.

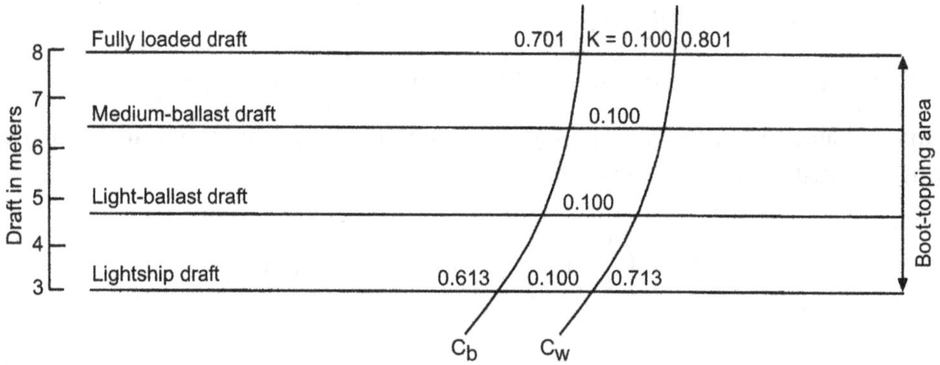

Figure 5.5: Variation of C_b and C_w Values with Draft.
Note how the two curves are parallel at a distance of 0.100 apart.

'K' is calculated for the fully loaded condition and is *held constant* for all remaining drafts down to the ship's lightship (empty ship) waterline.

Example

A ship is 64 m long, 10 m wide, and has a load draft of 4 m in salt water. Midship area = 39.40 square meters, W = 1968 tonnes, waterplane area = 533 square meters.

Calculate the design coefficients C_b, C_m, and C_p. Also calculate the TPC at this 4.00 m draft.

L	b	Loaded Draft	Density	Midship Area	W	WPA
64	10.00	4.00	1.025	39.4	1968	533

$C_b = 0.750$

$C_m = 0.985$

$C_p = 0.761$

TPC = 5.46 tonnes

Exercise 5

1. (a) Define 'coefficient of fineness of the waterplane'.
   (b) The length of a ship at the waterline is 100 m, the maximum beam is 15 m, and the coefficient of fineness of the waterplane is 0.800. Find the TPC at this draft.
2. (a) Define 'block coefficient of fineness of displacement'.
   (b) A ship's length at the waterline is 120 m when floating on an even keel at a draft of 4.5 m. The maximum beam is 20 m. If the ship's block coefficient is 0.750, find the displacement in tonnes at this draft in salt water.
3. A ship is 150 m long, has 20 m beam, load draft 8 m, and light draft 3 m. The block coefficient at the load draft is 0.766, and at the light draft is 0.668. Find the ship's deadweight.
4. A ship 120 m long × 15 m beam has a block coefficient of 0.700 and is floating at the load draft of 7 m in fresh water. Find how much more cargo can be loaded if the ship is to float at the same draft in salt water.
5. A ship 100 m long, 15 m beam, and 12 m deep is floating on an even keel at a draft of 6 m, block coefficient 0.800. The ship is floating in salt water. Find the cargo to discharge so that the ship will float at the same draft in fresh water.

CHAPTER 6

Discussion on LCB Position Relative to Amidships

LCB is the three-dimensional centroid of the underwater volume of displacement. It can be measured forward of the aft perp (foap). It can be measured forward or aft of amidships. This chapter uses the latter procedure.

Assume that each merchant ship is fully loaded, upright, and on even keel. Each vessel was assumed to be single screw fitted with a horizontal propeller shaft. Vessels fitted with azipod propulsion units are considered outside the remit of this chapter (see later note).

When fully loaded, the LCB ranges from amidships to being up to 3.0% L forward or aft of amidships. It depends upon ship type, such as supertankers to general cargo ships to container vessels. In other words, from full-form vessels to medium-form vessels to fine-form vessels (see Figure 6.1).

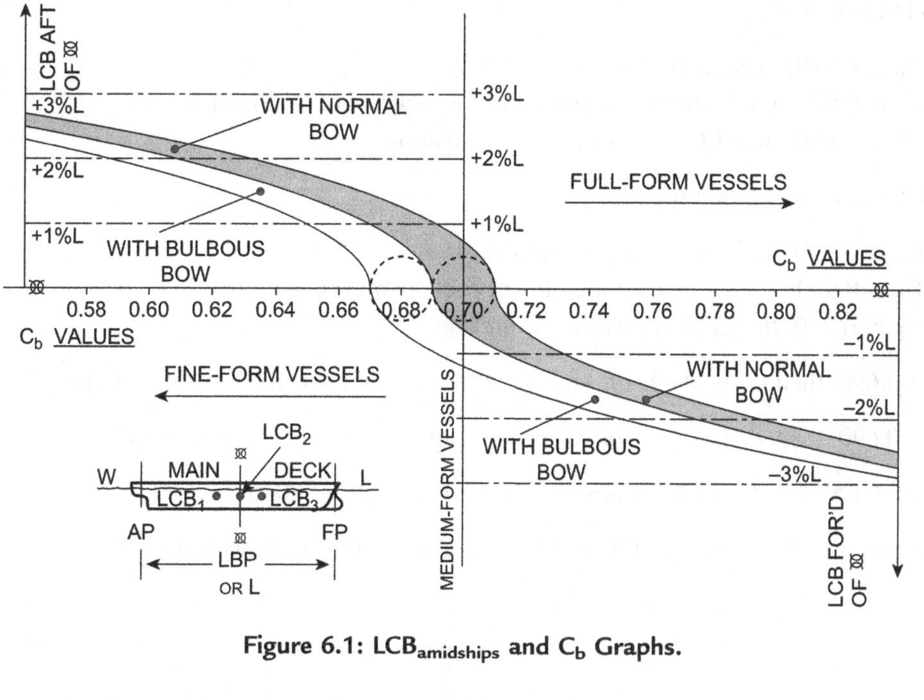

Figure 6.1: $LCB_{amidships}$ and C_b Graphs.

Ship Stability for Masters and Mates. http://dx.doi.org/10.1016/B978-0-08-097093-6.00006-2

It will also depend on several other factors:

1. LCB position will change depending upon the type of bow. Instead of a soft-nosed bow, it could be a bulbous bow or maybe a ram bow. This change in the forward underwater form will obviously change the position of LCB.
2. LCB will tend to move aft if (for a similar LBP) the ship's service speed is comparatively greater. C_b will generally have reduced in value.
3. LCB will have moved aft if (for a similar service speed) the ship's LBP is comparatively less.
4. LCB will move aft if $V/\sqrt{L}$ has increased.
5. LCB will change if LBP/(Br. Mld) is decreased. In the last two decades, this L/B for oil tankers has gradually reduced from being 6.25 to 6.00 to 5.75 to 5.00. Without loss in performance, modern merchant vessels tend to be shorter in length and wider in breadth molded.
6. LCB will change with type of propulsion system installed, e.g. azipod electro-diesel propulsion units rather than fitting a propeller at the end of a horizontal propeller shaft. Aft end contours will be very different. There will be no need for a sternframe, a solepiece, and a propeller aperture. This change in aft underwater form will obviously change the position of LCB.

Observations Regarding Figure 6.1

With a bulbous bow:

- If $C_b < 0.670$, then LCB is *aft* of amidships.
- If C_b is 0.670–0.690, the LCB can be *aft* or *forward* of amidships.
- If $C_b > 0.690$, then LCB is *forward* of amidships.

With a normal soft-nosed bow:

- If $C_b < 0.690$, then LCB is *aft* of amidships.
- If C_b is 0.690–0.710, then LCB can be *aft* or forward of amidships.
- If $C_b > 0.710$, the LCB is *forward* of amidships.

If $C_b = 0.690$, then C_b is 1% L *aft* amidships if a normal soft-nosed bow is fitted.

If $C_b = 0.690$, then C_b is 1% L *forward* amidships if a bulbous bow is fitted.

A range of 1.50% L for LCB positions is at C_b values of 0.680 and 0.700.

Lowest range of 0.20% L for LCB positions is at C_b values of 0.560 and 0.840.

Formulae Used for Figure 6.1

Normal bow:

$$\text{LCB} = 7 \times \sqrt{(0.700 - \text{actual } C_b)} \text{ for fine-body vessels.}$$
$$\text{LCB} = 7 \times \sqrt{(\text{actual } C_b - 0.700)} \text{ for full-body vessels.}$$

Bulbous bow:

$$\text{LCB} = 7 \times \sqrt{(0.680 - \text{actual } C_b)} \text{ for fine-body vessels.}$$
$$\text{LCB} = 7 \times \sqrt{(\text{actual } C_b - 0.680)} \text{ for full-body vessels.}$$

Fine-Body Boundary Lines Formulae (see Table 6.1)

$$\text{LCB} = 7 \times \sqrt{(0.710 - \text{actual } C_b)} \text{ for fine-body vessels.}$$
$$\text{LCB} = 7 \times \sqrt{(0.690 - \text{actual } C_b)} \text{ for fine-body vessels.}$$
$$\text{LCB} = 7 \times \sqrt{(0.670 - \text{actual } C_b)} \text{ for fine-body vessels.}$$

Full-Body Boundary Lines Formulae (see Table 6.1)

$$\text{LCB} = 7 \times \sqrt{(\text{actual } C_b - 0.710)} \text{ for full-body vessels.}$$
$$\text{LCB} = 7 \times \sqrt{(\text{actual } C_b - 0.690)} \text{ for full-body vessels.}$$
$$\text{LCB} = 7 \times \sqrt{(\text{actual } C_b - 0.670)} \text{ for full-body vessels.}$$

Observations and Formulae Regarding Figure 6.2

This show graphically the straight-line relationship between F_n and C_b for a range of hull forms. Let L = Ship's LBP in meters.

Naval Architects use the format of $V/\sqrt{L}$ in their work (V in kts and L in meters).

Ship model researchers use William Froude's formula (circa 1870), $F_n = (V/g\sqrt{L})$, where V is in m/s, g is acceleration due to gravity = 9.806 m/s^2, and L is in meters. F_n is a coefficient and thus dimensionless.

$$\text{Froude number } F_n = V/\sqrt{g \times L},\ C_b = \{1.20 - (F_n/0.421)\}$$
$$F_n = (1.20 - C_b) \times 0.421$$
$$\text{Global formula: } C_b = 1.20 - 0.39 \times (V/\sqrt{L}) \text{ for merchant ships.}$$

Table 6.1: Relationship between LCB about Amidships and C_b Values for Merchant Vessels.

C_b Values	LCB_1 (% L)	LCB_2 (% L)	LCB_3 (% L)	C_b Values	LCB_4 (% L)	LCB_5 (% L)	LCB_6 (% L)
	All Aft of Amidships				*All Forward of Amidships*		
0.560	2.71	2.52	2.32	0.820	−2.71	−2.52	−2.32
0.580	2.52	2.32	2.10	0.800	−2.52	−2.32	−2.10
0.600	2.32	2.10	1.85	0.780	−2.32	−2.10	−1.85
0.620	2.10	1.85	1.57	0.760	−2.10	−1.85	−1.57
0.640	1.85	1.57	1.21	0.740	−1.85	−1.57	−1.21
0.660	1.57	1.21	0.70	0.720	−1.57	−1.21	−0.70
0.662	1.53	1.17	0.63	0.718	−1.53	−1.17	−0.63
0.664	1.50	1.13	0.54	0.716	−1.50	−1.13	−0.54
0.666	1.47	1.08	0.44	0.714	−1.47	−1.08	−0.44
0.668	1.43	1.04	0.31	0.712	−1.43	−1.04	−0.31
0.670	1.40	0.99	0.00	0.710	−1.40	−0.99	0.00
0.672	1.36	0.94		0.708	−1.36	−0.94	
0.674	1.33	0.89		0.706	−1.33	−0.89	
0.676	1.29	0.83		0.704	−1.29	−0.83	
0.678	1.25	0.77		0.702	−1.25	−0.77	
0.680	1.21	0.70		0.700	−1.21	−0.70	
0.682	1.17	0.63		0.698	−1.17	−0.63	
0.684	1.13	0.54		0.696	−1.13	−0.54	
0.686	1.08	0.44		0.694	−1.08	−0.44	
0.688	1.04	0.31		0.692	−1.04	−0.31	
0.690	0.99	0.00		0.690	−0.99	0.00	
0.692	0.94			0.688	−0.94		
0.694	0.89			0.686	−0.89		
0.696	0.83			0.684	−0.83		
0.698	0.77			0.682	−0.77		
0.700	0.70			0.680	−0.70		
0.702	0.63			0.678	−0.63		
0.704	0.54			0.676	−0.54		
0.706	0.44			0.674	−0.44		
0.708	0.31			0.672	−0.31		
0.710	0.00			0.670	0.00		

LCB_1 (% L) $= 7 \times (0.710 - C_b)^{0.50}$.
LCB_2 (% L) $= 7 \times (0.690 - C_b)^{0.50}$.
LCB_3 (% L) $= 7 \times (0.670 - C_b)^{0.50}$.
LCB_4 (% L) $= 7 \times (C_b - 0.710)^{0.50}$.
LCB_5 (% L) $= 7 \times (C_b - 0.690)^{0.50}$.
LCB_6 (% L) $= 7 \times (C_b - 0.670)^{0.50}$.
0.00 is Station 5 positioned at amidships.

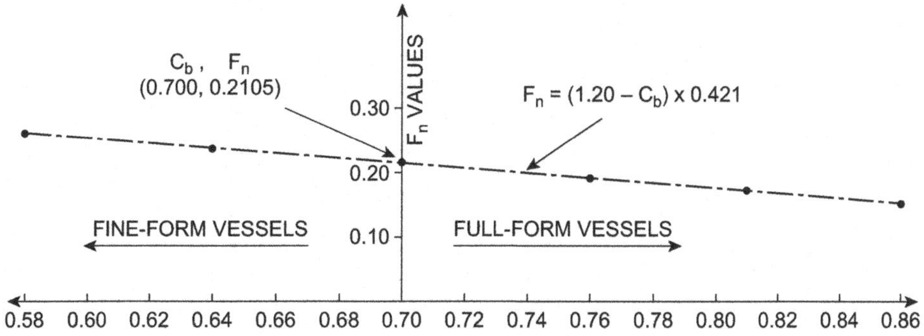

Figure 6.2: Relationship Between F_n and C_b Coefficients.

Observations and Formulae Regarding Figure 6.3

Figure 6.3 is basically Figure 6.2 superimposed onto a mean line plotted in Figure 6.1. Note the crossover points at zero LCB about amidships, of 0.700 for the C_b and 0.2105 for the Fn value. Select the variable C_b or F_n and insert into the appropriate formula to obtain the LCB about amidships in terms of a percentage L.

Example of Crosslink

$$LCB = 7\sqrt{(0.700 - \text{actual } C_b)} \text{ or } 10.75\sqrt{(\text{actual } F_n - 0.2105)} \text{ for fine-form ships.}$$

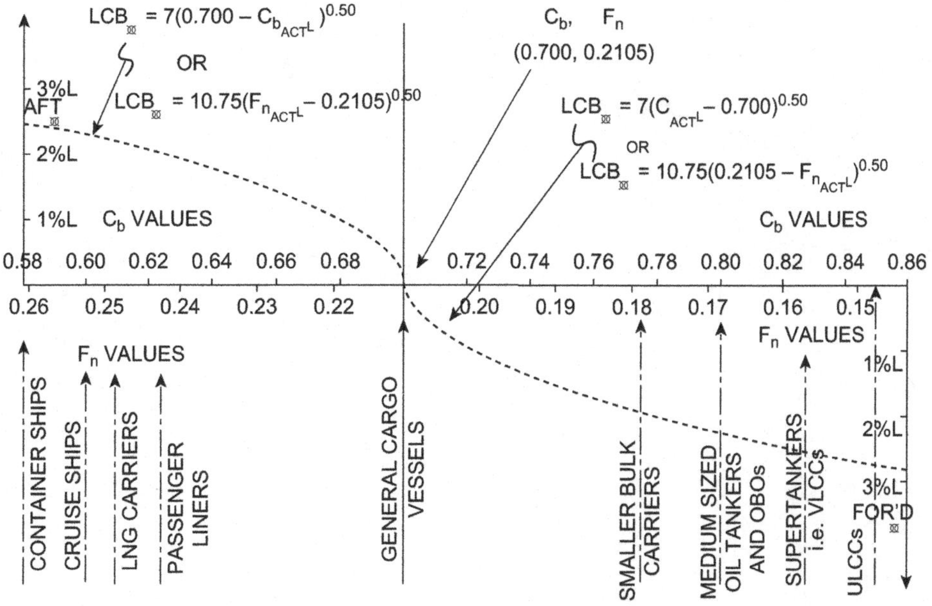

Figure 6.3: Relationship Between $LCB_{amidships}$ and (C_b, F_n) Coefficients (Mean Values).
See also Figure 6.2.

CHAPTER 7

Quadrature – Simpson's Rules for Areas and Centroids

Areas and Volumes

Simpson's Rules may be used to find the areas and volumes of irregular figures. The rules are based on the assumption that the boundaries of such figures are curves that follow a definite mathematical law. When applied to ships they give a good approximation of areas and volumes. The accuracy of the answers obtained will depend upon the spacing of the ordinates and upon how closely the curve follows the law.

Simpson's First Rule

This rule assumes that the curve is a parabola of the second order. A parabola of the second order is one whose equation, referred to coordinate axes, is of the form $y = a_0 + a_1x + a_2x^2$, where a_0, a_1, and a_2 are constants.

Let the curve in Figure 7.1 be a parabola of the second order. Let y_1, y_2, and y_3 be three ordinates equally spaced at 'h' units apart.

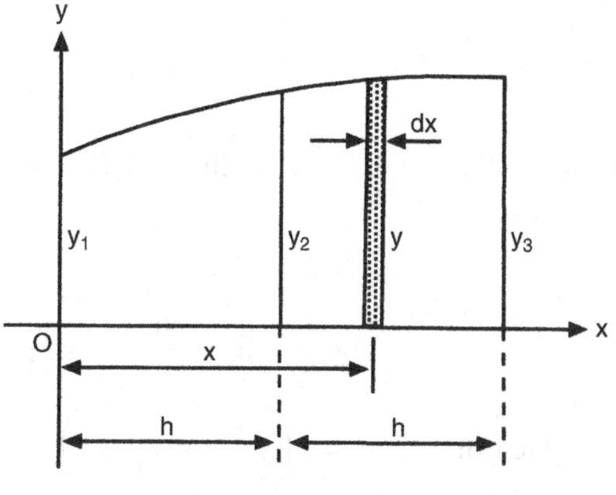

Figure 7.1

Ship Stability for Masters and Mates. http://dx.doi.org/10.1016/B978-0-08-097093-6.00007-4

The area of the elementary strip is y dx. Then the area enclosed by the curve and the axes of reference is given by:

$$\text{Area of figure} = \int_{O}^{2h} y\,dx$$

But

$$y = a_0 + a_1x + a_2x^2$$

$$\therefore \text{Area of figure} = \int_{O}^{2h} \left(a_0 + a_1x + a_2x^2\right) dx$$

$$= \left[a_0x + \frac{a_1x^2}{2} + \frac{a_2x^3}{3}\right]_0^{2h}$$

$$= 2a_0h + 2a_1h^2 + \frac{8}{3}a_2h^3$$

Assume that the area of figure $= Ay_1 + By_2 + Cy_3$

Using the equation of the curve and substituting 'x' for O, h, and 2h respectively:

$$\begin{aligned}\text{Area of figure} &= Aa_0 + B\left(a_0 + a_1h + a_2h^2\right)\\ &\quad + C\left(a_0 + 2a_1h + 4a_2h^2\right)\\ &= a_0(A + B + C) + a_1h(B + 2C)\\ &\quad + a_2h^2(B + 4C)\end{aligned}$$

$$\begin{aligned}\therefore 2a_0h + 2a_1h^2 + \frac{8}{3}a_2h^3 &= a_0(A + B + C) + a_1h(B + 2C)\\ &\quad + a_2h^2(B + 4C)\end{aligned}$$

Equating coefficients:

$$A + B + C = 2h, \qquad B + 2C = 2h, \qquad B + 4C = \frac{8}{3}h$$

From which:

$$A = \frac{h}{3} \quad B = \frac{4h}{3} \quad C = \frac{h}{3}$$

$$\therefore \text{Area of figure} = \frac{h}{3}\left(y_1 + 4y_2 + y_3\right)$$

This is Simpson's First Rule.

It should be noted that Simpson's First Rule can also be used to find the area under a curve of the third order, i.e. a curve whose equation, referred to the coordinate axes, is of the form $y = a_0 + a_1x + a_2x^2 + a_3x^3$, where a_0, a_1, a_2, and a_3 are constants.

Summary

A coefficient of 1/3 with multipliers of 1, 4, 1, etc.

Simpson's Second Rule

This rule assumes that the equation of the curve is of the third order, i.e. of a curve whose equation, referred to the coordinate axes, is of the form $y = a_0 + a_1x + a_2x^2 + a_3x^3$, where a_0, a_1, a_2, and a_3 are constants.

In Figure 7.2:

$$\text{Area of elementary strip} = y\,dx$$

$$\text{Area of the figure} = \int_0^{3h} y\,dx$$

$$= \int_0^{3h} \left(a_0 + a_1x + a_2x^2 + a_3x^3\right) dx$$

$$= \left[a_0x + \frac{1}{2}a_1x^2 + \frac{1}{3}a_2x^3 + \frac{1}{4}a_3x^4\right]_0^{3h}$$

$$= 3a_0h + \frac{9}{2}a_1h^2 + 9a_2h^3 + \frac{81}{4}a_3h^4$$

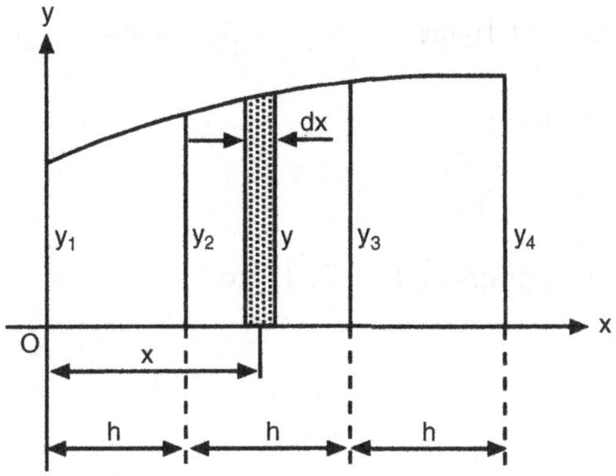

Figure 7.2

$$\begin{aligned}\text{Let the area of the figure} &= Ay_1 + By_2 + Cy_3 + Dy_4 \\ &= Aa_0 + B(a_0 + a_1h + a_2h^2 + a_3h^3) \\ &\quad + C(a_0 + 2a_1h + 4a_2h^2 + 8a_3h^3) \\ &\quad + D(a_0 + 3a_1h + 9a_2h^2 + 27a_3h^3) \\ &= a_0(A + B + C + D) + a_1h(B + 2C + 3D) \\ &\quad + a_2h^2(B + 4C + 9D) + a_3h^3(B + 8C + 27D)\end{aligned}$$

Equating coefficients:

$$\begin{aligned}A + B + C + D &= 3h \\ B + 2C + 3D &= \frac{9}{2}h \\ B + 4C + 9D &= 9\,h \\ B + 8C + 27D &= \frac{81}{4}h\end{aligned}$$

From which:

$$A = \frac{3}{8}h, \quad B = \frac{9}{8}h, \quad C = \frac{9}{8}h, \quad D = \frac{3}{8}h$$

$$\therefore \text{Area of figure} = \frac{3}{8}hy_1 + \frac{9}{8}hy_2 + \frac{9}{8}hy_3 + \frac{3}{8}hy_4$$

or

$$\text{Area of figure} = \frac{3}{8}h(y_1 + 3y_2 + 3y_3 + y_4)$$

This is Simpson's Second Rule.

Summary

A coefficient of 3/8 with multipliers of 1, 3, 3, 1, etc.

Simpson's Third Rule

In Figure 7.3:

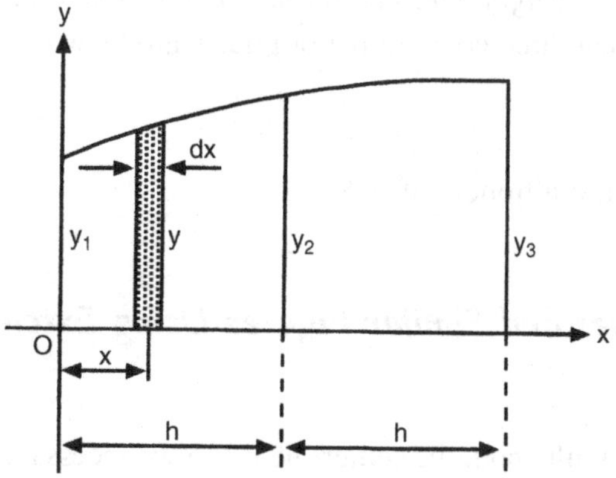

Figure 7.3

$$\text{Area of the elementary strip} = y\,dx$$

$$\text{Area between } y_1 \text{ and } y_2 \text{ in figure} = \int_O^h y\,dx$$

$$= a_0h + \frac{1}{2}a_1h^2 + \frac{1}{3}a_2h^3$$

$$\text{Let the area between } y_1 \text{ and } y_2 = Ay_1 + By_2 + Cy_3$$

$$\text{Then area} = Aa_0 + B(a_0 + a_1h + a_2h^2)$$
$$+ C(a_0 + 2a_1h + 4a_2h^2)$$
$$= a_0(A + B + C) + a_1h(B + 2C)$$
$$+ a_2h^2(B + 4C)$$

Equating coefficients:

$$A + B + C = h, \quad B + 2C = h/2, \quad B + 4C = h/3$$

From which:

$$A = \frac{5h}{12}, \quad B = \frac{8h}{12}, \quad C = -\frac{h}{12}$$

$$\therefore \text{Area of figure between } y_1 \text{ and } y_2 = \frac{5}{12}hy_1 + \frac{8}{12}hy_2 + \left(-\frac{1}{12}hy_3\right)$$

or

$$\text{Area} = \frac{h}{12}(5y_1 + 8y_2 - y_3)$$

This is the *Five/Eight (or five/eight minus one) Rule*, and is used to find the area between two consecutive ordinates when three consecutive ordinates are known.

Summary

A coefficient of 1/12 with multipliers of 5, 8, −1, etc.

Areas of Waterplanes and Similar Figures Using Extensions of Simpson's Rules

Since a ship is uniformly built about the centerline it is only necessary to calculate the area of half the waterplane and then double the area found to obtain the area of the whole waterplane.

Figure 7.4 represents the starboard side of a ship's waterplane area. To find the area, the centerline is divided into a number of equal lengths, each 'h' m long. The length 'h' is called the *common interval*. The half-breadths, a, b, c, d, etc., are then measured and each of these is called a *half-ordinate*.

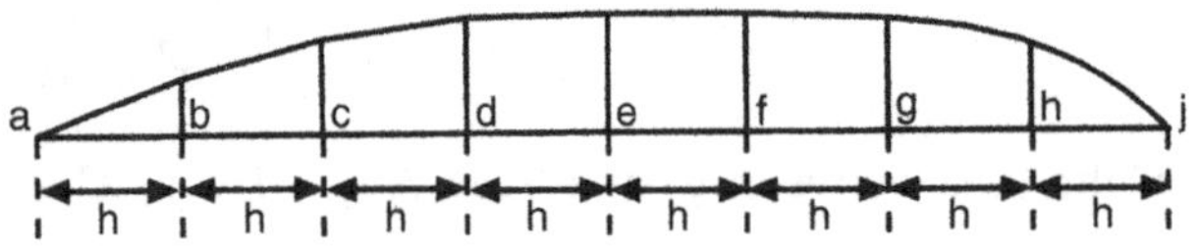

Figure 7.4

Using Simpson's First Rule

This rule can be used to find areas when there are an odd number of ordinates.

$$\text{Area of Figure 7.5(a)} = \frac{h}{13}(a + 4b + c)$$

If the common interval and the ordinates are measured in meters, the area found will be in square meters.

(a)

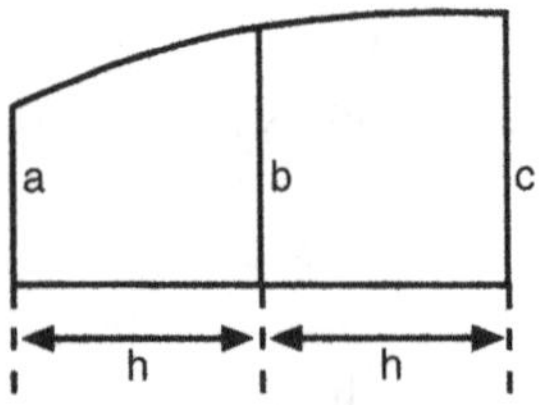

Figure 7.5(a)

Let this rule now be applied to a waterplane area such as that shown in Figure 7.5(b).

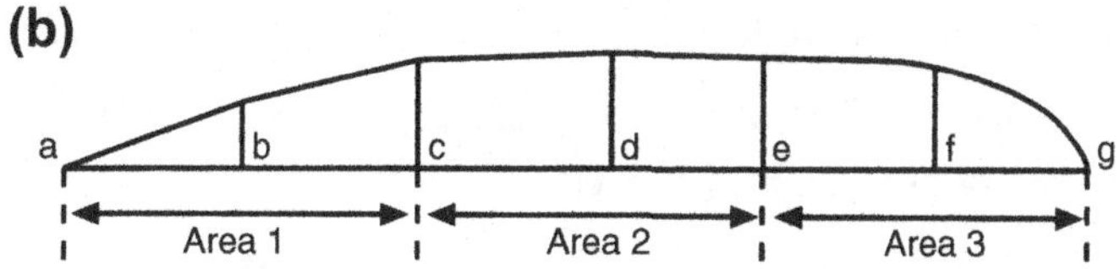

Figure 7.5(b)

	1	4	1		1	4	1
			+		+		
			1	4	1		
Combined multiplier	1	4	2	4	2	4	1

The waterplane is divided into three separate areas and Simpson's First Rule is used to find each separate area:

$$\text{Area 1} = h/3(a + 4b + c)$$

$$\text{Area 2} = h/3(c + 4d + e)$$

$$\text{Area 3} = h/3(e + 4f + g)$$

$$\text{Area of } \frac{1}{2}\,\text{WP} = \text{Area 1} + \text{Area 2} + \text{Area 3}$$

$$\therefore \text{Area of } \frac{1}{2}\,\text{WP} = h/3(a + 4b + c) + h/3(c + 4d + e) + h/3(e + 4f + g)$$

or

$$\text{Area of } \frac{1}{2}\,\text{WP} = h/3(a + 4b + 2c + 4d + 2e + 4f + g)$$

This is the form in which the formula should be used. Within the brackets the half-ordinates appear in their correct sequence from forward to aft. The coefficients of the half-ordinates are referred to as Simpson's Multipliers and they are in the form: 1424241. Had there been nine half-ordinates, the multipliers would have been: 142424241. It is usually much easier to set out that part of the problem within the brackets in tabular form. Note how Simpson's Multipliers begin and end with 1, as shown in Figure 7.5(b).

Example

A ship 120 meters long at the waterline has equidistantly spaced half-ordinates commencing from forward as follows:

0, 3.7, 5.9, 7.6, 7.5, 4.6, and 0.1 meters respectively.

Find the area of the waterplane and the TPC at this draft. Water density $= 1.025\ t/m^3$.

Note. There is an odd number of ordinates in the waterplane and therefore the First Rule can be used.

No.	1/2 Ord.	SM	Area Function
a	0	1	0
b	3.7	4	14.8
c	5.9	2	11.8
d	7.6	4	30.4
e	7.5	2	15.0
f	4.6	4	18.4
g	0.1	1	0.1
			$90.5 = \Sigma_1$

Σ_1 is used because it is a total; using Simpson's First Rule:

$$h = \frac{120}{6} = \text{the common interval CI}$$

$$\therefore \text{CI} = 20 \text{ meters}$$

$$\text{Area of WP} = \frac{1}{3} \times \text{CI} \times \Sigma_1 \times 2 \text{ (for both sides)}$$

$$= \frac{1}{3} \times 20 \times 90.5 \times 2$$

Ans. Area of WP = 1207 square meters.

$$\text{TPC}_{\text{SW}} = \frac{\text{WPA}}{97.56} = \frac{1207}{97.56}$$

Ans. TPC = 12.37 tonnes.

Note. If the half-ordinates are used in these calculations, then the area of half the waterplane is found. If, however, the whole breadths are used, the total area of the waterplane will be found. If half-ordinates are given and the WPA is requested, simply multiply by 2 at the end of the formula as shown above.

Using the Extension of Simpson's Second Rule

This rule can be used to find the area when the number of ordinates is such that if one is subtracted from the number of ordinates the remainder is divisible by 3.

$$\text{Area of Figure 7.6(a)} = \frac{3}{8}h(a + 3b + 3c + d)$$

Now consider a waterplane area that has been divided up using seven half-ordinates, as shown in Figure 7.6(b). The waterplane can be split into two sections as shown, each section having four ordinates.

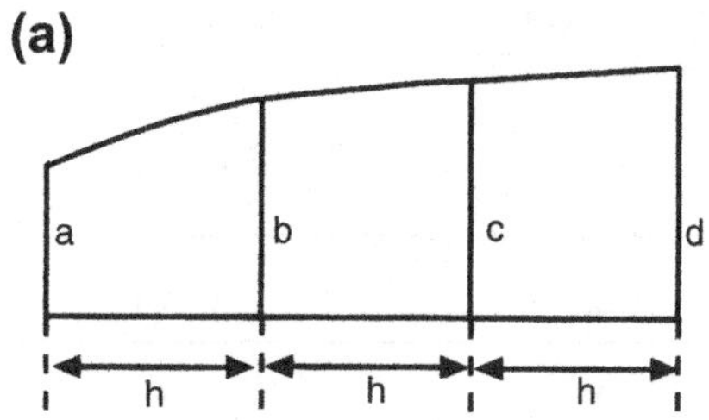

Figure 7.6(a)

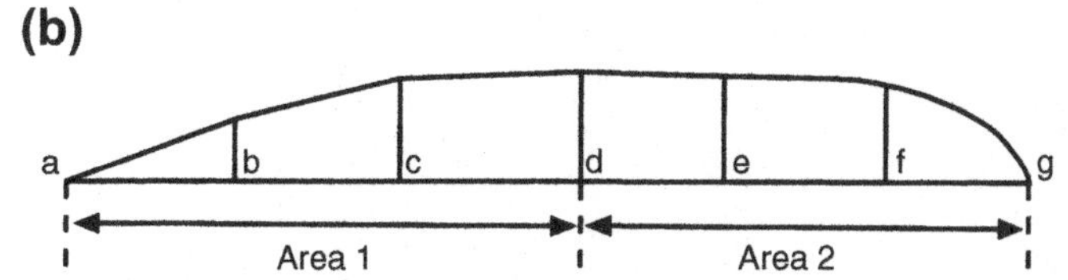

Figure 7.6(b)

	1	3	3	1			
				+			
				1	3	3	1
Combined multipliers	1	3	3	2	3	3	1

$$\text{Area 1} = \frac{3}{8}h(a + 3b + 3c + d)$$

$$\text{Area 2} = \frac{3}{8}h(d + 3e + 3f + g)$$

$$\text{Area of } \frac{1}{2}\text{WP} = \text{Area 1} + \text{Area 2}$$

$$\therefore \text{Area of } \frac{1}{2}\text{WP} = \frac{3}{8}h(a + 3b + 3c + d) + \frac{3}{8}h(d + 3e + 3f + g)$$

or

$$\text{Area of } \frac{1}{2}\text{ WP} = \frac{3}{8}\text{h}(\text{a} + 3\text{b} + 3\text{c} + 2\text{d} + 3\text{e} + 3\text{f} + \text{g})$$

This is the form in which the formula should be used. As before, all of the ordinates appear in their correct order within the brackets. The multipliers are now 1332331. Had there been 10 ordinates the multipliers would have been 1332332331. Note how Simpson's Multipliers begin and end with 1, as shown in Figure 7.6(b).

Example

Find the area of the waterplane described in the first example using Simpson's Second Rule.

Given information: Length of ship 120 m, number of intervals 6, common interval 20 m.

No.	1/2 Ord.	SM	Area Function
A	0	1	0
B	3.7	3	11.1
C	5.9	3	17.7
D	7.6	2	15.2
E	7.5	3	22.5
F	4.6	3	13.8
G	0.1	1	0.1
			$80.4 = \Sigma_2$

Σ_2 is used because it is a total; using Simpson's Second Rule:

$$\text{Area of WP} = \frac{3}{8} \times \text{CI} \times \Sigma_2 \times \ 2 \text{ (for both sides)}$$

$$= \frac{3}{8} \times 20 \times 80.4 \times 2$$

Ans. Area WP = 1206 square meters (compared to 1207 square meters previous answer).

The small difference in the two answers shows that the area found can only be a close approximation to the correct area.

The Five/Eight Rule (Simpson's Third Rule)

This rule may be used to find the area between two consecutive ordinates when three consecutive ordinates are known (see Figure 7.6(c) and (d)).

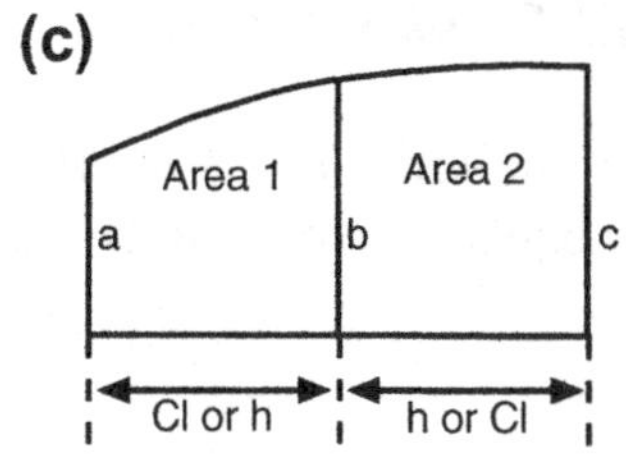

Figure 7.6(c)

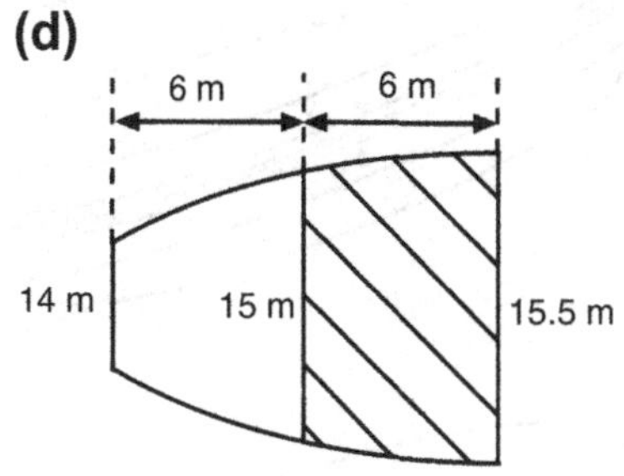

Figure 7.6(d)

The rule states that the area between two consecutive ordinates is equal to five times the first ordinate plus eight times the middle ordinate minus the external ordinate, all multiplied by 1/12 of the common interval.

$$\text{Thus: Area 1} = \frac{h}{12}(5a + 8b - c) \text{ or } \frac{1}{12} \times CI \times \Sigma_3$$

$$\text{Also: Area 2} = \frac{h}{12}(5c + 8b - a) \text{ or } \frac{1}{12} \times CI \times \Sigma_3$$

Σ_3 is used because it is a total, using Simpson's Third Rule. Consider the next example.

Example

Three consecutive ordinates in a ship's waterplane, spaced 6 meters apart, are 14, 15, and 15.5 m respectively. Find the area between the last two ordinates.

$$\text{Shaded area} = \frac{h}{12}(5a + 8b - c)$$

$$= \frac{6}{12}(77.5 + 120 - 14)$$

Ans. Area = 91.75 square meters.

Volumes of Ship Shapes and Similar Figures

Let the area of the elementary strip in Figure 7.7(a) and (b) be 'Y' square meters. Then the volume of the strip in each case is equal to Y dx and the volume of each ship is equal to $\int_{O}^{4h} Y\,dx$.

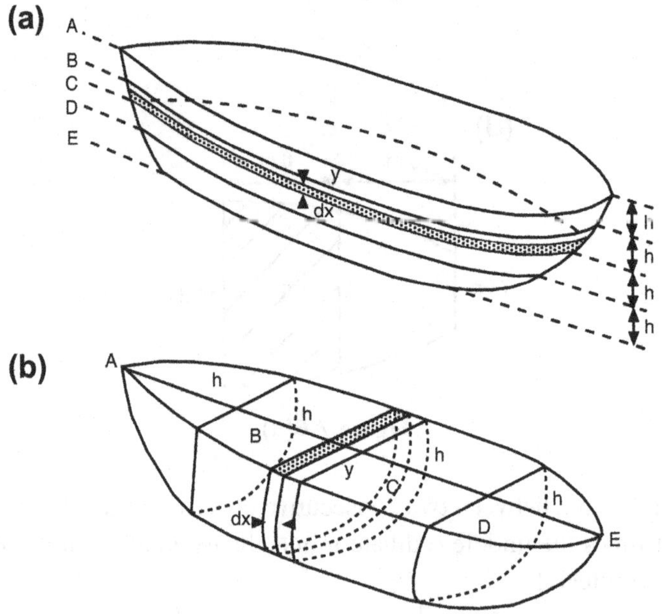

Figure 7.7

The value of the integral in each case is found by Simpson's Rules using the areas at equidistant intervals as ordinates, i.e.

$$\text{Volume} = \frac{h}{3}\,(A + 4B + 2C + 4D + E)$$

or

$$\frac{CI}{3} \times \Sigma_1$$

Thus the volume of displacement of a ship to any particular draft can be found first by calculating the areas of waterplanes or transverse areas at equidistant intervals and then using these areas as ordinates to find the volume by Simpson's Rules.

Example

The areas of a ship's waterplanes are as follows:

Draft (m)	0	1	2	3	4
Area of WP (square meters)	650	660	662	661	660

Calculate the ship's displacement in tonnes when floating in salt water at 4 meters draft. Also, if the ship's load draft is 4 meters, find the FWA.

Draft (m)	Area	SM	Volume Function
0 Keel	650	1	650
1	660	4	2640
2	662	2	1324
3	661	4	2644
4 SLWL	660	1	660
			$7918 = \Sigma_1$

Σ_1 is used because it is a total; using Simpson's First Rule:

$$\text{Underwater volume} = \frac{1}{3} \times \text{CI} \times \Sigma_1 = \frac{1}{3} \times 1.0 \times 7918$$

$$= 2639\ \frac{1}{3}\ \text{cubic meters}$$

$$\text{SW displacement} = 2639\ \frac{1}{3} \times 1.025\ \text{tonnes}$$

Ans. SW displacement = 2705 tonnes.

$$\text{Load TPC}_{SW} = \frac{\text{WPA}}{97.56}$$

$$= \frac{660}{97.56}$$

$$= 6.77\ \text{tonnes}$$

$$\text{FWA} = \frac{\text{Displacement}}{4 \times \text{TPC}}$$

$$= \frac{2705.3}{4 \times 6.77}$$

Ans. FWA = 999 mm or 9.99 cm, say 10 cm.

Appendages and Intermediate Ordinates

Appendages

It has been mentioned previously that areas and volumes calculated by the use of Simpson's Rules depend for their accuracy on the curvature of the sides following a definite mathematical law. It is very seldom that the ship's sides do follow one such curve. Consider the ship's waterplane area shown in Figure 7.8. The sides from the stern to the quarter form one curve but from this point to the stern is part of an entirely different curve. To obtain an answer that is as accurate as possible, the area from the stern to the quarter may be calculated by the use of Simpson's Rules and then the remainder of the area may be found by a second calculation. The remaining area mentioned above is referred to as an *appendage*.

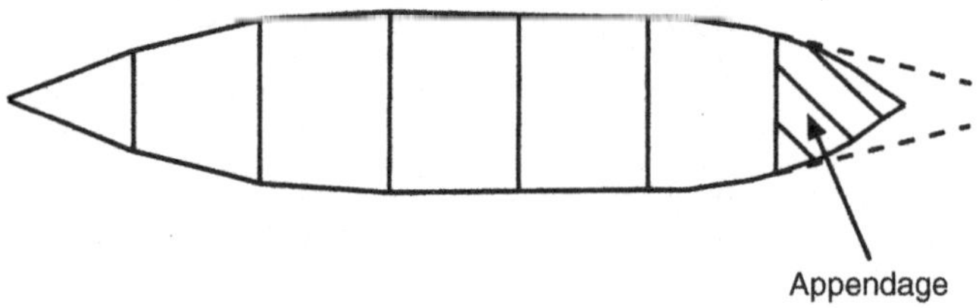

Figure 7.8

Similarly, in Figure 7.9 the side of the ship forms a reasonable curve from the waterline down to the turn of the bilge, but below this point the curve is one of a different form.

In this case the volume of displacement between the waterline (WL) and the waterplane XY could be found by use of Simpson's Rules and then the volume of the appendage found by means of a second calculation.

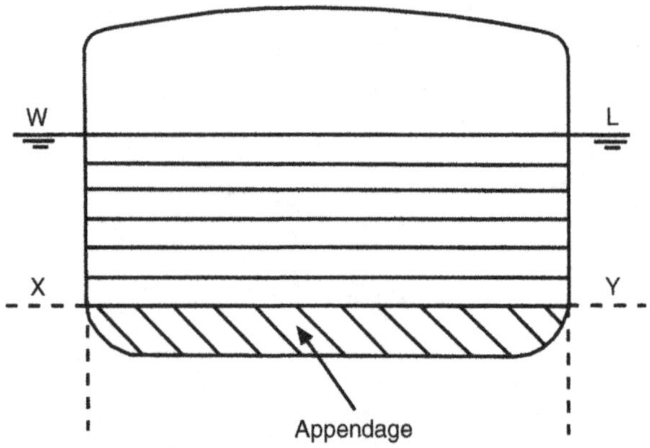

Figure 7.9

Example

A ship's breadths, at 9 m intervals commencing from forward, are as follows:

0, 7.6, 8.7, 9.2, 9.5, 9.4, and 8.5 meters respectively.

Abaft the last ordinate is an appendage of 50 square meters (see Figure 7.10). Find the total area of the waterplane.

Ord.	SM	Product for Area
0	1	0
7.6	4	30.4
8.7	2	17.4
9.2	4	36.8
9.5	2	19.0
9.4	4	37.6
8.5	1	8.5
		$149.7 = \Sigma_1$

$$\text{Area 1} = \frac{1}{3} \times \text{CI} \times \Sigma_1$$

$$\text{Area 1} = \frac{9}{3} \times 149.7$$

$$\text{Area 1} = 449 \text{ square meters}$$

$$\text{Appendage} = 50.0 \text{ square meters} = \text{Area 2}$$

Ans. Area of WP = 499 square meters.

Subdivided Common Intervals

The area or volume of an appendage may be found by the introduction of intermediate ordinates. Referring to the waterplane area shown in Figure 7.11, the length has been divided

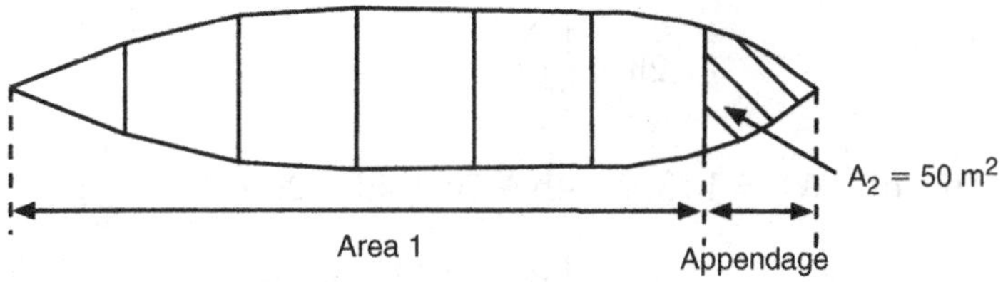

Figure 7.10

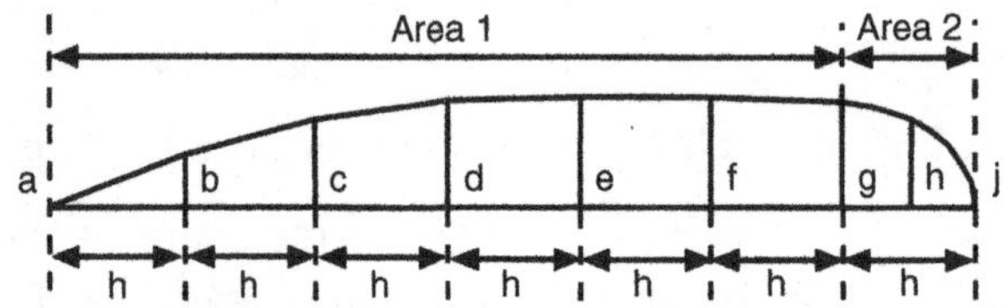

Figure 7.11

into seven equal parts and the half ordinates have been set up. Also, the side is a smooth curve from the stem to the ordinate 'g'.

If the area of the waterplane is found by putting the eight half-ordinates directly through the rules, the answer obtained will obviously be an erroneous one. To reduce the error the waterplane may be divided into two parts as shown in the figure.

Then,

$$\text{Area No. 1} = h/3\ (a + 4b + 2c + 4d + 2e + 4f + g)$$

To find area No. 2, an intermediate semi-ordinate is set up midway between the semi-ordinates g and j. The common interval for this area is h/2.

Then,

$$\text{Area No. 2} = h/2 \times \frac{1}{3} \times (g + 4h + j)$$

or

$$\text{Area No. 2} = h/3\left(\frac{1}{2}g + 2h + \frac{1}{2}j\right)$$

If CI is halved, then multipliers are halved, i.e. from 1, 4, 1 etc. to 1/2, 2, 1/2.

$$\begin{aligned}\text{Area of } \frac{1}{2}\text{WP} &= \text{Area 1} + \text{Area 2}\\ &= h/3(a + 4b + 2c + 4d + 2e + 4f + g)\\ &\quad + h/3\left(\frac{1}{2}g + 2h + \frac{1}{2}j\right)\\ &= h/3\left(a + 4b + 2c + 4d + 2e + 4f + g + \frac{1}{2}g\right.\\ &\quad \left. + 2h + \frac{1}{2}j\right)\end{aligned}$$

$$\therefore \text{Area of } \frac{1}{2}\text{WP} = h/3(a + 4b + 2c + 4d + 2e + 4f + 1\frac{1}{2}g$$
$$+ 2h + \frac{1}{2}j) \quad \text{or} \quad \frac{1}{3} \times h \times \Sigma_1$$

Example 1

The length of a ship's waterplane is 100 meters. The lengths of the half-ordinates commencing from forward are as follows:

0, 3.6, 6.0, 7.3, 7.7, 7.6, 4.8, 2.8, and 0.6 meters respectively.

Midway between the last two half-ordinates is one whose length is 2.8 meters. Find the area of the waterplane.

1/2 Ord.	SM	Area Function
0	1	0
3.6	4	14.4
6.0	2	12.0
7.3	4	29.2
7.7	2	15.4
7.6	4	30.4
4.8	1.5	7.2
2.8	2	5.6
0.6	0.5	0.3
		$114.5 = \Sigma_1$

$$\text{Area} = \frac{2}{3} \times \text{CI} \times \Sigma_1$$

$$\text{CI} = 100/7 = 14.29 \text{ m}$$

$$\text{Area of WP} = \frac{2}{3} \times 14.29 \times 114.5$$

Ans. Area of WP = 1090.5 square meters.

Note how the CI used was the *largest* CI in the ship's waterplane.

In some cases an even more accurate result may be obtained by dividing the waterplane into three separate areas as shown in Figure 7.12 and introducing intermediate semi-ordinates at the bow and the stern.

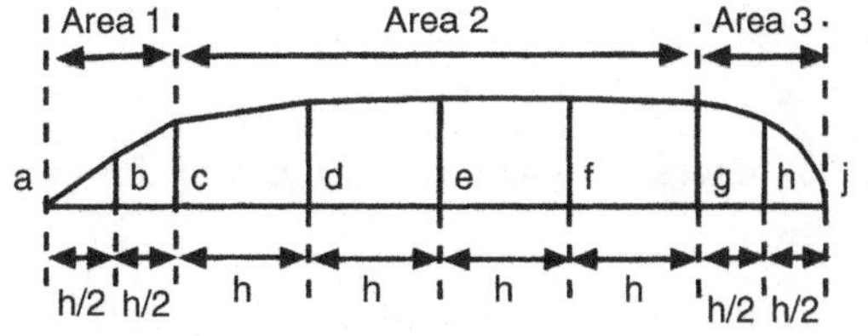

Figure 7.12

$$\text{Area 1} = h/2 \times \frac{1}{3} \times \left(a + 4b + c\right)$$

$$= h/3\left(\frac{1}{2}a + 2b + \frac{1}{2}c\right)$$

$$\text{Area 2} = h/3(c + 4d + 2e + 4f + g)$$

$$\text{Area 3} = h/2 \times \frac{1}{3}\left(g + 4h + j\right)$$

$$= h/3\left(\frac{1}{2}g + 2h + \frac{1}{2}j\right)$$

$$\text{Area}\ \frac{1}{2}\ \text{WP} - \text{Area 1} + \text{Area 2} + \text{Area 3}$$

$$= h/3\left(\frac{1}{2}a + 2b + \frac{1}{2}c\right) + c + 4d + 2e + 4f + g$$

$$+ \frac{1}{2}g + 2h + \frac{1}{2}j$$

$$\text{Area}\ \frac{1}{2}\ \text{WP} = h/3\left(\frac{1}{2}a + 2b + 1\frac{1}{2}c + 4d + 2e + 4f + 1\frac{1}{2}g\right.$$

$$\left. + 2h + \frac{1}{2}j\right) \quad \text{or} \quad h/3 \times \Sigma_1$$

	0.5	2	0.5		1	4	1		
			+		+		+		
			1	4	1		0.5	2	0.5
Combined multipliers	0.5	2	1.5	4	2	4	1.5	2	0.5

Example 2

A ship's waterplane is 72 meters long and the lengths of the half-ordinates commencing from forward are as follows:

0.2, 2.2, 4.4, 5.5, 5.8, 5.9, 5.9, 5.8, 4.8, 3.5, and 0.2 meters respectively.

1/2 Ord.	SM	Area Function
0.2	0.5	0.1
2.2	2	4.4
4.4	1.5	6.6
5.5	4	22.0
5.8	2	11.6
5.9	4	23.6
5.9	2	11.8
5.8	4	23.2
4.8	1.5	7.2
3.5	2	7.0
0.2	0.5	0.1
		$117.6 = \Sigma_1$

Largest common interval = 9.00 m.

The spacing between the first three and the last three half-ordinates is half of the spacing between the other half-ordinates. Find the area of the waterplane.

$$\text{Area} = \frac{1}{3} \times \text{CI} \times \Sigma_1 \times 2$$

$$\text{CI} = 72/8 = 9 \text{ m}$$

$$\text{Area of WP} = \frac{1}{3} \times 9 \times 117.6 \times 2$$

Ans. Area of WP = 705.6 square meters.

Note. It will be seen from this table that the effect of halving the common interval is to halve Simpson's Multipliers.

Σ_1 = is there because the solution has been obtained by using Simpson's First Rule.

■

Areas and Volumes Having an Awkward Number of Ordinates

Occasionally the number of ordinates used is such that the area or volume concerned cannot be found directly by use of either the First or the Second Rule. In such cases the area or volume should be divided into two parts, the area of each part being calculated separately, and the total area found by adding the areas of the two parts together.

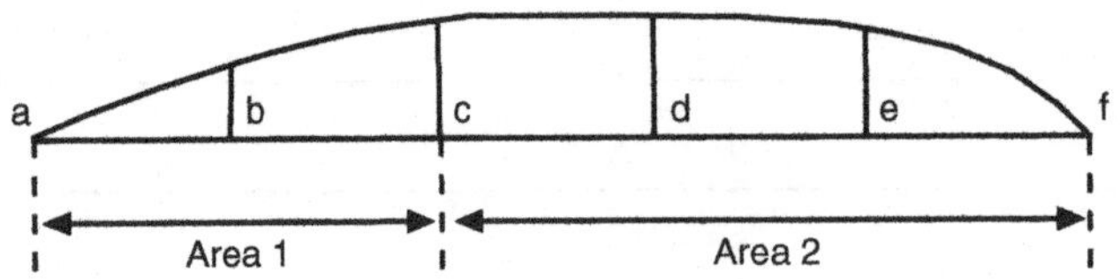

Figure 7.13

Example 1

Show how the area of a waterplane may be found when using six semi-ordinates. Neither the First nor the Second Rule can be applied directly to the whole area but the waterplane can be divided into two parts as shown in Figure 7.13. Area No. 1 can be calculated using the First Rule and area No. 2 by the Second Rule. The areas of the two parts may then be added together to find the total area.

An alternative method would be to find the area between the half-ordinates a and e by the First Rule and then find the area between the half-ordinates e and f by the 'Five/Eight' Rule.

Example 2

Show how the area may be found when using eight semi-ordinates.

Divide the area up as shown in Figure 7.14. Find area No. 1 using the Second Rule and area No. 2 using the First Rule.

An alternative method is again to find the area between the half-ordinates a and g by the first rule and the area between the half-ordinates g and h by the 'Five/Eight' Rule.

In practice, the naval architect divides the ship's length into 10 stations and then subdivides the forward and aft ends in order to obtain extra accuracy with the calculations.

In doing so, the calculations can be made using Simpson's First and Second Rules, perhaps as part of a computer package.

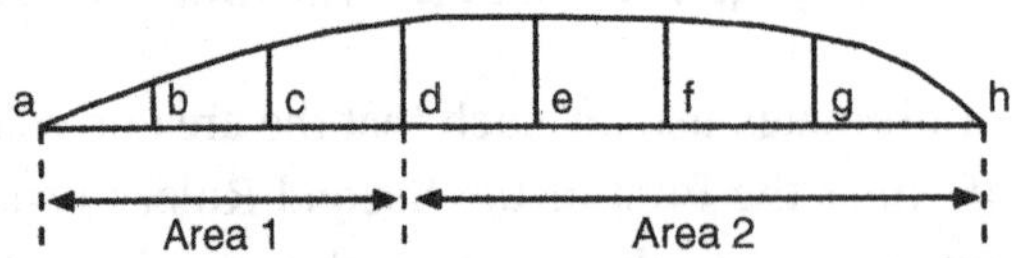

Figure 7.14

Centroids and Centers of Gravity

To Find the Center of Flotation

The *center of flotation* is the center of gravity or *centroid* of the waterplane area, and is the point about which a ship heels and trims. It must lie on the longitudinal centerline but may be slightly forward or aft of amidships (up to say 3% L forward of amidships for oil tankers and up to say 3% L aft of amidships for container ships).

To find the area of a waterplane by Simpson's Rules, the half-breadths are used as ordinates. If the moments of the half-ordinates about any point are used as ordinates, then the total moment of the area about that point will be found. If the total moment is now divided by the total area, the quotient will give the distance of the centroid of the area from the point about which the moments were taken. This may be shown as follows.

In Figure 7.15:

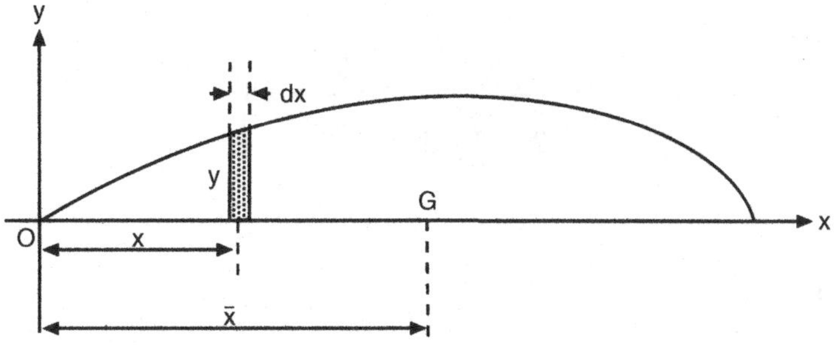

Figure 7.15

$$\text{Area of strip} = y\,dx$$

$$\text{Area of the } \frac{1}{2}\text{ WP} = \int_O^L y\,dx$$

$$\text{Area of the WP} = 2 \times \int_O^L y\,dx$$

The value of the integral is found using the formula:

$$\int_O^L y\,dx = \frac{h}{3}(a + 4b + 2c + 4d + e)$$

Thus, the value of the integral is found by Simpson's Rules using values of the variable y as ordinates:

$$\text{Moment of strip about OY} = x\,y\,dx$$

$$\text{Moment of } \frac{1}{2} \text{ WP about OY} = \int_O^L x\,y\,dx$$

$$\text{Moment of WP about OY} = 2 \times \int_O^L x\,y\,dx$$

The value of this integral is found by Simpson's Rules using values of the product xy as ordinates.

Let the distance of the center of flotation be $\overline{X}$ from OY, then:

$$\overline{X} = \frac{\text{Moment}}{\text{Area}}$$

$$= \frac{2 \times \int_O^L x\,y\,dx}{2 \times \int_O^L y\,dx} = \frac{\Sigma_2}{\Sigma_1} \times CI$$

Example 1

A ship 150 meters long has half-ordinate commencing from aft as follows (see Figure 7.16):

0, 5, 9, 9, 9, 7, and 0 meters respectively.

Find the distance of the center of flotation from forward.

Note. To avoid using large numbers the levers used are in terms of CI, the common interval. It is more efficient than using levers in meters.

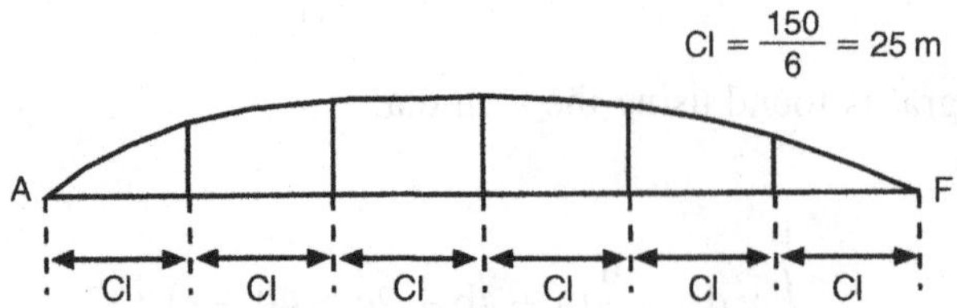

Figure 7.16

$$\text{Area of the waterplane} = \frac{2}{3} \times \text{CI} \times \Sigma_1$$

$$= \frac{2}{3} \times 25 \times 120 \text{ square meters} = 2000 \text{ m}^2$$

$$\text{Distance of C.F. from aft} = \frac{\Sigma_2}{\Sigma_1} \times \text{CI}$$

$$= \frac{376}{120} \times \text{CI}$$

$$= 78.33 \text{ m}$$

Ans. C.F. is 78.33 m from aft.

1/2 Ord.		SM	Products for Area	Levers from A*	Moment Function
Aft	0	1	0	0	0
	5	4	20	1	20
	9	2	18	2	36
Amidships	9	4	36	3	108
	9	2	18	4	72
	7	4	28	5	140
Forward	0	1	0	6	0
			$120 = \Sigma_1$		$376 = \Sigma_2$

Σ_1, because it is the first total.
Σ_2, because it is the second total.
*The levers are in terms of the number of CI from the aft ordinate through to the foremost ordinate.

This problem can also be solved by taking the moments about amidships as in the following example:

Example 2

A ship 75 m long has half-ordinates at the load waterplane commencing from aft as follows:

$$0, 1, 2, 4, 5, 5, 5, 4, 3, 2, \text{ and } 0 \text{ meters respectively.}$$

The spacing between the first three semi-ordinates and the last three semi-ordinates is half of that between the other semi-ordinates (see Figure 7.17). Find the position of the center of flotation relative to amidships.

Use positive sign for levers and moments *aft* of amidships.

Use negative sign for levers and moments *forward* of amidships.

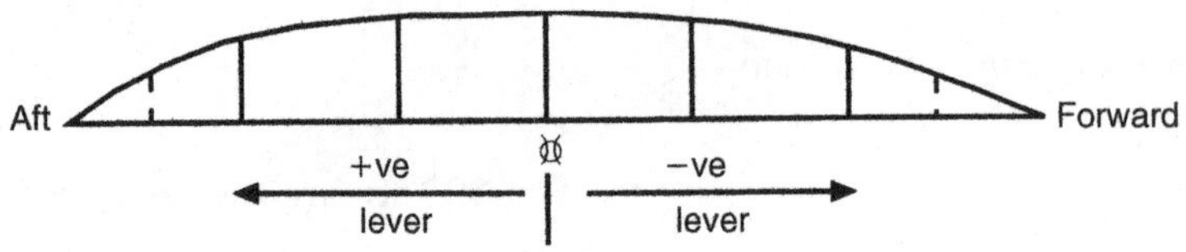

Figure 7.17

1/2 Ord.		SM	Area Function	Levers	Moment Function
Aft	0	0.5	0	+4	0
	1	2	2	+3.5	+7
	2	1.5	3	+3	+9
	4	4	16	+2	+32
	5	2	10	+1	+10
Amidships	5	4	20	0	0
	5	2	10	−1	−10
	4	4	16	−2	−32
	3	1.5	4.5	−3	−13.5
	2	2	4.0	−3.5	−14
Forward	0	0.5	0	−4	0
			$85.5 = \Sigma_1$		$-11.5 = \Sigma_2$

$$\text{Largest common interval} = \frac{75}{8} = 9.375 \text{ m}$$

Σ_1 denotes first total.

Σ_2 denotes second algebraic total.

The point having a lever of zero is the fulcrum point. All other levers positive and negative are then relative to this point.

$$\text{Distance of C.F. from amidships} = \frac{\Sigma_2}{\Sigma_1} \times \text{CI}$$

$$= \frac{-11.5}{85.5} \times 9.375$$

$$= -1.26 \text{ m}$$

The negative sign shows C.F. is forward of amidships (a positive sign indicates C.F. is aft of amidships).

Ans. C.F. is 1.26 meters forward of amidships.

To Find the KB

The center of buoyancy is the three-dimensional center of gravity of the underwater volume and Simpson's Rules may be used to determine its height above the keel.

First, the areas of waterplanes are calculated at equidistant intervals of draft between the keel and the waterline. Then the volume of displacement is calculated by using these areas as ordinates in the rules. The moments of these areas about the keel are then taken to find the total moment of the underwater volume about the keel. The KB is then found by dividing the total moment about the keel by the volume of displacement.

It will be noted that this procedure is similar to that for finding the position of the center of flotation, which is the two-dimensional center of gravity of each waterplane.

Example 1

A ship is floating upright on an even keel at 6.0 m draft F and A (see Figure 7.18). The areas of the waterplanes are as follows:

Draft (m)	0	1	2	3	4	5	6
Area (square meters)	5000	5600	6020	6025	6025	6025	6025

Find the ship's KB at this draft.

Waterplane	Area	SM	Volume Function	Levers	Moment Function
A	6025	1	6025	6	36,150
B	6025	4	24,100	5	120,500
C	6025	2	12,050	4	48,200
D	6025	4	24,100	3	72,300

Continued

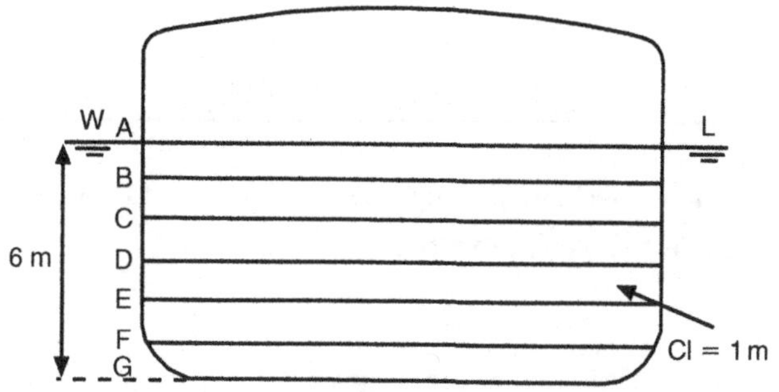

Figure 7.18

Waterplane	Area	SM	Volume Function	Levers	Moment Function
E	6020	2	12,040	2	24,080
F	5600	4	22,400	1	22,400
G	5000	1	5000	0	0
			$105{,}715 = \Sigma_1$		$323{,}630 = \Sigma_2$

$$KB = \frac{\text{Moment about keel}}{\text{Volume of displacement}}$$

$$\therefore KB = \frac{\Sigma_2}{\Sigma_1} \times \text{CI}$$

$$= \frac{323{,}630}{105{,}715} \times 1.0$$

Ans. $\underline{KB} = \underline{3.06 \text{ meters}}$

$$= 0.51 \times d \text{ approximately.}$$

The lever of zero was at the keel so the final answer was relative to this point, i.e. above base.

If we Simpsonize 1/2 ords we will obtain *areas*.

If we Simpsonize areas we will obtain *volumes*.

Example 2

A ship is floating upright in S.W. on an even keel at 7 m draft F and A. The TPCs are as follows:

Draft (m)	1	2	3	4	5	6	7
TPC (tonnes)	60.0	60.3	60.5	60.5	60.5	60.5	60.5

The volume between the outer bottom and 1 m draft is 3044 cubic meters, and its center of gravity is 0.5 m above the keel. Find the ship's KB.

In Figure 7.19:

Let KY represent the height of the center of gravity of volume A above the keel, and KZ represent the height of the center of gravity of volume B above the keel.

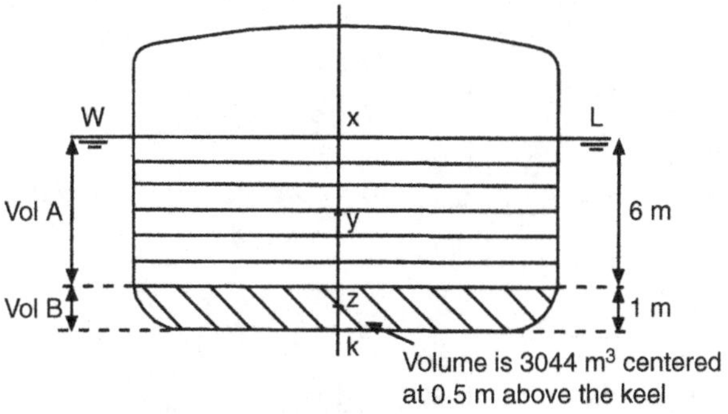

Figure 7.19

Let X = 100/1.025, then the area of each waterplane is equal to TPC × X square meters.

Σ_1 denotes the first total
Σ_2 denotes the second total } see table below

Draft	Area	SM	Volume Function	Levers 1 m	Moment Function
7	60.5	1	60.5	0	0
6	60.5	4	242.0	1	242.0
5	60.5	2	121.0	2	242.0
4	60.5	4	242.0	3	726.0
3	60.5	2	121.0	4	484.0
2	60.3	4	241.2	5	1206.0
1	60.0	1	60.0	6	360.0
			$1087.7 = \Sigma_1$		$3260.0 = \Sigma_2$

$$\text{Volume A} = \frac{1}{3} \times \text{CI} \times \Sigma_1 \times X$$

$$= \frac{1}{3} \times 1.0 \times 1087.7 \times \frac{100}{1.025} = 35,372$$

Volume A = 35,372 cubic meters

Volume B = +3044 cubic meters

Total volume = 38,416 cubic meters

$$XY = \frac{\Sigma_2}{\Sigma_1} \times CI = \frac{\text{Moment}}{\text{Volume A}}$$

$$= \frac{3260}{1087.7} \times 1.0 = 3 \text{ m below 7 m waterline}$$

$$XY = 3 \text{ m}$$

$$KX = 7 \text{ m}$$

$$KX - XY = KY, \text{ so } \underline{KY = 4 \text{ m}}$$

Volume	KG_{keel}	Moments About Keel
35,372	4	141,488
+3044	0.5	+1522
38,416		143,010

Moments About the Keel

$$KB = \frac{\text{Total moment}}{\text{Total volume}} = \frac{143{,}010}{38{,}416} = \underline{3.72 \text{ meters}}$$

$$= 0.531 \times d$$

Summary

When using Simpson's Rules for ship calculations always use the following procedure:

1. Make a sketch using the given information.
2. Insert values into a *table* as shown in worked examples.
3. Use tabulated summations to finally *calculate* requested values.

Exercise 7

1. A ship's load waterplane is 60 m long. The lengths of the half-ordinates commencing from forward are as follows:

   0.1, 3.5, 4.6, 5.1, 5.2, 5.1, 4.9, 4.3, and 0.1 m respectively.

   Calculate the area of the waterplane, the TPC in salt water, and the position of the center of flotation, from amidships.

2. The half-ordinates of a ship's waterplane, which is 60 m long, commencing from forward, are as follows:

   0, 3.8, 4.3, 4.6, 4.7, 4.7, 4.5, 4.3, and 1 m respectively.

   Find the area of the waterplane, the TPC, the coefficient of fineness of the waterplane area, and the position of the center of flotation, from amidships.
3. The breadths at the load waterplane of a ship 90 meters long, measured at equal intervals from forward, are as follows:

   0, 3.96, 8.53, 11.58, 12.19, 12.5, 11.58, 5.18, 3.44, and 0.30 m respectively.

   If the load draft is 5 meters and the block coefficient is 0.6, find the FWA and the position of the center of flotation, from amidships.
4. The areas of a ship's waterplanes, commencing from the load draft of 24 meters, and taken at equal distances apart, are:

   2000, 1950, 1800, 1400, 800, 400, and 100 square meters respectively.

   The lower area is that of the ship's outer bottom. Find the displacement in salt water, the Fresh Water Allowance, and the height of the center of buoyancy above the keel.
5. The areas of vertical transverse sections of a forward hold, spaced equidistantly between bulkheads, are as follows:

   800, 960, 1100, and 1120 square meters respectively.

   The length of the hold is 20 m. Find how many tonnes of coal (stowing at 4 cubic meters per tonne) it will hold.
6. A ship 90 meters long is floating on an even keel at 6 m draft. The half-ordinates, commencing from forward, are as follows:

   0, 4.88, 6.71, 7.31, 7.01, 6.40, and 0.9 m respectively.

   The half-ordinates 7.5 meters from bow and stern are 2.13 and 3.35 m respectively. Find the area of the waterplane and the change in draft if 153 tonnes of cargo are loaded with its center of gravity vertically over the center of flotation. Find also the position of the center of flotation.
7. The areas of a ship's waterplanes commencing from the load waterplane and spaced at equidistant intervals down to the inner bottom are:

   2500, 2000, 1850, 1550, 1250, 900, and 800 square meters respectively.

   Below the inner bottom is an appendage 1 meter deep that has a mean area of 650 square meters. The load draft is 7 meters. Find the load displacement in salt water, the Fresh Water Allowance, and the height of the center of buoyancy above the keel.

8. A ship's waterplane is 80 meters long. The breadths commencing from forward are as follows:

   0, 3.05, 7.1, 9.4, 10.2, 10.36, 10.3, 10.0, 8.84, 5.75, and 0 m respectively.

   The space between the first three and the last three ordinates is half of that between the other ordinates. Calculate the area of the waterplane and the position of the center of flotation.

9. Three consecutive ordinates in a ship's waterplane area are:

   6.30, 3.35, and 0.75 m respectively.

   The common interval is 6 m. Find the area contained between the last two ordinates.

10. The transverse horizontal ordinates of a ship's amidships section commencing from the load waterline and spaced at 1-meter intervals are as follows:

    16.30, 16.30, 16.30, 16.00, 15.50, 14.30, and 11.30 m respectively.

    Below the lowest ordinate there is an appendage of 8.5 square meters. Find the area of the transverse section.

11. The following table gives the area of a ship's waterplane at various drafts:

Draft (m)	6	7	8
Area (square meters)	700	760	800

    Find the volume of displacement and approximate mean TPC between the drafts of 7 and 8 m.

12. The areas of a ship's waterplanes, commencing from the load waterplane and spaced 1 meter apart, are as follows:

    800, 760, 700, 600, 450, and 10 square meters respectively.

    Midway between the lowest two waterplanes the area is 180 square meters. Find (a) the load displacement in salt water and (b) the height of the center of buoyancy above the keel.

■

CHAPTER 8

Quadrature – Simpson's Rules for Moments of Inertia

The second moment of an element of an area about an axis is equal to the product of the area and the square of its distance from the axis. In some textbooks, this second moment of area is called the 'moment of inertia'. Let dA in Figure 8.1 represent an element of an area and let y be its distance from the axis AB. The second moment of the element about AB is equal to $dA \times y^2$.

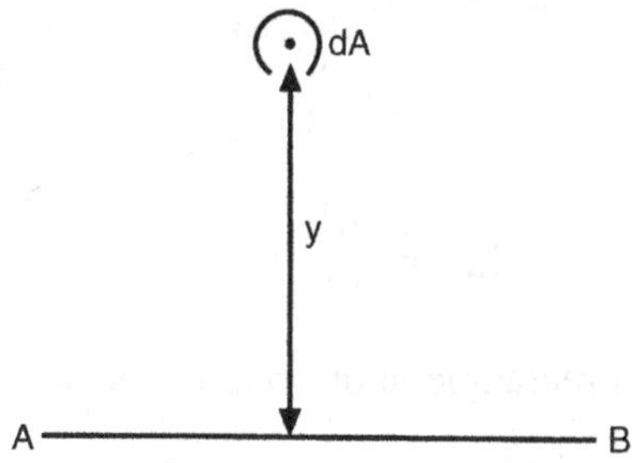

Figure 8.1

To find the second moment of a rectangle about an axis parallel to one of its sides and passing through the centroid.

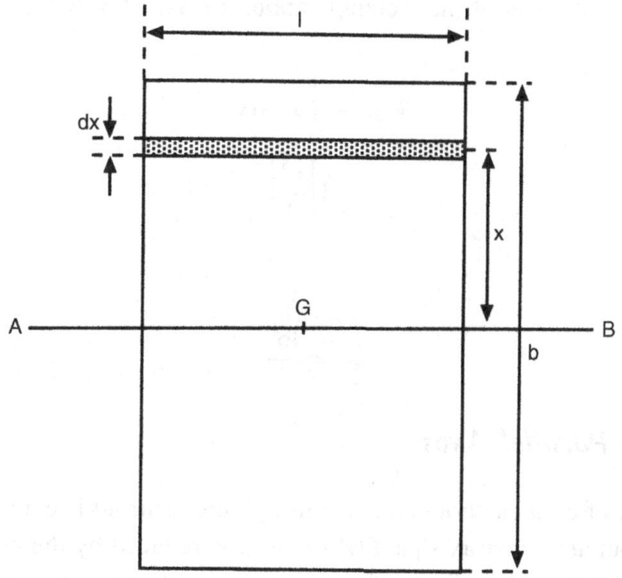

Figure 8.2

Ship Stability for Masters and Mates. http://dx.doi.org/10.1016/B978-0-08-097093-6.00008-6

In Figure 8.2, l represents the length of the rectangle and b represents the breadth. Let G be the centroid and let AB, an axis parallel to one of the sides, pass through the centroid. Consider the elementary strip that is shown shaded in the figure. The second moment (i) of the strip about the axis AB is given by the equation:

$$i = l\mathrm{d}x \times x^2$$

Let I_{AB} be the second moment of the whole rectangle about the axis AB, then:

$$I_{AB} = \int_{-b/2}^{+b/2} lx^2 \mathrm{d}x$$

$$I_{AB} = l \int_{-b/2}^{+b/2} x^2 \mathrm{d}x$$

$$= l\left[\frac{x^3}{3}\right]_{-b/2}^{+b/2}$$

$$I_{AB} = \frac{lb^3}{12}$$

To find the second moment of a rectangle about one of its sides:

Consider the second moment (i) of the elementary strip shown in Figure 8.3 about the axis AB:

$$i = l\mathrm{d}x \times x^2$$

Let I_{AB} be the second moment of the rectangle about the axis AB, then:

$$I_{AB} = \int_{O}^{b} lx^2 \mathrm{d}x$$

$$= l\left[\frac{x^3}{3}\right]_{O}^{b}$$

or

$$I_{AB} = \frac{lb^3}{3}$$

The Theorem of Parallel Axes

The second moment of an area about an axis through the centroid is equal to the second moment about any other axis parallel to the first reduced by the product of the area and the square of the perpendicular distance between the two axes. Thus,

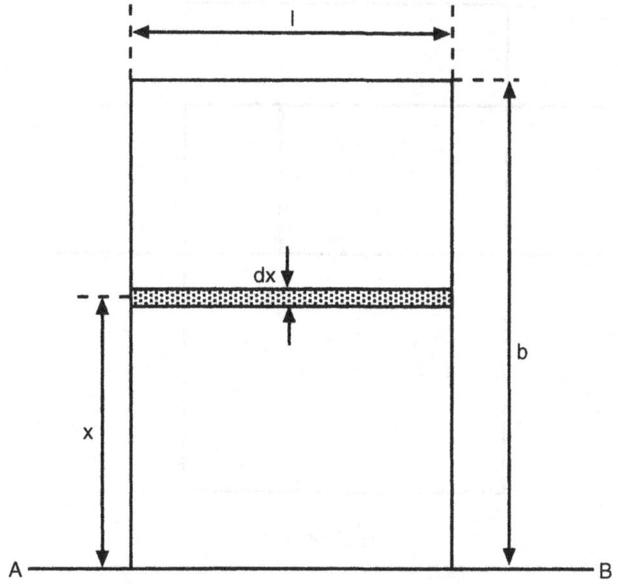

Figure 8.3

in Figure 8.4, if G represents the centroid of the area (A) and the axis OZ is parallel to AB, then:

$$I_{OZ} = I_{AB} - Ay^2 = \text{parallel axis theorem equation}$$

To find the second moment of a ship's waterplane area about the centerline:

In Figure 8.5,

$$\text{Area of elementary strip} = ydx$$

$$\text{Area of waterplane} = \int_O^L ydx$$

It has been shown in Chapter 7 that the area under the curve can be found by Simpson's Rules, using the values of y, the half-breadths, as ordinates.

The second moment of a rectangle about one end is given by $lb^3/3$, and therefore the second moment of the elementary strip about the centerline is given by $(y^3dx)/3$ and the second moment of the half waterplane about the centerline is given by:

$$\int_O^L \frac{y^3}{3}\,dx$$

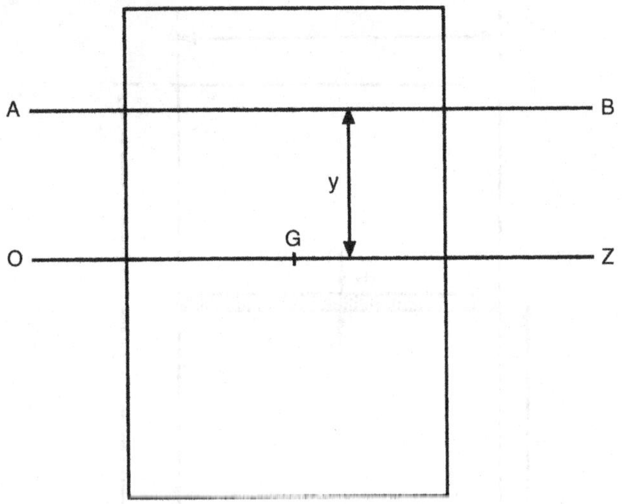

Figure 8.4

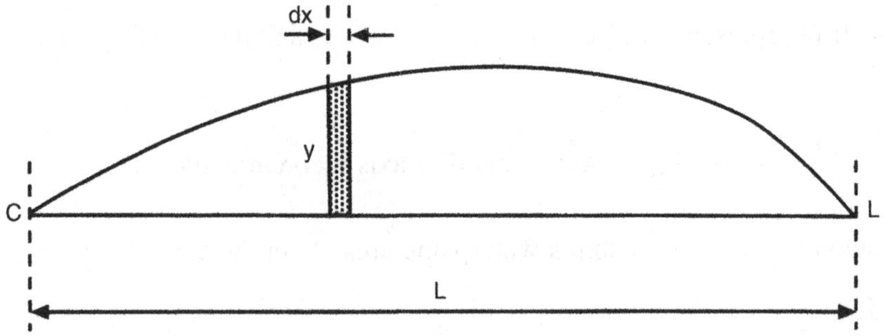

Figure 8.5

Therefore, if I_{CL} is the second moment of the whole waterplane area about the centerline, then:

$$I_{CL} = \frac{2}{3}\int_{O}^{L} y^3 dx$$

The integral part of this expression can be evaluated by Simpson's Rules using the values of y^3 (i.e. the half-breadths cubed) as ordinates, and I_{CL} is found by multiplying the result by 2/3. I_{CL} is also known as the 'moment of inertia about the centerline'.

Example 1

A ship's waterplane is 18 meters long. The half-ordinates at equal distances from forward are as follows:

0, 1.2, 1.5, 1.8, 1.8, 1.5, and 1.2 meters respectively

Find the second moment of the waterplane area about the centerline.

Given information: Length 18 m, number of intervals 6, common interval 3 m.

1/2 Ord.	1/2 Ord.^{3}	SM	Inertia Function
0	0	1	0
1.2	1.728	4	6.912
1.5	3.375	2	6.750
1.8	5.832	4	23.328
1.8	5.832	2	11.664
1.5	3.375	4	13.500
1.2	1.728	1	1.728
			$63.882 = \Sigma_1$

$$I_{CL} = \frac{2}{9} \times CI \times \Sigma_1$$

$$I_{CL} = \frac{2}{9} \times \frac{18}{6} \times 63.882$$

Moment of inertia about the centerline $=$ $\underline{42.588\ m^4}$

This moment of inertia was solved via Simpson's First Rule.

To find the second moment of the waterplane area about a transverse axis through the center of flotation:

$$\text{Area of elementary strip} = y\,dx$$

$$I_{AB}\text{ of the elementary strip} = x^2 y\,dx$$

$$I_{AB}\text{ of the waterplane area} = 2\int_O^L x^2 y\,dx$$

Once again the integral part of this expression can be evaluated by Simpson's Rules using the values of x^2y as ordinates and the second moment about AB is found by multiplying the result by 2.

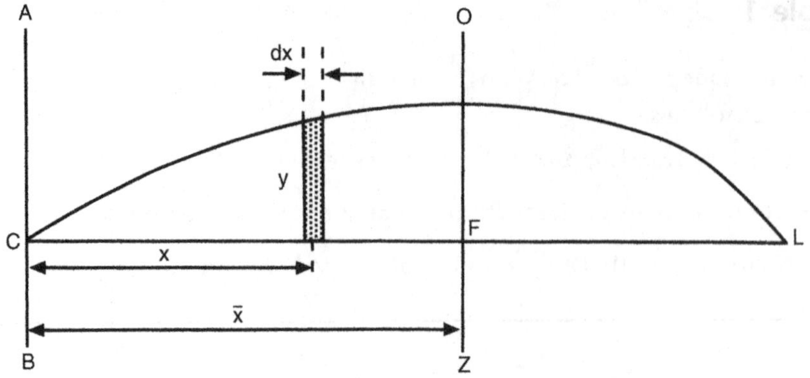

Figure 8.6

Let OZ be a transverse axis through the center of flotation. The second moment about OZ can then be found by the theorem of parallel axes shown in Figure 8.6, i.e.

$$I_{OZ} = I_{AB} - A\bar{X}^2$$

Example 2

A ship's waterplane is 18 meters long. The half-ordinates at equal distances from forward are as follows:

0, 1.2, 1.5, 1.8, 1.8, 1.5, and 1.2 meters respectively

Find the second moment of the waterplane area about a transverse axis through the center of flotation.

1/2 Ord.	SM	Area Function	Lever	Moment Function	Lever	Inertia Function
0	1	0	0	0	0	0
1.2	4	4.8	1	4.8	1	4.8
1.5	2	3.0	2	6.0	2	12.0
1.8	4	7.2	3	21.6	3	64.8
1.8	2	3.6	4	14.4	4	57.6
1.5	4	6.0	5	30.0	5	50.0
1.2	1	1.2	6	7.2	6	43.2
		$25.8 = \Sigma_1$		$84.0 = \Sigma_2$		$332.4 = \Sigma_3$

$$\text{Area of waterplane} = \frac{1}{3} \times \text{CI} \times \Sigma_1 \times 2$$
$$= \frac{1}{3} \times \frac{18}{6} \times 25.8 \times 2$$
$$= 51.6 \text{ square meters}$$

$$\text{Distance of the center of flotation from forward} = \frac{\Sigma_2}{\Sigma_1} \times \text{CI}$$
$$= \frac{84}{25.8} \times \frac{18}{6}$$
$$= 9.77 \text{ m}$$
$$= 0.77 \text{ m aft of amidships}$$

$$I_{AB} = \frac{1}{3} \times (\text{CI})^3 \times \Sigma_3 \times 2$$
$$= \frac{1}{3} \times \left(\frac{18}{6}\right)^3 \times 332.4 \times 2 = 5983 \text{ m}^4$$
$$I_{OZ} = I_{AB} - A\overline{X}^2$$
$$= 5983 - 51.6 \times 9.77^2$$
$$= 5983 - 4925$$

Ans. I_{OZ} = 1058 meters^4.

There is a quicker and more efficient method of obtaining the solution to the above problem. Instead of using the foremost ordinate at the datum, use the midship ordinate. Proceed as follows:

1/2 Ord.	SM	Area Function	Lever	Moment Function	Lever	Inertia Function
0	1	0	−3	0	−3	0
1.2	4	4.8	−2	−9.6	−2	+19.2
1.5	2	3.0	−1	−3.0	−1	+3.0
1.8	4	7.2	0	0	0	0
1.8	2	3.6	+1	+3.6	+1	+3.6
1.5	4	6.0	+2	+12.0	+2	+24.0
1.2	1	1.2	+3	+3.6	+3	+10.8
		$25.8 = \Sigma_1$		$+6.6 = \Sigma_2$		$60.6 = \Sigma_3$

Common interval 3.00 m

$$\text{Area of waterplane} = \frac{1}{3} \times \Sigma_1 \times h \times 2 \qquad h = \frac{18}{6} = 3 \text{ m}$$

$$= \frac{1}{3} \times 25.8 \times 3 \times 2$$

$$= 51.6 \text{ m}^2 \text{ (as before)}$$

Reconsider $\Sigma_2 = +6.6$

The positive sign shows the center of flotation is in the aft body.

$$\text{Center of flotation from amidships} = \frac{\Sigma_2}{\Sigma_1} \times h$$

$$= +\frac{6.6}{25.8} \times 3$$

$\therefore$ Center of flotation $= +0.77$ m or 0.77 m aft amidships (as before).

$$I_{amidships} = \frac{1}{3} \times \Sigma_3 \times h^3 \times 2 = \frac{1}{3} \times 60.6 \times 3^3 \times 2$$

$$\therefore I_{amidships} = 1090.8 \text{ m}^4$$

But, using the parallel axis theorem:

$$I_{LCF} = I_{amidships} - A\bar{y} = 1090.8 - (51.6 \times 0.77^2)$$

$$= 1090.8 - 30.6$$

$$= 1060 \text{ m}^4$$

i.e. very close to the previous answer of 1058 m^4.

With this improved method the levers are much lower in value. Consequently, the error is decreased when predicting $LCF_{amidships}$ and I_{LCF}. $I_{amidships}$ is also known as the 'moment of inertia about amidships'. I_{LCF} is also known as the 'moment of inertia about the LCF'.

■

Summary

When using Simpson's Rules for second moments of area the procedure should be as follows:

1. Make a sketch from the given information.
2. Use a moment table and insert values.

3. Using summations obtained in the table proceed to calculate area, LCF, $I_{amidships}$, I_{LCF}, I_{CL}, etc.
4. Remember: sketch, table, calculation.

Exercise 8

1. A large square has a smaller square cut out of its center such that the second moment of the smaller square about an axis parallel to one side and passing through the centroid is the same as that of the portion remaining about the same axis. Find what proportion of the area of the original square is cut out.
2. Find the second moment of a square of side 2a about its diagonals.
3. Compare the second moment of a rectangle 40 cm × 30 cm about an axis through the centroid and parallel to the 40 cm side with the second moment about an axis passing through the centroid and parallel to the 30 cm side.
4. An H-girder is built from 5-cm-thick steel plate. The central web is 25 cm high and the overall width of each of the horizontal flanges is 25 cm. Find the second moment of the end section about an axis through the centroid and parallel to the horizontal flanges.
5. A ship's waterplane is 36 m long. The half-ordinates, at equidistant intervals, commencing from forward, are as follows:

   0, 4, 5, 6, 6, 5, and 4 m respectively

   Calculate the second moment of the waterplane area about the centerline and also about a transverse axis through the center of flotation.
6. A ship's waterplane is 120 meters long. The half-ordinates at equidistant intervals from forward are as follows:

   0, 3.7, 7.6, 7.6, 7.5, 4.6, and 0.1 m respectively

   Calculate the second moment of the waterplane area about the centerline and about a transverse axis through the center of flotation.
7. A ship of 12,000 tonnes displacement is 150 meters long at the waterline. The half-ordinates of the waterplane at equidistant intervals from forward are as follows:

   0, 4.0, 8.5, 11.6, 12.2, 12.5, 12.5, 11.6, 5.2, 2.4, and 0.3 m respectively

   Calculate the longitudinal and transverse BM.
8. The half-ordinates of a ship's waterplane at equidistant intervals from forward are as follows:

   0, 1.3, 5.2, 8.3, 9.7, 9.8, 8.3, 5.3, and 1.9 m respectively

   If the common interval is 15.9 meters, find the second moment of the waterplane area about the centerline and a transverse axis through the center of flotation.

9. A ship's waterplane is 120 meters long. The half-ordinates commencing from aft are as follows:

   0, 1.3, 3.7, 7.6, 7.6, 7.5, 4.6, 1.8, and 0.1 m respectively

   The spacing between the first three and the last three half-ordinates is half of that between the other half-ordinates. Calculate the second moment of the waterplane area about the centerline and about a transverse axis through the center of flotation.

10. A ship's waterplane is 90 meters long between perpendiculars. The half-ordinates at this waterplane are as follows:

Station	AP	0.5	1	2	3	4	5	5.5	FP
1/2 Ords (m)	0	2.00	4.88	6.71	7.31	7.01	6.40	2.00	0.90

Calculate the second moment of the waterplane area about the centerline and also about a transverse axis through the center of flotation.

■

CHAPTER 9

Quadrature – Simpson's Rules for Centers of Pressure on Transverse Bulkheads

Centers of Pressure by Simpson's Rules

The center of pressure is the point at which the resultant thrust on an immersed surface may be considered to act. Its position can be found using Simpson's Rules. The procedures are outlined below.

Using Horizontal Ordinates

Referring to Figure 9.1:

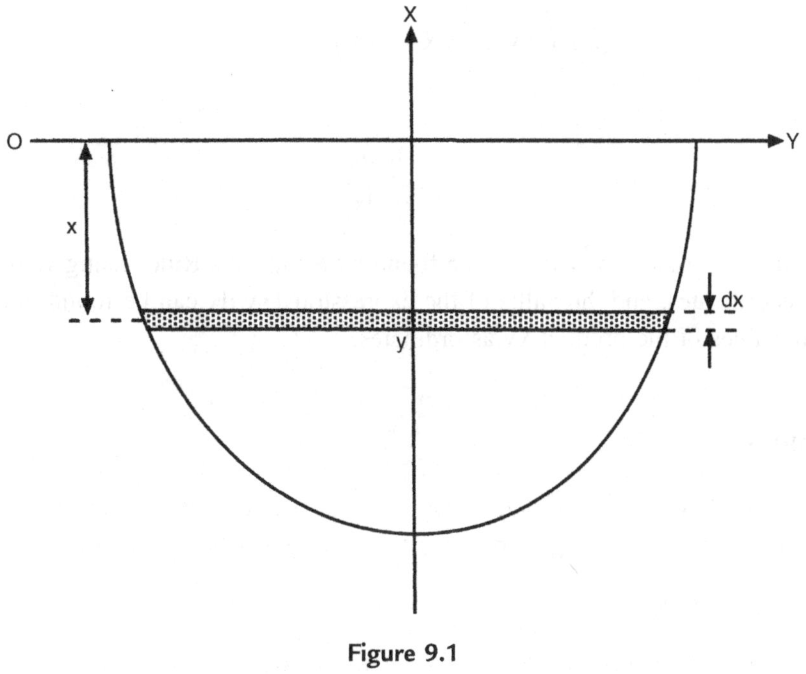

Figure 9.1

Ship Stability for Masters and Mates. http://dx.doi.org/10.1016/B978-0-08-097093-6.00009-8

$$\text{Thrust on the element} = \text{wgxy dx}$$

$$\text{Moment of the thrust about OY} = \text{wgx}^2\text{y dx}$$

$$\text{Moment of total thrust about OY} = \int \text{wgx}^2\text{y dx}$$

$$= \text{wg}\int \text{x}^2\text{y dx}$$

$$\text{Total thrust} = \text{wgA} \times \text{Depth of centroid}$$

$$= \text{wg}\int \text{y dx}\frac{\int \text{xy dx}}{\int \text{y dx}}$$

$$= \text{wg}\int \text{xy dx}$$

$$\text{Moment of total thrust about OY} = \text{Total thrust} \times \bar{x}$$

where

$$\bar{x} = \text{Depth of center of pressure below the surface}$$

$$\therefore \text{Moment of total thrust about OY} = \text{wg}\int \text{xy dx} \times \bar{x}$$

or

$$\text{wg}\int \text{xy dx} \times \bar{x} = \text{wg}\int \text{x}^2\text{y dx}$$

and

$$\bar{x} = \frac{\int \text{x}^2\text{y dx}}{\int \text{xy dx}}$$

The value of the expression $\int x^2y\,dx$ can be found by Simpson's Rules using values of the product x^2y as ordinates, and the value of the expression $\int xy\,dx$ can be found in a similar manner using values of the product xy as ordinates.

Example 1

A lower hold bulkhead is 12 meters deep. The transverse widths of the bulkhead, commencing at the upper edge and spaced at 3-m intervals, are as follows:

15.4, 15.4, 15.4, 15.3, and 15 m respectively

Find the depth of the center of pressure below the waterplane when the hold is flooded to a depth of 2 meters above the top of the bulkhead.

Ord.	SM	Area Function	Lever	Moment Function	Lever	Inertia Function
15.40	1	15.40	0	0.00	0	0
15.40	4	61.60	1	61.60	1	61.6
15.40	2	30.80	2	61.60	2	123.2
15.30	4	61.20	3	183.60	3	550.8
15.00	1	15.00		60.00		240.0
		$184.00 = \Sigma_1$		$366.80 = \Sigma_2$		$975.6 = \Sigma_3$

$$\text{Area} = \frac{1}{3} \times h \times \Sigma_1 = \frac{3}{3} \times 184.0$$

$$= 184 \text{ square meters}$$

VCG below the topmost 15.40 m ordinate = 5.98 m

VCG below the flooded waterline = 7.98 m

Center of pressure below topmost 15.40 m ordinate = 7.98 m

Referring to Figure 9.2:

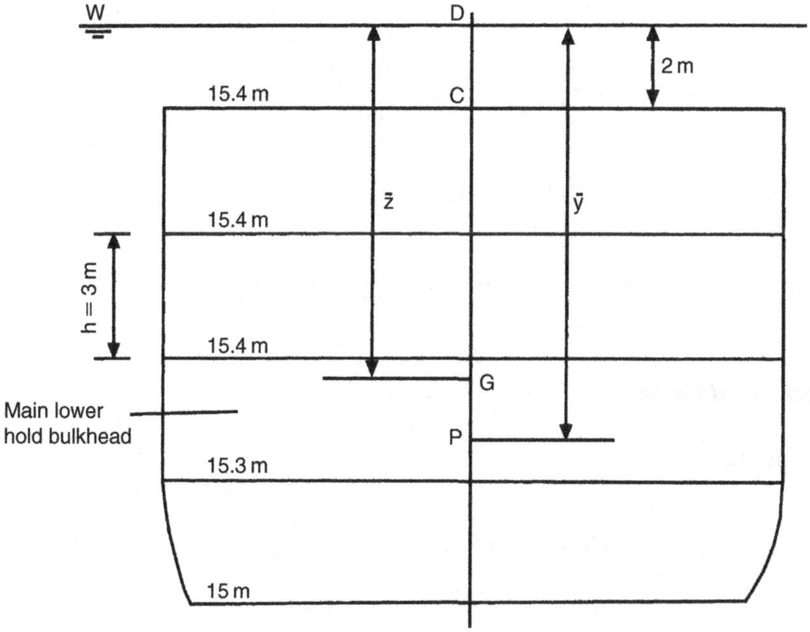

Figure 9.2

$$CG = \frac{\Sigma_2}{\Sigma_1} \times h$$

$$= \frac{366.8}{184} \times 3$$

$$= 5.98 \text{ m}$$

$$+CD = 2.00 \text{ m}$$

$$\bar{z} = 7.98 \text{ m}$$

$$I_{OZ} = \frac{1}{3} \times h^3 \times \Sigma_3 = \frac{1}{3} \times 3^3 \times 975.6$$

$$= 8780 \text{ m}^4$$

$$I_{CG} = I_{OZ} - A(CG)^2, \text{ i.e. parallel axis theorem}$$

$$I_{WL} = I_{CG} + A\bar{z}^2$$

$$= I_{OZ} - A(CG^2 - \bar{z}^2)$$

$$I_{WL} = 8780 - 184(5.98^2 - 7.98^2)$$

$$= 13{,}915 \text{ m}^4$$

$$\bar{y} = \frac{I_{WL}}{A\bar{z}}$$

$$= \frac{13{,}915}{184 \times 7.98}$$

$$\bar{y} = 9.48 \text{ m}$$

Ans. The center of pressure is 9.48 m below the waterline.

■

Using Vertical Ordinates

Referring to Figure 9.3:

$$\text{Thrust on the element} = wg\frac{y}{2}y\,dx$$

$$= \frac{wgy^2}{2}dx$$

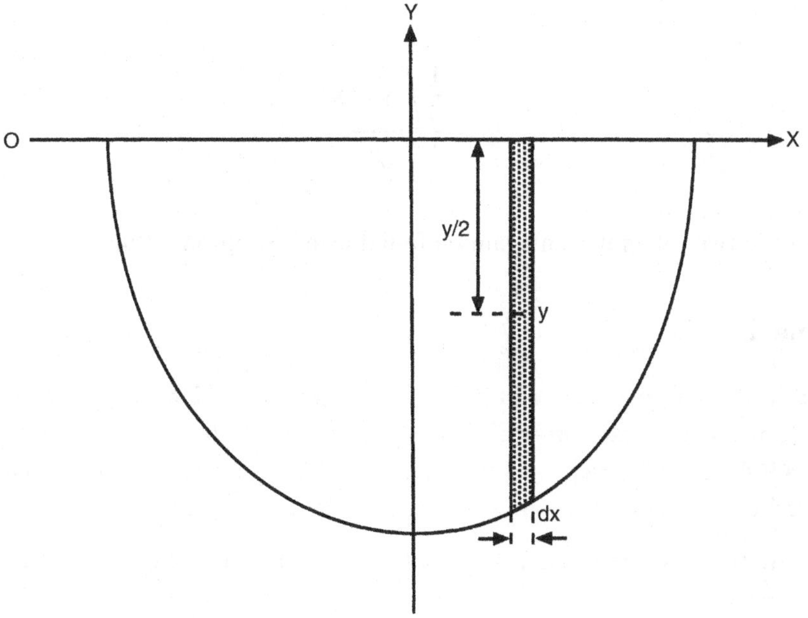

Figure 9.3

$$\text{Moment of the thrust about OX} = \frac{wgy^2}{2}\,dx \times \frac{2}{3}y$$

$$= \frac{wgy^3}{3}\,dx$$

$$\text{Moment of total thrust about OX} = \frac{w}{3} \times g \times \int y^3\,dx$$

$$\text{Total thrust} = wgA \times \text{Depth at center of gravity}$$

$$= wg \int y\,dx \frac{\frac{1}{2}\int y^2\,dx}{\int y\,dx}$$

$$= \frac{w}{2} g \int y^2\,dx$$

Let $\bar{y}$ be the depth of the center of pressure below the surface, then:

$$\text{Moment of total thrust about OX} = \text{Total thrust} \times \bar{y}$$

$$\frac{wg}{3}\int y^3\,dx = \frac{wg}{2}\int y^2\,dx \times \bar{y}$$

or

$$\bar{y} = \frac{\frac{1}{3}\int y^3\,dx}{\frac{1}{2}\int y^2\,dx}$$

The values of the two integrals can again be found using Simpson's Rules.

Example 2

The breadth of the upper edge of a deep tank bulkhead is 12 m. The vertical heights of the bulkhead at equidistant intervals across it are 0, 3, 5, 6, 5, 3, and 0 m respectively (see Figure 9.4). Find the depth of the center of pressure below the waterline when the tank is filled to a head of 2 m above the top of the tank.

Given information: Common interval = 2.00 m, waterline above top bulkhead = 2.00 m.

$$\text{Area} = \frac{1}{3} \times \text{CI} \times \Sigma_1$$

$$\text{Area} = \frac{1}{3} \times 2 \times 68$$

$$= 45.33 \text{ square meters}$$

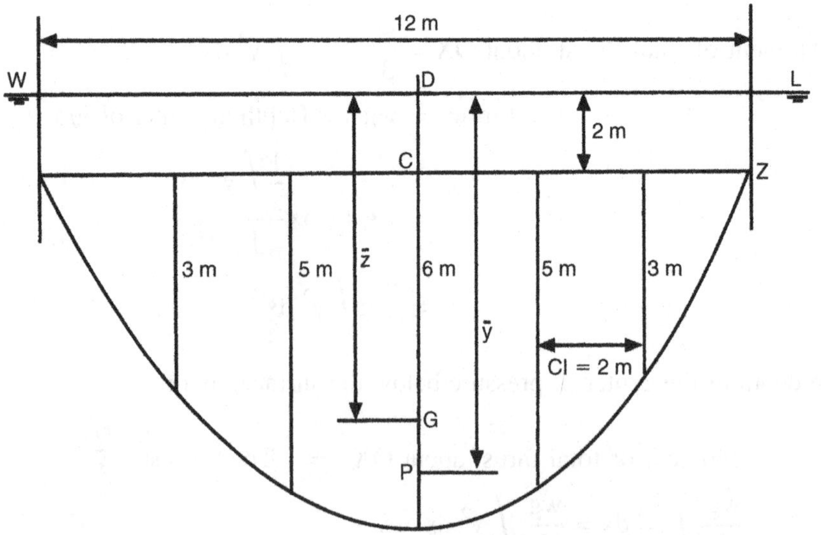

Figure 9.4

Ord.	SM	Area Function	Lever	Moment Function	Lever	Inertia Function
0.00	1	0.00	0	0.00	0	0
3.00	4	12.00	3	36.00	3	108
5.00	2	10.00	5	50.00	5	250
6.00	4	24.00	6	144.00	6	864
5.00	2	10.00	5	50.00	5	250
3.00	4	12.00	3	36.00	3	108
0.00	1	0.00	0	0.00	0	0
		$68.00 = \Sigma_1$		$316.00 = \Sigma_2$		$1580 = \Sigma_3$

Referring to Figure 9.4:

$$CG = \frac{\Sigma_2}{\Sigma_1} \times \frac{1}{2}$$
$$= \frac{316}{68} \times \frac{1}{2}$$
$$= 2.324 \text{ m}$$
$$CD = \underline{2.000 \text{ m}}$$
$$DG = \underline{4.324 \text{ m}} = \bar{z}$$

$$I_{OZ} = \frac{1}{9} \times CI \times \Sigma_3$$
$$= \frac{1}{9} \times 2 \times 1580 = 351 \text{m}^4$$
$$I_{CG} = I_{OZ} - A(CG)^2$$
$$I_{WL} = I_{CG} + A\bar{z}^2$$
$$= I_{OZ} - A(CG)^2 + A\bar{z}^2$$
$$= I_{OZ} - A(CG^2 - \bar{z}^2)$$
$$= 351 - 45.33(2.324^2 - 4.324^2)$$
$$I_{WL} = 954 \text{ m}^4$$
$$\bar{y} = \frac{I_{WL}}{A\bar{z}}$$
$$= \frac{953.75}{45.33 \times 4.324}$$
$$\bar{y} = 4.87 \text{ m}$$

Ans. The center of pressure is 4.87 m below the waterline.

■

Summary

When using Simpson's Rules to estimate the area of a bulkhead under liquid pressure together with the VCG and center of pressure, the procedure should be as follows:

1. Make a sketch from the given information.
2. Make a table and insert the relevant ordinates and multipliers.
3. Calculate the area of the bulkhead's plating.
4. Estimate the ship's VCG below the stipulated datum level.
5. Using the parallel axis theorem, calculate the requested center of pressure.
6. Remember: sketch, table, calculation.

Exercise 9

1. A forepeak tank bulkhead is 7.8 m deep. The widths at equidistant intervals from its upper edge to the bottom are as follows:

   16, 16.6, 17, 17.3, 16.3, 15.3, and 12 m respectively

   Find the load on the bulkhead and the depth of the center of pressure below the top of the bulkhead when the fore peak is filled with salt water to a head of 1.3 m above the crown of the tank.
2. A deep tank transverse bulkhead is 30 m deep. Its width at equidistant intervals from the top to the bottom is:

   20, 20.3, 20.5, 20.7, 18, 14, and 6 m respectively

   Find the depth of the center of pressure below the top of the bulkhead when the tank is filled to a head of 4 m above the top of the tank.
3. The transverse end bulkhead of a deep tank is 18 m wide at its upper edge. The vertical depths of the bulkhead at equidistant intervals across it are as follows:

   0, 3.3, 5, 6, 5, 3.3, and 0 m respectively

   Find the depth of the center of pressure below the top of the bulkhead when the tank is filled with salt water to a head of 2 m above the top of the bulkhead. Find also the load on the bulkhead.
4. A forepeak bulkhead is 18 m wide at its upper edge. Its vertical depth at the centerline is 3.8 m. The vertical depths on each side of the centerline at 3-m intervals are 3.5, 2.5, and 0.2 m respectively. Calculate the load on the bulkhead and the depth of the center of pressure below the top of the bulkhead when the forepeak tank is filled with salt water to a head of 4.5 m above the top of the bulkhead.

5. The vertical ordinates across the end of a deep tank transverse bulkhead measured downwards from the top at equidistant intervals are:

   4, 6, 8, 9.5, 8, 6, and 4 m respectively

   Find the distance of the center of pressure below the top of the bulkhead when the tank is filled with salt water.

■

CHAPTER 10

KB, BM, and KM Calculations and Graphics on Metacentric Diagrams

The method used to determine the final position of the center of gravity is examined in Chapter 7. To ascertain the GM for any condition of loading it is necessary also to calculate the KB and BM (i.e. KM) for any draft.

To Find KB

The center of buoyancy is the center of gravity of the underwater volume.

For a box-shaped vessel on an even keel, the underwater volume is rectangular in shape and the center of buoyancy will be at the half-length, on the centerline, and at half the draft, as shown in Figure 10.1(a). Therefore, for a box-shaped vessel on an even keel: KB = 1/2 draft.

For a vessel that is in the form of a triangular prism, as shown in Figure 10.1(b), the underwater section will also be in the form of a triangular prism. The centroid of a triangle is at two-thirds of the median from the apex. Therefore, the center of buoyancy will be at the half-length, on the centerline, but the KB = 2/3 draft.

For an ordinary ship the KB may be found fairly accurately by Simpson's Rules, as explained in Chapter 7. The approximate depth of the center of buoyancy of a ship *below* the waterline usually lies between 0.44 × draft and 0.49 × draft. A closer approximation of this depth can be obtained by using Morrish's formula, which states:

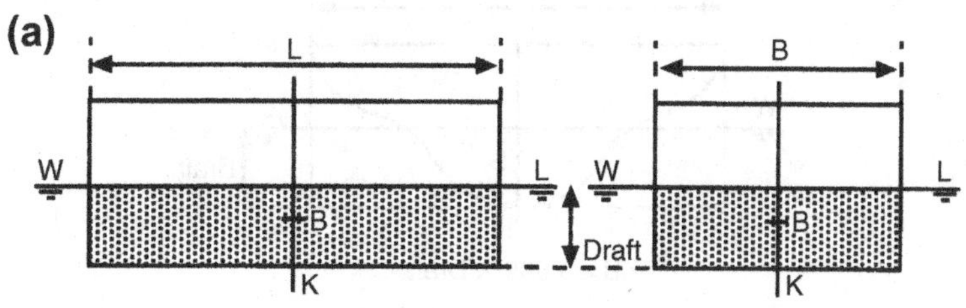

Figure 10.1(a):
Box-Shaped Vessel.

Ship Stability for Masters and Mates. http://dx.doi.org/10.1016/B978-0-08-097093-6.00010-4

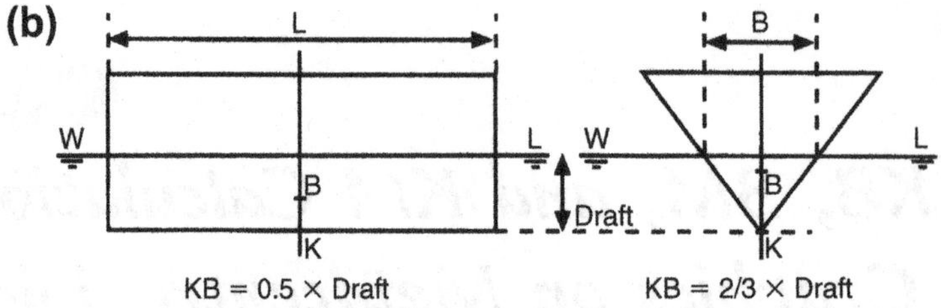

Figure 10.1(b):
Triangular-Shaped Vessel.

$$\text{Depth of center of buoyancy } \textit{below} \text{ waterline} = \frac{1}{3}\left(\frac{d}{2} + \frac{V}{A}\right)$$

where d = mean draft, V = volume of displacement, and A = area of the waterplane.

The derivation of this formula is as follows:

In Figure 10.2, let ABC be the curve of waterplane areas plotted against drafts to the load waterline. Let DE = V/A and draw EG parallel to the base, cutting the diagonal FD in H.

It must first be shown that area DAHC is equal to area DABC:

$$\begin{aligned} \text{Rectangle AH} &= \text{Rectangle HC} \\ \therefore \text{Triangle AGH} &= \text{Triangle HEC} \end{aligned}$$

and

$$\begin{aligned} \text{Area AHCD} &= \text{Area AGED} \\ \text{Area AGED} &= V/A \times A \\ &= V \end{aligned}$$

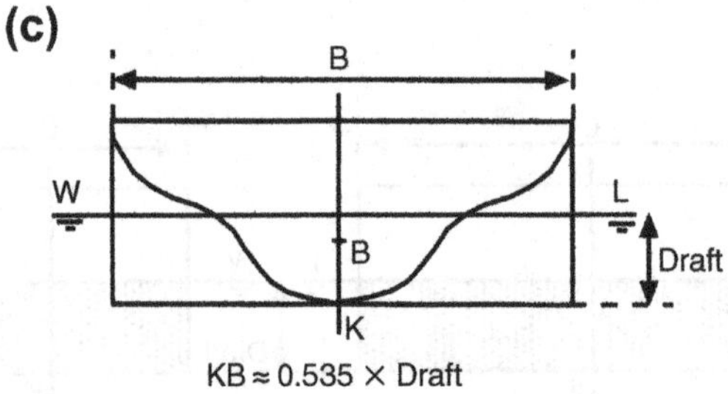

Figure 10.1(c):
Ship-Shaped Vessel.

but

$$\text{Area DABC} = V$$
$$\therefore \text{Area DAHC} = \text{Area DABC}$$

The distance of the centroid of DABC below AD is the distance of the center of buoyancy below the load waterline. It is now assumed that the centroid of the area DAHC is the same distance below the load waterline as the centroid of area DABC.

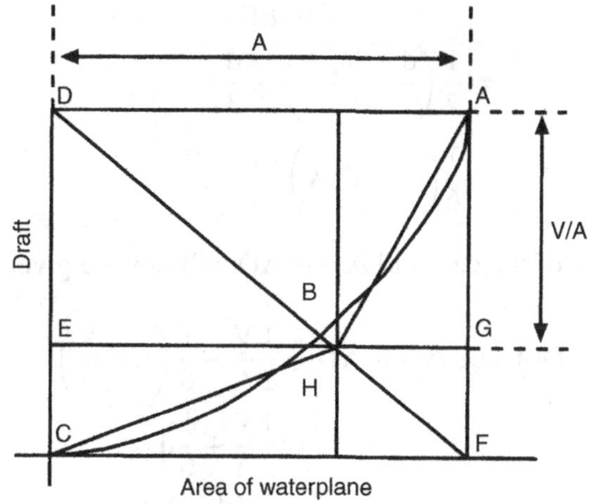

Figure 10.2

To find the distance of the centroid of area DAHC below AD:

$$\frac{\text{Area AGH}}{\text{Area AGED}} = \frac{\frac{1}{2}\text{AG} \times \text{GH}}{\text{AG} \times \text{AD}}$$
$$= \frac{1}{2}\frac{\text{GH}}{\text{AD}}$$
$$= \frac{1}{2}\frac{\text{GH}}{\text{AF}}$$
$$= \frac{1}{2}\frac{\text{AF} - \text{AG}}{\text{AF}}$$
$$= \frac{1}{2}\left(\frac{\text{d} - \text{AG}}{\text{d}}\right)$$
$$\therefore \text{Area AGH} = \frac{1}{2}\frac{(\text{d} - \text{V/A})}{\text{d}} \times \text{Area AGED}$$

The centroid of AGED is V/2A from AD.

Now let triangle AGH be shifted to HEC. The centroid of AGED will move parallel to the shift of the centroid of AGH and the vertical component of this shift (x) is given by:

$$x = \frac{AGH \times d/3}{AGED}$$
$$= \frac{\frac{1}{2}\left(\frac{d - V/A}{d}\right) \times \frac{d}{3} \times AGED}{AGED}$$
$$= \frac{1}{2}\left(\frac{d - V/A}{d}\right) \times \frac{d}{3}$$
$$= \frac{1}{6}\left(d - V/A\right)$$

The new vertical distance of the centroid *below* AD will now be given by:

$$\text{Distance } \textit{below} \text{ AD} = \frac{1}{2}\frac{V}{A} + \frac{1}{6}\left(d - \frac{V}{A}\right)$$
$$= \frac{1}{3}\frac{V}{A} + \frac{1}{6}d$$
$$= \frac{1}{3}\left(\frac{d}{2} + \frac{V}{A}\right)$$

Therefore, the distance of the center of buoyancy *below* the load waterline is given by the formula:

$$\text{Distance } \textit{below} \text{ LWL} = \frac{1}{3}\left(\frac{d}{2} + \frac{V}{A}\right)$$

This is known as *Morrish's* or *Normand's formula* and will give very good results for merchant ships.

To Find Transverse BM

The transverse BM is the height of the transverse metacenter above the center of buoyancy and is found by using the formula:

$$BM = \frac{1}{V}$$

where 1 = the second moment of the waterplane area about the centerline and V = the ship's volume of displacement.

The derivation of this formula is as follows:

Consider a ship inclined to a small angle (θ), as shown in Figure 10.3(a). Let 'y' be the half-breadth.

(a)

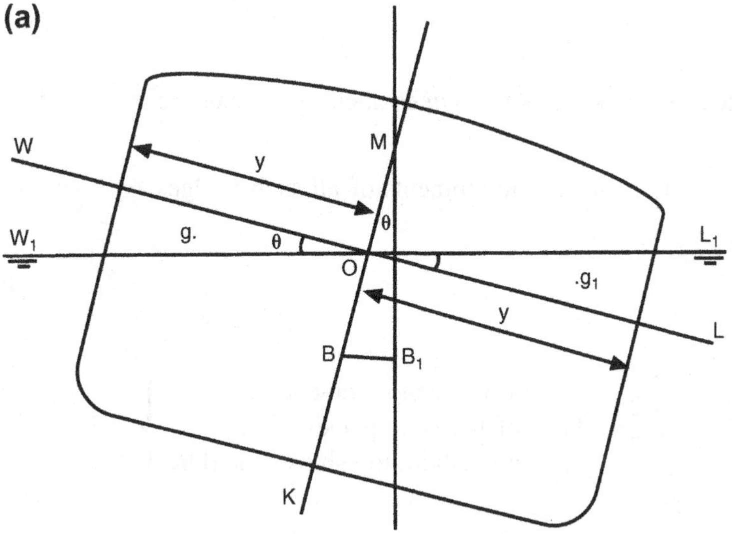

Figure 10.3(a)

$$\text{Since } \theta \text{ is a small angle then arc } WW_1 = \text{arc } LL_1 = \theta y$$

Also:

$$\text{Area of wedge } WOW_1 = \text{Area of wedge } LOL_1 = \frac{1}{2}\theta y^2$$

Consider an elementary wedge of longitudinal length dx as in Figure 10.3(b).

(b)

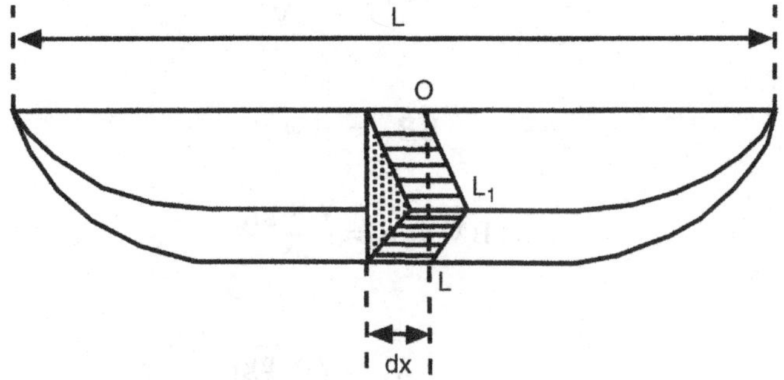

Figure 10.3(b)

$$\text{The volume of this wedge} = \frac{1}{2}\theta y^2\, dx$$

$$\text{The moment of the wedge about the centerline} = \frac{1}{2}\theta y^2\, dx \times \frac{2}{3}y$$

$$= \frac{1}{3}\theta y^3\, dx$$

$$\text{The total moment of both wedges about the centerline} = \frac{2}{3}\theta y^3\, dx$$

$$\text{The sum of the moments of all such wedges} = \int_O^L \frac{2}{3}\theta y^3\, dx$$

$$= \theta \int_O^L \frac{2}{3} y^3\, dx$$

But

$$\int_O^L \frac{2}{3} y^3\, dx = \left.\begin{array}{l}\text{The second moment}\\ \text{of the waterplane}\\ \text{area about the ships centerline}\end{array}\right\} = I$$

$$\therefore \text{The sum of the moments of the wedges} = I \times \theta$$

$$\text{But the sum of the moments} = v \times gg_1$$

where v is the volume of the immersed or emerged wedge.

$$\therefore I \times \theta = v \times gg_1$$

or

$$I = \frac{v \times gg_1}{\theta} \qquad \text{(I)}$$

Now:

$$BB_1 = \frac{v \times gg_1}{V}$$

and

$$BB_1 = BM \times \theta$$

$$\therefore BM \times \theta = \frac{v \times gg_1}{V}$$

or

$$BM \times V = \frac{v \times gg_1}{\theta}$$

Substituting in (I) above:

$$BM \times V = I$$

$$\therefore BM = \frac{I}{V}$$

For a rectangular waterplane area the second moment about the centerline is found by the formula:

$$I = \frac{LB^3}{12}$$

where L = the length of the waterplane and B = the breadth of the waterplane.

Thus, for a vessel having a rectangular waterplane area:

$$BM = \frac{LB^3}{12V}$$

For a box-shaped vessel:

$$\begin{aligned} BM &= \frac{1}{V} \\ &= \frac{LB^3}{12V} \\ &= \frac{L \times B^3}{12 \times L \times B \times \text{draft}} \\ \therefore BM &= \frac{B^2}{12d} \end{aligned}$$

where B = the beam of the vessel, d = any draft of the vessel, B = constant, and d = variable.

For a triangular-shaped prism:

$$\begin{aligned} BM &= \frac{1}{V} \\ &= \frac{LB^3}{12V} \\ &= \frac{L \times B^3}{12\left(\frac{1}{2} \times L \times B \times \text{draft}\right)} \\ \therefore BM &= \frac{B^2}{6d} \end{aligned}$$

where B = the breadth *at the waterline*, d = the corresponding draft, and B, d are variables.

Example 1

A box-shaped vessel is 24 m × 5 m × 5 m and floats on an even keel at 2 m draft. KG = 1.5 m. Calculate the initial metacentric height.

$$KB = \frac{1}{2}\text{draft} \qquad KB = 1\text{ m}$$

$$BM = \frac{B^2}{12d} \qquad BM = \frac{5^2}{12 \times 2} \qquad BM = 1.04\text{ m}$$

KB	=	1.00 m
BM	=	+ 1.04 m
KM	=	2.04 m
KG	=	− 1.50 m
GM	=	0.54 m

Ans. GM = +0.54 m.

Example 2

A vessel is in the form of a triangular prism 32 m long, 8 m wide at the top, and 5 m deep. KG = 3.7 m. Find the initial metacentric height when floating on even keel at 4 m draft.

Let 'x' be the half-breadth *at the waterline,* as shown in Figure 10.4. Then:

$$\frac{x}{4} = \frac{4}{5}$$

$$x = \frac{16}{5}$$

$$x = 3.2\text{ m}$$

$$\therefore \text{The breadth at the waterline} = 6.4\text{ m}$$

$$KB = 2/3\text{ draft} = 2/3 \times 4 \qquad KB = 2.67\text{ m}$$

$$BM = \frac{B^2}{6d} = \frac{6.4 \times 6.4}{6 \times 4} \qquad BM = 1.71\text{ m}$$

KB	=	2.67 m
BM	=	+ 1.71 m
KM	=	4.38 m
KG	=	− 3.70 m
GM	=	0.68 m

Ans. GM = +0.68 m.

Note how the breadth 'B' would decrease at the lower drafts (see also Figure 10.6(b)).

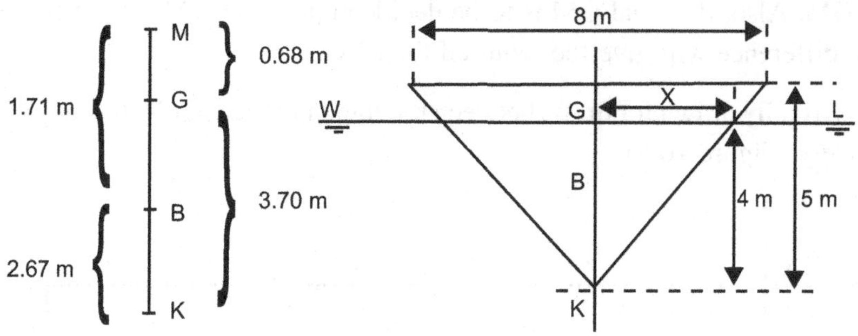

Figure 10.4

Example 3

The second moment of a ship's waterplane area about the centerline is 20,000 m^4 units. The displacement is 7000 tonnes whilst floating in dock water of density 1008 kg per cubic meter. KB = 1.9 m and KG = 3.2 m. Calculate the initial metacentric height.

$$\text{Volume of water displaced} = \frac{7000 \times 1000}{1008} \text{ cubic meters} = 6944 \text{ cubic meters}$$

$$BM = \frac{I}{V}$$

$$\therefore BM = \frac{20{,}000}{6944}$$

$$BM = 2.88 \text{ m}$$

$$KB = \underline{+1.90 \text{ m}}$$

$$KM = 4.78 \text{ m}$$

$$KG = 3.20 \text{ m}$$

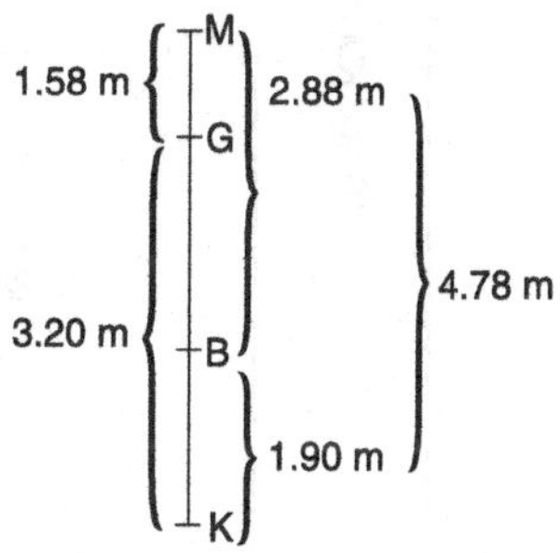

Ans. GM = +1.58 m.

Metacentric Diagrams

It has been mentioned in Chapter 2 that the officer responsible for loading a ship should aim to complete the loading with a GM that is neither too large nor too small (for typical GM values for merchant ships when fully loaded, see Table 2.1 in Chapter 2). A metacentric diagram is a figure in graphical form from which the KB, BM, and thus the KM can be found for any draft by inspection. If the KG is known and the KM is found from the diagram, the difference will give the GM. Also, if a final GM is to be decided upon, the KM can be taken from the graph and the difference will give the required final KG.

The diagram is usually drawn for drafts between the light and loaded displacements, i.e. 3 and 13 m respectively (Figure 10.5).

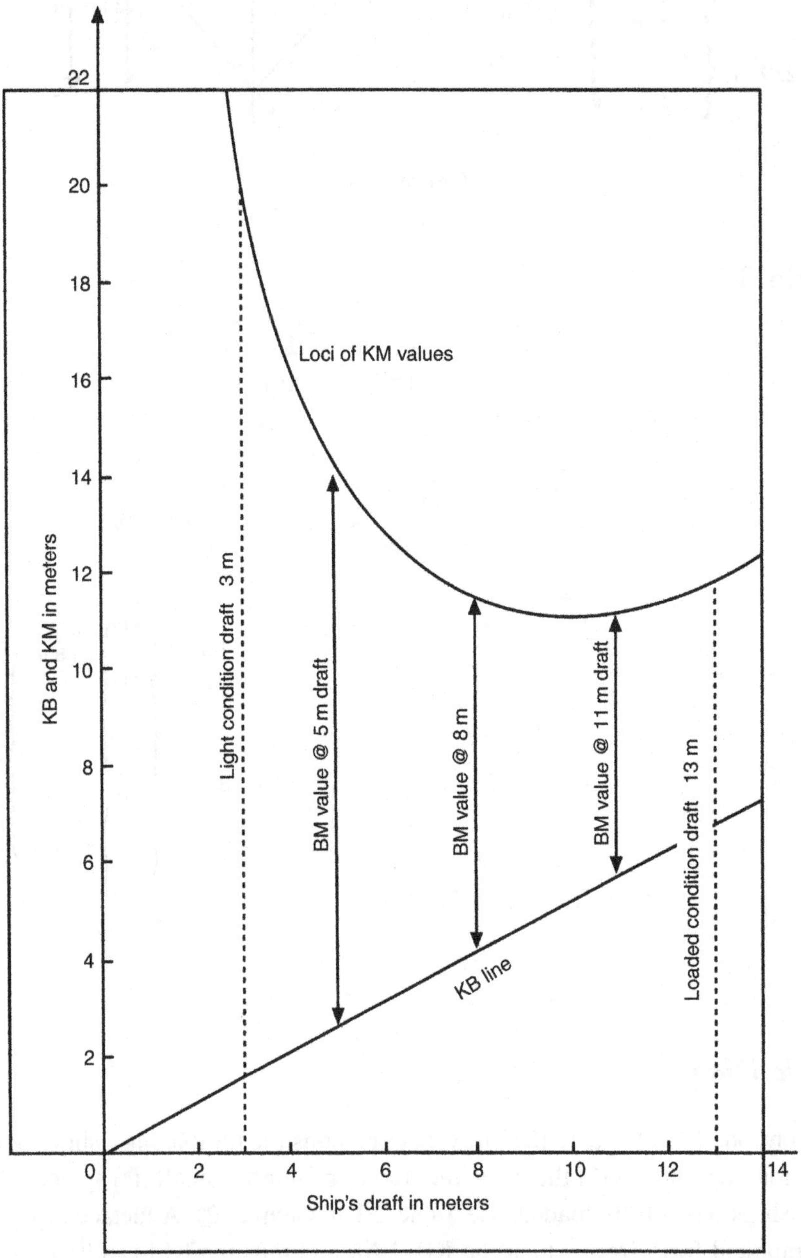

Figure 10.5:
Metacentric Diagram for a Ship-Shaped Vessel.

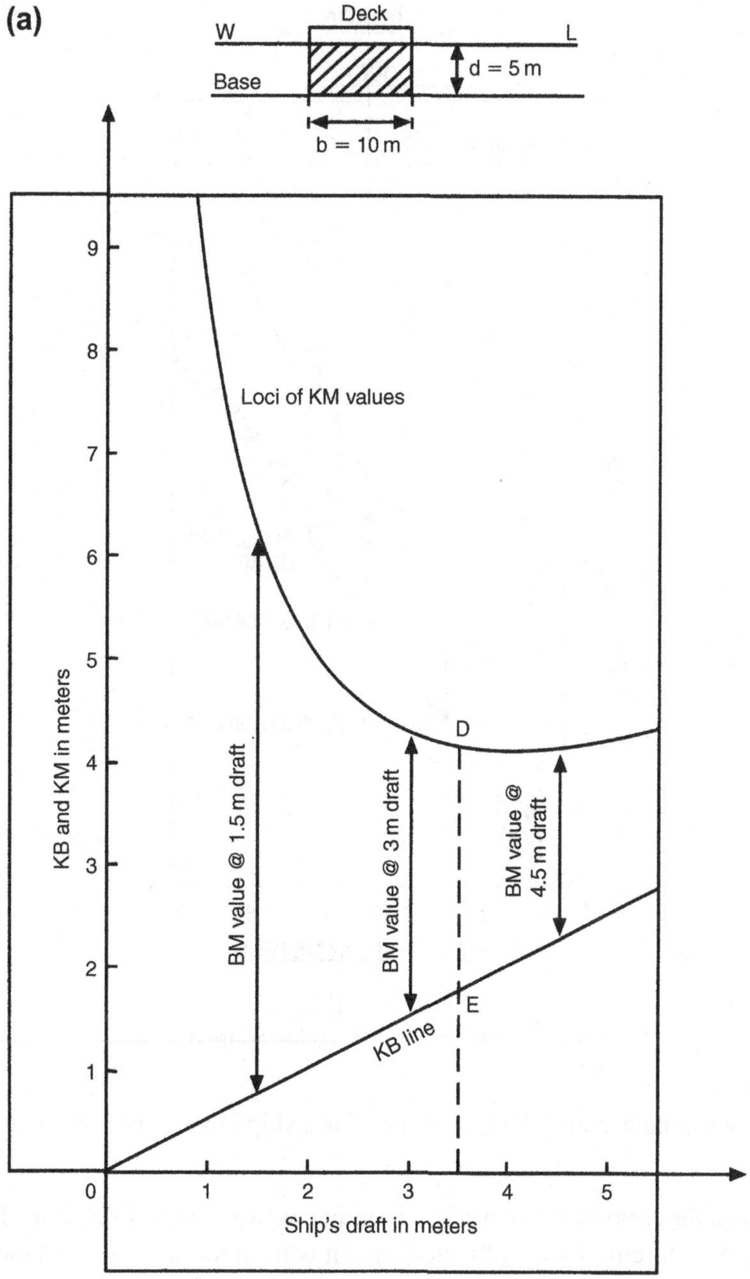

Figure 10.6(a):
Metacentric Diagram for a Box-Shaped Vessel.

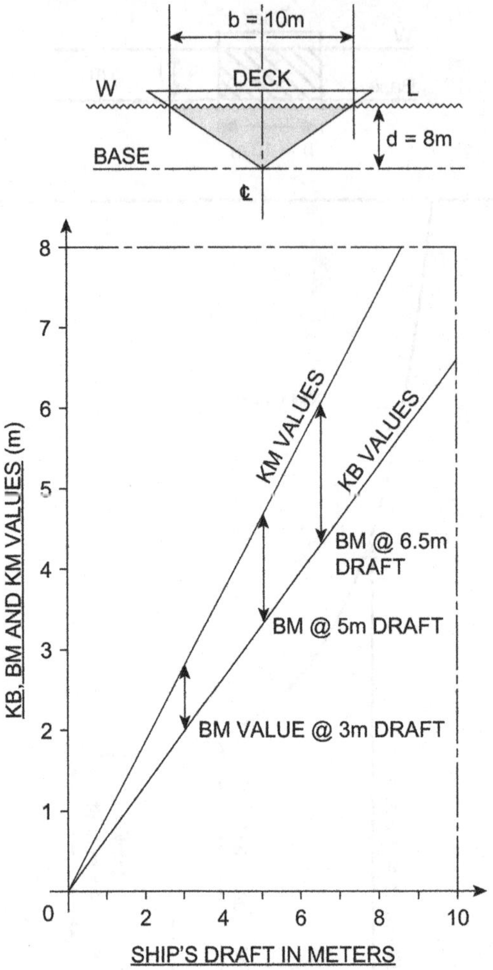

Figure 10.6(b):
Metacentric Diagram for Triangular-Shaped Vessel.

Figure 10.5 shows a metacentric diagram drawn for a ship having the particulars given in Table 10.1.

The following is a description of the method used in constructing this diagram. The scale on the left-hand side represents a scale of meters, and it is from this scale that all measurements are to be taken.

First, the curve of the centers of buoyancy is plotted. For each draft plot the corresponding KB. For example, plot 6.65 m at 13 m, 6.13 m at 12 m draft and so on to 1.55 m at 3 m draft. Join these points together to form the KB curve. In practice it will be very close to being a straight line because the curvature will be so small (see Figure 10.5).

Table 10.1:

Draft (m)	KB (m)	KM (m)
13	6.65	11.60
12	6.13	11.30
11	5.62	11.14
10	5.11	11.10
9	4.60	11.15
8	4.10	11.48
7	3.59	11.94
6	3.08	12.81
5	2.57	14.30
4	2.06	16.63
3	1.55	20.54
–	–	–

Next is the KM curve or locus of metacenters. For each draft plot the corresponding KM value given in Table 10.1. At 13 m plot 11.60 m. At 12 m plot 11.30 m and so on down to plotting 20.54 m KM at 3 m draft. These points are then joined by a smooth curve, as shown in Figure 10.5.

Note how it is possible for two different drafts to have the same value of KM in the range of drafts from 7 to 13 m approximately.

For any draft being considered, the vertical distance between the KB line and the KM curve gives the BM value. To find the KBs and KMs the vertical distances are measured from the baseline to the curves.

Example 1

Construct the metacentric diagram for a box-shaped vessel 64 m long, 10 m beam, and 6 m deep, for even keel drafts at 0.5-m intervals between the light draft 1 m and the load draft 5 m. Also, from the diagram find:

(a) The minimum KM and the draft at which it occurs.
(b) The BM at 3.5 m.

Draft (m)	KB = 1/2 Draft (m)	$BM_T = B^2/12d$ (m)	KM_T = KB + BM (m)
1.00	0.50	8.33	8.833
1.50	0.75	5.56	6.306
2.00	1.00	4.17	5.167
2.50	1.25	3.33	4.583
3.00	1.50	2.78	4.278

Continued

Draft (m)	KB = 1/2 Draft (m)	$BM_T = B^2/12d$ (m)	KM_T = KB + BM (m)
3.50	1.75	2.38	4.131
4.00	2.00	2.08	4.083
4.50	2.25	1.85	4.102
5.00	2.50	1.67	4.167

See Figure 10.6(a) for KB and KM plotted against draft.

Explanation. To find the minimum KM, draw a horizontal tangent to the lowest point of the curve of metacenters, i.e. through A. The point where the tangent cuts the scale will give the minimum KM and the draft at which it occurs.

Note. It is shown below that for a box-shaped vessel the minimum KM and the draft at which it occurs are both given by $B/\sqrt{6}$, where B is the beam.

Therefore, the answer to part (a) of the question is:

$$\text{Minimum KM} = 4.08 \text{ m occurring at } 4.08 \text{ m draft}$$

To find the BM at 3.5 m draft, measure the distance DE on the scale and it will give the BM (2.38 m).

Therefore, the answer to part (b) of the question is:

$$\text{BM at } 3.5 \text{ m draft} = 2.38 \text{ m}$$

To show that, for a box-shaped vessel, the minimum KM and the draft at which it occurs are both given by the expression $B/\sqrt{6}$, where B is equal to the vessel's beam:

$$KM = KB + BM$$

For a box-shaped vessel:

$$KM = \frac{d}{2} + \frac{B^2}{12d} \qquad (I)$$

$$\frac{\partial KM}{\partial d} = \frac{1}{2} + \frac{B^2}{12d^2}$$

For minimum KM:

$$\frac{\partial KM}{\partial d} = O$$

$$\therefore O = \frac{1}{2} + \frac{B^2}{12d^2}$$

$$B^2 = 6d^2$$

and

$$d = B/\sqrt{6}$$

Substituting in equation (I) above:

$$\text{Minimum KM} = \frac{B}{2\sqrt{6}} + \frac{B^2\sqrt{6}}{12B}$$

$$= \frac{6B + 6B}{12\sqrt{6}}$$

$$\text{Minimum KM} = B/\sqrt{6}$$

	WL Breadth (m)	KB (m)	BM (m)	KM (m)
Breadth at 100% draft	10.00	5.33	2.08	7.42
Breadth at 75% draft	7.50	4.00	1.56	5.56
Breadth at 50% draft	5.00	2.67	1.04	3.71
Breadth at 25% draft	2.50	1.33	0.52	1.85
Breadth at 0% draft	0.00	0.00	0.00	0.00

Draft (m)	KB (m)	KM (m)
8.00	5.33	7.42
6.00	4.00	5.56
4.00	2.67	3.71
2.00	1.33	1.85
0.00	0.00	0.00

Figure 10.6(b) shows a metacentric diagram for a triangular-shaped underwater form with apex at the base. Note how the KM values have produced a straight line instead of the parabolic curve of the rectangular hull form. Note also how BM increases with *every increase* in *draft*. If d = 8, 6, 4, 2, and 0 m, then B = 10, 7.5, 5, 2.5, and 0 m.

Exercise 10

1. A box-shaped vessel 75 m long, 12 m beam, and 7 m deep is floating on even keel at 6 m draft. Calculate the KM.
2. Compare the initial metacentric heights of two barges, each 60 m long, 10 m beam at the waterline, 6 m deep, floating upright on even keel at 3 m draft and having

KG = 3 m. One barge is in the form of a rectangular prism and the other is in the form of a triangular prism, floating apex downwards.

3. Two box-shaped vessels are each 100 m long, 4 m deep, float at 3 m draft, and have KG = 2.5 m. Compare their initial metacentric heights if one has 10 m beam and the other has 12 m beam.
4. Will a homogeneous log of square cross-section and relative density 0.7 have a positive initial metacentric height when floating in fresh water with one side parallel to the waterline? Verify your answer by means of a calculation.
5. A box-shaped vessel 60 m × 12 m × 5 m is floating on even keel at a draft of 4 m. Construct a metacentric diagram for drafts between 1 and 4 m. From the diagram find:

   (a) the KMs at drafts of 2.4 and 0.9 m;
   (b) the draft at which the minimum KM occurs.

6. Construct a metacentric diagram for a box-shaped vessel 65 m × 12 m × 6 m for drafts between 1 and 6 m. From the diagram find:

   (a) the KMs at drafts of 1.2 and 3.6 m;
   (b) the minimum KM and the draft at which it occurs.

7. Construct a metacentric diagram for a box-shaped vessel 70 m long and 10 m beam, for drafts between 1 and 6 m. From the diagram find:

   (a) the KMs at drafts of 1.5 and 4.5 m;
   (b) the draft at which the minimum KM occurs.

8. A box-shaped vessel is 60 m long, 13.73 m wide, and floats at 8 m even-keel draft in salt water.

   (a) Calculate the KB, BM, and KM values for drafts 3–8 m at intervals of 1 m. From your results draw the metacentric diagram.
   (b) At 3.65 m draft even keel, it is known that the VCG is 4.35 m above base. Using your diagram, estimate the transverse GM for this condition of loading.
   (c) At 5.60 m draft even keel, the VCG is also 5.60 m above base. Using your diagram, estimate the GM for this condition of loading. What state of equilibrium is the ship in?

Draft (m)	3	4	5	6	7	8
KM (m)	6.75	5.94	5.64	5.62	5.75	5.96

CHAPTER 11

Final KG Plus 20 Reasons for Rise in KG

When a ship is completed by the builders, certain written stability information must be handed over to the shipowner with the ship. Details of the information required are contained in the 1998 load line rules, parts of which are reproduced in Chapter 58. The information includes details of the ship's lightweight, the lightweight VCG and LCG, and also the positions of the centers of gravity of cargo and bunker spaces. This gives an initial condition from which the displacement and KG for any condition of loading may be calculated. The final KG is found by taking the moments of the weights loaded or discharged, about the keel. For convenience, when taking the moments, consider the ship to be on her beam ends.

In Figure 11.1(a), KG represents the original height of the center of gravity above the keel and W represents the original displacement. The original moment about the keel is therefore W × KG.

Now load a weight w_1 with its center of gravity at g_1 and discharge w_2 from g_2. This will produce moments about the keel of $w_1 \times Kg_1$ and $w_2 \times Kg_2$ in the directions indicated in the figure. The final moment about the keel will be equal to the original moment plus the moment of the weight added minus the moment of the weight discharged. But the final moment must also be equal to the final displacement multiplied by the final KG, as shown in Figure 11.1(b), i.e.

$$\text{Final moment} = \text{Final KG} \times \text{Final displacement}$$

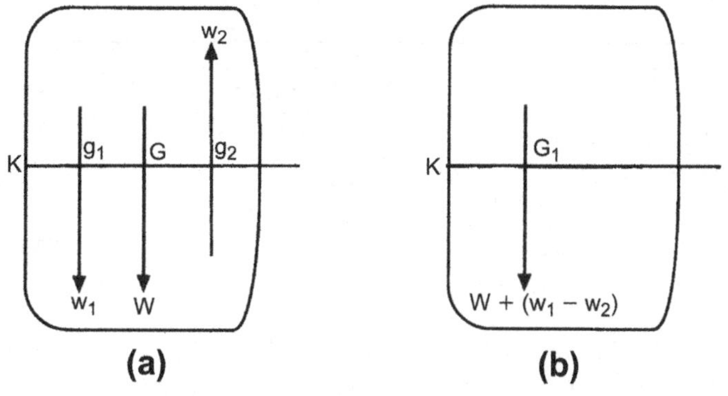

Figure 11.1

Ship Stability for Masters and Mates. http://dx.doi.org/10.1016/B978-0-08-097093-6.00011-6

or

$$\text{Final KG} = \frac{\text{Final moment}}{\text{Final displacement}}$$

Example 1

A ship of 6000 tonnes displacement has KG = 6 m and KM = 7.33 m. The following cargo is then loaded:

1000 tonnes KG = 2.5 m
500 tonnes KG = 3.5 m
750 tonnes KG = 9.0 m

The following is then discharged:

450 tonnes of cargo KG = 0.6 m

and

800 tonnes of cargo KG = 3.0 m

Find the final GM.

Weight	KG	Moment About the Keel
+6000	6.0	+36,000
+1000	2.5	+2500
+500	3.5	+1750
+750	9.0	+6750
+8250		+47,000
−450	0.6	−270
−800	3.0	−2400
+7000		+44,330

$$\text{Final KG} = \frac{\text{Final moment}}{\text{Final displacement}}$$

$$= \frac{44,330}{7000} = 6.33 \text{ m}$$

$$\text{GM} = \text{KM} - \text{KG}$$

$$\text{KM} = 7.33 \text{ m, as given}$$

$$\text{Final KG} = \underline{6.33 \text{ m}}, \text{ as calculated}$$

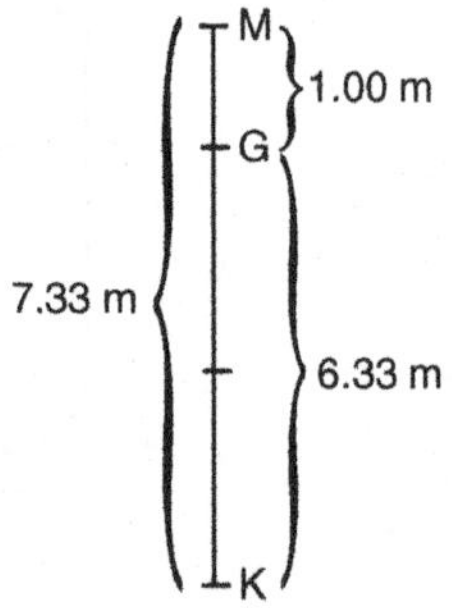

Ans. Final GM = 1.00 m.

Note. KM was assumed to have a similar value at 6000 tonnes and 7000 tonnes displacement. This is feasible. As can be seen in Figure 2.2, it is possible to have the same KM value at two different drafts.

Example 2

A ship of 5000 tonnes displacement has KG = 4.5 m and KM = 5.3 m. The following cargo is then loaded:

2000 tonnes KG = 3.7 m and 1000 tonnes KG = 7.5 m

Find how much deck cargo (KG = 9 m) may now be loaded if the ship is to sail with a minimum GM of 0.3 m.

Let 'x' tonnes of deck cargo be loaded, so that the vessel sails with GM = 0.3 m.

Final KM = 5.3 ms
Final GM = 0.3 m

Final KG = 5.0 m

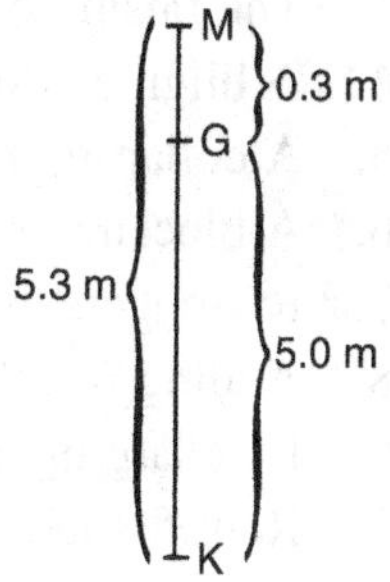

$$\text{Final KG} = \frac{\text{Final moment}}{\text{Final displacement}} = 5.0\text{ m}$$

$$\therefore 5 = \frac{37{,}400 + 9x}{8000 + x} = \frac{\Sigma_v}{\Sigma_1}$$

$$40{,}000 + 5x = 37{,}400 + 9x$$

$$2600 = 4x$$

$$x = 650 \text{ tonnes}$$

Ans. Maximum to load = 650 tonnes.

Weight	KG	Moment About the Keel
5000	4.5	22,500
2000	3.7	7400
1000	7.5	7500
x	9.0	9x
$(8000 + x) = \Sigma_1$		$(37{,}400 + 9x) = \Sigma_v$

Twenty Reasons for a Rise in G

When the vertical center of gravity G rises, there will normally be a loss in the ship's stability. G may even rise above the transverse metacenter M to make the ship unstable. The master and mate onboard ship must be aware of changes in a ship that would cause such a rise in G. The following list gives reasons for such a rise:

1. Free-surface effects in partially filled tanks.
2. Collapse of a longitudinal division/bulkhead in a partially filled tank of liquid.
3. Icing up of superstructures.
4. Loading cargo in upper reaches of the vessel.
5. Water entering the ship through badly maintained hatches on upper deck and flooding the tween decks.
6. Hatches or bow doors inadvertently left open on the main deck.
7. Water landing on the deck from the sea in heavy weather conditions.
8. Raising of a weight from a deck using a mast and derrick.
9. Raising a weight low down in the ship to a higher position within the ship.
10. Timber deck cargo becoming saturated due to bad weather conditions.
11. Vessel making first contact with keel blocks in a dry dock at the stern.
12. A ship's first contact with a raised shelf or submerged wreck.
13. The raising of the sails on a yacht.
14. A bilging situation, causing free-surface effects.
15. A collapse of grainboards or fishboards.
16. A blockage of freeing ports or scuppers on the upper deck.
17. Passengers crowding on superstructure decks at time of departure or arrival.
18. Adding weight at a point *above* the ship's initial overall VCG.
19. Discharging a weight at a point *below* the ship's initial overall VCG.
20. Retrofits in accommodation decks and navigation spaces.

Exercise 11

1. A ship has a displacement of 1800 tonnes and KG = 3 m. She loads 3400 tonnes of cargo (KG = 2.5 m) and 400 tonnes of bunkers (KG = 5.0 m). Find the final KG.
2. A ship has a light displacement of 2000 tonnes and light KG = 3.7 m. She then loads 2500 tonnes of cargo (KG = 2.5 m) and 300 tonnes of bunkers (KG = 3 m). Find the new KG.
3. A ship sails with displacement 3420 tonnes and KG = 3.75 m. During the voyage bunkers were consumed as follows: 66 tonnes (KG = 0.45 m) and 64 tonnes (KG = 2 m). Find the KG at the end of the voyage.
4. A ship has displacement 2000 tonnes and KG = 4 m. She loads 1500 tonnes of cargo (KG = 6 m), 3500 tonnes of cargo (KG = 5 m), and 1520 tonnes of bunkers

(KG = 1 m). She then discharges 2000 tonnes of cargo (KG = 2.5 m) and consumes 900 tonnes of oil fuel (KG = 0.5 m) during the voyage. Find the final KG on arrival at the port of destination.

5. A ship has a light displacement of 2000 tonnes (KG = 3.6 m). She loads 2500 tonnes of cargo (KG = 5 m) and 300 tonnes of bunkers (KG = 3 m). The GM is then found to be 0.15 m. Find the GM with the bunkers empty.
6. A ship has a displacement of 3200 tonnes (KG = 3 m and KM = 5.5 m). She then loads 5200 tonnes of cargo (KG = 5.2 m). Find how much deck cargo having a KG = 10 m may now be loaded if the ship is to complete loading with a positive GM of 0.3 m.
7. A ship of 5500 tonnes displacement has KG = 5 m, and she proceeds to load the following cargo:

1000 tonnes	KG = 6 m
700 tonnes	KG = 4 m
300 tonnes	KG = 5 m

   She then discharges 200 tonnes of ballast, KG = 0.5 m. Find how much deck cargo (KG = 10 m) can be loaded so that the ship may sail with a positive GM of 0.3 meters. The load KM is 6.3 m.
8. A ship of 3500 tonnes light displacement and light KG = 6.4 m has to load 9600 tonnes of cargo. The KG of the lower hold is 4.5 m and that of the tween deck is 9 m. The load KM is 6.2 m and, when loading is completed, the righting moment at 6 degrees of heel is required to be 425 tonnes m. Calculate the amount of cargo to be loaded into the lower hold and tween deck respectively. (Righting moment = W × GM × sin heel.)
9. A ship arrives in port with displacement 6000 tonnes and KG = 6 m. She then discharges and loads the following quantities:

Discharge	1250 tonnes of cargo	KG = 4.5 meters
	675 tonnes of cargo	KG = 3.5 meters
	420 tonnes of cargo	KG = 9.0 meters

Load	980 tonnes of cargo	KG = 4.25 meters
	550 tonnes of cargo	KG = 6.0 meters
	700 tonnes of bunkers	KG = 1.0 meter
	70 tonnes of FW	KG = 12.0 meters

   During the stay in port, 30 tonnes of oil (KG = 1 m) are consumed. If the final KM is 6.8 m, find the GM on departure.
10. A ship has light displacement 2800 tonnes and light KM = 6.7 m. She loads 400 tonnes of cargo (KG = 6 m) and 700 tonnes (KG = 4.5 m). The KG is then found to be 5.3 m. Find the light GM.

11. A ship's displacement is 4500 tonnes and KG = 5 m. The following cargo is loaded:

450 tonnes	KG = 7.5 m
120 tonnes	KG = 6.0 m
650 tonnes	KG = 3.0 m

Find the amount of cargo to load in a tween deck (KG = 6 m) so that the ship sails with a GM of 0.6 m. (The load KM is 5.6 m.)

12. A ship of 7350 tonnes displacement has KG = 5.8 m and GM = 0.5 m. Find how much deck cargo must be loaded (KG = 9 m) if there is to be a metacentric height of not less than 0.38 m when loading is completed.
13. A ship is partly loaded and has a displacement of 9000 tonnes, KG = 6 m, and KM = 7.3 m. She is to make a 19-day passage consuming 26 tonnes of oil per day (KG = 0.5 m). Find how much deck cargo she may load (KG = 10 m) if the GM on arrival at the destination is to be not less than 0.3 m.

■

CHAPTER 12

Angle of List Considerations — Text, Calculations, and Graphics

Consider a ship floating upright as shown in Figure 12.1. The centers of gravity and buoyancy are on the centerline. The resultant force acting on the ship is zero, and the resultant moment about the center of gravity is zero.

Now let a weight already on board the ship be shifted transversely such that G moves to G_1 as in Figure 12.2(a). This will produce a listing moment of W × GG_1, and the ship will list until G_1 and the center of buoyancy are in the same vertical line, as in Figure 12.2(b).

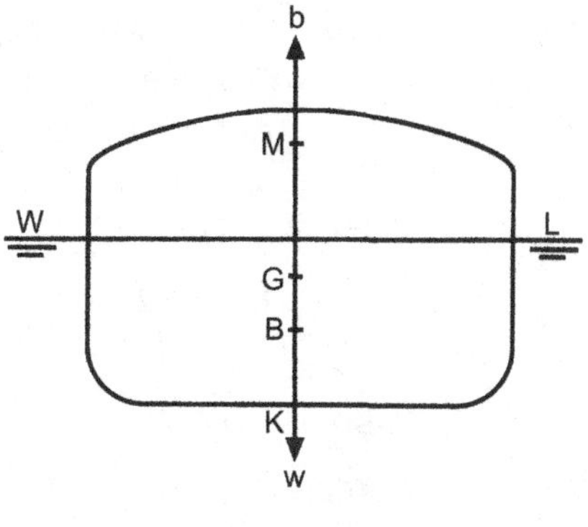

Figure 12.1

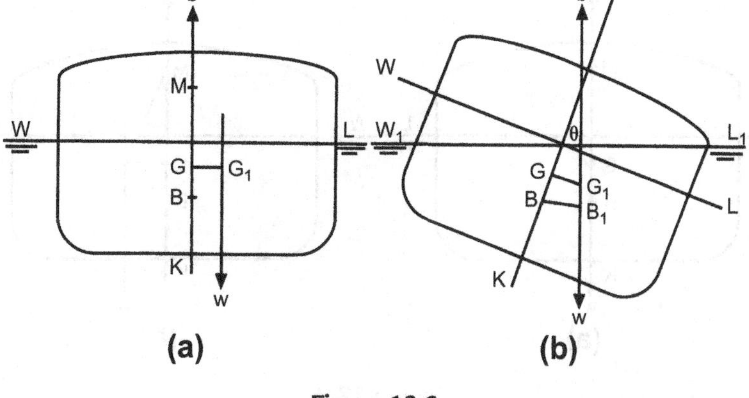

Figure 12.2

Ship Stability for Masters and Mates. http://dx.doi.org/10.1016/B978-0-08-097093-6.00012-8

In this position G_1 will also lie vertically under M so long as the angle of list is small. Therefore, if the final positions of the metacenter and the center of gravity are known, the final list can be found, using trigonometry, in the triangle GG_1M, which is right-angled at G.

The final position of the center of gravity is found by taking moments about the keel and about the centerline.

Note. It will be found more convenient in calculations, when taking moments, to consider the ship to be upright throughout the operation.

Example 1

A ship of 6000 tonnes displacement has KM = 7.3 m and KG = 6.7 m, and is floating upright. A weight of 60 tonnes already on board is shifted 12 m transversely. Find the resultant list.

Figure 12.3(a) shows the initial position of G before the weight was shifted and Figure 12.3(b) shows the final position of G after the weight has been shifted.

When the weight is shifted transversely the ship's center of gravity will also shift transversely, from G to G_1. The ship will then list θ degrees to bring G_1 vertically under M, the metacenter:

$$GG_1 = \frac{w \times d}{W}$$

$$= \frac{60 \times 12}{6000}$$

$$GG_1 = 0.12 \text{ m}$$

$$GM = KM - KG = 7.3 - 6.7 = 0.6 \text{ m}$$

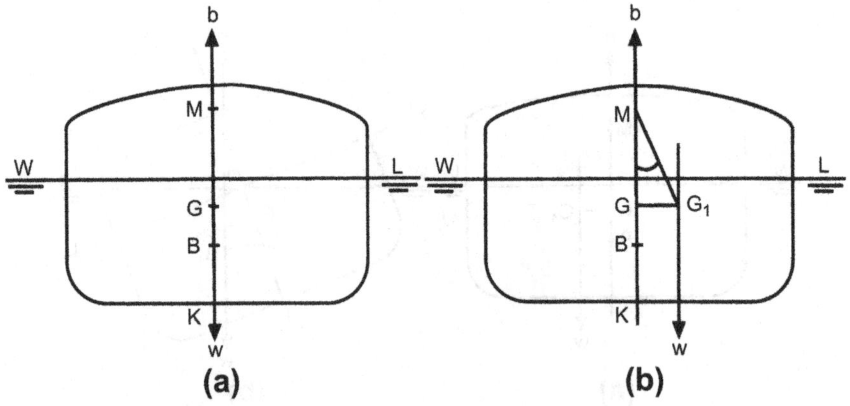

Figure 12.3

In triangle GG_1M:

$$\tan\theta = \frac{GG_1}{GM} = \frac{0.12}{0.60} = 0.20$$

Ans. List = 11° 18.5′.

Example 2

A ship of 8000 tonnes displacement has KM = 8.7 m and KG = 7.6 m. The following weights are then loaded and discharged:

Load 250 tonnes cargo KG = 6.1 m and center of gravity 7.6 m to starboard of the centerline.

Load 300 tonnes fuel oil KG = 0.6 m and center of gravity 6.1 m to port of the centerline.

Discharge 50 tonnes of ballast KG = 1.2 m and center of gravity 4.6 m to port of the centerline.

Find the final list. (See Figure 12.3).

Note. In this type of problem find the final KG by taking moments about the keel, and the final distance of the center of gravity from the centerline by taking moments about the centerline.

Moments About the Keel

Weight	KG	Moments About the Keel
8000	7.6	60,800
250	6.1	1525
300	0.6	180
8550		62,505
−50	1.2	−60
8500		62,445

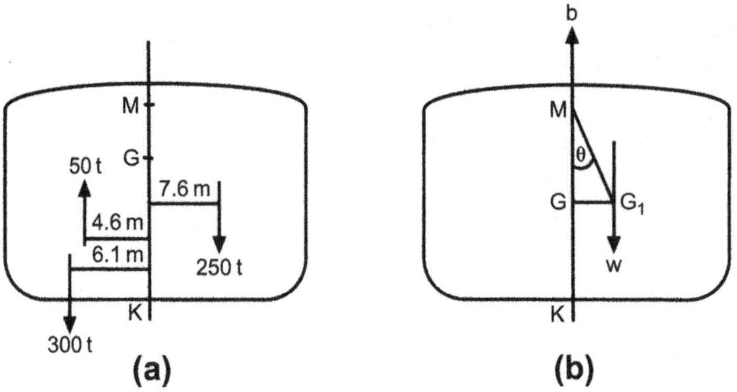

Figure 12.4

$$\text{Final KG} = \frac{\text{Final moment}}{\text{Final displacement}} = \frac{62{,}445}{8500}$$

KM = 8.70 m
Final KG = − 7.35 m
Final GM = 1.35 m

Final KG = 7.35 m.

Moments About the Centerline (as in Figure 12.4(a))

For levers to port, use positive sign.

For levers to starboard, use negative sign.

w	D	Listing Moment	
		To Port Positive	To Starboard Negative
+250	−7.6	—	−1900
−50	+4.6	—	−230
+300	+6.1	+1830	—
		+1830	−2130
			+1830
		Final moment	−300

Let the final position of the center of gravity be as shown in Figure 12.4(b):

$$\therefore \text{ Final listing moment} = W \times GG_1$$

or

$$GG_1 = \frac{\text{Final moment}}{\text{Final displacement}}$$

$$= \frac{-300}{8500} = -0.035\text{ m}$$

$GG_1 = 0.035$ m to starboard, because of the negative sign used in the table.

Since the final position of the center of gravity must lie vertically under M, it follows that the ship will list $\theta°$ to starboard:

$$\tan\theta = \frac{GG_1}{GM}, \text{ as in Figure12.4(b)}$$

$$= \frac{-0.035}{1.35} = -0.0259$$

$$\therefore \theta = 1°29'$$

Ans. Final list = 1° 29′ to starboard.

Example 3

A ship of 8000 tonnes displacement has a GM = 0.5 m. A quantity of grain in the hold, estimated at 80 tonnes, shifts and, as a result, the center of gravity of this grain moves 6.1 m horizontally and 1.5 m vertically. Find the resultant list.

Referring to Figure 12.5, let the center of gravity of the grain shift from g to g_2. This will cause the ship's center of gravity to shift from G to G_2 in a direction parallel to gg_2. The horizontal components of these shifts are g to g_1 and G to G_1 respectively, whilst the vertical components are g_1g_2 and G_1G_2.

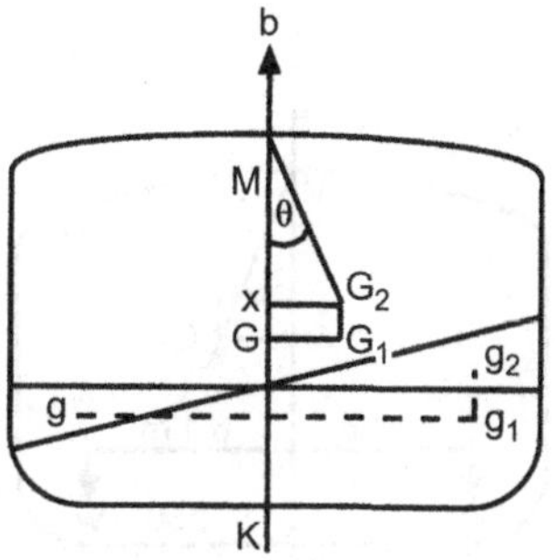

Figure 12.5

$$GG_1 = \frac{w \times d}{W} \qquad G_1G_2 = \frac{w \times d}{W}$$

$$= \frac{80 \times 6.1}{8000} \qquad = \frac{80 \times 1.5}{8000}$$

$$GG_1 = 0.061 \text{ m} \qquad G_1G_2 = 0.015 \text{ m}$$

In Figure 12.5:

$$GX = G_1G_2$$
$$GX = 0.015 \text{ m}$$
$$GM = 0.500 \text{ m}$$
$$XM = 0.485 \text{ m}$$

$$XG_2 = GG_1$$
$$XG_2 = 0.061 \text{ m}$$

$$\tan\theta = \frac{XG_2}{MX}$$
$$\tan\theta = \frac{0.061}{0.485} = 0.126$$
$$\tan\theta = 0.126$$

Ans. List = 7°12′.

Example 4

A ship of 13,750 tonnes displacement, GM = 0.75 m, is listed 2.5 degrees to starboard and has yet to load 250 tonnes of cargo. There is space available in each side of No. 3 between deck (center of gravity 6.1 m out from the centerline). See Figure 12.6. Find how much cargo to load on each side if the ship is to be upright on completion of loading.

Load 'w' tonnes to port and (250 − w) tonnes to starboard.

In triangle GG_1M:

$$GG_1 = GM \tan\theta$$
$$= 0.75 \tan 2.5°$$
$$= 0.75 \times 0.0437$$
$$GG_1 = 0.0328 \text{ m}$$

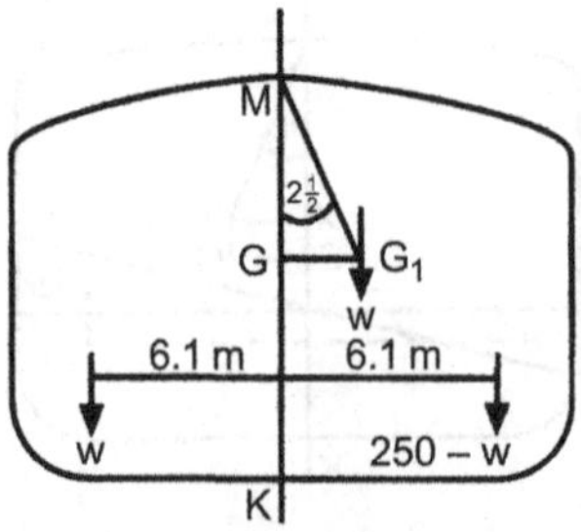

Figure 12.6

Moments About the Centerline

w	D	Listing Moment	
		To Port	To Starboard
w	6.1	6.1w	–
13,750	0.0328	–	451
250 – w	6.1	–	1525 – 6.1w
		6.1w	1976 – 6.1w

If the ship is to complete loading upright, then:

$$\text{Moment to port} = \text{Moment to starboard}$$
$$6.1\ w = 1976 - 6.1\ w$$
$$w = 161.97\ \text{tonnes}$$

Ans. Load 161.97 tonnes to port and 88.03 tonnes to starboard.

Note. 161.97 + 88.03 = 250 tonnes of cargo (as given in question).

■

Summary

1. Always make a *sketch* from the given information.
2. Use a moment of weight table.
3. Use values from the table to *calculate* the final requested data.

Exercise 12

1. A ship of 5000 tonnes displacement has KG = 4.2 m and KM = 4.5 m, and is listed 5° to port. Assuming that the KM remains constant, find the final list if 80 tonnes of bunkers are loaded in No. 2 starboard tank whose center of gravity is 1 meter above the keel and 4 meters out from the centerline.
2. A ship of 4515 tonnes displacement is upright, and has KG = 5.4 m and KM = 5.8 m. It is required to list the ship 2° to starboard and a weight of 15 tonnes is to be shifted transversely for this purpose. Find the distance through which it must be shifted.
3. A ship of 7800 tonnes displacement has a mean draft of 6.8 m and is to be loaded to a mean draft of 7 meters. GM = 0.7 m and TPC = 20 tonnes. The ship is at present listed 4° to starboard. How much more cargo can be shipped in the port

and starboard tween deck, centers of gravity 6 m and 5 m respectively from the centerline, for the ship to complete loading and finish upright?

4. A ship of 1500 tonnes displacement has KB = 2.1 m, KG = 2.7 m and KM = 3.1 m, and is floating upright in salt water. Find the list if a weight of 10 tonnes is shifted transversely across the deck through a distance of 10 meters.
5. A weight of 12 tonnes, when moved transversely across the deck through a distance of 12 m, causes a ship of 4000 tonnes displacement to list 3.8° to starboard. KM = 6 m. Find the KG.
6. A quantity of grain, estimated at 100 tonnes, shifts 10 m horizontally and 1.5 m vertically in a ship of 9000 tonnes displacement. If the ship's original GM was 0.5 m, find the resulting list.
7. A ship of 7500 tonnes displacement has KM = 8.6 m, KG = 7.8 m, and 20 m beam. A quantity of deck cargo is lost from the starboard side (KG = 12 m, and center of gravity is 6 m in from the rail). If the resulting list is 3° 20′ to port, find how much deck cargo was lost.
8. A ship of 12,500 tonnes displacement, KM = 7 m and KG = 6.4 m, has a 3° list to starboard and has yet to load 500 tonnes of cargo. There is space available in the tween decks, centers of gravity 6 m each side of the centerline. Find how much cargo to load on each side if the ship is to complete loading upright.
9. A ship is listed 2.5° to port. The displacement is 8500 tonnes, KM = 5.5 m, and KG = 4.6 m. The ship has yet to load a locomotive of 90 tonnes mass on deck on the starboard side (center of gravity 7.5 m from the centerline), and a tender of 40 tonnes. Find how far from the centerline the tender must be placed if the ship is to complete loading upright, and also find the final GM (KG of the deck cargo is 7 m).
10. A ship of 9500 tonnes displacement is listed 3.5° to starboard, and has KM = 9.5 m and KG = 9.3 m. She loads 300 tonnes of bunkers in No. 3 double-bottom tank port side (KG = 0.6 m and center of gravity 6 m from the centerline), and discharges two parcels of cargo each of 50 tonnes from the port side of No. 2 shelter deck (KG = 11 m and center of gravity 5 m from the centerline). Find the final list.
11. A ship of 6500 tonnes displacement is floating upright and has GM = 0.15 m. A weight of 50 tonnes, already on board, is moved 1.5 m vertically downwards and 5 m transversely to starboard. Find the list.
12. A ship of 5600 tonnes displacement is floating upright. A weight of 30 tonnes is lifted from the port side of No. 2 tween deck to the starboard side of No. 2 shelter deck (10 m horizontally). Find the weight of water to be transferred into No. 3 double-bottom tank from starboard to port to keep the ship upright. The distance between the centers of gravity of the tanks is 6 m.

CHAPTER 13

Angle of Heel – Effects of Suspended Weights

The center of gravity of a body is the point through which the force of gravity may be considered to act vertically downwards. Consider the center of gravity of a weight suspended from the head of a derrick as shown in Figure 13.1.

It can be seen from Figure 13.1 that whether the ship is upright or inclined in either direction, the point in the ship through which the force of gravity may be considered to act

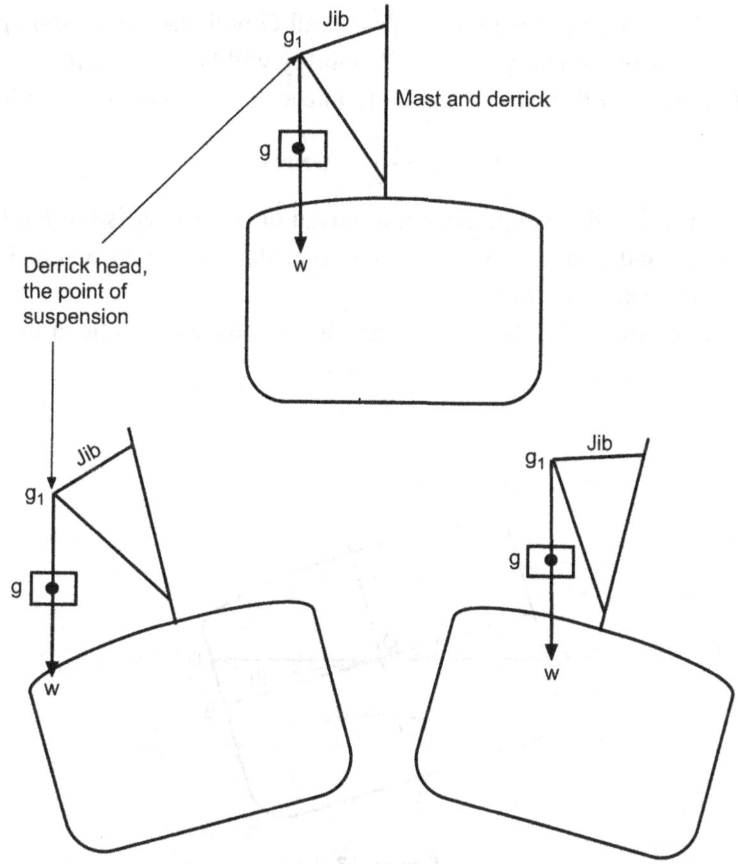

Figure 13.1

Ship Stability for Masters and Mates. http://dx.doi.org/10.1016/B978-0-08-097093-6.00013-X

vertically downwards is g_1, the point of suspension. Thus, the center of gravity of a suspended weight is considered to be at the point of suspension.

Conclusions

1. The center of gravity of a body will move directly *towards* the center of gravity of any *weight added.*
2. The center of gravity of a body will move directly *away* from the center of gravity of any *weight removed.*
3. The center of gravity of a body will *move parallel* to the shift of the center of gravity of any *weight moved* within the body.
4. No matter where the weight 'w' was initially in the ship relative to G, when this weight is moved *downwards* in the ship, then the ship's overall G will also be moved *downwards* to a *lower* position. Consequently, the ship's stability will be *improved.*
5. No matter where the weight 'w' was initially in the ship relative to G, when this weight is moved *upwards* in the ship, then the ship's overall G will also be moved *upwards* to a *higher* position. Consequently, the ship's stability will be *decreased.*
6. The *shift of the center of gravity* of the body in each case is given by the formula:

$$GG_1 = \frac{w \times d}{W} \text{ meters}$$

   where w is the mass of the weight added, removed or shifted, W is the *final* mass of the body, and d is, in 1 and 2, the distance between the centers of gravity, and in 3, the distance through which the weight is shifted.
7. When a weight is *suspended* its center of gravity is considered to be at the *point of suspension.*

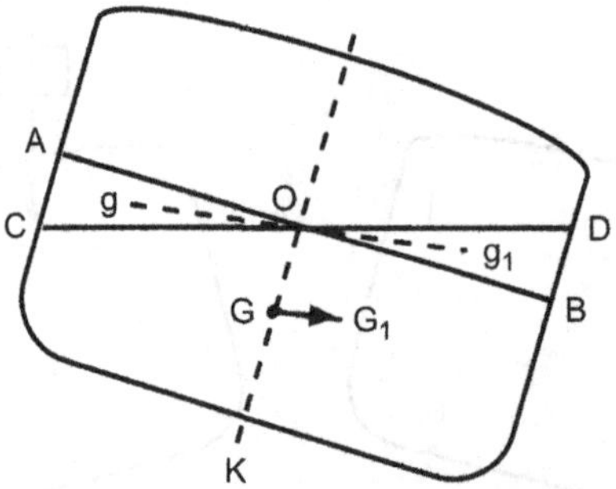

Figure 13.2

Example 1

A hold is partly filled with a cargo of bulk grain. During the loading, the ship takes a list and a quantity of grain shifts so that the surface of the grain remains parallel to the waterline. Show the effect of this on the ship's center of gravity.

In Figure 13.2, G represents the original position of the ship's center of gravity when upright. AB represents the level of the surface of the grain when the ship was upright and CD the level when inclined. A wedge of grain AOC with its center of gravity at g has shifted to ODB with its center of gravity at g_1. The ship's center of gravity will shift from G to G_1, such that GG_1 is parallel to gg_1, and the distance

$$GG_1 = \frac{w \times d}{W} \text{ meters}$$

■

Example 2

A ship is lying starboard side to a quay. A weight is to be discharged from the port side of the lower hold by means of the ship's own derrick. Describe the effect on the position of the ship's center of gravity during the operation.

Note. When a weight is suspended from a point, the center of gravity of the weight appears to be at the point of suspension regardless of the distance between the point of suspension and the weight. Thus, as soon as the weight is clear of the deck and is being borne at the derrick head, the center of gravity of the weight appears to move from its original position to the derrick head. For example, it does not matter whether the weight is 0.6 or 6.0 meters above the deck, or whether it is being raised or lowered, its center of gravity will appear to be at the derrick head.

In Figure 13.3, G represents the original position of the ship's center of gravity, and g represents the center of gravity of the weight when lying in the lower hold. As soon as the weight is raised clear of the deck, its center of gravity will appear to move vertically upwards to g_1. This will cause the ship's center of gravity to move upwards from G to G_1, parallel to gg_1. The centers of gravity will remain at G_1 and g_1 respectively during the whole of the time the weight is being raised. When the derrick is swung over the side, the derrick head will move from g_1 to g_2, and since the weight is suspended from the derrick head, its center of gravity will also appear to move from g_1 to g_2. This will cause the ship's center of gravity to move from G_1 to G_2. If the weight is now landed on the quay it is in effect being discharged from the derrick head and the ship's center of

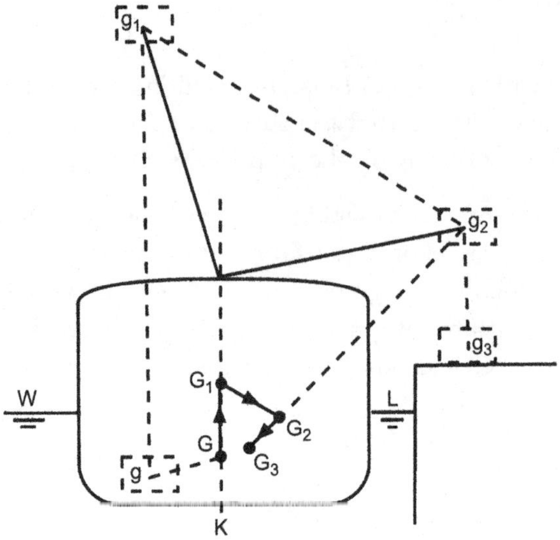

Figure 13.3

gravity will move from G_2 to G_3 in a direction directly away from g_2. G_3 is therefore the final position of the ship's center of gravity after discharging the weight.

From this it can be seen that the net effect of discharging the weight is a shift of the ship's center of gravity from G to G_3, directly away from the center of gravity of the weight discharged.

Note. The only way in which the position of the center of gravity of a ship can be altered is by changing the distribution of the weights within the ship, i.e. by *adding, removing,* or *shifting* weights.

Students find it hard sometimes to accept that the weight, when suspended from the derrick, acts at its point of suspension.

However, it can be proved, by experimenting with ship models or observing full-size ship tests. When measured, the final angle of heel verifies that this assumption is indeed correct. For the method of working, see Example 3 below.

Example 3

A ship of 9900 tonnes displacement has KM = 7.3 m and KG = 6.4 m. She has yet to load two 50-tonne lifts with her own gear and the first lift is to be placed on deck on the inshore side (KG = 9 m and center of gravity 6 m out from the centerline). When the derrick plumbs the quay its head is 15 m above the keel and 12 m out from the centerline. Calculate the maximum list during the operation.

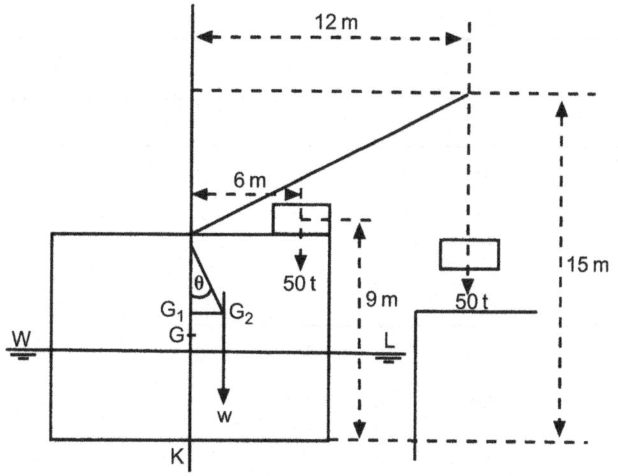

Figure 13.4

Note. The maximum list will obviously occur when the first lift is in place on the deck and the second weight is suspended over the quay, as shown in Figure 13.4.

Moments About the Keel

Weight	KG	Moment
9900	6.4	63,360
50	9.0	450
50	15.0	750
10,000		64,560

$$\text{Final KG} = \frac{\text{Final moment}}{\text{Final displacement}}$$

$$= \frac{64{,}560}{10{,}000}$$

$$\text{Final KG} = 6.456 \text{ m } (KG_1)$$

$$\text{So } GG_1 = 0.056 \text{ m}$$

That is, a rise of 0.056 m above the original KG of 6.4 m.

Moments About the Centerline

Weight	d	Listing Moment	
		To Port	To Starboard
9900	0	0	0
50	12	0	600
50	6	0	300
10,000			900

$$\text{Listing moment} = 900 \text{ tonnes m}$$

$$\text{But listing moment} = W \times G_1G_2$$

$$\therefore W \times G_1G_2 = 900$$

$$G_1G_2 = \frac{900}{10{,}000}$$

$$G_1G_2 = 0.09 \text{ m}$$

$$GG_1 = KG_1 - KG = 6.456 - 6.400 = 0.056 \text{ m}$$

$$\text{Original GM} = KM - KG = 7.3 - 6.4$$

$$\therefore GM = 0.9 \text{ m}$$

In triangle G_1G_2M:

$$\text{New GM} = G_1M = GM - GG_1$$

$$= 0.9 \text{ m} - 0.056 \text{ m}$$

$$G_1M = 0.844 \text{ m}$$

$$\tan\theta = \frac{G_1G_2}{G_1M}$$

$$= \frac{0.09}{0.844}$$

$$\tan\theta = 0.1066$$

Ans. Maximum list = 6° 5′.

Summary

1. Always make a *sketch* from the given information.
2. Use a moment of weight table.
3. Use values from the table to *calculate* the final requested data.

Exercise 13

1. A ship of 2000 tonnes displacement has KG = 4.5 m. A heavy lift of 20 tonnes mass is in the lower hold and has KG = 2 m. This weight is then raised 0.5 meters clear of the tank top by a derrick whose head is 14 meters above the keel. Find the new KG of the ship.
2. A ship has a displacement of 7000 tonnes and KG = 6 m. A heavy lift in the lower hold has KG = 3 m and mass 40 tonnes. Find the new KG when this weight is raised through 1.5 meters vertically and is suspended by a derrick whose head is 17 meters above the keel.
3. Find the shift in the center of gravity of a ship of 1500 tonnes displacement when a weight of 25 tonnes mass is shifted from the starboard side of the lower hold to the port side on deck through a distance of 15 meters.
4. A ship is just about to lift a weight from a jetty and place it on board. Using the data given below, calculate the angle of heel after the weight has just been lifted from this jetty. Weight to be lifted is 140 t with an outreach of 9.14 m. Displacement of ship prior to the lift is 10,060 tonnes. Prior to lift-off, the KB is 3.4 m. KG is 3.66 m, TPC_{SW} is 20, I_{NA} is 22,788 m^4, and draft is 6.7 m in salt water. Height to derrick head is 18.29 m above the keel.

■

CHAPTER 14

Angle of List Due to Bilging of Side Compartments

When a compartment in a ship is bilged the buoyancy provided by that compartment is lost. This causes the center of buoyancy of the ship to move directly away from the center of the lost buoyancy and, unless the center of gravity of the compartment is on the ship's centerline, a listing moment will be created, $b = w$.

Let the ship in Figure 14.1 float upright at the waterline, WL. G represents the position of the ship's center of gravity and B the center of buoyancy.

Now let a compartment that is divided at the centerline be bilged on the starboard side, as shown in the figure. To make good the lost buoyancy the ship will sink to the waterline W_1L_1. That is, the lost buoyancy is made good by the layer between WL and W_1L_1.

The center of buoyancy will move from B to B_1, directly away from the center of gravity of the lost buoyancy, and the distance BB_1 is equal to $(w \times d)/W$, where w represents the lost buoyancy and d represents the distance between the ship's center of buoyancy and the center of the lost buoyancy.

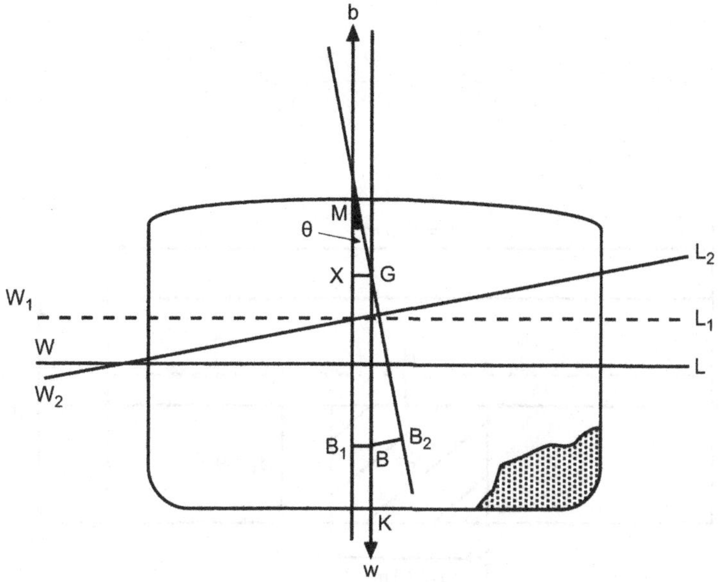

Figure 14.1

Ship Stability for Masters and Mates. http://dx.doi.org/10.1016/B978-0-08-097093-6.00014-1

The shift in the center of buoyancy produces a listing moment. Let θ be the resultant list. Then:

$$\tan\theta = \frac{GX}{XM} = \frac{BB_1}{XM}$$

where XM represents the initial metacentric height for the bilged condition.

Example 1

A box-shaped vessel, of length 100 m and breadth 18 m, floats in salt water on an even keel at 7.5 m draft. KG = 4 m. The ship has a continuous centerline bulkhead that is watertight (see Figure 14.2). Find the list if a compartment amidships, which is 15 m long and is empty, is bilged on one side.

(a) *Find the new mean draft:*

$$\text{Bodily increase in draft} = \frac{\text{Volume of lost buoyancy}}{\text{Area of intact W.P.}}$$

$$= \frac{15 \times 9 \times 7.5}{(100 \times 18) - (15 \times 9)} = 0.61$$

$$\text{New draft} = 7.50 + 0.61$$

$$\therefore \text{New draft} = 8.11 \text{ m} = \text{Draft } d_2$$

(b) *Find the shift of the center of buoyancy:*

$$BB_1 = \frac{a \times B/4}{LB - a}$$

$$= \frac{15 \times 9 \times 18/4}{(100 \times 18) - (15 \times 9)} = \frac{607.5}{1665}$$

$$= 0.365 \text{ m}$$

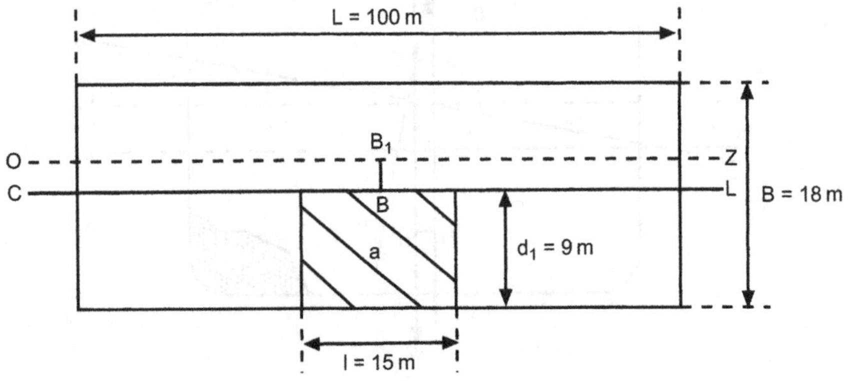

Figure 14.2

(c) *To find* I_{OZ}:

$$I_{CL} = \left(\frac{B}{2}\right)^3 \times \frac{L}{3} + \left(\frac{B}{2}\right)^3 \times \frac{(L-l)}{3}$$

$$I_{CL} = \frac{9^3 \times 100}{3} + \frac{9^3 \times 85}{3} = 24{,}300 + 20{,}655$$

$$= 44{,}955 \quad m^4$$

$$I_{OZ} = I_{CL} - A \times BB_1^2 \qquad A = \text{Intact area of waterplane}$$

$$= 44{,}955 - \{(100 \times 18) - (15 \times 9)\} \times 0.365^2$$

$$= 44{,}955 - 222$$

$$= 44{,}733 \quad m^4$$

(d) *To find GM*:

$$BM = \frac{I_{OZ}}{V}$$

$$= \frac{44{,}733}{100 \times 18 \times 7.5}$$

$$= 3.31 \text{ m}$$

$$+$$

$$KB = \frac{d_2}{2} \quad \therefore KB = \underline{4.06 \text{ m}}$$

$$KM = 7.37 \text{ m}$$

$$-$$

$$KG = \underline{4.00 \text{ m}} \text{ as before bilging}$$

$$\text{After bilging, } GM = \underline{3.37 \text{ m}}$$

(e) *To find the list*:

$$\tan \text{List} = \frac{BB_1}{GM}$$

$$= \frac{0.365}{3.37} = 0.1083$$

Ans. List = 6° 11′.

Example 2

A box-shaped vessel, 50 m long × 10 m wide, floats in salt water on an even keel at a draft of 4 m. A centerline longitudinal watertight bulkhead extends from end to end and for the full depth of the vessel. A compartment amidships on the starboard side is 15 m long and contains cargo with permeability 'μ' of 30% (see Figure 14.3). Calculate the list if this compartment is bilged. KG = 3 m.

(a) *Find the new mean draft:*

$$\text{Volume of bilged tank} = 90\ \text{m}^3$$

$$\text{Area of intact waterplane} = 477.5\ \text{m}^2$$

$$\text{Bodily increase in draft} = \frac{\text{Volume of lost buoyancy}}{\text{Area of intact W.P.}}$$

$$= \frac{\frac{30}{100} \times 15 \times 5 \times 4}{\left(50 \times 10\right) - \left(\frac{30}{100} \times 15 \times 5\right)} = \frac{90}{477.5}$$

$$= 0.19\ \text{m}$$

$$\therefore \text{New draft} = 4.00 + 0.19 = 4.19\ \text{m say draft}\ d_2$$

(b) *Find the shift of the center of buoyancy:*

$$BB_1 = \frac{\mu a \times \frac{B}{4}}{LB - \mu a}$$

$$= \frac{\frac{30}{100} \times 15 \times 5 \times \frac{10}{4}}{\left(50 \times 10\right) - \left(\frac{30}{100} \times 15 \times 5\right)} = \frac{56.25}{477.5}$$

$$= 0.12\ \text{m}$$

(c) *To find* I_{OZ}:

$$I_{CL} = \frac{LB^3}{12} - \frac{\mu l b^3}{3}$$

$$= \left(\frac{50 \times 10^3}{12}\right) - \left(\frac{30}{100} \times \frac{15 \times 5^3}{3}\right)$$

$$= 4166.7 - 187.5$$

$$= 3979\ \text{m}^4$$

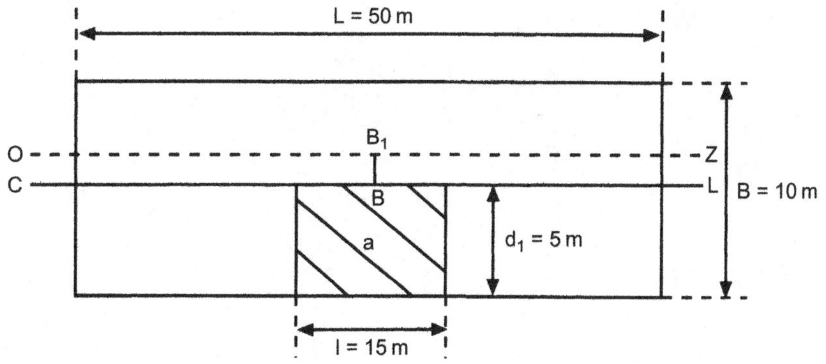

Figure 14.3

$$
\begin{aligned}
I_{OZ} &= I_{CL} - A \times BB_1^2 \\
&= 3979 - (477.5 \times 0.12^2) \\
&= 3979 - 7 \\
&= 3972 \text{ m}^4
\end{aligned}
$$

(d) *To find GM*:

$$BM_2 = \frac{I_{OZ}}{V}$$

$$BM_2 = \frac{3972}{50 \times 10 \times 4}$$

$$\therefore BM_2 = 1.99 \text{ m}$$

$$+$$

$$KB_2 = \underline{2.09 \text{ m}}$$

$$KM_2 = \underline{4.08 \text{ m}}$$

$$-$$

$$\text{as before bilging, } KG = 3.00 \text{ m}$$

$$GM_2 = \underline{1.08 \text{ m}}$$

(e) *To find the list*:

$$\tan \text{List} = \frac{BB_1}{GM_2}$$

$$= \frac{0.12}{1.08} = 0.109023$$

Ans. List = 6° 17′ to starboard.

Note. When $\mu = 100\%$, then:

$$I_{CL} = \left(\frac{B}{2}\right)^3 \times \frac{L}{3} + \left(\frac{B}{2}\right)^3 \left(\frac{L-l}{3}\right) m^4$$

or

$$I_{CL} = \frac{LB^3}{12} - \frac{lb^3}{3} m^4$$

Both formulae give the same answer.

Summary

1. Make a sketch from the given information.
2. Calculate the mean bodily increase in draft.
3. Calculate the shift in the center of buoyancy.
4. Estimate the second moment of area in the bilged condition with the use of the parallel axis theorem.
5. Evaluate the new KB, BM, KM, and GM.
6. Finally calculate the requested angle of list.

Exercise 14

1. A box-shaped tanker barge is 100 m long, 15 m wide, and floats in salt water on an even keel at 5 m draft. KG = 3 m. The barge is divided longitudinally by a centerline watertight bulkhead. An empty compartment amidships on the starboard side is 10 m long. Find the list if this compartment is bilged.
2. A box-shaped vessel, 80 m long and 10 m wide, is floating on an even keel at 5 m draft. Find the list if a compartment amidships, 15 m long, is bilged on one side of the centerline bulkhead. KG = 3 m.
3. A box-shaped vessel, 120 m long and 24 m wide, floats on an even keel in salt water at a draft of 7 m. KG = 7 m. A compartment amidships is 12 m long and is divided at the centerline by a full depth watertight bulkhead. Calculate the list if this compartment is bilged.
4. A box-shaped vessel is 50 m long, 10 m wide, and is divided longitudinally at the centerline by a watertight bulkhead. The vessel floats on an even keel in salt water at a draft of 4 m. KG = 3 m. A compartment amidships is 12 m long and contains cargo of permeability 30%. Find the list if this compartment is bilged.
5. A box-shaped vessel 68 m long and 14 m wide has KG = 4.7 m, and floats on an even keel in salt water at a draft of 5 m. A compartment amidships 18 m long is divided longitudinally at the centerline and contains cargo of permeability 30%. Calculate the list if this compartment is bilged.

CHAPTER 15

Heel Due to Turning

When a body moves in a circular path there is an acceleration towards the center equal to v^2/r, where v represents the velocity of the body and r represents the radius of the circular path. The force required to produce this acceleration, called a 'centripetal' force, is equal to Mv^2/r, where M is the mass of the body.

In the case of a ship turning in a circle, the centripetal force is produced by the water acting on the side of the ship away from the center of the turn. The force is considered to act at the center of lateral resistance which, in this case, is the centroid of the underwater area of the ship's side away from the center of the turn. The centroid of this area is considered to be at the level of the center of buoyancy. For equilibrium there must be an equal and opposite force, called the 'centrifugal' force, and this force is considered to act at the center of mass (G).

When a ship's rudder is put over to port, the forces on the rudder itself will cause the ship to develop a small angle of heel initially to port, say $\alpha°_1$.

However, the underwater form of the ship and centrifugal force on it cause the ship to heel to starboard, say $\alpha°_2$.

In this situation $\alpha°_2$ is always greater than $\alpha°_1$. Consequently, for port rudder helm, the final angle of heel due to turning will be to starboard and vice versa.

It can be seen from Figure 15.1 that these two forces produce a couple that tends to heel the ship away from the center of the turn, i.e.

$$\text{Heeling couple} = \frac{Mv^2}{r} \times B_1Z$$

Equilibrium is produced by a righting couple equal to W × GZ, where W is equal to the weight of the ship, the weight being a unit of force, i.e. W = Mg.

$$\therefore\ MgGZ = \frac{Mv^2}{r} \times B_1Z$$

or

Ship Stability for Masters and Mates. http://dx.doi.org/10.1016/B978-0-08-097093-6.00015-3

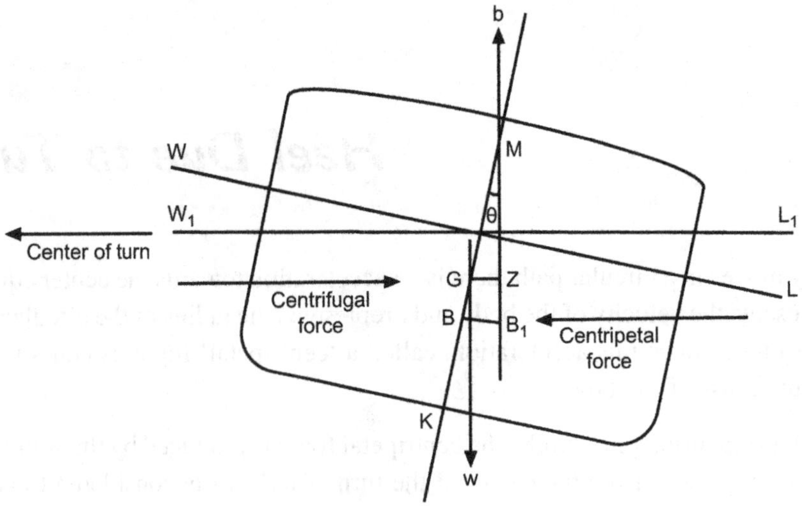

Figure 15.1

$$GZ = \frac{v^2}{gr} \times B_1Z$$

but at a small angle

$$GZ = GM \sin \theta$$

$$B_1Z = BG \cos \theta$$

$$\therefore GM \sin \theta = \frac{v^2}{gr} BG \cos \theta$$

and

$$\tan \theta = \frac{v^2 \times BG}{grGM}$$

Example

A ship turns to port in a circle of radius 100 m at a speed of 15 knots. The GM is 0.67 m and BG is 1 m. If $g = 9.81\ m/s^2$ and 1 knot is equal to 1852 km/hour, find the heel due to turning.

$$\text{Ship speed v in m/s} = 15 \times \frac{1852}{3600}$$

$$v = 7.72 \text{ m/s}$$

$$\tan\theta = \frac{v^2 \times BG}{grGM}$$

$$= \frac{7.72^2 \times 1.0}{9.81 \times 100 \times 0.67}$$

$$\tan\theta = 0.0906$$

Ans. Heel = 5° 11′ to starboard, due to centrifugal forces only.

In practice, this angle of heel will be slightly smaller. Forces on the rudder will have produced an angle of heel, say 1° 17′ to port. Consequently the overall angle of heel due to turning will be:

$$\text{Heel} = 5°\ 11' - 1°\ 17' = 3°\ 54' \text{ or } 3.9° \text{to starboard}$$

Exercise 15

1. A ship's speed is 12 knots. The helm is put hard over and the ship turns in a circle of radius 488 m. GM = 0.3 m and BG = 3 m. Assuming that 1 knot is equal to 1852 km/hour, find heel due to turning.
2. A ship steaming at 10 knots turns in a circle of radius 366 m. GM = 0.24 m, BM = 3.7 m. Calculate the heel produced.
3. A ship turns in a circle of radius 100 m at a speed of 15 knots. BG = 1 m. Find the heel if the GM = 0.6 m.
4. A ship with a transverse metacentric height of 0.40 m has a speed of 21 knots. The center of gravity is 6.2 m above keel whilst the center of lateral resistance is 4.0 m above keel. The rudder is put hard over to port and the vessel turns in a circle of 550 m radius. Considering only the centrifugal forces involved, calculate the angle of heel as this ship turns at the given speed.

CHAPTER 16

Angle of Loll

When a ship with negative initial metacentric height is inclined to a small angle, the righting lever is negative, resulting in a capsizing moment. This effect is shown in Figure 16.1(a) and it can be seen that the ship will tend to heel still further.

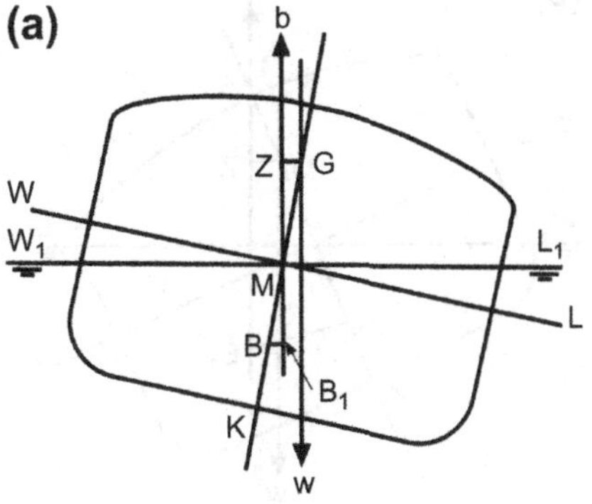

Figure 16.1(a)

At a large angle of heel the center of buoyancy will have moved further out the low side and the force of buoyancy can no longer be considered to act vertically upwards through M, the initial metacenter. If, by heeling still further, the center of buoyancy can move out far enough to lie vertically under G, the center of gravity, as in Figure 16.1(b), the righting lever and thus the righting moment will be zero.

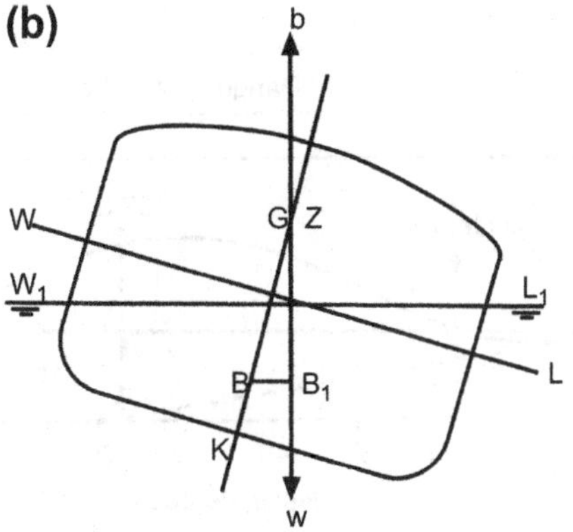

Figure 16.1(b)

Ship Stability for Masters and Mates. http://dx.doi.org/10.1016/B978-0-08-097093-6.00016-5

The angle of heel at which this occurs is referred to as the *angle of loll* and may be defined as the angle to which a ship with negative initial metacentric height will lie at rest in still water.

If the ship should now be inclined to an angle greater than the angle of loll, as shown in Figure 16.1(c), the righting lever will be positive, giving a moment to return the ship to the angle of loll.

From this it can be seen that the ship will oscillate about the angle of loll instead of the upright.

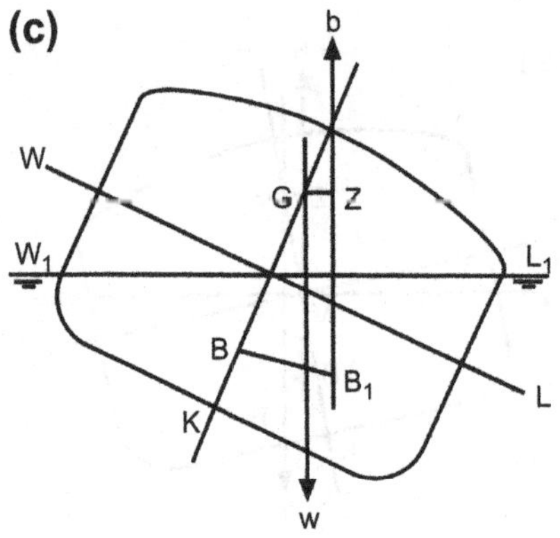

Figure 16.1(c)

The curve of statical stability for a ship in this condition of loading is illustrated in Figure 16.2. Note from the figure that the GZ at the angle of loll is zero. At angles of heel less than the angle of loll the righting levers are negative, whilst beyond the angle of loll the righting levers are positive up to the angle of vanishing stability.

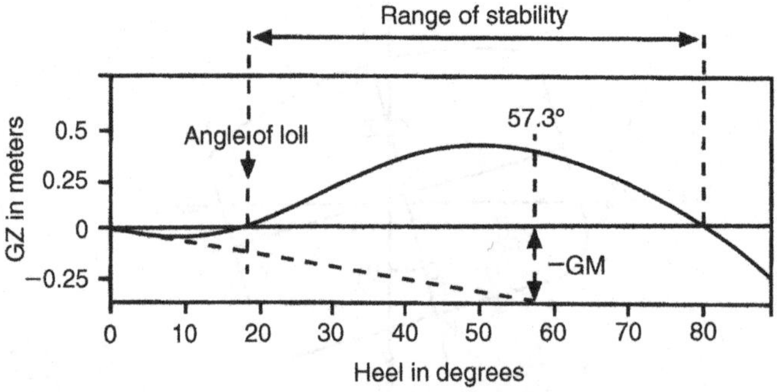

Figure 16.2

Note how the range of stability in this case is measured from the angle of loll and not from the 'o–o' axis.

To Calculate the Angle of Loll

When the vessel is 'wall-sided' between the upright and inclined waterlines, the GZ may be found using the formula:

$$GZ = \sin\theta(GM + \frac{1}{2}BM\tan^2\theta)$$

At the angle of loll:

$$GZ \qquad = o$$

$$\therefore \text{either } \sin\theta = o$$

or

$$(GM + \frac{1}{2}BM\tan^2\theta) = o$$

If

$$\sin\theta = o$$

then

$$\theta = o$$

But then the angle of loll cannot be zero, therefore:

$$(GM + \frac{1}{2}BM\tan^2\theta) = o$$

$$\frac{1}{2}BM\tan^2\theta = -GM$$

$$BM\tan^2\theta = -2GM$$

$$\tan^2\theta = \frac{-2GM}{BM}$$

$$\tan\theta = \sqrt{\frac{-2GM}{BM}}$$

The angle of loll is caused by a negative GM, therefore:

$$\tan\theta = \sqrt{\frac{-2(-GM)}{BM}}$$

or

$$\tan\theta = \sqrt{\frac{2GM}{BM}}$$

where

θ = the angle of loll

GM = a negative initial metacentric height

BM = the BM when upright

Example

Will a homogeneous log 6 m × 3 m × 3 m and relative density 0.4 float in fresh water with a side perpendicular to the waterline? If not, what will be the angle of loll?

Since the log is homogeneous the center of gravity must be at half-depth, i.e. KG = 1.5 m (see Figure 16.3).

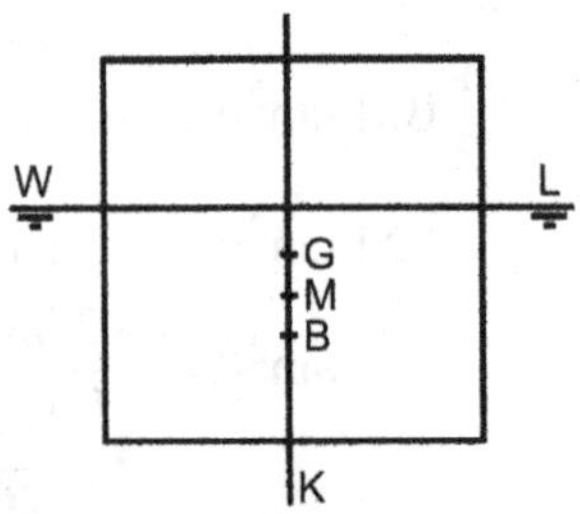

Figure 16.3

$$\frac{\text{Draft of log}}{\text{Depth of log}} = \frac{\text{SG of log}}{\text{SG of water}}$$

$$\text{Draft of log} = \frac{3 \times 0.4}{1}$$

$$d = 1.2 \text{ m}$$

$$KB = \frac{1}{2} \text{ draft}$$

$$KB = 0.600 \text{ m}$$

$$BM = B^2/12d$$

$$= \frac{3 \times 3}{12 \times 1.2}$$

$$BM = 0.625 \text{ m}$$

$$+$$

$$KB = \underline{0.600 \text{ m}}$$

$$KM = 1.225 \text{ m}$$

$$KG = \underline{1.500 \text{ m}}$$

$$GM = -0.275 \text{ m}$$

Therefore, the log is unstable and will take up an angle of loll:

$$\tan\theta = \sqrt{\frac{2GM}{BM}}$$

$$= \sqrt{\frac{0.55}{0.625}} = 0.9381$$

$$\theta = 43^\circ\ 10'$$

Ans. The angle of loll = 43°10′.

Question: What exactly is angle of list and angle of loll? List the differences/characteristics.

Angle of List

'G', the centroid of the loaded weight, has *moved off the centerline* due to a shift of cargo or bilging effects, say to the port side.

GM is positive, i.e. 'G' is below 'M'. In fact, GM will *increase* at the angle of list compared to GM when the ship is upright. The ship is in *stable equilibrium.*

In still water conditions the ship will remain at this *fixed* angle of heel. She will list to one side only, i.e. the same side as movement of weight.

In heavy weather conditions the ship will roll about this angle of list, say 3° P, but will not stop at 3° S. See comment below.

To bring the ship back to upright, load weight on the other side of the ship, for example if she lists 3° P add weight onto the starboard side of ship.

Angle of Loll

KG = KM so GM is zero. 'G' remains *on the centerline* of the ship.

The ship is in *neutral equilibrium*. She is in a *more dangerous situation* than a ship with an angle of list, because once 'G' goes above 'M' she will capsize.

Angle of loll may be *3° P or 3° S* depending upon external forces such as wind and waves acting on her structure. She may suddenly flop over from 3° P to 3° S and then back again to 3° P.

To improve this condition 'G' must be brought below 'M'. This can be done by moving weight downwards towards the keel, adding water ballast in double-bottom tanks or removing weight above the ship's 'G'. Beware of free surface effects when moving, loading, and discharging liquids.

With an angle of list or an angle of loll the calculations must be carefully made *prior* to any changes in loading being made.

Exercise 16

1. Will a homogeneous log of square cross-section and relative density 0.7 be stable when floating in fresh water with two opposite sides parallel to the waterline? If not, what will be the angle of loll?
2. A box-shaped vessel 30 m × 6 m × 4 m floats in salt water on an even keel at 2 m, draft F and A. KG = 3 m. Calculate the angle of loll.
3. A ship is upright and is loaded with a full cargo of timber with timber on deck. During the voyage the ship develops a list, even though stores, fresh water, and bunkers have been consumed evenly from each side of the centerline. Discuss the probable cause of the list and the method that should be used to bring the ship to the upright.

4. A ship loaded with a full cargo of timber and timber on deck is alongside a quay and has taken up an angle of loll away from the quay. Describe the correct method of discharging the deck cargo and the precautions that must be taken during this process.

■

CHAPTER 17

Moments of Statical Stability

When a ship is inclined by an external force, such as wind and wave action, the center of buoyancy moves out to the low side, parallel to the shift of the center of gravity of the immersed and emerged wedges, to the new center of gravity of the underwater volume. The force of buoyancy is considered to act vertically upwards through the center of buoyancy, whilst the weight of the ship is considered to act vertically downwards through the center of gravity. These two equal and opposite forces produce a moment or couple that may tend to right or capsize the ship. The moment is referred to as the *moment of statical stability* and may be defined as the moment to return the ship to the initial position when inclined by an external force.

A ship that has been inclined by an external force is shown in Figure 17.1.

The center of buoyancy has moved from B to B_1 parallel to gg_1, and the force of buoyancy (W) acts vertically upwards through B_1. The weight of the ship (W) acts vertically downwards through the center of gravity (G). The perpendicular distance between the lines of action of the forces (GZ) is called the *righting lever*. Taking moments about the center of gravity, the moment of statical stability is equal to the product of the righting lever and the displacement, or:

$$\text{Moment of statical stability} = W \times GZ$$

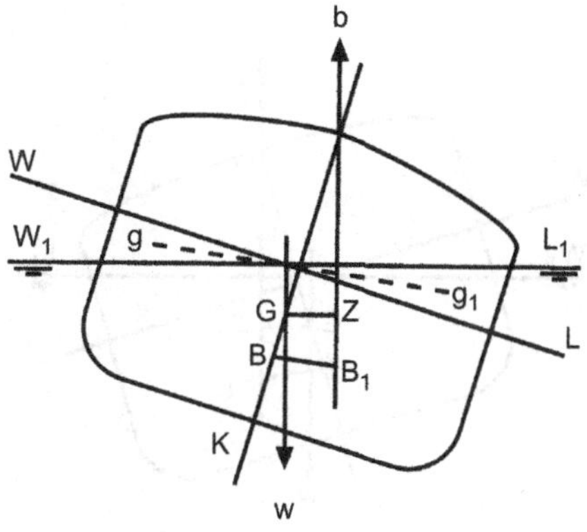

Figure 17.1

Ship Stability for Masters and Mates. http://dx.doi.org/10.1016/B978-0-08-097093-6.00017-7

The Moment of Statical Stability at a Small Angle of Heel

At small angles of heel the force of buoyancy may be considered to act vertically upwards through a fixed point called the initial metacenter (M). This is shown in Figure 17.2, in which the ship is inclined to a small angle (θ degrees).

$$\begin{aligned} \text{Moment of statical stability} &= W \times GZ \\ \text{But in triangle GZM : } GZ &= GM \sin \theta^\circ \\ \therefore \text{Moment of statical stability} &= W \times GM \times \sin \theta^\circ \end{aligned}$$

From this formula it can be seen that for any particular displacement at small angles of heel, the righting moments will vary directly as the initial metacentric height (GM). Hence, if the ship has a comparatively large GM she will tend to be 'stiff', whilst a small GM will tend to make her 'tender'. It should also be noticed, however, that the stability of a ship depends not only upon the size of the GM or GZ, but also upon the displacement. Thus, two similar ships may have identical GMs, but if one is at the light displacement and the other at the load displacement, their respective states of stability will be vastly different. The ship that is at the load displacement will be much more 'stiff' than the other.

Example 1

A ship of 4000 tonnes displacement has KG = 5.5 m and KM = 6.0 m. Calculate the moment of statical stability when heeled 5°.

$$\begin{aligned} GM = KM - KG &= 6.0 - 5.5 = 0.5 \text{ m} \\ \text{Moment of statical stability} &= W \times GM \times \sin \theta^\circ \\ &= 4000 \times 0.5 \times \sin 5^\circ \end{aligned}$$

Ans. Moment of statical stability = 174.4 tonnes m.

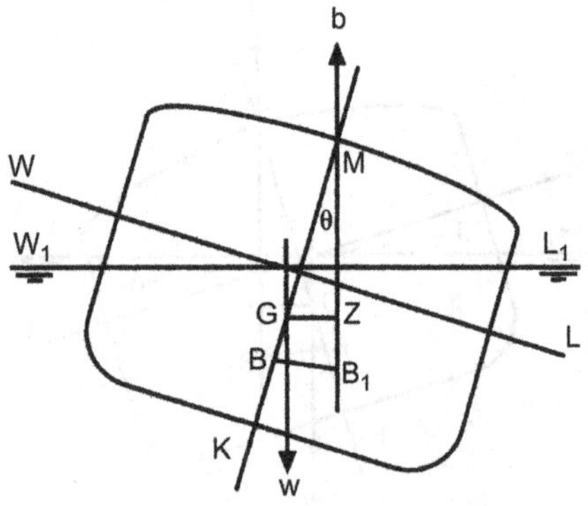

Figure 17.2

Example 2

When a ship of 12,000 tonnes displacement is heeled 6.5° the moment of statical stability is 600 tonnes m. Calculate the initial metacentric height.

$$\text{Moment of statical stability} = W \times GM \times \sin \theta^\circ$$

$$\therefore GM = \frac{\text{Moment of statical stability}}{W \times \sin \theta^\circ}$$

$$= \frac{600}{12{,}000 \sin 6.5^\circ}$$

Ans. GM = 0.44 m.

The Moment of Statical Stability at a Large Angle of Heel

At a large angle of heel the force of buoyancy can no longer be considered to act vertically upwards through the initial metacenter (M). This is shown in Figure 17.3, where the ship is heeled to an angle of more than 15°. The center of buoyancy has moved further out to the low side, and the vertical through B_1 no longer passes through (M), the initial metacenter. The righting lever (GZ) is once again the perpendicular distance between the vertical through G and the vertical through B_1, and the moment of statical stability is equal to W × GZ.

However, GZ is no longer equal to GM sin θ°. Up to the angle at which the deck edge is immersed, it may be found by using a formula known as the *wall-sided formula*, i.e.

$$GZ = (GM + \frac{1}{2} BM \tan^2 \theta) \sin \theta$$

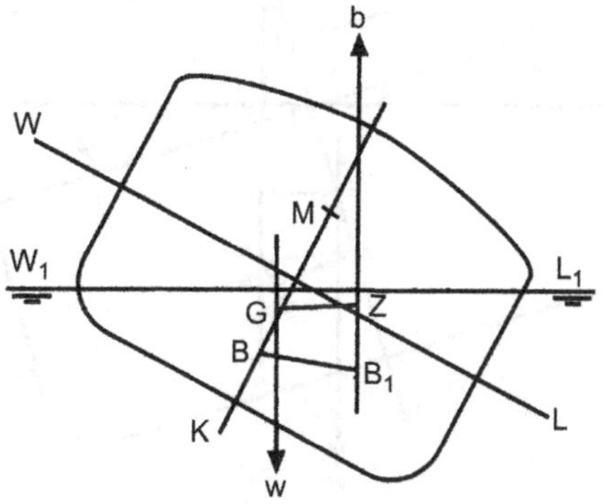

Figure 17.3

The derivation of this formula is as follows:

Refer to the ship shown in Figure 17.4. When inclined the wedge WOW_1 is transferred to LOL_1 such that its center of gravity shifts from g to g_1. This causes the center of buoyancy to shift from B to B_1. The horizontal components of these shifts are hh_1 and BB_2 respectively, the vertical components being $(gh + g_1h_1)$ and B_1B_2 respectively.

Let BB_2 be 'a' units and let B_1B_2 be 'b' units (see Figure 17.5(a)). Now consider the wedge LOL_1:

$$\text{Area} = \frac{1}{2} y^2 \tan\theta$$

Consider an elementary strip longitudinally of length dx as in Figure 17.5(b):

$$\text{Volume} = \frac{1}{2} y^2 \tan\theta \, dx$$

The total horizontal shift of the wedge (hh_1) is 2/3 × 2y or 4/3 × y.

$$\therefore \text{Moment of shifting this wedge} = \frac{4}{3} y \times \frac{1}{2} y^2 \tan\theta \, dx$$
$$= \frac{2}{3} y^3 \tan\theta \, dx$$

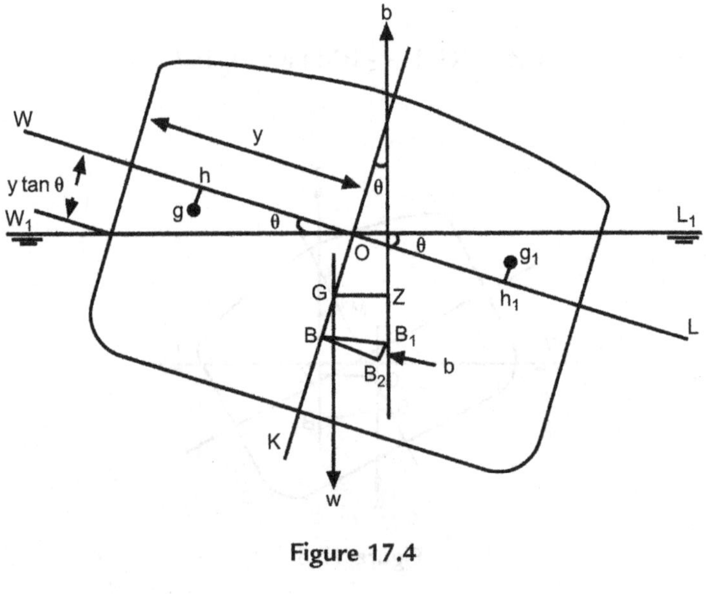

Figure 17.4

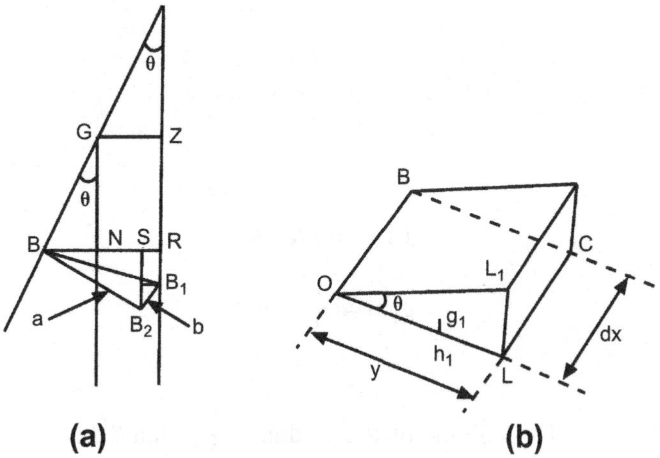

Figure 17.5

$$\text{The sum of the moment of all such wedges} = \int_O^L \frac{2}{3} y^3 \tan\theta \, dx$$

$$= \tan\theta \int_O^L \frac{2}{3} y^3 \, dx$$

$$\text{But the second moment of the waterplane area about the centerline (I)} = \int_O^L \frac{2}{3} y^3 \, dx$$

$$\therefore \text{Sum of the moment of all such wedges} = I \tan\theta$$

$$BB_2 = \frac{v \times hh_1}{V}$$

or

$$V \times BB_2 = v \times hh_1$$

$$\text{But the sum of the moments of the wedges} = v \times hh_1$$

$$\therefore V \times BB_2 = I \tan\theta$$

$$BB_2 = \frac{I}{V} \tan\theta$$

$$BB_2 = BM \tan\theta \qquad \text{(a)}$$

$$\text{The vertical shift of the wedge} = gh + g_1h_1$$

$$= 2gh$$

$$\therefore \text{The vertical moment of the shift} = v \times 2gh$$

$$= 2vgh$$

In Figure 17.5(b):

$$OL = y \quad \text{and} \quad Oh_1 = \frac{2}{3}y$$

But

$$LL_1 = y \tan \theta$$

$$\therefore\ g_1h_1 = \frac{1}{3}y \tan \theta$$

$$\text{The volume of the wedge} = \frac{1}{2}y^2 \tan \theta\, dx$$

$$\text{The moment of the vertical shift} = \frac{1}{2}y^2 \tan \theta\, dx \times \frac{2}{3}y \tan \theta$$

$$= \frac{1}{3}y^3 \tan^2 \theta\, dx$$

$$\text{The vertical moment of all such wedges} = \int_O^L \frac{1}{3}y^3 \tan^2 \theta\, dx$$

$$= \frac{1}{2}I \tan^2 \theta$$

$$\therefore\ \text{The moment of the vertical shift} = \frac{1}{2}I \tan^2 \theta$$

Also:

$$B_1B_2 = \frac{v \times 2gh}{V}$$

or

$$V \times b = 2vgh$$

but

$$2vgh = \text{The vertical moment of the shift}$$

$$\therefore\ V \times b = \frac{1}{2}I \tan^2 \theta$$

or

$$b = \frac{I}{V} \times \frac{\tan^2 \theta}{2}$$

$$B_1B_2 = \frac{BM \tan^2 \theta}{2} \qquad \text{(b)}$$

Referring to Figure 17.5(a),

$$\begin{aligned} GZ &= NR \\ &= BR - BN \\ &= (BS + SR) - BN \\ &= a \cos\theta + b \sin\theta - BG \sin\theta \\ &= BM \tan\theta \cos\theta + \frac{1}{2} BM \tan^2\theta \sin\theta - BG \sin\theta \qquad \text{(from (a) and (b))} \\ &= BM \sin\theta + \frac{1}{2} BM \tan^2\theta \sin\theta - BG \sin\theta \\ &= \sin\theta\,(BM + \frac{1}{2} BM \tan^2\theta - BG) \\ GZ &= \sin\theta\,(GM + \frac{1}{2} BM \tan^2\theta) \quad (\text{for } \theta \text{ up to } 25^\circ) \end{aligned}$$

This is the *wall-sided formula.*

Note. This formula may be used to obtain the GZ at any angle of heel so long as the ship's side at WW_1 is parallel to LL_1, but for small angles of heel (θ up to 5°), the term ½BM $\tan^2\theta$ may be omitted.

Example 1

A ship of 6000 tonnes displacement has KB = 3 m, KM = 6 m, and KG = 5.5 m. Find the moment of statical stability at 25° heel.

$$\begin{aligned} GZ &= (GM + \frac{1}{2} BM \tan^2\theta)\sin\theta \\ &= (0.5 + \frac{1}{2} \times 3 \times \tan^2 25^\circ)\sin 25^\circ \\ &= 0.8262 \sin 25^\circ \\ GZ &= 0.35\ \text{m} \\ \text{Moment of statical stability} &= W \times GZ \\ &= 6000 \times 0.35 \end{aligned}$$

Ans. Moment of statical stability = 2100 tonnes m.

Example 2

A box-shaped vessel 65 m × 12 m × 8 m depth has KG = 4 m, and is floating in salt water upright on an even keel at 4 m draft. Calculate the moments of statical stability at (a) 5° and (b) 25° heel.

$$W = L \times B \times \text{draft} \times 1.025$$
$$= 65 \times 12 \times 4 \times 1.025 \text{ tonnes}$$
$$W = 3198 \text{ tonnes}$$

$$KB = \frac{1}{2}\text{draft}$$
$$KB = 2 \text{ m}$$

$$BM = \frac{B^2}{12d}$$
$$= \frac{12 \times 12}{12 \times 4}$$
$$BM = 3 \text{ m}$$

$$\begin{aligned} KB &= 2 \text{ m} \\ BM &= +\underline{3 \text{ m}} \\ KM &= 5 \text{ m} \\ KG &= -\underline{4 \text{ m}} \\ GM &= 1 \text{ m} \end{aligned}$$

At 5° heel:

$$\begin{aligned} GZ &= GM\sin\theta \\ &= 1 \times \sin 5^\circ \\ GZ &= 0.0872 \\ \text{Moment of statical stability} &= W \times GZ \\ &= 3198 \times 0.0872 \\ &= 278.9 \text{ tonnes m} \end{aligned}$$

At 25° heel

$$\begin{aligned} GZ &= (GM + \frac{1}{2}BM\ \tan^2\theta)\sin\theta \\ &= (1 + \frac{1}{2} \times 3 \times \tan^2 25^\circ)\sin 25^\circ \\ &= (1 + 0.3262)\sin 25^\circ \\ &= 1.3262 \sin 25^\circ \\ GZ &= 0.56 \text{ meters} \\ \text{Moment of statical stability} &= W \times GZ \\ &= 3198 \times 0.56 \\ &= 1790.9 \text{ tonnes m} \end{aligned}$$

Ans. (a) 278.9 tonnes m and (b) 1790.9 tonnes m.

The moment of statical stability at a large angle of heel may also be calculated using a formula known as *Atwood's formula*, i.e.

$$\text{Moment of statical stability} = W\left(\frac{v \times hh_1}{V} - BG\sin\theta\right)$$

The derivation of this formula is as follows:

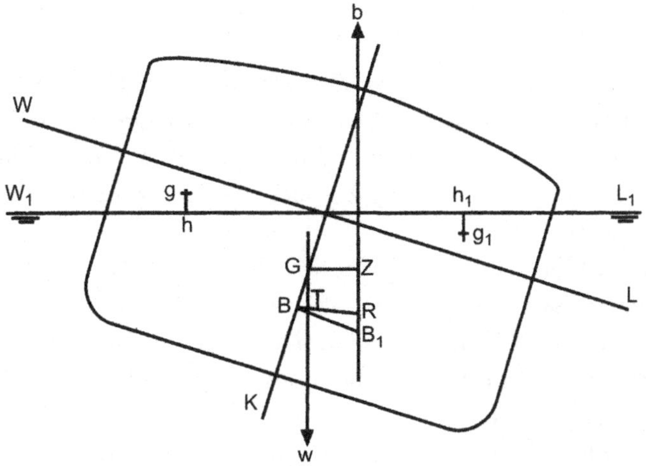

Figure 17.6

With reference to Figure 17.6,

$$\begin{aligned}\text{Moment of statical stability} &= W \times GZ\\ &= W(BR - BT)\end{aligned}$$

Let:

v = the volume of the immersed or emerged wedge

hh_1 = the horizontal component of the shift of the center of gravity of the wedge

V = the underwater volume of the ship

BR = the horizontal component of the shift of the center of buoyancy.

$$BT = BG\sin\theta$$

also

$$BR = \frac{v \times hh_1}{V}$$

$$\therefore \text{Moment of statical stability} = W\left(\frac{v \times hh_1}{V} - BG\sin\theta\right)$$

Exercise 17

1. A ship of 10,000 tonnes displacement has GM = 0.5 m. Calculate the moment of statical stability when the ship is heeled 7.75°.
2. When a ship of 12,000 tonnes displacement is heeled 5.25° degrees the moment of statical stability is 300 tonnes m, KG = 7.5 m. Find the height of the metacenter above the keel.
3. Find the moment of statical stability when a ship of 10,450 tonnes displacement is heeled 6° if the GM is 0.5 m.
4. When a ship of 10,000 tonnes displacement is heeled 15°, the righting lever is 0.2 m, KM = 6.8 m. Find the KG and the moment of statical stability.
5. A box-shaped vessel 55 m × 7.5 m × 6 m has KG = 2.7 m, and floats in salt water on an even keel at 4 m draft F and A. Calculate the moments of statical stability at (a) 6° heel and (b) 24° heel.
6. A ship of 10,000 tonnes displacement has KG = 5.5 m, KB = 2.8 m, and BM = 3 m. Calculate the moments of statical stability at (a) 5° heel and (b) 25° heel.
7. A box-shaped vessel of 3200 tonnes displacement has GM = 0.5 m and beam 15 m, and is floating at 4 m draft. Find the moments of statical stability at 5° and 25° heel.
8. A ship of 11,000 tonnes displacement has a moment of statical stability of 500 tonnes m when heeled 5°. Find the initial metacentric height.
9. (a) Write a brief description on the characteristics associated with an 'angle of loll'.
   (b) For a box-shaped barge, the breadth is 6.4 m and the draft is 2.44 m even keel, with a KG of 2.67 m.
   Using the given wall-sided formula, calculate the GZ ordinates up to an angle of heel of 20°, in 4° increments. From the results construct a statical stability curve up to 20° angle of heel. Label the important points on this constructed curve:

$$\mathrm{GZ} = \sin\theta\left(\mathrm{GM} + \frac{1}{2}\,\mathrm{BM}\tan^2\theta\right)$$

CHAPTER 18

Aspects of Trim — The Main Factors Involved

Trim may be considered as the longitudinal equivalent of list. Trim is also known as 'longitudinal stability'. It is in effect transverse stability turned through 90°. Instead of trim being measured in degrees it is measured as the difference between the drafts forward and aft. If the difference is zero then the ship is on even keel. If forward draft is greater than aft draft, the vessel is trimming *by the bow.* If aft draft is greater than forward draft, the vessel is trimming *by the stern.*

Consider a ship to be floating at rest in still water and on an even keel, as shown in Figure 18.1.

The center of gravity (G) and the center of buoyancy (B) will be in the same vertical line and the ship will be displacing her own weight of water. So W = b.

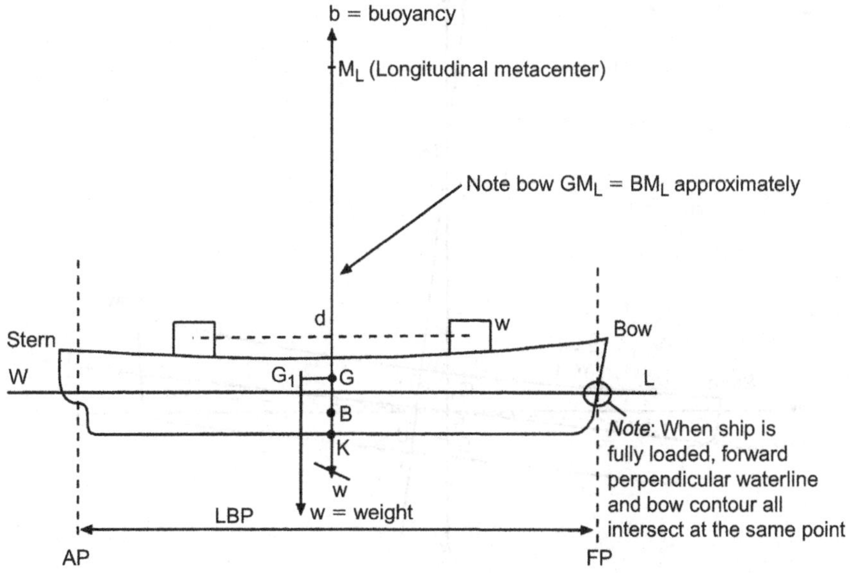

Figure 18.1

Ship Stability for Masters and Mates. http://dx.doi.org/10.1016/B978-0-08-097093-6.00018-9

Now let a weight 'w', already on board, be shifted aft through a distance 'd', as shown in Figure 18.1. This causes the center of gravity of the ship to shift from G to G_1, parallel to the shift of the center of gravity of the weight shifted, so that:

$$GG_1 = \frac{w \times d}{W}$$

or

$$W \times GG_1 = w \times d$$

A trimming moment of $W \times GG_1$ is thereby produced. However,

$$W \times GG_1 = w \times d$$

$$\therefore \text{The trimming moment} = w \times d$$

The ship will now trim until the centers of gravity and buoyancy are again in the same vertical line, as shown in Figure 18.2. When trimmed, the wedge of buoyancy LFL_1 emerges and the wedge WFW_1 is immersed. Since the ship, when trimmed, must displace the same weight of water as when on an even keel, the volume of the immersed wedge must be equal to the

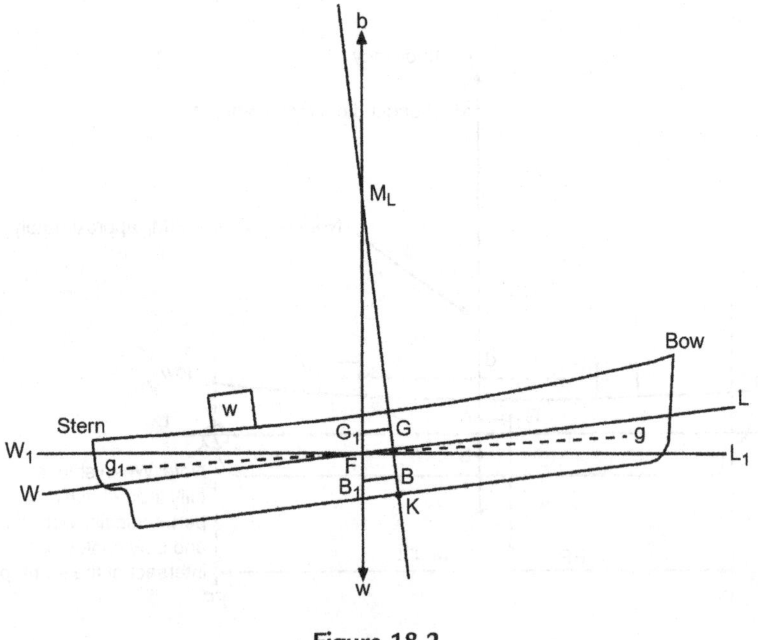

Figure 18.2

volume of the emerged wedge and F, the point about which the ship trims, is the center of gravity of the waterplane area. The point F is called the 'center of flotation' or 'tipping center'.

A vessel with a rectangular waterplane has its center of flotation on the centerline amidships but, on a ship, it may be a little forward or abaft amidships, depending on the shape of the waterplane. In trim problems, unless stated otherwise, it is to be assumed that the center of flotation is situated amidships.

Trimming moments are taken about the center of flotation since this is the point about which rotation takes place.

The longitudinal metacenter (M_L) is the point of intersection between the verticals through the longitudinal positions of the centers of buoyancy. The vertical distance between the center of gravity and the longitudinal metacenter (GM_L) is called the longitudinal metacentric height.

BM_L is the height of the longitudinal metacenter above the center of buoyancy and is found for any shape of vessel by the formula:

$$BM_L = \frac{I_L}{V}$$

where

I_L = the longitudinal second moment of the waterplane about the center of flotation
V = the vessels volume of displacement

The derivation of this formula is similar to that for finding the transverse BM.

For a rectangular waterplane area:

$$I_L = \frac{BL^3}{12}$$

where

L = the length of the waterplane
B = the breadth of the waterplane

Thus, for a vessel having a rectangular waterplane:

$$BM_L = \frac{BL^3}{12V}$$

For a box-shaped vessel:

$$BM_L = \frac{I_L}{V}$$

$$= \frac{BL^3}{12V}$$

$$= \frac{BL^3}{12 \times L \times B \times d}$$

$$BM_L = \frac{L^2}{12d}$$

where

$$\left.\begin{array}{l} L = \text{the length of the vessel} \\ d = \text{the draft of the vessel} \end{array}\right\} \begin{array}{l} \text{Hence, } BM_L \text{ is independent} \\ \text{of ship's Br. Mld} \end{array}$$

For a triangular prism:

$$BM_L = \frac{I_L}{V}$$

$$= \frac{BL^3}{12 \times \frac{1}{2} \times L \times B \times d}$$

$$BM_L = \frac{L^2}{6d}, \text{ so again is independent of Br. Mld}$$

It should be noted that the distance BG is small when compared with BM_L or GM_L and, for this reason, BM_L may, without appreciable error, be substituted for GM_L in the formula for finding MCT 1 cm.

The Moment to Change Trim 1 cm (MCT 1 cm or MCTC)

The MCT 1 cm, or MCTC, is the moment required to change trim by 1 cm, and may be calculated using the formula:

$$\text{MCT 1 cm} = \frac{W \times GM_L}{100L}$$

where

$$W = \text{the vessel's displacement in tonnes}$$
$$GM_L = \text{the longitudinal metacentric height in meters}$$
$$L = \text{the vessel's length in meters}$$

The derivation of this formula is as follows:

Consider a ship floating on an even keel as shown in Figure 18.3(a). The ship is in equilibrium. Now shift the weight 'w' forward through a distance of 'd' meters. The ship's center of gravity will shift from G to G_1, causing a trimming moment of $W \times GG_1$, as shown in Figure 18.3(b). The ship will trim to bring the centers of buoyancy and gravity into the same vertical line, as shown in Figure 18.3(c). The ship is again in equilibrium.

Let the ship's length be L meters and let the tipping center (F) be l meters from aft. The longitudinal metacenter (M_L) is the point of intersection between the verticals through the center of buoyancy when on an even keel and when trimmed.

$$GG_1 = \frac{w \times d}{W} \text{ and } GG_1 = GM_L \tan\theta$$

$$\therefore \tan\theta = \frac{w \times d}{W \times GM_L}$$

but

$$\tan\theta = \frac{t}{L} \text{ (see Figure 18.4(b))}$$

Let the change of trim due to shifting the weight be 1 cm. Then $w \times d$ is the moment to change trim 1 cm.

$$\therefore \tan\theta = \frac{1}{100L}$$

but

$$\tan\theta = \frac{w \times d}{W \times GM_L}$$

$$\therefore \tan\theta = \frac{\text{MCT 1 cm}}{W \times GM_L}$$

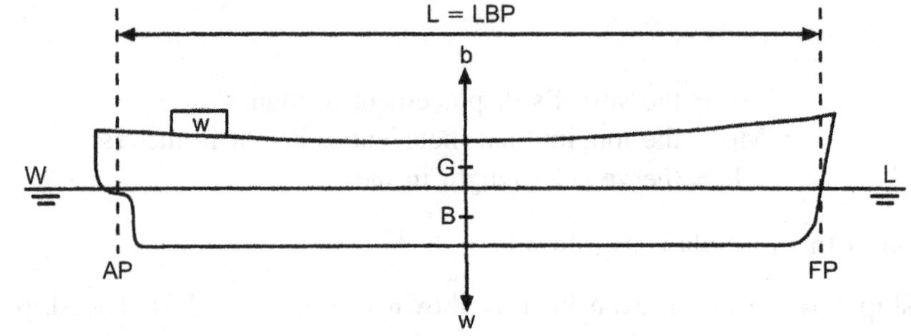

Figure 18.3a

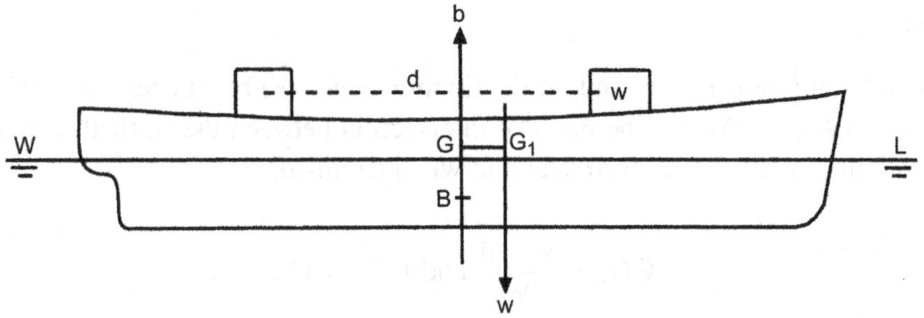

Figure 18.3b

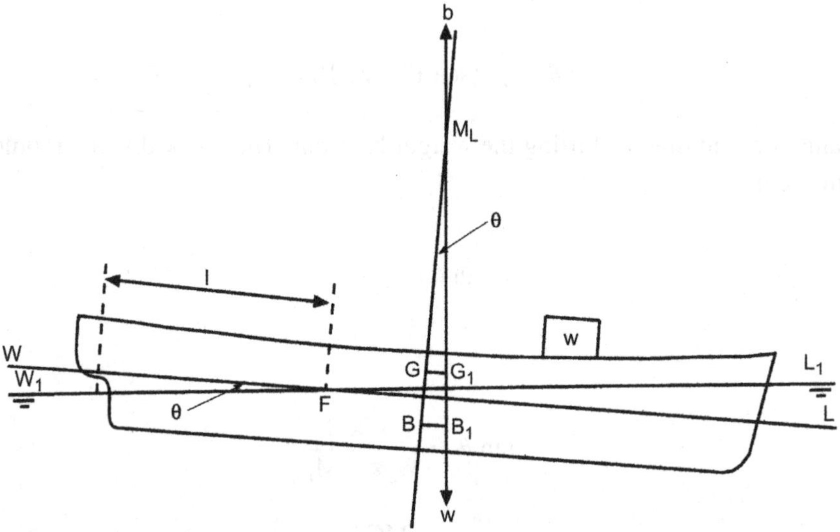

Figure 18.3c

or

$$\frac{\text{MCT 1 cm}}{\text{W} \times \text{GM}_\text{L}} = \frac{1}{100\text{L}}$$

and

$$\text{MCT 1 cm} = \frac{\text{W} \times \text{GM}_\text{L}}{100\text{L}} \text{ tonnes m/cm}$$

To Find the Change of Draft Forward and Aft Due to Change of Trim

When a ship changes trim it will obviously cause a change in the drafts forward and aft. One of these will be increased and the other decreased. A formula must now be found that will give the change in drafts due to change of trim.

Consider a ship floating upright, as shown in Figure 18.4(a). F_1 represents the position of the center of flotation, which is *l* meters from aft. The ship's length is L meters and a weight 'w' is on deck forward. Let this weight now be shifted aft a distance of 'd' meters. The ship will trim about F_1 and change the trim 't' cm by the stern, as shown in Figure 18.4(b). W_1C is a line drawn parallel to the keel. 'A' represents the new draft aft and 'F' the new draft forward. The trim is therefore equal to A − F and, since the original trim was zero, this must also be equal to the change of trim.

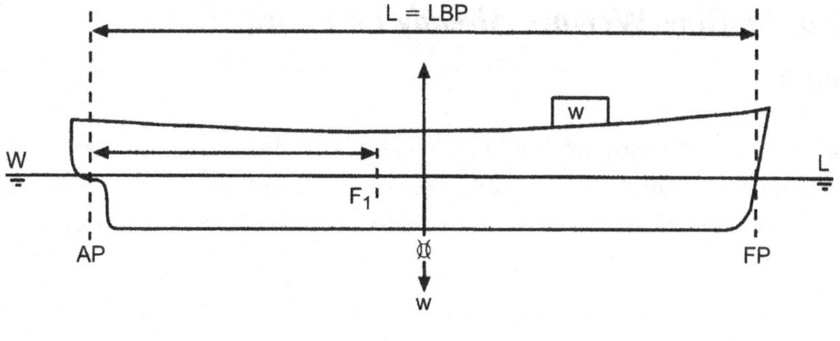

Figure 18.4a

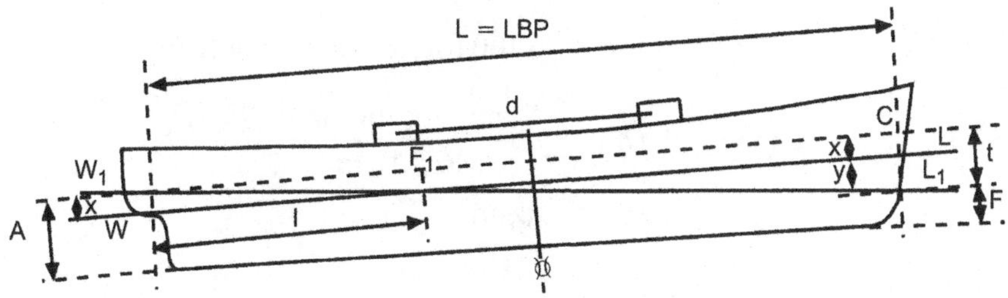

Figure 18.4b

Let 'x' represent the change of draft aft due to the change of trim and let 'y' represent the change forward. In the triangles WW_1F_1 and W_1L_1C, using the property of similar triangles:

$$\frac{\text{x cm}}{l\text{ m}} = \frac{\text{t cm}}{\text{L m}}$$

or

$$\text{x cm} = \frac{l\text{ m} \times \text{t cm}}{\text{L m}}$$

$$\therefore \text{Change of draft aft in cm} = \frac{l}{\text{L}} \times \text{Change of trim in cm}$$

where

l = the distance of center of flotation from aft in meters and
L = the ships length in meters

It will also be noticed that $x + y = t$.

$$\therefore \text{Change of draft F in cm} = \text{Change of trim} - \text{Change of draft A}$$

The Effect of Shifting Weights Already on Board

Example 1

A ship 126 m long is floating at drafts of 5.5 m F and 6.5 m A. The center of flotation is 3 m aft of amidships. MCT 1 cm = 240 tonnes m. Displacement = 6000 tonnes. Find the new drafts if a weight of 120 tonnes already on board is shifted forward a distance of 45 meters.

$$\begin{aligned}\text{Trimming moment} &= \text{w} \times \text{d}\\ &= 120 \times 45\\ &= 5400 \text{ tonnes m by the head}\end{aligned}$$

$$\begin{aligned}\text{Change of trim} &= \frac{\text{Trimming moment}}{\text{MCT 1 cm}}\\ &= \frac{5400}{240}\\ &= 22.5 \text{ cm by the head}\end{aligned}$$

$$\text{Change of draft aft} = \frac{l}{L} \times \text{Change of trim}$$

$$= \frac{60}{126} \times 22.5$$

$$= 10.7\text{cm}$$

$$\text{Change of draft forward} = \frac{66}{126} \times 22.5$$

$$= 11.8\text{cm}$$

Original drafts	6.500 m A	5.500 m F
Change due to trim	−0.107 m	+0.118 m
Ans. New drafts	6.393 m A	5.618 mF

Example 2

A box-shaped vessel 90 m × 10 m × 6 m depth floats in salt water on an even keel at 3 m draft (see Figure 18.5). Find the new drafts if a weight of 64 tonnes already on board is shifted a distance of 40 meters aft.

$$BM_L = \frac{L^2}{12d}$$

$$= \frac{90 \times 90}{12 \times 3}$$

$$BM_L = 225\text{ m}$$

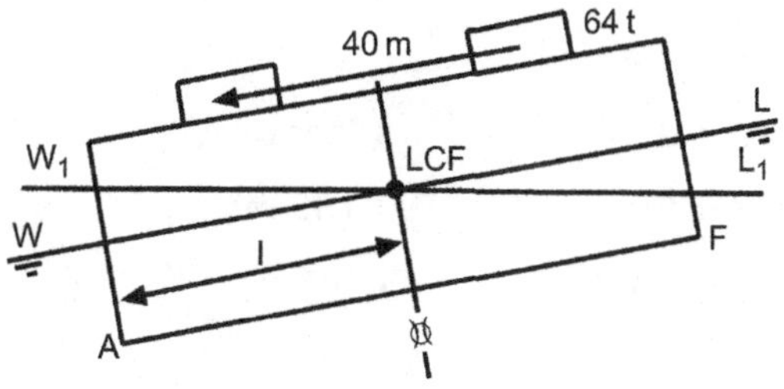

Figure 18.5

$$W = L \times B \times d \times 1.025$$

$$= 90 \times 10 \times 3 \times 1.025$$

$$W = 2767.5 \text{ tonnes}$$

$$\text{MCT 1 cm} = \frac{W \times GM_L}{100L}$$

Since BG is small compared with GM_L, BM_L can be used instead of GM_L:

$$\text{MCT 1 cm} \simeq \frac{W \times BM_L}{100L}$$

$$= \frac{2767.5 \times 225}{100 \times 90}$$

$$\text{MCT 1 cm} = 69.19 \text{ tonnes m/cm}$$

$$\text{Change of trim} = \frac{w \times d}{\text{MCT 1 cm}}$$

$$= \frac{64 \times 40}{69.19}$$

$$\text{Change of trim} = 37 \text{ cm by the stern}$$

$$\text{Change of draft aft} = \frac{l}{L} \times \text{Change of trim}$$

$$= \frac{1}{2} \times 37\text{cm}$$

$$\text{Change of draft aft} = 18.5\text{cm}$$

$$\text{Change of draft forward} = 18.5\text{cm}$$

Original drafts	3.000 m A	3.000 m F
Change due to trim	+0.185 m	−0.185 m
Ans. New drafts	3.185 m A	2.815 m F

■

Exercise 18

1. A ship of 8500 tonnes displacement has TPC = 10 tonnes, MCT 1 cm = 100 tonnes m, and the center of flotation is amidships. She is completing loading under coal tips. Nos. 2 and 3 holds are full, but space is available in No. 1 hold (center of gravity 50 m forward of amidships) and in No. 4 hold (center of gravity 45 m aft of amidships). The present drafts are 6.5 m F and 7 m A, and the load draft is 7.1 m. Find how much cargo is to be loaded in each of the end holds so as to put the ship down to the load draft and complete loading on an even keel.
2. An oil tanker 150 m long, displacement 12,500 tonnes, and MCT 1 cm = 200 tonnes m leaves port with drafts 7.2 m F and 7.4 m A. There are 550 tonnes of fuel oil in the foward deep tank (center of gravity 70 m forward of the center of flotation) and 600 tonnes in the after deep tank (center of gravity 60 m aft of center of flotation). The center of flotation is 1 m aft of amidships. During the sea passage 450 tonnes of oil are consumed from aft. Find how much oil must be transferred from the forward tank to the after tank if the ship is to arrive on an even keel.
3. A ship 100 m long, and with a displacement of 2200 tonnes, has longitudinal metacentric height 150 m. The present drafts are 5.2 m F and 5.3 m A. Center of flotation is 3 m aft of amidships. Find the new drafts if a weight of 5 tonnes already on board is shifted aft through a distance of 60 meters.
4. A ship is floating at drafts of 6.1 m F and 6.7 m A. The following cargo is then loaded:

   20 tonnes in a position whose center of gravity is 30 meters forward of amidships.
   45 tonnes in a position whose center of gravity is 25 meters forward of amidships.
   60 tonnes in a position whose center of gravity is 15 meters aft of amidships.
   30 tonnes in a position whose center of gravity is 3 meters aft of amidships.
   The center of flotation is amidships, MCT 1 cm = 200 tonnes m, and TPC = 35 tonnes. Find the new drafts forward and aft.
5. A ship arrives in port trimmed 0.3 m by the stern and is to discharge 4600 tonnes of cargo from four holds; 1800 tonnes of the cargo is to be discharged from No. 2 and 800 tonnes from No. 3 hold. Center of flotation is amidships, MCT 1 cm = 250 tonnes m.

   The center of gravity of No. 1 hold is 45 m forward of amidships.
   The center of gravity of No. 2 hold is 25 m forward of amidships.
   The center of gravity of No. 3 hold is 20 m aft of amidships.
   The center of gravity of No. 4 hold is 50 m aft of amidships.
   Find the amount of cargo that must be discharged from Nos. 1 and 4 holds if the ship is to sail on an even keel.
6. A ship is 150 m long, displacement 12,000 tonnes, and is floating at drafts of 7 m F and 8 m A. The ship is to enter port from an anchorage with a maximum draft of

7.6 m. Find the minimum amount of cargo to discharge from a hold whose center of gravity is 50 m aft of the center of flotation (which is amidships), TPC 15 tonnes, and MCT 1 cm = 300 tonnes m.

7. A ship 150 m × 20 m floats on an even keel at 10 m draft, and has a block coefficient of fineness 0.8 and LGM of 200 meters. If 250 tonnes of cargo are discharged from a position 32 m from the center of flotation, find the resulting change of trim.
8. A ship is floating in salt water at drafts of 6.7 m F and 7.3 m A. MCT 1 cm = 250 tonnes m. TPC = 10 tonnes. Length of ship = 120 meters. The center of flotation is amidships; 220 tonnes of cargo are then discharged from a position 24 m forward of the center of flotation. Find the weight of cargo that must now be shifted from 5 m aft of the center of flotation to a position 20 m forward of the center of flotation, to bring the draft aft to 7 meters. Find also the final draft forward.

■

CHAPTER 19

Trim Calculations — Changing Conditions of Loading

The Effect of Loading, Discharging, and Moving Weights

When a weight is loaded at the center of flotation it will produce no trimming moment, but the ship's drafts will increase uniformly so that the ship displaces an extra weight of water equal to the weight loaded. If the weight is now shifted forward or aft away from the center of flotation, it will cause a change of trim. From this it can be seen that when a weight is loaded away from the center of flotation, it will cause both a bodily sinkage and a change of trim.

Similarly, when a weight is being discharged, if the weight is first shifted to the center of flotation it will produce a change of trim, and if it is then discharged from the center of flotation the ship will rise bodily. Thus, both a change of trim and bodily rise must be considered when a weight is being discharged away from the center of flotation.

Example 1

A ship 90 m long is floating at drafts 4.5 m F and 5.0 m A. The center of flotation is 1.5 m aft of amidships, TPC = 10 tonnes, and MCT 1 cm = 120 tonnes m (see Figure 19.1). Find the new drafts if a total weight of 450 tonnes is loaded in a position 14 m forward of amidships.

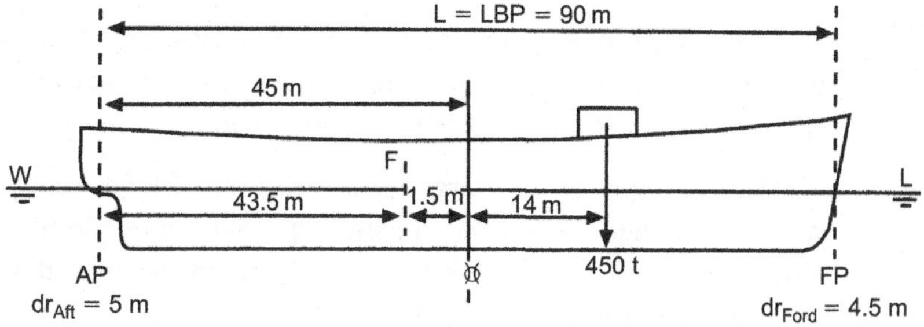

Figure 19.1

Ship Stability for Masters and Mates. http://dx.doi.org/10.1016/B978-0-08-097093-6.00019-0

$$\text{Bodily sinkage} = \frac{w}{\text{TPC}}$$

$$= \frac{450}{10}$$

$$\text{Bodily sinkage} = 45\ \text{cm}$$

$$\text{Change of trim} = \frac{\text{Trim moment}}{\text{MCT 1 cm}}$$

$$= \frac{450 \times 15.5}{120}$$

$$\text{Change of trim} = 58.12\ \text{cm by the head}$$

$$\text{Change of draft aft} = \frac{l}{L} \times \text{Change of trim}$$

$$= \frac{43.5}{90} \times 58.12$$

$$\text{Change of draft aft} = 28.09\ \text{cm}$$

$$\text{Change of draft forward} = \frac{46.5}{90} \times 58.12$$

$$\text{Change of draft forward} = 30.03\ \text{cm}$$

Original drafts	5.000 m A	4.500 m F
Bodily sinkage	+0.450 m	+0.450 m
	5.450 m	4.950 m
Change due trim	−0.281 m	+0.300 m
Ans. New drafts	5.169 m A	5.250 m F

Note. In the event of more than one weight being loaded or discharged, the net weight loaded or discharged is used to find the net bodily increase or decrease in draft, and the resultant trimming moment is used to find the change of trim.

Also, when the net weight loaded or discharged is large, it may be necessary to use the TPC and MCT 1 cm at the original draft to find the approximate new drafts, and then rework the problem using the TPC and MCT 1 cm for the mean of the old and the new drafts to find a more accurate result.

Example 2

A box-shaped vessel 40 m × 6 m × 3 m depth is floating in salt water on an even keel at 2 m draft (see Figure 19.2). Find the new drafts if a weight of 35 tonnes is discharged from a position 6 m from forward. MCT 1 cm = 8.4 tonnes m.

$$\text{TPC} = \frac{\text{WPA}}{97.56}$$

$$= \frac{40 \times 6}{97.56}$$

$$\text{TPC} = 2.46 \text{ tonnes}$$

$$\text{Bodily rise} = \frac{w}{\text{TPC}}$$

$$= \frac{35}{2.46}$$

$$\text{Bodily rise} = 14.2 \text{ cm}$$

$$\text{Change of trim} = \frac{w \times d}{\text{MCT 1 cm}}$$

$$= \frac{35 \times 14}{8.4}$$

$$\text{Change of trim} = 58.3 \text{ cm by the stern}$$

$$\text{Change of draft aft} = \frac{l}{L} \times \text{Change of trim}$$

$$= \frac{1}{2} \times 58.3 \text{ cm}$$

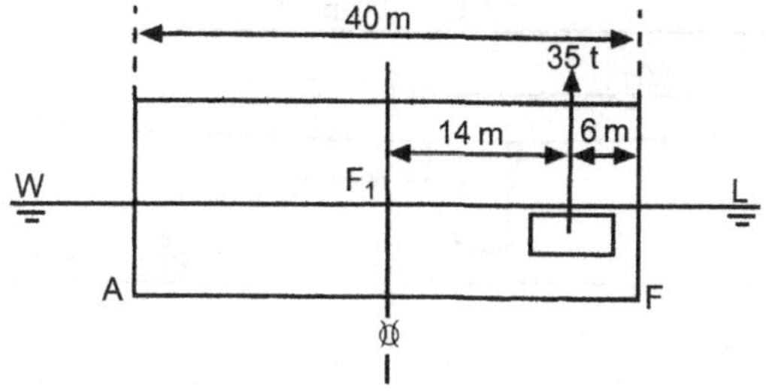

Figure 19.2

$$\text{Change of draft aft} = 29.15 \text{ cm}$$

$$\text{Change of draft forward} = \frac{1}{2} \times 58.3$$

$$\text{Change of draft forward} = 29.15 \text{ cm}$$

Original drafts	2.000 m A	2.00 m F
Bodily rise	− 0.140 m	− 0.140 m
	1.860 m	1.860 m
Change due trim	+ 0.290 m	− 0.290 m
Ans. New drafts	2.150 m A	1.570 m F

Example 3

A ship 100 m long arrives in port with drafts 3 m F and 4.3 m A. TPC = 10 tonnes. MCT 1 cm = 120 tonnes m. The center of flotation is 3 m aft of amidships (see Figure 19.3). If 80 tonnes of cargo are loaded in a position 24 m forward of amidships and 40 tonnes of cargo are discharged from 12 m aft of amidships, what are the new drafts?

Given information:

LBP	Draft Aft	Draft Forward	MCTC	TPC	LCF
100.00	4.30	3.00	120	10.00	3.00
wgt 1	lcg 1	wgt 2	lcg 2		
80.00	24.00	−40.00	12.00		

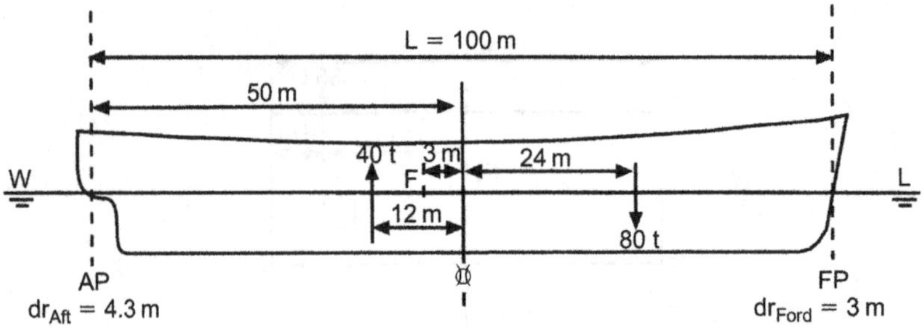

Figure 19.3

If LCF is aft of amidships, use plus sign in above table.

If LCF is forward of amidships, use negative sign in above table.

lcg 1 and lcg 2 are given about amidships and must be adjusted for position of LCF.

If weight 1 is to be added then use positive sign.

If weight 2 is to be discharched then use negative sign.

Cargo loaded	80 tonnes
Cargo discharged	40 tonnes
Net loaded	40 tonnes

$$\text{Bodily sinkage} = \frac{w}{TPC}$$

$$= \frac{40}{10}$$

$$\text{Bodily sinkage} = 4\text{ cm} = 0.04\text{ m}$$

To find the change of trim take moments about the center of flotation.

Weight	Lever	Moment of Weight
+80.00	−27.00	−2160
−40.00	+9.00	−360
		−2520

Negative sign indicates trim by the head, positive sign indicates trim by the stern.

Lever aft 47 m, lever forward 53 m.

$$\text{Change of trim} = \frac{\text{Trim moment}}{\text{MCT 1 cm}}$$

$$= \frac{2520}{120}$$

$$\text{Change of trim} = 21\text{ cm by the head}$$

$$\text{Change of draft aft} = \frac{l}{L} \times \text{Change of trim}$$

$$= \frac{47}{100} \times 21$$

$$\text{Change of draft aft} = 9.87\text{ cm}$$

$$\text{Change of draft forward} = \frac{53}{100} \times 21$$

$$\text{Change of draft forward} = 11.13\text{ cm}$$

Summary

	Aft	Forward	Units
Original drafts	4.300	3.000	meters
Bodily sinkage	+0.040	+0.040	meters
New drafts	4.340	3.040	meters
Change due trim	−0.099	+0.111	meters
Final end drafts	4.241	3.151	meters

Example 4

A ship of 6000 tonnes displacement has drafts 7 m F and 8 m A. MCT 1 cm = 100 tonnes m, TPC = 20 tonnes, center of flotation is amidships; 500 tonnes of cargo are then discharged from each of the following four holds (see Figure 19.4):

No. 1 hold, center of gravity 40 m forward of amidships
No. 2 hold, center of gravity 25 m forward of amidships
No. 3 hold, center of gravity 20 m aft of amidships
No. 4 hold, center of gravity 50 m aft of amidships

The following bunkers are also loaded:

150 tonnes at 12 m forward of amidships
50 tonnes at 15 m aft of amidships

Find the new drafts forward and aft.

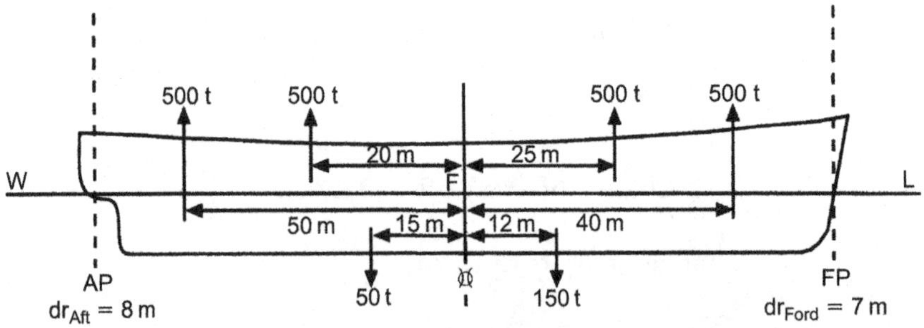

Figure 19.4

Total cargo discharged	2000	tonnes
Total bunkers loaded	200	tonnes
Net weight discharged	1800	tonnes

$$\text{Bodily rise} = \frac{w}{\text{TPC}} = \frac{1800}{20}$$

$$\text{Bodily rise} = 90\ \text{cm}$$

Assume levers and moments aft of LCF are positive.

Assume levers and moments forward of LCF are negative.

Weight	Distance from LCF	Moments
−500	−40	+20,000
−500	−25	+12,500
−500	+20	−10,000
−500	+50	−25,000
+150	−12	−1800
+50	+15	+750
		−3550

Resultant moment 3550 tonnes m by the head because of the negative sign.

$$\text{Change of trim} = \frac{\text{Trim moment}}{\text{MCT 1 cm}} = \frac{3550}{100}$$

$$\text{Change of trim} = 35.5\ \text{cm by the head}$$

Since the center of flotation is amidships,

$$\text{Change of draft aft} = \text{Change of draft forward} = \frac{1}{2}\ \text{change of trim} = 17.75\ \text{cm, say } 0.18\ \text{m}$$

Original drafts	8.000 m A	7.000 m F
Bodily rise	− 0.900 m	− 0.900 m
	7.100 m	6.100 m
Change due trim	− 0.180 m	+ 0.180 m
Ans. New drafts	6.920 m A	6.280 m F

■

Example 5

A ship arrives in port trimmed 25 cm by the stern. The center of flotation is amidships. MCT 1 cm = 100 tonnes m. A total of 3800 tonnes of cargo is to be discharged from four holds, and 360 tonnes of bunkers loaded in No. 4 double-bottom tank; 1200 tonnes of the cargo are to be discharged from No. 2 hold and 600 tonnes from No. 3 hold (see Figure 19.5). Find the remaining amount to be discharged from Nos. 1 and 4 holds if the ship is to complete on an even keel.

Center of gravity of No. 1 hold is 50 m forward of the center of flotation

Center of gravity of No. 2 hold is 30 m forward of the center of flotation

Center of gravity of No. 3 hold is 20 m abaft of the center of flotation

Center of gravity of No. 4 hold is 45 m abaft of the center of flotation

Center of gravity of No. 4 DB tank is 5 m abaft of the center of flotation

Total cargo to be discharged from four holds	3800 tonnes
Total cargo to be discharged from Nos. 2 and 3	1800 tonnes
Total cargo to be discharged from Nos. 1 and 4	2000 tonnes

Let 'x' tonnes of cargo be discharged from No. 1 hold

Let (2000 − x) tonnes of cargo be discharged from No. 4 hold

Take moments about the center of flotation, or as shown in Figure 19.5.

Original trim = 25 cm by the stern, i.e. + 25 cm

Required trim = 0

Change of trim required = 25 cm by the head, i.e. − 25 cm

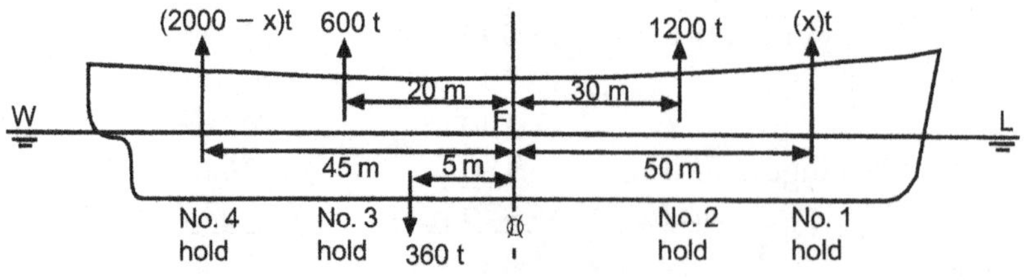

Figure 19.5

Weight	Distance from C.F.	Moments	
		Negative	Positive
−x	−50	–	+50x
−1200	−30	–	+36,000
−600	+20	−12,000	–
−(2000 − x)	+45	−(90,000 − 45x)	–
+360	+5	–	+1800
		−102,000 + 45x	+37,800 + 50x

$$\text{Trimming moment required} = \text{Change of trim} \times \text{MCT 1 cm}$$

$$= -25 \times 100 = -2500$$

$$\text{Trimming moment required} = 2500 \text{ tonnes m by the head}$$

$$\text{Resultant moment} = \text{Moment to change trim by head} - \text{MCT by stern}$$

$$\therefore\ -2500 = -102,000 + 45x + 37,800 + 50x$$

$$2500 = 64,200 - 95x$$

or

$$95x = 61,700$$

$$\underline{x = 649.5 \text{ tonnes}}$$

and

$$\underline{2000 - x = 1350.5 \text{ tonnes}}$$

Ans. Discharge 649.5 tonnes from No. 1 hold and 1350.5 tonnes from No. 4 hold.

■

Using Trim to Find the Position of the Center of Flotation

■ Example

A ship arrives in port floating at drafts of 4.50 m A and 3.80 m F. The following cargo is then loaded:

100 tonnes in a position 24 m aft of amidships
30 tonnes in a position 30 m forward of amidships
60 tonnes in a position 15 m forward of amidships

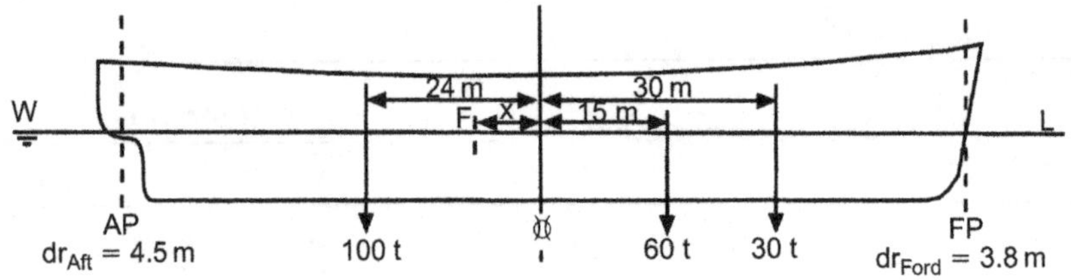

Figure 19.6

The drafts are then found to be 5.10 m A and 4.40 m F (see Figure 19.6). Find the position of the longitudinal center of flotation aft of amidships.

Original drafts 4.50 m A, 3.80 m F give 0.70 m trim by the stern, i.e. +70 cm.

New drafts 5.10 m A, 4.40 m F give 0.70 m trim by the stern, i.e. +70 cm.

Therefore, there has been no change in trim, which means that:

Moment to change trim by the head = Moment to the change trim by the stern

Let the center of flotation be 'x' meters aft of amidships. Taking moments, then

$$100(24 - x) = 30(30 + x) + 600(15 + x)$$
$$2400 - 100x = 900 + 30x + 900 + 60x$$
$$190x = 600$$
$$x = 3.16 \text{ m}$$

Ans. Center of flotation is 3.16 meters aft of amidships.

Note. In this type of question it is usual to assume that the center of flotation is aft of amidships, but this may not be the case. Had it been assumed that the center of flotation was aft of amidships when in actual fact it was forward, then the answer obtained would have been minus.

Remember. Levers, moments, and trim by the stern all have a positive sign. Levers, moments, and trim by the head all have a negative sign.

Exercise 19

1. A ship floats in salt water on an even keel displacing 6200 tonnes. KG = 5.5 m, KM = 6.3 m, and there are 500 tonnes of cargo yet to load. Space is available in No. 1 tween deck (KG = 7.6 m, center of gravity 40 m forward of the center of flotation) and in No. 4 lower hold (KG = 5.5 m, center of gravity 30 m aft of the center of flotation). Find how much cargo to load in each space to complete loading trimmed 0.6 m by the stern, and also find the final GM. MCT 1 cm = 200 tonnes m.

2. A ship, floating at drafts of 7.7 m F and 7.9 m A, sustains damage in an end-on collision and has to lift the bow to reduce the draft forward to 6.7 m. The ship is about to enter a port in which the maximum permissible draft is 8.3 m. To do this it is decided to discharge cargo from No. 1 hold (center of gravity 75 m forward of amidships) and No. 4 hold (center of gravity 45 m aft of amidships). MCT 1 cm = 200 tonnes m, TPC = 15 tonnes. Center of flotation is amidships. Find the minimum amount of cargo to discharge from each hold.
3. A ship 100 m long has center of flotation 3 m aft of amidships and is floating at drafts 3.2 m F and 4.4 m A. TPC = 10 tonnes, MCT 1 cm = 150 tonnes m; 30 tonnes of cargo are then discharged from 20 m forward of amidships and 40 tonnes are discharged from 12 m aft of amidships. Find the final drafts.
4. A ship 84 meters long is floating on an even keel at a draft of 5.5 meters; 45 tonnes of cargo are then loaded in a position 30 m aft of amidships. The center of flotation is 1 m aft of amidships. TPC = 15 tonnes, MCT 1 cm = 200 tonnes m. Find the final drafts.
5. A ship arrives in port with drafts 6.8 m F and 7.2 m A; 500 tonnes of cargo are then discharged from each of four holds.

   The center of gravity of No. 1 hold is 40 m forward of amidships
   The center of gravity of No. 2 hold is 25 m forward of amidships
   The center of gravity of No. 3 hold is 20 m aft of amidships
   The center of gravity of No. 4 hold is 50 m aft of amidships
   Also, 50 tonnes of cargo are loaded in a position whose center of gravity is 15 m aft of amidships, and 135 tonnes of cargo center of gravity 40 m forward of amidships. TPC = 15 tonnes, MCT 1 cm = 400 tonnes m. The center of flotation is amidships. Find the final drafts.
6. A ship is floating at drafts 5.5 m F and 6.0 m A. The following cargo is then loaded:

   97 tonnes center of gravity 8 m forward of amidships
   20 tonnes center of gravity 40 m aft of amidships
   28 tonnes center of gravity 20 m aft of amidships
   The draft is now 5.6 m F and 6.1 m A. Find the position of the center of flotation relative to amidships.
7. Find the position of the center of flotation of a ship if the trim remains unchanged after loading the following cargo:

   100 tonnes center of gravity 8 meters forward of amidships
   20 tonnes center of gravity 40 meters aft of amidships
   28 tonnes center of gravity 20 meters aft of amidships
8. A ship arrives in port with drafts 6.8 m F and 7.5 m A. The following cargo is discharged:

   90 tonnes center of gravity 30 m forward of amidships

40 tonnes center of gravity 25 m aft of amidships
50 tonnes center of gravity 50 m aft of amidships
The drafts are now 6.7 m F and 7.4 m A. Find the position of the center of flotation relative to amidships.

■

CHAPTER 20

Trim Calculations — Satisfying Prescribed Requirements for End Drafts

Loading a Weight to Keep the After Draft Constant

When a ship is being loaded it is usually the aim of those in charge of the operation to complete loading with the ship trimmed by the stern. Should the ship's draft on sailing be restricted by the depth of water over a dock-sill or by the depth of water in a channel, then the ship will be loaded in such a manner as to produce this draft aft and be trimmed by the stern.

Assume now that a ship loaded in this way is ready to sail. It is then found that the ship has to load an extra weight. The weight must be loaded in such a position that the draft aft is not increased and also that the maximum trim is maintained.

If the weight is loaded at the center of flotation, the ship's drafts will increase uniformly and the draft aft will increase by a number of centimeters equal to w/TPC. The draft aft must now be decreased by this amount.

Now let the weight be shifted through a distance of 'd' meters forward. The ship will change trim by the head, causing a reduction in the draft aft by a number of centimeters equal to l/L × Change of trim.

Therefore, if the same draft is to be maintained aft, the above two quantities must be equal, i.e.

$$\frac{l}{L} \times \text{Change of trim} = \frac{w}{TPC}$$

So:

$$\text{Change of trim} = \frac{w}{TPC} \times \frac{L}{l} \qquad \text{(I)}$$

But

$$\text{Change of trim} = \frac{w \times d}{\text{MCT 1 cm}} \qquad \text{(II)}$$

Equating (I) and (II):

$$\therefore \frac{w \times d}{\text{MCT 1 cm}} = \frac{w}{TPC} \times \frac{L}{l}$$

Ship Stability for Masters and Mates. http://dx.doi.org/10.1016/B978-0-08-097093-6.00020-7

or

$$d = \frac{L \times MCT\ 1\ cm}{l \times TPC}$$

where d = the distance forward of the center of flotation to load a weight to keep the draft aft constant, L = the ship's length, LBP, and l = the distance of the center of flotation to the stern.

Example

A box-shaped vessel 60 m long, 10 m beam, and 6 m deep is floating in salt water at drafts 4 m F and 4.4 m A. Find how far forward of amidships a weight of 30 tonnes must be loaded if the draft aft is to remain at 4.4 m.

$$TPC_{SW} = \frac{WPA}{97.56}$$

$$= \frac{60 \times 10}{97.56}$$

$$TPC_{SW} = 6.15 \text{ tonnes}$$

$$W = L \times B \times d \times \rho_{SW} \text{ tonnes}$$

$$= 60 \times 10 \times 4.2 \times 1.025$$

$$W = 2583 \text{ tonnes}$$

$$BM_L = \frac{L^2}{12d}$$

$$= \frac{60 \times 60}{12 \times 4.2}$$

$$BM_L = 71.42 \text{ meters}$$

$$MCT\ 1\ cm \simeq \frac{W \times BM_L}{100L} \text{ because } GM_L \simeq BM_L$$

$$MCT\ 1\ cm = \frac{2583 \times 71.42}{100 \times 60}$$

$$MCT\ 1\ cm = 30.75 \text{ t m/cm}$$

$$d = \frac{L \times MCT\ 1\ cm}{l \times TPC_{SW}}$$

$$= \frac{60}{30} \times 30.75 \times \frac{1}{6.15}$$

$$d = 10 \text{ meters from LCF}$$

LCF is at amidships.

Ans. Load the weight 10 meters forward of amidships.

Loading a Weight to Produce a Required Draft

Example

A ship 150 meters long arrives at the mouth of a river with drafts 5.5 m F and 6.3 m A. MCT 1 cm = 200 tonnes m, TPC = 15 tonnes. Center of flotation is 1.5 m aft of amidships. The ship has then to proceed up the river where the maximum draft permissible is 6.2 m. It is decided that SW ballast will be run into the forepeak tank to reduce the draft aft to 6.2 m (see Figure 20.1). If the center of gravity of the forepeak tank is 60 meters forward of the center of flotation, find the minimum amount of water that must be run in and also find the final draft forward.

(a) *Load 'w' tonnes at the center of flotation*

$$\text{Bodily sinkage} = \frac{w}{TPC}$$

$$= \frac{w}{15}\ \text{cm}$$

$$\text{New draft aft} = 6.3\ \text{m} + \frac{w}{15}\ \text{cm} \qquad \text{(I)}$$

$$\text{Required draft aft} = 6.2\ \text{m} \qquad \text{(II)}$$

$$\text{Equations(I)} - \text{(II)} = \underline{\text{Reduction required}} = 0.1\ \text{m} + \frac{w}{15}\ \text{cm}$$

$$= 10\ \text{cm} + \frac{w}{15}\ \text{cm}$$

$$= \left(10 + \frac{w}{15}\right)\ \text{cm} \qquad \text{(III)}$$

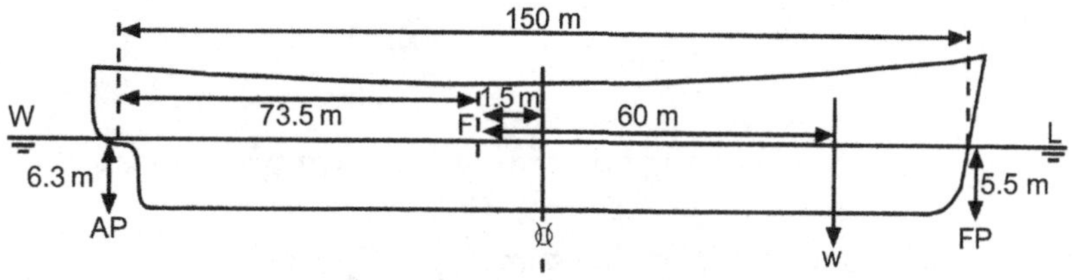

Figure 20.1

(b) *Shift 'w' tonnes from the center of flotation to the forepeak tank*

$$\text{Change of trim} = \frac{w \times d}{\text{MCT 1 cm}} = \frac{60w}{200}$$

$$\text{Change of trim} = \frac{3w}{10} \text{ cm by the head}$$

$$\text{Change of draft aft due to trim} = \frac{l}{L} \times \text{Change of trim} = \frac{73.5}{150} \times \frac{3w}{10}$$

$$\text{Change of draft aft due to trim} = 0.147w \text{ cm}$$

$$\text{But change of draft required aft} = \left(10 + \frac{w}{15}\right) \text{ cm as per equation (III)}$$

$$0.147w = 10 + \frac{w}{15}, \text{ i.e. equation (IV)} = \text{equation (III)}$$

$$2.205w = 150 + w$$

$$1.205w = 150$$

$$w = 124.5 \text{ tonnes}$$

Therefore, by loading 124.5 tonnes in the forepeak tank the draft aft will be reduced to 6.2 meters.

(c) *To find the new draft forward*

$$\text{Bodily sinkage} = \frac{w}{\text{TPC}} = \frac{124.5}{15}$$

$$\text{Bodily sinkage} = 8.3 \text{ cm}$$

$$\text{Change of trim} = \frac{w \times d}{\text{MCT 1 cm}} = \frac{124.5 \times 60}{200}$$

$$\text{Change of trim} = 37.35 \text{ cm by the head}$$

$$\text{Change of draft aft due trim} = \frac{l}{L} \times \text{Change of trim}$$

$$\text{Change of draft aft due trim} = \frac{73.5}{150} \times 37.35$$

$$= 18.3 \text{ cm}$$

$$\text{Change of draft forward due trim} = \text{Change of trim} - \text{Change of draft aft}$$

$$= 37.35 - 18.3 \text{ cm} = 19.05 \text{ cm, or}$$

$$\text{Change of draft forward due trim} = \frac{76.5 \times 37.35}{150} = 19.05 \text{ cm}$$

Original drafts	6.300 m A	5.500 m F
Bodily sinkage	+ 0.080 m	+ 0.080 m
	6.380 m	5.580 m
Change due trim	− 0.180 m	+ 0.190 m
New drafts	6.200 m A	5.770 m F

Ans. Load 124.5 tonnes in forepeak tank. Final draft forward is 5.770 meters.

■

Using Change of Trim to Find the Longitudinal Metacentric Height (GM_L)

Earlier it was shown that, when a weight is shifted longitudinally within a ship, it will cause a change of trim. It will now be shown how this effect may be used to determine the longitudinal metacentric height.

Consider Figure 20.2(a), which represents a ship of length 'L' at the waterline, floating upright on an even keel with a weight on deck forward. The center of gravity is at G, the center of buoyancy at B, and the longitudinal metacenter at M_L. The longitudinal metacentric height is therefore GM_L.

Now let the weight be shifted aft horizontally, as shown in Figure 20.2(b). The ship's center of gravity will also shift horizontally, from G to G_1, producing a trimming moment of $W \times GG_1$ by the stern.

The ship will now trim to bring G_1 under M_L, as shown in Figure 20.2(c). In Figure 20.2(c), W_1L_1 represents the new waterline, F the new draft forward, and A the new draft aft. It was shown in Figure 18.4(b), and by the associated notes, that F – A is equal to the new trim (t) and since the ship was originally on an even keel, then 't' must also be equal to the change of trim.

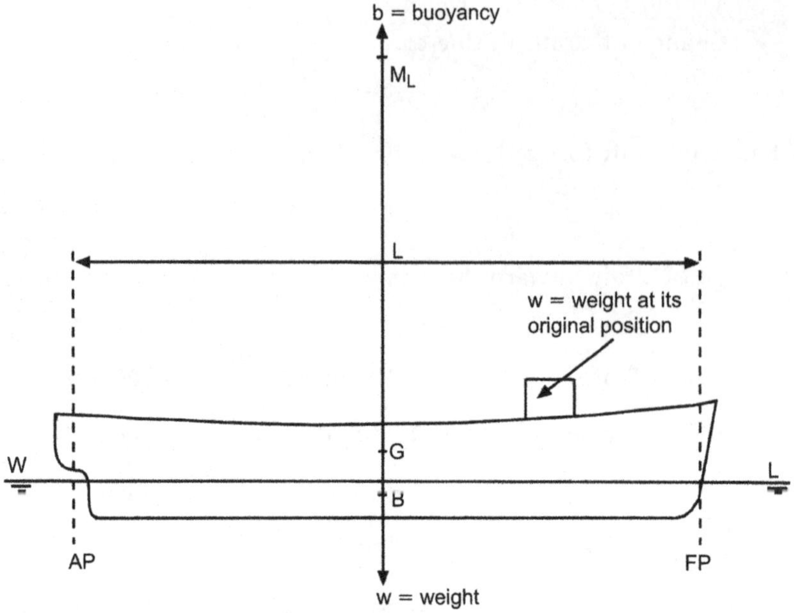

Figure 20.2a

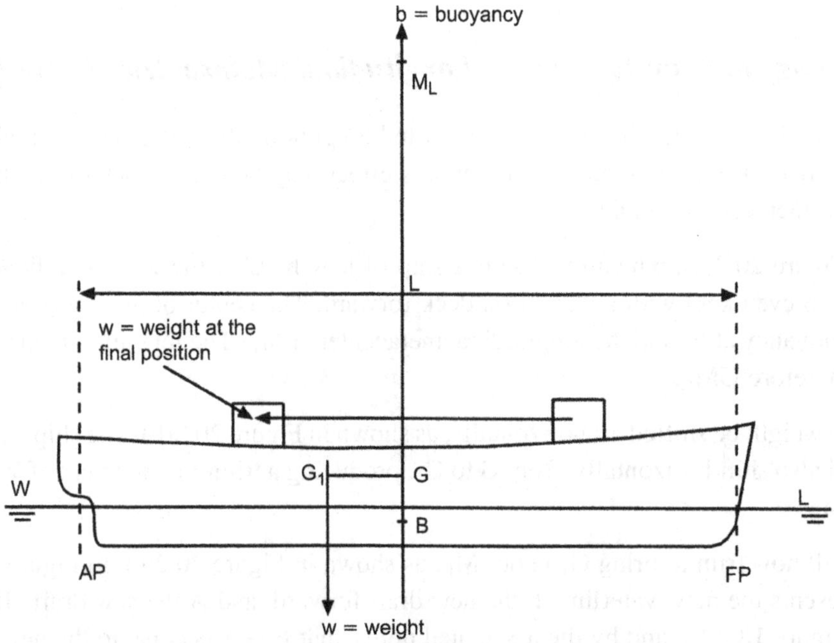

Figure 20.2b

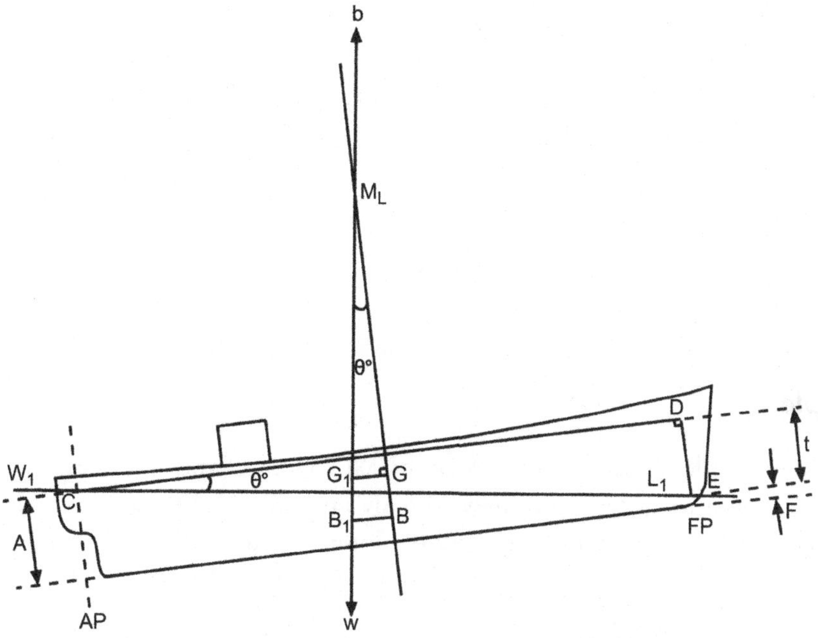

Figure 20.2c

If the angle between the new and old verticals is equal to θ, then the angle between the new and old horizontals must also be equal to θ (the angle between two straight lines being equal to the angle between their normals).

It will also be seen in Figure 20.2(c) that the triangles GG_1M_L and CDE are similar triangles.

$$\therefore \frac{GM_L}{GG_1} = \frac{L}{t}$$

or

$$GM_L = \frac{L}{t} \times GG_1$$

(All measurements are in meters.)

Example 1

When a weight is shifted aft in a ship 120 meters long, it causes the ship's center of gravity to move 0.2 meters horizontally and the trim to change by 0.15 meters. Find the longitudinal metacentric height.

$$\frac{GM_L}{GG_1} = \frac{L}{t}$$

$$\therefore GM_L = \frac{L}{t} \times GG_1$$

$$= \frac{120 \times 0.2}{0.15}$$

Ans. GM_L = 160 meters.

Example 2

A ship 150 meters long has a displacement of 7200 tonnes, and is floating upright on an even keel. When a weight of 60 tonnes, already on board, is shifted 24 meters forward, the trim is changed by 0.15 meters. Find the longitudinal metacentric height.

$$\frac{GM_L}{GG_1} = \frac{L}{t}$$

$$GM_L = GG_1 \times \frac{L}{t}, \qquad GG_1 = \frac{w \times d}{W}$$

$$\text{So } GM_L = \frac{w \times d}{W} \times \frac{L}{t}$$

$$GM_L = \frac{60 \times 24}{7200} \times \frac{150}{0.15}$$

Ans. GM_L = 200 meters.

Which Way Will the Ship Trim?

As previously discussed, when a weight 'w' is placed onto a ship the LCG moves to a new position. This creates a temporary gap between the LCG and the LCB.

To obtain equilibrium, and for the ship to settle down at new end drafts, the vessel will have a change of trim. It will continue to do so until the LCB is once again vertically in line and below the new LCG position.

$$\text{Change of trim} = \frac{W \times \{LCB_{foap} - LCG_{foap}\}}{MCTC} \text{cm}$$

$$\text{Change of trim also} = \frac{w \times d}{MCTC}\,\text{cm}$$

$$\text{foap} = \text{forward of after perpendicular}$$

$$W \times \{LCB_{foap} - LCG_{foap}\} = w \times d \text{ also } = \text{Trimming moment in tm}$$

Observations

- Many naval architects measure LCG and LCB from amidships.
- SQA/MCA advocate measuring LCG and LCB forward of aft perp (foap).
- Both procedures will end up giving the same change of trim.
- LCG is the longitudinal overall center of gravity for the lightweight plus all of the deadweight items on board the ship.
- LCB is the three-dimensional longitudinal centroid of the underwater form of the vessel.
- LCF is the two-dimensional center of each waterplane.
- W = ship's displacement in tonnes, that is lightweight + deadweight.
- d = the horizontal lever from the LCF to the lcg of the added weight 'w'.
- MCTC = moment to change trim 1 cm.

Always remember that:

- If $\{LCB_{foap} - LCG_{foap}\}$ is *positive* in value, ship will trim by the **stern**.
- If $\{LCB_{foap} - LCG_{foap}\}$ is *negative* in value, ship will trim by the **bow**.

Summary

1. Make a sketch from the given information.
2. Estimate the mean bodily sinkage.
3. Calculate the change of trim using levers measured from LCF.
4. Evaluate the trim ratio forward and aft at FP and AP, from the LCF position.
5. Collect the above calculated values to estimate the final end drafts.
6. In the solutions shown in the text, these final end drafts have been calculated to three decimal places. In practice, naval architects and ship officers round off the drafts to two decimal places only. This gives acceptable accuracy.
7. Note how the formulae were written in *letters* first and then the *figures* were put in. In the event of a mathematical error, marks will be given for a correct formula and for a correct sketch.

Exercise 20

1. A ship is 150 m long, MCT 1 cm = 400 tonnes m, TPC = 15 tonnes. The center of flotation is 3 m aft of amidships. Find the position in which to load a mass of 30 tonnes, with reference to the center of flotation, so as to maintain a constant draft aft.
2. A ship 120 m long, with maximum beam 15 m, is floating in salt water at drafts 6.6 m F and 7 m A. The block coefficient and coefficient of fineness of the waterplane is 0.75. Longitudinal metacentric height = 120 m. Center of flotation is amidships. Find how much more cargo can be loaded and in what position relative to amidships if the ship is to cross a bar with a maximum draft of 7 m F and A.
3. A ship 120 m long floats in salt water at drafts 5.8 m F and 6.6 m A. TPC = 15 tonnes, MCT 1 cm = 300 tonnes m. Center of flotation is amidships. What is the minimum amount of water ballast required to be taken into the forepeak tank (center of gravity 60 m forward of the center of flotation) to reduce the draft aft to 6.5 meters? Find also the final draft forward.
4. A ship leaves port with drafts 7.6 m F and 7.9 m A; 400 tonnes of bunkers are burned from a space whose center of gravity is 15 m forward of the center of flotation, which is amidships. TPC = 20 tonnes, MCT 1 cm = 300 tonnes m. Find the minimum amount of water that must be run into the forepeak tank (center of gravity 60 m forward of the center of flotation) in order to bring the draft aft to the maximum of 7.7 m. Find also the final draft forward.
5. A ship 100 m long has MCT 1 cm = 300 tonnes m, requires 1200 tonnes of cargo to complete loading, and is at present floating at drafts of 5.7 m F and 6.4 m A. She loads 600 tonnes of cargo in a space whose center of gravity is 3 m forward of amidships. The drafts are then 6.03 m F and 6.67 m A. The remainder of the cargo is to be loaded in No. 1 hold (center of gravity 43 m forward of amidships) and in No. 4 hold (center of gravity 37 m aft of amidships). Find the amount that must be loaded in each hold to ensure that the draft aft will not exceed 6.8 meters. LCF is at amidships.
6. A ship 100 m long is floating in salt water at drafts 5.7 m F and 8 m A. The center of flotation is 2 m aft of amidships. Find the amount of water to run into the forepeak tank (center of gravity 48 m forward of amidships) to bring the draft aft to 7.9 m. TPC = 30 tonnes, MCT 1 cm = 300 tonnes m.
7. A ship 140 m long arrives off a port with drafts 5.7 m F and 6.3 m A. The center of flotation is 3 m aft of amidships. TPC = 30 tonnes, MCT 1 cm = 420 tonnes m. It is required to reduce the draft aft to 6.2 m by running water into the forepeak tank (center of gravity 67 m forward of amidships). Find the minimum amount of water to load and also give the final draft forward.

8. A ship 150 m long is floating upright on an even keel. When a weight already on board is shifted aft, it causes the ship's center of gravity to shift 0.2 m horizontally and the trim to change by 0.15 m. Find the longitudinal metacentric height.
9. A ship of 5000 tonnes displacement is 120 m long and floats upright on an even keel. When a mass of 40 tonnes, already on board, is shifted 25 m forward horizontally, it causes the trim to change by 0.1 m. Find the longitudinal metacentric height.
10. A ship is 130 m LBP and is loaded ready for departure as shown in the table below. From her hydrostatic curves at 8 m even-keel draft in salt water it is found that: MCTC is 150 tm/cm, LCF is 2.5 m forward [symbol], W is 12,795 tonnes, and LCB is 2 m forward [symbol]. Calculate the final end drafts for this vessel. What is the final value for the trim? What is the Dwt value for this loaded condition?

Item	Weight in Tonnes	LCG from Amidships
Lightweight	3600	2.0 m aft
Cargo	8200	4.2 m forward
Oil fuel	780	7.1 m aft
Stores	20	13.3 m forward
Fresh water	100	20.0 m aft
Feed water	85	12.0 m aft
Crew and effects	10	At amidships

■

CHAPTER 21

Large-Angle Stability Considerations – GZ and KN Cross Curves of Stability

Cross Curves of Stability

GZ Cross Curves of Stability

These are a set of curves from which the righting lever about an assumed center of gravity for any angle of heel at any particular displacement may be found by inspection. The curves are plotted for an assumed KG and, if the actual KG of the ship differs from this, a correction must be applied to the righting levers taken from the curves.

Figure 21.1 shows a set of Stability Cross Curves plotted for an imaginary ship called M.V. 'Tanker', assuming the KG to be 9 meters. A scale of displacements is shown along the bottom margin and a scale of righting levers (GZs) in meters on the left-hand margin. The GZ scale extends from +4.5 m through 0 to −1 m. The curves are plotted at 15° intervals of heel between 15° and 90°.

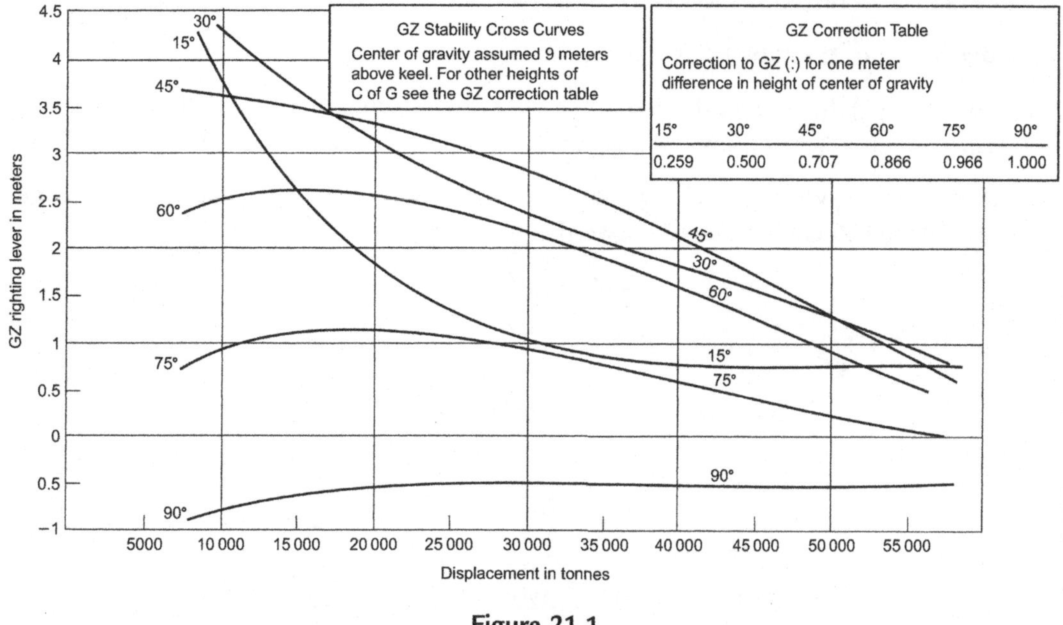

Figure 21.1

Ship Stability for Masters and Mates. http://dx.doi.org/10.1016/B978-0-08-097093-6.00021-9

To find the GZs for any particular displacement locate the displacement concerned on the bottom scale and, through this point, erect a perpendicular to cut all the curves. Translate the intersections with the curves horizontally to the left-hand scale and read off the GZs for each angle of heel.

Example 1

Using the Stability Cross Curves for M.V. 'Tanker', find the GZs at 15° intervals between 0° and 90° heel when the displacement is 35,000 tonnes and KG = 9 meters.

Erect a perpendicular through 35,000 tonnes on the displacement scale and read off the GZs from the left-hand scale as follows:

Angle of heel	0°	15°	30°	45°	60°	75°	90°
GZ in meters	0	0.86	2.07	2.45	1.85	0.76	−0.5

Should the KG of the ship be other than 9 meters, a correction must be applied to the GZs taken from the curves to obtain the correct GZs. The corrections are tabulated in the block on the top right-hand side of Figure 21.1 and are given for each 1 meter difference between 9 meters and the ship's actual KG. To find the correction to the GZ, multiply the correction taken from the table for the angle of heel concerned by the difference in KGs. To apply the correction: when the ship's KG is *greater* than 9 meters the ship is less stable and the correction must be *subtracted*, but when the KG is *less* than 9 meters the ship is more stable and the correction is to be *added*.

The derivation of the table is as follows:

In Figure 21.2(a), KG is 9 m, this being the KG for which this set of curves is plotted, and GZ represents the righting lever, as taken from the curves for this particular angle of heel.

Consider the case when the KG is greater than 9 m (KG_1 in Figure 21.2(a)). The righting lever is reduced to G_1Z_1. Let G_1X be perpendicular to GZ. Then:

$$\begin{aligned} G_1Z_1 &= XZ \\ &= GZ - GX \end{aligned}$$

or

$$\text{Corrected GZ} = \text{Tabulated GZ} - \text{Correction}$$

Also, in triangle GXG_1:

$$GX = GG_1 \sin \theta^\circ$$

or

$$\text{Correction} = GG_1 \sin \theta^\circ, \text{ where } \theta^\circ \text{ is the angle of heel}$$

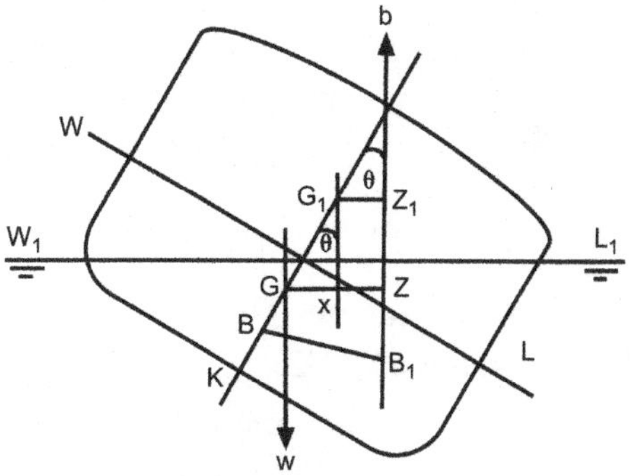

Figure 21.2a

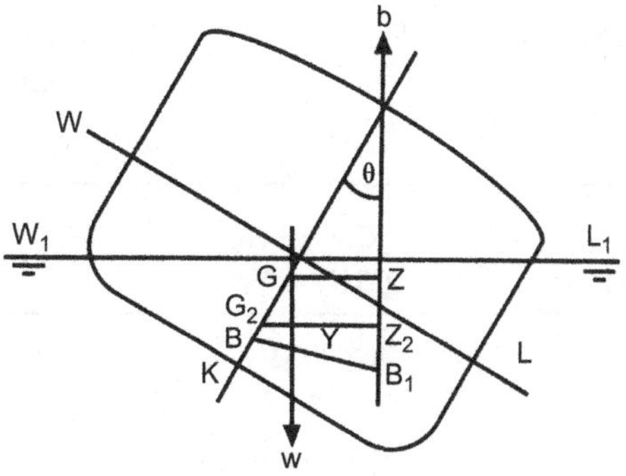

Figure 21.2b

But GG_1 is the difference between 9 m and the ship's actual KG. Therefore, the corrections shown in the table on the Cross Curves for each 1 meter difference of KG are simply the sines of the angle of heel.

Now consider the case where KG is less than 9 m (KG_2 in Figure 21.2(b)). The length of the righting lever will be increased to G_2Z_2.

Let GY be perpendicular to G_2Z_2, then:

$$G_2Z_2 = YZ_2 + G_2Y$$

but

$$YZ_2 = GZ$$

therefore

$$G_2Z_2 = GZ + G_2Y$$

or

$$\text{Corrected GZ} = \text{Tabulated GZ} + \text{Correction}$$

Also, in triangle GG_2Y:

$$G_2Y = GG_2 \sin\theta^\circ$$

or

$$\text{Correction} = GG_2 \sin \text{Heel}$$

It will be seen that this is similar to the previous result except that in this case the correction is to be *added* to the tabulated GZ.

Example 2

Using the Stability Cross Curves for M.V. 'Tanker' (Figure 21.1), find the GZs at 15° intervals between 0° and 90° when the displacement is 38,000 tonnes and the KG is 8.5 meters.

Heel	GZ Ords (KG = 9 m)	Correction ($GG_1 \sin\theta^\circ$)	New GZ (KG = 8.5 m)
0°	0	$0.5 \times 0 = 0$	$0 + 0 = 0$
15°	0.81	$0.5 \times 0.259 = 0.13$	$0.81 + 0.13 = 0.94$
30°	1.90	$0.5 \times 0.500 = 0.25$	$1.90 + 0.25 = 2.15$
45°	2.24	$0.5 \times 0.707 = 0.35$	$2.24 + 0.35 = 2.59$
60°	1.70	$0.5 \times 0.866 = 0.43$	$1.70 + 0.43 = 2.13$
75°	0.68	$0.5 \times 0.966 = 0.48$	$0.68 + 0.48 = 1.16$
90°	−0.49	$0.5 \times 1.000 = 0.50$	$-0.49 + 0.50 = 0.01$

KN Cross Curves of Stability

It has already been shown that the Stability Cross Curves for a ship are constructed by plotting the righting levers for an assumed height of the center of gravity above the keel. In some cases the curves are constructed for an assumed KG of zero. The curves are then referred to as KN curves, KN being the righting lever measured from the keel. Figure 21.3(a) shows the KN curves for an imaginary ship called the M.V. 'Cargo-Carrier'.

To obtain the righting levers for a particular displacement and KG, the values of KN are first obtained from the curves by inspection at the displacement concerned. The correct righting

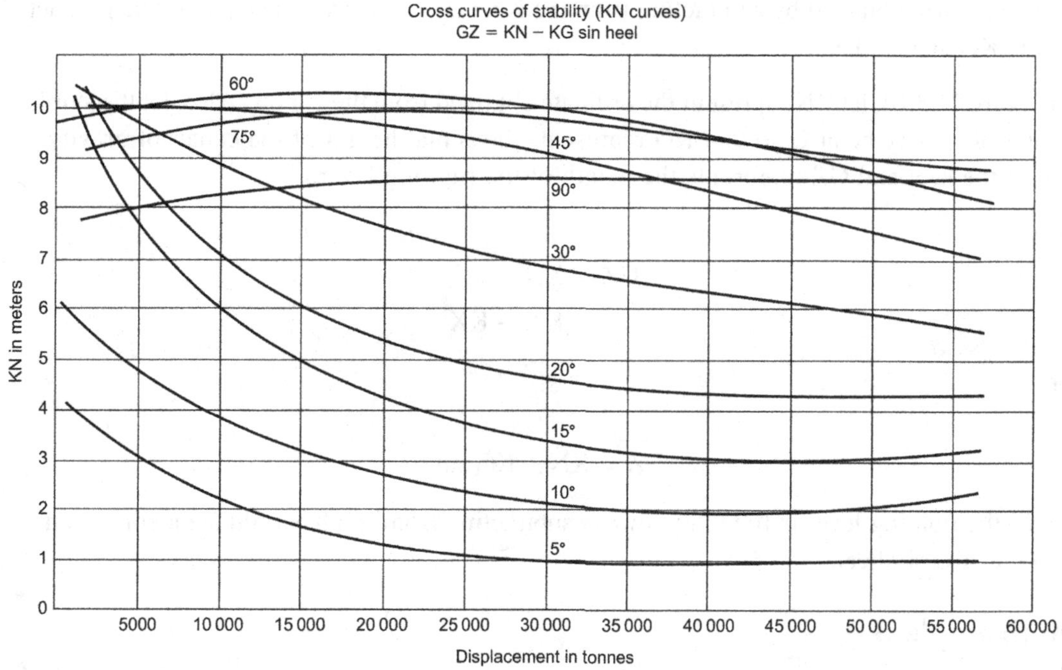

Figure 21.3a:
M.V. 'Cargo-Carrier'.

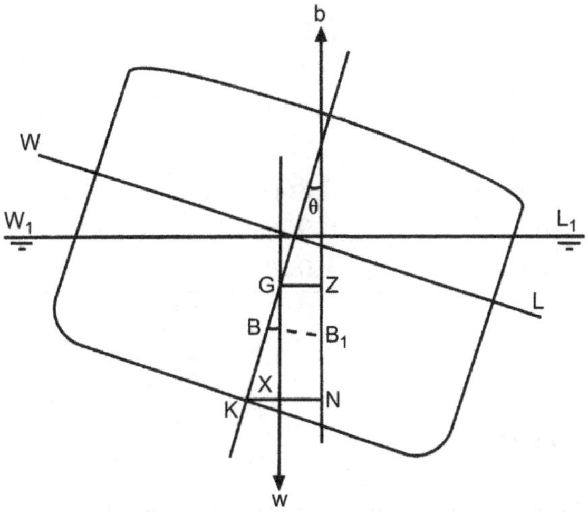

Figure 21.3b

levers are then obtained by subtracting from the KN values a correction equal to the product of the KG and sin heel.

In Figure 21.3(b), let KN represent the ordinate obtained from the curves. Also, let the ship's center of gravity be at G so that KG represents the actual height of the center of gravity above the keel and GZ represents the length of the righting lever.

Now

$$\begin{aligned} GZ &= XN \\ &= KN - KX \end{aligned}$$

or

$$GZ = KN - KG \sin\theta$$

Thus, the righting lever is found by *always* subtracting from the KN ordinate a correction equal to KG sin heel.

Example 3

Find the righting levers for M.V. 'Cargo-Carrier' when the displacement is 40,000 tonnes and the KG is 10 meters.

Heel (θ)	KN Ords	sin θ	Correction −KG sin θ	New GZ
5°	0.90	0.087	−0.871	0.03
10°	1.92	0.174	−1.736	0.18
15°	3.11	0.259	−2.588	0.52
20°	4.25	0.342	−3.420	0.83
30°	6.30	0.500	−5.000	1.30
45°	8.44	0.707	−7.071	1.37
60°	9.39	0.866	−8.660	0.73
75°	9.29	0.966	−9.659	−0.37
90°	8.50	1.000	−10.00	−1.50

Statical Stability Curves

The curve of statical stability for a ship in any particular condition of loading is obtained by plotting the righting levers against angle of heel, as shown in Figures 21.4 and 21.5.

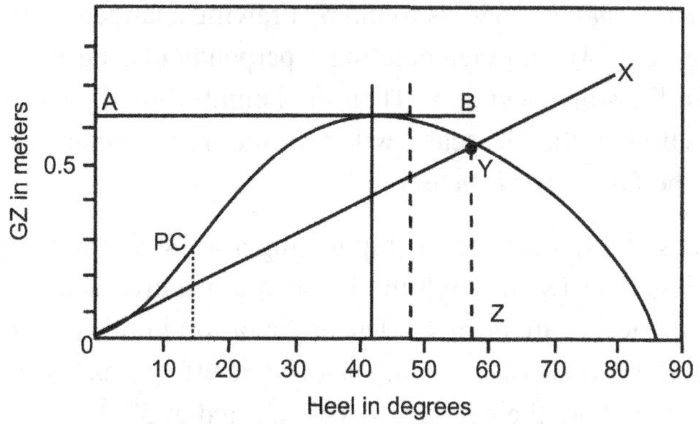

Figure 21.4:
Curve for a Ship with Positive Initial Metacentric Height.

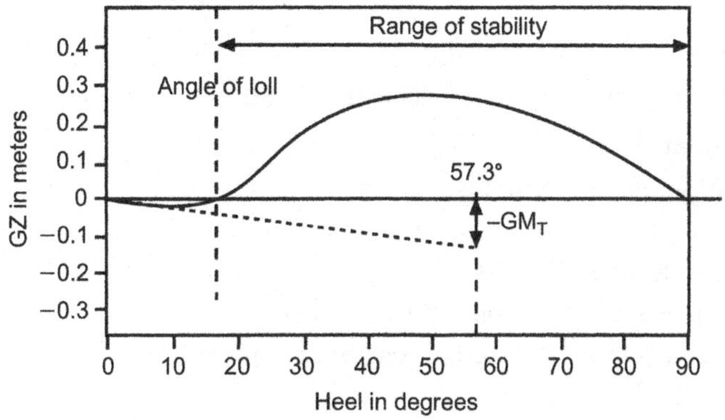

Figure 21.5:
Curve for a Ship with Negative Initial Metacentric Height.

From this type of graph a considerable amount of stability information may be found by inspection:

- PC = point of contraflexure on the curve. It is where the corner of the deck of a heeled ship becomes level with the waterline. Beyond angle α, the deck becomes flooded.
- *The range of stability* is the range over which the ship has positive righting levers. In Figure 21.4 the range is from 0° to 86°.
- *The angle of vanishing stability* is the angle of heel at which the righting lever returns to zero, or is the angle of heel at which the sign of the righting levers changes from positive to negative. The angle of vanishing stability in Figure 21.4 is 86°.
- *The maximum GZ* is obtained by drawing a tangent to the highest point in the curve. In Figure 21.4, AB is the tangent and this indicates a maximum GZ of 0.63 meters. If a perpendicular is dropped from the point of tangency, it cuts the heel scale at the angle of heel at which the maximum GZ occurs. In the present case the maximum GZ occurs at 42° heel.

- *The initial metacentric height* (GM) is found by drawing a tangent to the curve through the origin (OX in Figure 21.4), and then erecting a perpendicular through an angle of heel of 57.3°. Let the two lines intersect at Y. Then the height of the intersection above the base (YZ), when measured on the GZ scale, will give the initial metacentric height. In the present example the GM is 0.54 meters.

Figure 21.5 shows the stability curve for a ship having a negative initial metacentric height. At angles of heel of less than 18° the righting levers are negative, whilst at angles of heel between 18° and 90° the levers are positive. The angle of loll in this case is 18°, the range of stability is 18–90°, and the angle of vanishing stability is 90° (for an explanation of angle of loll, see Chapter 16). Note how the negative GM is plotted at 57.3°.

Example 1

Using the Stability Cross Curves for M.V. 'Tanker', plot the curve of statical stability when the displacement is 33,500 tonnes and KG = 9.3 meters. From the curve find the following:

(a) The range of stability.
(b) The angle of vanishing stability.
(c) The maximum righting lever and the angle of heel at which it occurs.
(d) The initial metacentric height.
(e) The moment of statical stability at 25° heel.
(f) Angle of heel at which deck edge just becomes immersed.

Use the GZ Cross Curves of Stability shown in Figure 21.1.

Heel	Tabulated GZ (KG = 9 m)	Correction to GZ (GG_1 sin Heel)	Required GZ (KG = 9.3 m)
0°	0	0 × 0 = 0	= 0
15°	0.90	0.3 × 0.259 = 0.08	0.90 − 0.08 = 0.82
30°	2.15	0.3 × 0.500 = 0.15	2.15 − 0.15 = 2.00
45°	2.55	0.3 × 0.707 = 0.21	2.55 − 0.21 = 2.34
60°	1.91	0.3 × 0.866 = 0.26	1.91 − 0.26 = 1.65
75°	0.80	0.3 × 0.966 = 0.29	0.80 − 0.29 = 0.51
90°	−0.50	0.3 × 1.000 = 0.30	−0.50 − 0.30 = −0.80

For the graph see Figure 21.6(a).

Answers from the curve in Figure 21.6(a):

(a) Range of stability 0–81°.

(b) Angle of vanishing stability 81°.
(c) Maximum GZ = 2.35 m occurring at 43° heel.
(d) GM is 2.30 m.
(e) GZ at 25° heel = 1.64 m.
(f) Deck edge just becomes immersed when angle of heel is 20°.

$$\begin{aligned}\text{Moment of statical stability} &= W \times GZ \\ &= 33{,}500 \times 1.64 \\ &= 54{,}940 \text{ tonnes m}\end{aligned}$$

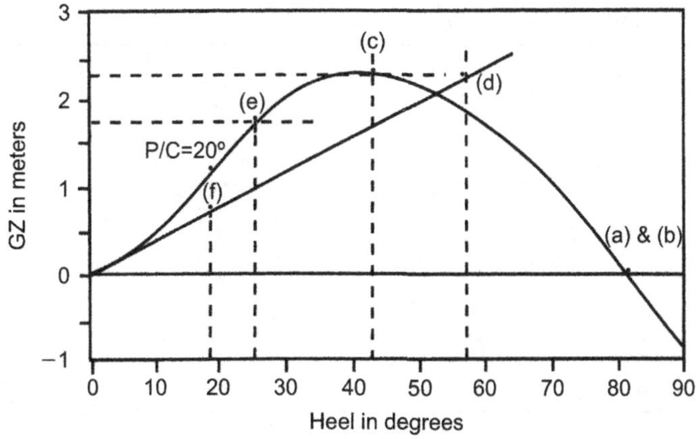

Figure 21.6a

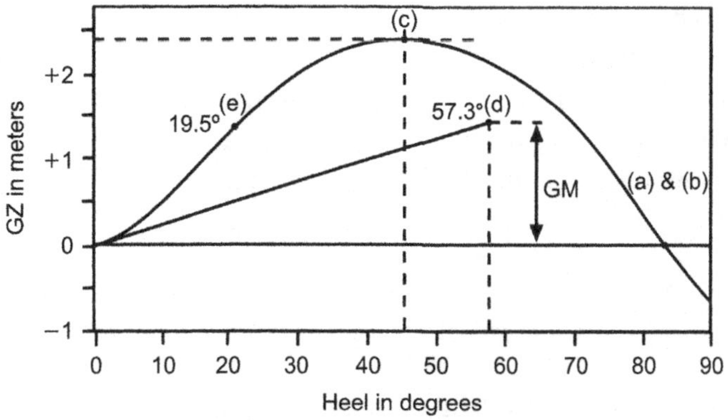

Figure 21.6b:
The Curve of Statical Stability.

Example 2

Construct the curve of statical stability for the M.V. 'Cargo-Carrier' when the displacement is 35,000 tonnes and KG is 9 meters. From the curve you have constructed find the following:

(a) The range of stability.
(b) The angle of vanishing stability.
(c) The maximum righting lever and the angle of the heel at which it occurs.
(d) The approximate initial metacentric height.
(e) Angle of heel at which deck edge just becomes immersed.

From the Stability Cross Curves:

Heel (θ)	KN	sin θ	KG sin θ	GZ = KN − KG sin θ
5°	0.9	0.087	0.783	0.12
10°	2.0	0.174	1.566	0.43
15°	3.2	0.259	2.331	0.87
20°	4.4	0.342	3.078	1.32
30°	6.5	0.500	4.500	2.00
45°	8.75	0.707	6.363	2.39
60°	9.7	0.866	7.794	1.91
75°	9.4	0.966	8.694	0.71
90°	8.4	1.000	9.000	−0.60

Use the KN Cross Curves of Stability shown in Figure 21.3(a).

Answers from the curve in Figure 21.6(b):

(a) Range of stability 0–83.75°.
(b) Angle of vanishing stability 83.75°.
(c) Maximum GZ = 2.39 m occurring at 45° heel.
(d) Approximate GM = 1.4 m.
(e) Deck edge just becomes immersed when angle of heel = 19.5°.

Exercise 21

1. Plot the curve of stability for M.V. 'Tanker' (Figure 21.1) when the displacement is 34,500 tonnes and KG = 9 m. From this curve find the approximate GM, the range of stability, the maximum GZ, and the angle of heel at which it occurs.

2. Plot the curve of statical stability for M.V. 'Tanker' (Figure 21.1) when the displacement is 23,400 tonnes and KG = 9.4 m. From this curve find the approximate GM, the maximum moment of statical stability, and the angle of heel at which it occurs. Find also the range of stability.
3. The displacement of M.V. 'Tanker' (Figure 21.1) is 24,700 tonnes and KG = 10 m. Construct a curve of statical stability and state what information may be derived from it. Find also the moments of statical stability at 10° and 40° heel.
4. Using the cross curves of stability for M.V. 'Tanker' (Figure 21.1):
   (a) Draw a curve of statical stability when the displacement is 35,000 tonnes, KG = 9.2 m, and KM = 11.2 m.
   (b) From this curve determine the moment of statical stability at 10° heel and state what other information may be obtained from the curve.
5. Plot the stability curve for M.V. 'Tanker' (Figure 21.1) when the displacement is 46,800 tonnes and KG = 8.5 m. From this curve find the approximate GM, the range of stability, the maximum GZ, and the angle of heel at which it occurs.
6. Construct the curve of statical stability for M.V. 'Cargo-Carrier' (Figure 21.3(a)) when the displacement is 35,000 tonnes and KG is 8 m. From this curve find:
   (a) The range of stability
   (b) The angle of vanishing stability
   (c) The maximum GZ and the heel at which it occurs.
7. Construct the curve of statical stability for M.V. 'Cargo-Carrier' (Figure 21.3(a)) when the displacement is 28,000 tonnes and the KG is 10 m. From the curve find:
   (a) The range of stability
   (b) The angle of vanishing stability
   (c) The maximum GZ and the heel at which it occurs.
8. A vessel is loaded up ready for departure. KM is 11.9 m. KG is 9.52 m with a displacement of 20,550 tonnes. From the ship's Cross Curves of Stability, the GZ ordinates for a displacement of 20,550 tonnes and a VCG of 8 m above base are as follows:

Angle of heel (θ)	0°	15°	30°	45°	60°	75°	90°
GZ ordinate (m)	0	1.10	2.22	2.60	2.21	1.25	0.36

   Using this information, construct the ship's statical stability curve for this condition of loading and determine the following:
   (a) Maximum righting lever GZ
   (b) Angle of heel at which this maximum GZ occurs
   (c) Angle of heel at which the deck edge just becomes immersed
   (d) Range of stability.

9. Using the table of KN ordinates below, calculate the righting levers for a ship when her displacement is 40,000 tonnes and her actual KG is 10 m. Draw the resulting Statical Stability Curve and from it determine:
   (a) Maximum GZ value
   (b) Approximate GM value
   (c) Righting moment at an angle of heel of 25°
   (d) Range of stability.

Angle of heel (θ)	0°	5°	10°	15°	20°	30°	45°	60°
KN ordinates (m)	0	0.90	1.92	3.11	4.25	6.30	8.44	9.39
Angle of heel (θ)	75°	90°						
KN ordinates (m)	9.29	8.50						

■

Effects of Beam and Freeboard on Stability

To investigate the effect of beam and freeboard on stability, it will be necessary to assume the stability curve for a particular vessel in a particular condition of loading. Let curve A in Figure 22.1 represent the curve of stability for a certain box-shaped vessel whose deck edge becomes immersed at about 17° heel.

The Effect of Increasing the Beam

Let the draft, freeboard, and KG remain unchanged, but increase the beam and consider the effect this will have on the stability curve.

For a ship-shaped vessel BM = I/V and for a box-shaped vessel $BM = B^2/12d$. Therefore, an increase in beam will produce an increase in BM. Hence the GM will also be increased, as will the righting levers at all angles of heel. The range of stability is also increased. The new curve of stability would appear as curve B in Figure 22.1.

It will be noticed that the curve, at small angles of heel, is much steeper than the original curve, indicating the increase in GM. Also, the maximum GZ and the range of stability have been increased whilst the angle of heel at which the deck edge becomes immersed has been reduced. The reason for the latter change is shown in Figure 22.2. Angle θ is reduced from 17° to 12°.

Figure 22.2(a) represents the vessel in her original condition with the deck edge becoming immersed at about 17°. The increase in the beam, as shown in Figure 22.2(b), will result in the

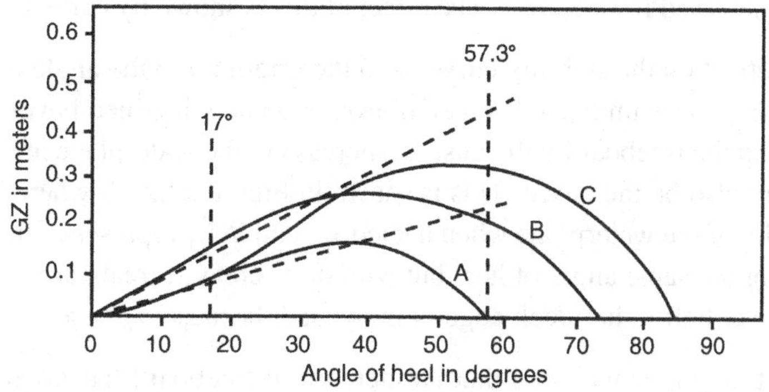

Figure 22.1

Ship Stability for Masters and Mates. http://dx.doi.org/10.1016/B978-0-08-097093-6.00022-0

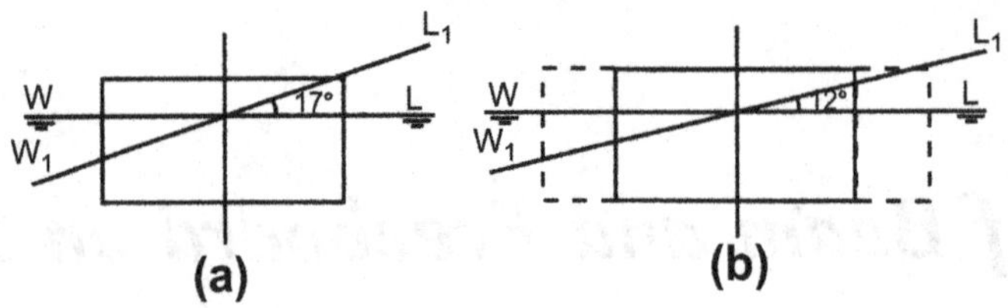

Figure 22.2 a,b

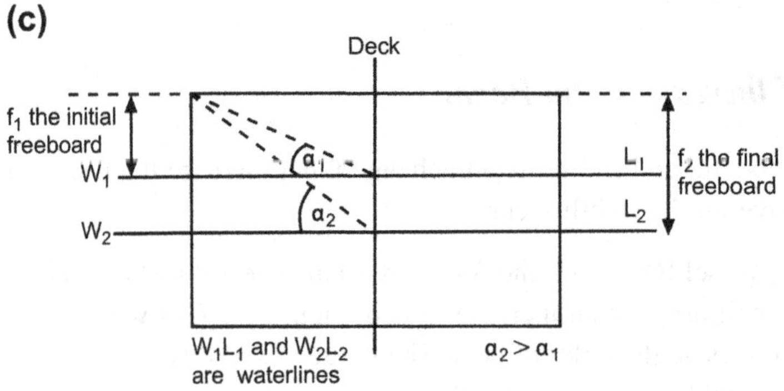

Figure 22.2c

deck edge becoming immersed at a smaller angle of heel. When the deck edge becomes immersed, the breadth of the waterplane will decrease and this will manifest itself in the curve by a reduction in the rate of increase of the GZs with increase in heel.

The Effect of Increasing the Freeboard

Now return to the original vessel. Let the draft, KG, and the beam remain unchanged, but let the freeboard be increased from f_1 to f_2. The effect of this is shown by curve C in Figure 22.1.

There will be no effect on the stability curve from the origin up to the angle of heel at which the original deck edge was immersed. When the vessel is now inclined beyond this angle of heel, the increase in the freeboard will cause an increase in the waterplane area and, thus, the righting levers will also be increased. This is shown in Figure 22.2(c), where WL represents the original breadth of the waterplane when heeled x°, and WL_1 represents the breadth of the waterplane area for the same angle of heel but with the increased freeboard. Thus, the vessel can heel further over before her deck edge is immersed, because $\alpha_2 > \alpha_1$.

From the above it may be concluded that an increase in freeboard has no effect on the stability of the vessel up to the angle of heel at which the original deck edge became immersed, but beyond this angle of heel all of the righting levers will be increased in length. The maximum GZ and the angle at which it occurs will be increased, as also will be the range of stability.

Summary

With increased beam:

- GM_T and GZ increase.
- Range of stability increases.
- Deck edge immerses earlier.
- KB remains similar.

With increased freeboard:

- GM_T and GZ increase.
- Range of stability increases.
- Deck edge immerses later at greater θ.
- KB decreases.

Dynamical Stability Relating to Statical Stability Curves

Dynamical stability is defined as the work done in inclining a ship. Consider the ship shown in Figure 23.1. When the ship is upright the force 'W' acts upwards through B and downwards through G. These forces act throughout the inclination, b = w.

$$\text{Work done} = \text{Weight} \times \text{Vertical separation of G and B}$$

or

$$\text{Dynamical stability} = W \times (B_1Z - BG)$$

$$= W \times (B_1R + RZ - BG)$$

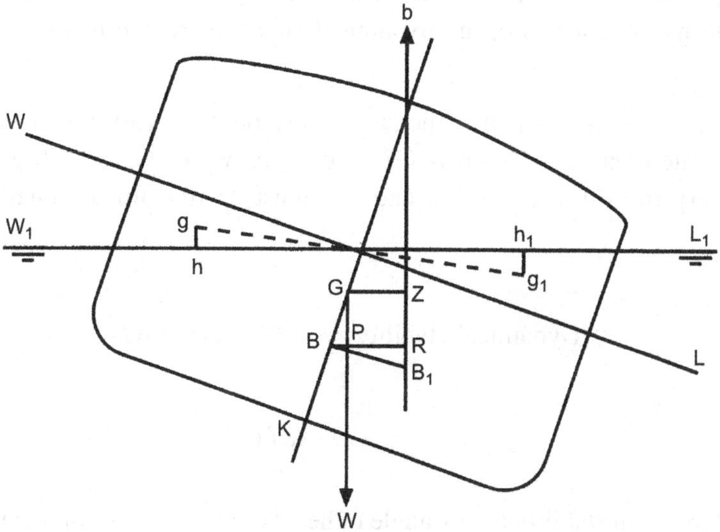

Figure 23.1

Ship Stability for Masters and Mates. http://dx.doi.org/10.1016/B978-0-08-097093-6.00023-2

$$= W \times \left[\frac{v(gh + g_1h_1)}{V} + PG - BG\right]$$

$$= W \times \left[\frac{v(gh + g_1h_1)}{V} + BG\cos\theta - BG\right]$$

$$\text{Dynamical stability} = W\left[\frac{v(gh + g_1h_1)}{V} - BG(1 - \cos\theta)\right]$$

This is known as *Moseley's formula* for dynamical stability.

If the curve of statical stability for a ship has been constructed the dynamical stability to any angle of heel may be found by multiplying the area under the curve to the angle concerned by the vessel's displacement, i.e.

$$\text{Dynamical stability} = W \times \text{Area under the stability curve}$$

The derivation of this formula is as follows:

Consider Figure 23.2(a), which shows a ship heeled to an angle θ. Now let the ship be heeled through a further very small angle $d\theta$. The center of buoyancy B_1 will move parallel to W_1L_1 to the new position B_2, as shown in Figure 23.2(b).

B_2Z_1 is the new vertical through the center of buoyancy and GZ_1 is the new righting arm. The vertical separation of Z and Z_1 is therefore $GZ \times d\theta$. But this is also the vertical separation of B and G. Therefore, the dynamical stability from θ to $(\theta + d\theta)$ is $W \times (GZ \times d\theta)$.

Refer now to Figure 23.2(c), which is the curve of statical stability for the ship. At θ the ordinate is GZ. The area of the strip is $GZ \times d\theta$. But $W \times (GZ \times d\theta)$ gives the dynamical stability from θ to $(\theta + d\theta)$, and this must be true for all small additions of inclination:

$$\therefore \text{Dynamical stability} = \int_0^\theta W \times GZ \times d\theta$$

$$= W\int_0^\theta GZ\, d\theta$$

Therefore, the dynamical stability to any angle of heel is found by multiplying the area under the stability curve to that angle by the displacement.

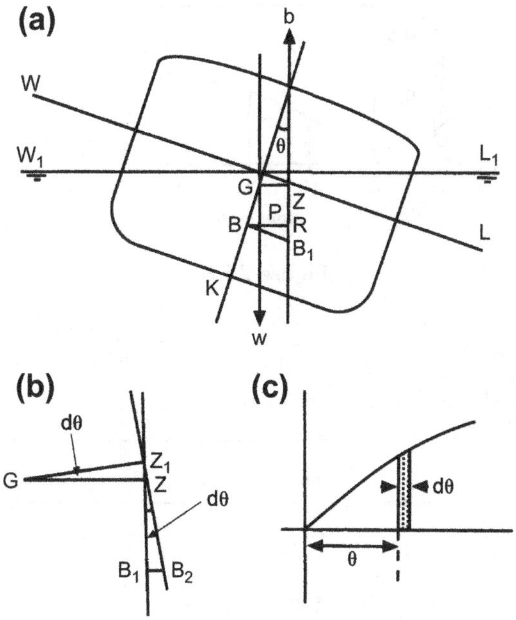

Figure 23.2

It should be noted that in finding the area under the stability curve by the use of Simpson's Rules, the common interval must be expressed in *radians*:

$$57.3^\circ = 1 \text{ radian}$$

$$1^\circ = \frac{1}{57.3} \text{ radians}$$

or

$$x^\circ = \frac{x}{57.3} \text{ radians}$$

Therefore, to convert degrees to radians, simply divide the number of degrees by 57.3.

Example 1

A ship of 5000 tonnes displacement has righting levers as follows (see Figure 23.3):

Angle of heel	0°	10°	20°	30°	40°
GZ (meters)	0	0.21	0.33	0.40	0.43

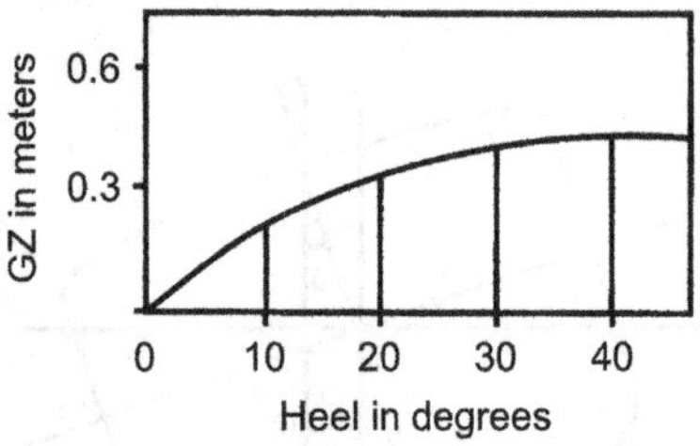

Figure 23.3

Calculate the dynamical stability to 40° heel.

GZ	SM	Functions of Area
0	1	0
0.21	4	0.84
0.33	2	0.66
0.40	4	1.60
0.43	1	0.43
		$3.53 = \Sigma_1$

$$h = 10°$$

$$h = \frac{10}{57.3} \text{ radians} = \text{Common interval (CI)}$$

$$\text{The area under the stability curve} = \frac{1}{3} \times \text{CI} \times \Sigma_1$$

$$= \frac{1}{3} \times \frac{10}{57.3} \times 3.53$$

$$= 0.2053 \text{ meter-radians}$$

$$\text{Dynamical stability} = \text{W} \times \text{Area under the stability curve}$$

$$= 5000 \times 0.2053$$

Ans. Dynamical stability = 1026.5 meter tonnes.

Example 2

A box-shaped vessel 45 m × 10 m × 6 m is floating in salt water at a draft of 4 m F and A. GM = 0.6 m (see Figure 23.4). Calculate the dynamical stability to 20° heel.

$$BM = \frac{B^2}{12d} \qquad \text{Displacement} = 45 \times 10 \times 4 \times 1.025 \text{ tonnes}$$

$$= \frac{10 \times 10}{12 \times 4} \qquad \text{Displacement} = 1845 \text{ tonnes}$$

$$BM = 2.08 \text{ m}$$

Note. When calculating the GZs 10° may be considered a small angle of heel, but 20° is a large angle of heel, and therefore the wall-sided formula must be used to find the GZ.

GZ	SM	Products for Area
0	1	0
0.104	4	0.416
0.252	1	0.252
		$0.668 = \Sigma_1$

GZ ordinates are derived below.

At 10° heel:

$$GZ = GM \times \sin\theta$$

$$= 0.6 \times \sin 10°$$

$$GZ = 0.104 \text{ m}$$

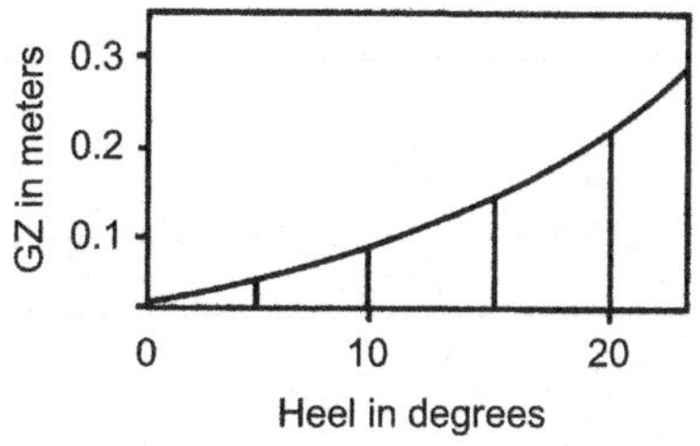

Figure 23.4

At 20° heel:

$$GZ = \left(GM + \frac{1}{2}BM\ \tan^2\theta\right)\ \sin\theta$$

$$= \left(0.6 + \frac{1}{2} \times 2.08 \times \tan^2 20°\right)\ \sin 20°$$

$$= (0.6 + 0.138)\ \sin 20°$$

$$= 0.738 \sin 20°$$

$$GZ = 0.252\ \text{m}$$

$$\text{Area under the curve} = \frac{1}{3} \times CI \times \Sigma_1$$

$$= \frac{1}{3} \times \frac{10}{57.3} \times 0.668$$

$$\text{Area under the curve} = 0.0389\ \text{meter-radians}$$

$$\text{Dynamical stability} = W \times \text{Area under the curve}$$

$$= 1845 \times 0.0389$$

Ans. Dynamical stability = 71.77 m tonnes.

Exercise 23

1. A ship of 10,000 tonnes displacement has righting levers as follows:

Heel	10°	20°	30°	40°
GZ (m)	0.09	0.21	0.30	0.33

   Calculate the dynamical stability to 40° heel.

2. When inclined, a ship of 8000 tonnes displacement has the following righting levers:

Heel	15°	30°	45°	60°
GZ (m)	0.20	0.30	0.32	0.24

Calculate the dynamical stability to 60° heel.

3. A ship of 10,000 tonnes displacement has the following righting levers when inclined:

Heel	0°	10°	20°	30°	40°	50°
GZ (m)	0.0	0.02	0.12	0.21	0.30	0.33

Calculate the dynamical stability to 50° heel.

4. A box-shaped vessel 42 m × 6 m × 5 m is floating in salt water on an even keel at 3 m draft and has KG = 2 m. Assuming that the KM is constant, calculate the dynamical stability to 15° heel.
5. A box-shaped vessel 65 m × 10 m × 6 m is floating upright on an even keel at 4 m draft in salt water. GM = 0.6 m. Calculate the dynamical stability to 20° heel.

■

CHAPTER 24

Changes in Statical Stability Relating to Wave Profiles — Loss of Quasi-Static Stability

Figure 24.1(a) shows a Statical Stability curve for a trimmed vessel in calm water conditions. The stability values are very satisfactory. The seaworthiness looks to be good.

Figure 24.1(b) shows a vessel with wave crests at the ends of the vessel. There is a wave trough at or very near to amidships. The stability of this condition is that it is better than the previous condition in calm water. The larger GZ values and the greater range of stability give indications of this.

Figure 24.1(c) shows a vessel with wave troughs at the ends of the vessel. There is a wave crest at or very near to amidships. The stability of this condition is that it is less safe than the two conditions considered previously. The much lower GZ values and the much smaller range of stability give indications of this. It would take very little to capsize this vessel.

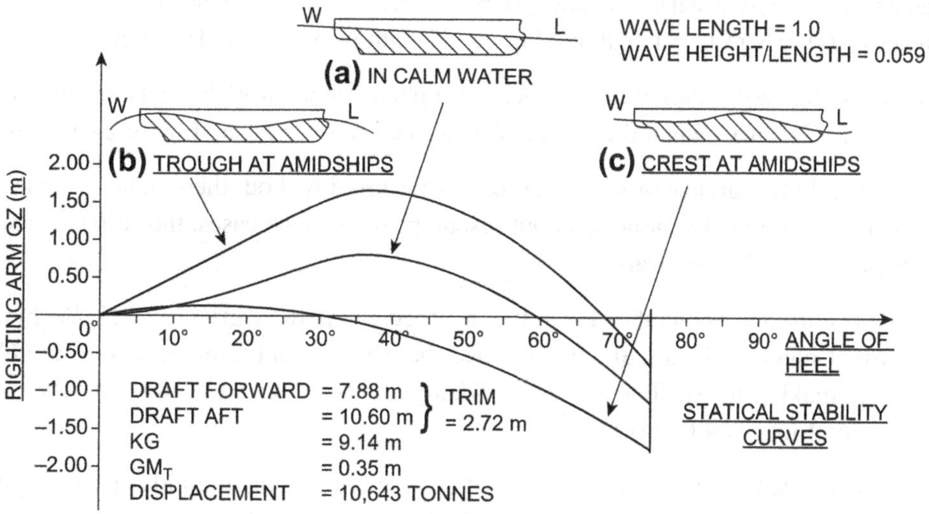

Figure 24.1:
GZ Curves in Calm Water and with Wave Passing the Ship.

Ship Stability for Masters and Mates. http://dx.doi.org/10.1016/B978-0-08-097093-6.00024-4

Table 24.1: Statical Stability curve readings for Figure 24.1

Type of Wave Profile	Maximum GZ (m)	Range of Stability	Angle of Heel when Max GZ Occurs	Deck Edge Flooded Angle of Heel
Calm water	0.78	0–59.0°	37°	23°
Trough at amidships	1.63	0–68.5°	40°	25°
Crest at amidships	0.12	0–29.5°	15°	5°

Loss of stability typically occurs in following and stern quartering seas with a low encounter frequency, where the ship's speed lies between the group and the phase velocities of the waves.

When a wave of a certain length and steepness is slowly overtaking the ship, is at or very near the amidships position, and the ship's ends are in a wave trough, the transverse moment of inertia of the waterline will decrease (see Figure 24.1(c)). This causes a lowering of the BM_T. Consequently, there is a reduction in the righting lever GZ, a reduction in GM_T, and reductions in accompanying large heel angles. If this reduction in GZ is such that GZ becomes negative, then a sudden capsize may result. The opposite applies when the wave trough is at amidships; the ship's stability will increase (see Figure 24.1(b)).

Critical waves tend to be fairly steep, having amplitude similar to the ship's freeboard and with a wavelength similar to the ship's length.

Capsize due to quasi-steady loss of stability is usually preceded by moderate rolling motions. This can occur suddenly when the wavelength is between 0.8L and 2.0L and the wave steepness H/λ exceeds 0.04, where L = length of ship, λ = length of wave, and H = height of wave.

Some stability standards attempt to consider these phenomena through the assessment of the decrease in the GZ curve when the vessel is balanced on a wave (see Figure 24.1(a)–(c)).

In the past, the disappearance of several ships has occurred without there being time to send out an emergency signal, let alone send out rescue boats. In such cases, the ship was usually lost along with everyone on board.

Such a case occurred in September 1980. It concerned a bulk ore carrier, the *MV Derbyshire*, en route from Canada to Japan. She had radioed that she was in trouble in The South China Seas and was making for shallow waters. Suddenly all communication stopped. The vessel sank with a total loss of 44 persons.

The question arises: Could this be the result of a wave crest being at amidships and causing a very sudden large decrease in the vessel's Statical Stability? See Table 24.1 and compare the values therein.

CHAPTER 25

Hydrostatic Curves and Values for Vessels Initially on Even Keel

Hydrostatic Curves

Hydrostatic information is usually supplied to the ship's officer in the form of a table or a graph. Figure 25.1 shows the hydrostatic curves for the imaginary ship M.V. 'Tanker'. The various items of hydrostatic information are plotted against draft.

When information is required for a specific draft, first locate the draft on the scale on the left-hand margin of the figure. Then draw a horizontal line through the draft to cut all of the curves

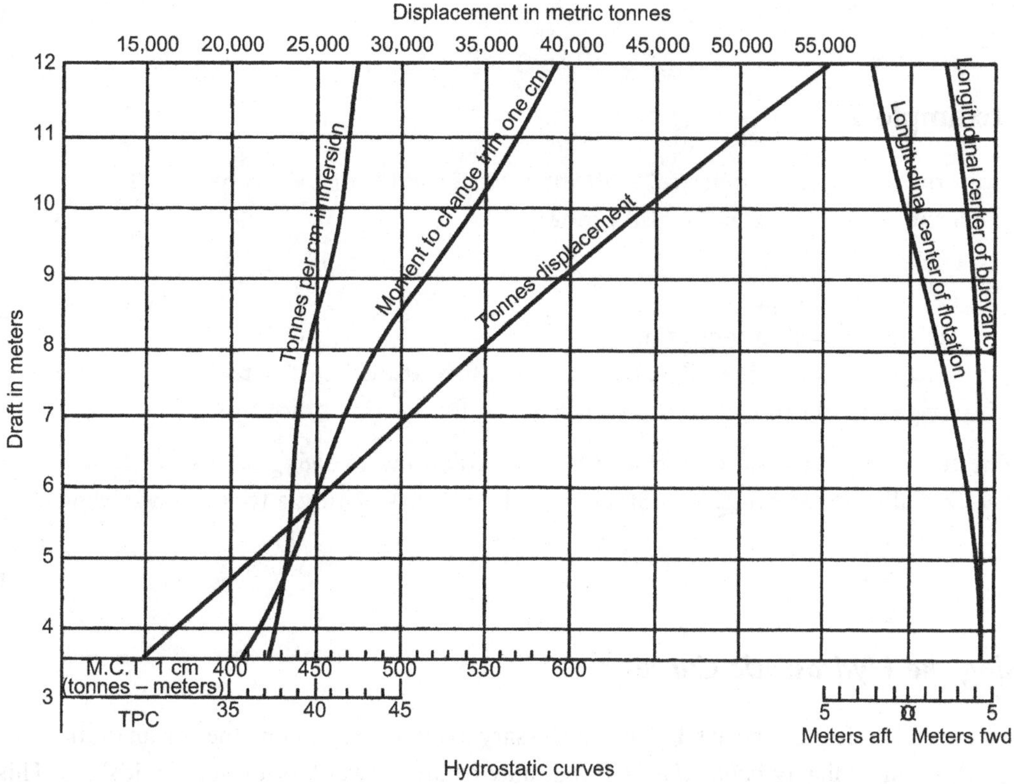

Figure 25.1

Ship Stability for Masters and Mates. http://dx.doi.org/10.1016/B978-0-08-097093-6.00025-6

on the figure. Next draw a perpendicular through the intersections of this line with each of the curves in turn and read off the information from the appropriate scale.

Example 1

Using the hydrostatic curves for M.V. 'Tanker' in Figure 25.1, take off all of the information possible for the ship when the mean draft is 7.6 meters.

1. TPC = 39.3 tonnes.
2. MCT 1 cm = 475 tonnes m.
3. Displacement = 33,000 tonnes.
4. Longitudinal center of flotation is 2.2 m forward of amidships.
5. Longitudinal center of buoyancy is 4.0 m forward of amidships.

When information is required for a specific displacement, locate the displacement on the scale along the top margin of the figure and drop a perpendicular to cut the curve marked 'Displacement'. Through the intersection draw a horizontal line to cut all of the other curves and the draft scale. The various quantities can then be obtained as before.

Example 2

From the hydrostatic curves take off the information for M.V. 'Tanker' in Figure 25.1 when the displacement is 37,500 tonnes.

1. Draft = 8.55 m.
2. TPC = 40 tonnes.
3. MCT 1 cm = 500 tonnes m.
4. Longitudinal center of flotation is 1.2 m forward of amidships.
5. Longitudinal center of buoyancy is 3.7 m forward of amidships.

The curves themselves are produced from calculations involving Simpson's Rules. These involve half-ordinates, areas, moments, and moments of inertia for each waterline under consideration.

Using the Hydrostatic Curves

After the end drafts have been taken it is necessary to interpolate to find the 'mean draft'. This is the draft immediately *below the LCF* that may be aft, forward, or even at amidships. This draft can be labeled d_H.

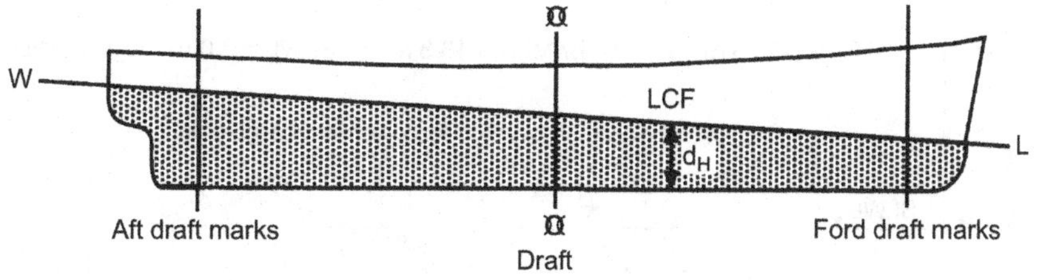

Figure 25.2

If d_H is taken as being simply the average of the two end drafts, then in large full-form vessels (supertankers) and fine-form vessels (container ships) an appreciable error in the displacement can occur (see Figure 25.2).

Let us assume the true mean draft 'd_H' is 6 m. The Naval Architect or mate on board ship draws a horizontal line parallel to the SLWL at 6 m on the vertical axis right across all of the hydrostatic curves. At each intersection with a curve and this 6 m line, he or she projects downwards and reads off on the appropriate scale on the 'x' axis.

From the hydrostatic curves in Figure 25.3, at a mean draft of 6 m, for example, we would obtain the following:

TPC = 19.70 t	Displacement = 10,293 t
MCTC = 152.5 tm/cm	$LCF_{amidships}$ = 0.05 m forward amidships
$LCB_{amidships}$ = 0.80 m forward amidships	KM_L = 207.4 m
KM_T = 7.46 m	

These values can then be used to calculate the *new end drafts* and *transverse stability*, if weights are *added* to the ship, *discharged* from the ship, or simply *moved* longitudinally or transversely within the ship.

$LCF_{amidships}$ and $LCB_{amidships}$ are distances measured from amidships, or forward of the aft perp (foap).

Nowadays these values can be put on a spreadsheet in a computer package. When the hydrostatic draft d_H is keyed, the hydrostatic values appertaining to this draft are then displayed, ready for use.

A set of hydrostatic values has been calculated for a 135.5 m general cargo ship of about 10,000 tonnes deadweight. These are shown in Table 25.1. From those values a set of hydrostatic curves were drawn. These are shown in Figure 25.3.

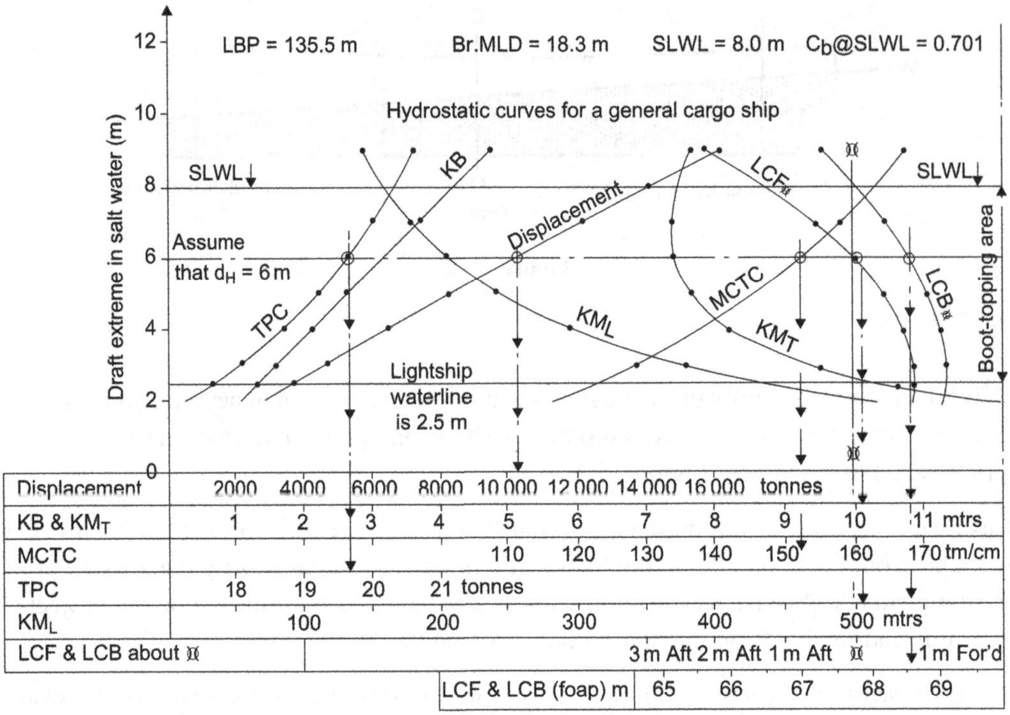

Figure 25.3:
Hydrostatic Curves Based on Values in Table 25.1.

Table 25.1: Hydrostatic values for Figure 25.3 (these are for a 135.5 m LBP general cargo ship).

Draft d_H (see below) (m)	TPC (tonnes)	KB (m)	Displacement (tonnes)	KM_L (m)	MCTC (tm/cm)	KM_T (m)	$LCF_{amidships}$ (m)	$LCB_{amidships}$ (m)
9	20.64	4.80	16,276	146.5	167.4	7.71	2.10 aft	0.45 aft
8	20.36	4.27	14,253	161.8	162.9	7.54	1.20 aft	Amidships
7	20.06	3.74	12,258	181.5	158.1	7.44	0.50 aft	0.45 forward
6	19.70	3.21	10,293	207.4	152.5	7.46	0.05 forward	0.80 forward
5	19.27	2.68	8361	243.2	145.9	7.70	0.42 forward	1.05 forward
4	18.76	2.15	6486	296.0	138.3	8.28	0.70 forward	1.25 forward
3	18.12	1.61	4674	382.3	129.1	9.56	0.83 forward	1.30 forward
2.5	17.69	1.35	3785	449.0	123.0	10.75	0.85 forward	1.30 forward

Note. LCF and LCB are now measured forward of aft perp (foap).

Exercise 25

1. From the hydrostatic curves for M.V. 'Tanker', find the mean draft when entering a dock where the density of the water is 1009 kg per cubic meter, if the mean draft in salt water is 6.5 m.
2. Find the mean draft of M.V. 'Tanker' in dock water of density 1009 kg per cubic meter when the displacement is 35,400 tonnes.
3. M.V. 'Tanker' is 200 m long and is floating at the maximum permissible draft aft. There is one more parcel of heavy cargo to load on deck to put her down to her marks at the load displacement of 51,300 tonnes. Find the position, relative to amidships, to load the parcel so as to maintain the maximum possible trim by the stern.
4. Using the hydrostatic curves for M.V. 'Tanker', find the displacement, MCT 1 cm, and the TPC when the ship is floating in salt water at a mean draft of 9 m. Also find the new mean draft if she now enters a dock where the density of the water is 1010 kg per cubic meter.
5. M.V. 'Tanker' is 200 m long and has a light displacement of 16,000 tonnes. She has on board 100 tonnes of cargo, 300 tonnes of bunkers, and 100 tonnes of fresh water and stores. The ship is trimmed 1.5 m by the stern. Find the new drafts if the 100 tonnes of cargo already on board are now shifted 50 m forward.

CHAPTER 26

Hydrostatic Curves and Values for Vessels Initially Having Trim by the Bow or by the Stern

Hydrostatics for the design condition of the vessel *MOANA* are considered:

DELFTload demo vessel	DELFT ship
Designer	M. van Engeland
Created by	Used for demonstration purposes
Filename	DELFTload demo.fbm

Design length	84.950 m	Midship location	42.475 m
Length overall	89.011 m	Relative water density	1.025
Design beam	15.250 m	Mean shell thickness	0.0110 m
Maximum beam	15.251 m	Appendage coefficient	1.0005
Design draft	5.633 m	General cargo/container vessel.	

The general particulars of the *MOANA* are listed in Table 26.1.

$$C_b = (\text{Displacement in tonnes/density})/(L \times B \times d)$$

$$C_b = (95645.2/1)/(84.95 \times 15.25 \times 5.633) = 0.7547 \text{ as given above.}$$

The *MOANA*'s displacement and KG are assumed to remain at constant values for conditions of even keel and trimmed situations.

If when static, the vessel is *trimming by the stern*, then when compared to being on even keel:

- LCB and LCF move forward
- KM_T decreases, so stability decreases
- MCTC decreases
- TPC decreases.

As trim by the stern was increased from 0.50 to 1.00 m, LCB and LCF both moved forward. KM_T, MCTC, and TCP values all decreased (see Tables 26.2–26.10 and Figure 26.1).

Ship Stability for Masters and Mates. http://dx.doi.org/10.1016/B978-0-08-097093-6.00026-8

Table 26.1: General particulars for the *MOANA*.

Molded volume	5483.8 m^3
Total displaced volume	5507.5 m^3
Displacement	5645.2 tonnes
Block coefficient C_b	0.7547
Prismatic coefficient C_p	0.7645
Midship area	84.802 m^2
Wetted surface area	1900.5 m^2
Longitudinal center of buoyancy	41.593 m foap
Longitudinal center of buoyancy	−1.021% L
GM_T	6.892 m
Vertical center of buoyancy	3.063 m
Length on waterline	86.460 m
Beam on waterline	15.251 m
Entrance angle	88.492°
Waterplane area (WPA)	1192.2 m^2
WPA coefficient	0.9203
Midship area coefficient C_m	0.9872
Transverse moment of inertia	21,003 m^4
Longitudinal moment of inertia	626,867 m^4
LCF position foap	39.118 m
GM_L	117.37 m

foap denotes forward of aft perp.

If when static, the vessel is *trimming by the bow*, then when compared to being on even keel:

- LCB and LCF move aft
- KM_T increases, so stability increases

Table 26.2: Hydrostatic particulars for the MOANA when on even keel at zero speed.

Draft (meters)	Mld Volume (m^3)	Extreme Volume (m^3)	Displacement in FW (tonnes)	Displacement in SW (tonnes)	LCB foap (m)	LCF foap (m)	KM_T (meters)	MCTC (t-m/cm)	TPC in SW (tonnes)
2.00	1612	1625	1625	1666	43.44	43.12	10.221	40.845	9.60
2.05	1659	1672	1672	1714	43.43	43.10	10.052	41.167	9.63
2.10	1706	1719	1719	1762	43.42	43.08	9.890	41.480	9.67
2.15	1753	1766	1766	1810	43.41	43.06	9.737	41.788	9.70
2.20	1800	1814	1814	1859	43.40	42.95	9.622	42.404	9.76
2.25	1848	1862	1862	1908	43.39	42.92	9.487	42.750	9.80
2.30	1896	1910	1910	1957	43.37	42.89	9.358	43.096	9.83
2.35	1944	1958	1958	2007	43.36	42.86	9.235	43.446	9.87
2.40	1992	2006	2006	2056	43.35	42.82	9.119	43.801	9.90
2.45	2040	2054	2054	2106	43.34	42.78	9.007	44.161	9.94

foap denotes forward of aft perp.

- MCTC increases
- TPC increases.

As trim by the bow was increased from 0.50 to 1.00 m, LCB and LCF both moved aft. KM_T, MCTC, and TCP values all increased (see Tables 26.2–26.10 and Figure 26.1).

Table 26.3: Hydrostatic particulars for the *MOANA* when trimming 0.50 m by the stern at zero speed.

Draft (meters)	Mld Volume (m^3)	Extreme Volume (m^3)	Displacement in FW (tonnes)	Displacement in SW (tonnes)	LCB foap (m)	LCF foap (m)	KM_T (meters)	MCTC (t-m/cm)	TPC in SW (tonnes)
2.00	1612	1625	1625	1666	44.66	43.46	10.126	40.752	9.57
2.05	1659	1672	1672	1714	44.62	43.43	9.962	41.075	9.61
2.10	1706	1719	1719	1762	44.59	43.40	9.806	41.389	9.65
2.15	1753	1766	1766	1810	44.56	43.38	9.658	41.695	9.68
2.20	1800	1814	1814	1859	44.53	43.35	9.517	41.991	9.71
2.25	1848	1862	1862	1908	44.50	43.32	9.381	42.281	9.74
2.30	1896	1910	1910	1957	44.47	43.30	9.253	42.563	9.77
2.35	1944	1958	1958	2007	44.44	43.19	9.156	43.130	9.83
2.40	1992	2006	2006	2056	44.41	43.16	9.042	43.447	9.86
2.45	2040	2054	2054	2106	44.38	43.12	8.934	43.758	9.90

foap denotes forward of aft perp.

Table 26.4: Hydrostatic particulars for the *MOANA* when trimming 1.00 m by the stern at zero speed.

Draft (meters)	Mld Volume (m^3)	Extreme Volume (m^3)	Displacement in FW (tonnes)	Displacement in SW (tonnes)	LCB foap (m)	LCF foap (m)	KM_T (meters)	MCTC (t-m/cm)	TPC in SW (tonnes)
2.00	1612	1625	1625	1666	45.87	43.81	10.028	40.493	9.54
2.05	1659	1672	1672	1714	45.81	43.77	9.869	40.820	9.58
2.10	1706	1719	1719	1762	45.75	43.73	9.718	41.141	9.61
2.15	1753	1766	1766	1810	45.70	43.69	9.575	41.452	9.65
2.20	1800	1814	1814	1859	45.64	43.65	9.438	41.755	9.68
2.25	1848	1862	1862	1908	45.59	43.62	9.308	42.043	9.71
2.30	1896	1910	1910	1957	45.54	43.58	9.184	42.319	9.75
2.35	1944	1958	1958	2007	45.50	43.55	9.065	42.584	9.78
2.40	1992	2006	2006	2056	45.45	43.51	8.951	42.840	9. 804
2.45	2040	2054	2054	2106	45.40	43.48	8.843	43.088	9.83

foap denotes forward of aft perp.

It is important to note that at the lower drafts, the changes in KM_T were greater than at higher drafts for the *MOANA*. Consequently, depending on which way the vessel was trimming, the stability was increased or decreased at these lower drafts.

It must be stressed that these results and conclusions are only for the *MOANA*. Other ship forms may yield different conclusions. However, they are indicative that trimming a vessel does change, to some degree, the stability safety of a ship.

Table 26.5: Hydrostatic particulars for the *MOANA* when trimming 0.50 m by the bow at zero speed.

Draft (meters)	Mld Volume (m^3)	Extreme Volume (m^3)	Displacement in FW (tonnes)	Displacement in SW (tonnes)	LCB foap (m)	LCF foap (m)	KM_T (meters)	MCTC (t-m/cm)	TPC in SW (tonnes)
2.00	1612	1625	1625	1666	42.21	42.59	10.341	41.113	9.64
2.05	1659	1672	1672	1714	42.22	42.67	10.171	41.488	9.68
2.10	1706	1719	1719	1762	42.23	42.64	10.010	41.865	9.72
2.15	1753	1766	1766	1810	42.24	42.61	9.857	42.248	9.76
2.20	1800	1814	1814	1859	42.25	42.58	9.711	42.635	9.80
2.25	1848	1862	1862	1908	42.26	42.54	9.572	43.027	9.83
2.30	1896	1910	1910	1957	42.27	42.51	9.441	43.428	9.87
2.35	1944	1958	1958	2007	42.27	42.47	9. 315	43.833	9.91
2.40	1992	2006	2006	2056	42.28	42.43	9.196	44.245	9.95
2.45	2040	2054	2054	2106	42.28	42.39	9.082	44.665	9.99

foap denotes forward of aft perp.

Table 26.6: Hydrostatic particulars for the *MOANA* when trimming 1.00 m by the bow at zero speed.

Draft (meters)	Mld Volume (m^3)	Extreme Volume (m^3)	Displacement in FW (tonnes)	Displacement in SW (tonnes)	LCB foap (m)	LCF foap (m)	KM_T (meters)	MCTC (t-m/cm)	TPC in SW (tonnes)
2.00	1612	1625	1625	1666	40.97	42.28	10.440	41.327	9.67
2.05	1659	1672	1672	1714	41.01	42.25	10. 267	41.666	9.71
2.10	1706	1719	1719	1762	41.04	42.22	10.103	42.098	9.75
2.15	1753	1766	1766	1810	41.07	42.18	9. 947	42.536	9.80
2.20	1800	1814	1814	1859	41.10	42.15	9.798	42.982	9.84
2.25	1848	1862	1862	1908	41.13	42.11	9.657	43.432	9.88
2.30	1896	1910	1910	1957	41.15	42.07	9.523	43.890	9.92
2.35	1944	1958	1958	2007	41.18	42.02	9. 395	44.356	9.96
2.40	1992	2006	2006	2056	41.20	41.98	9.274	44.832	10.00
2.45	2040	2054	2054	2106	41.21	41.93	9.158	45.314	10.05

Table 26.7: KM_T values for the *MOANA* when trimming 1.00 m by the stern on even keel.

Ship speed (knots)	KM_T (meters)	KM_T (meters)	Difference (meters)
2.00	10.221	10.028	0.193
2.05	10.052	9.869	0.183
2.10	9.890	9.718	0.172
2.15	9.737	9.575	0.162
2.20	9.622	9.438	0.184
2.25	9.487	9.308	0.179
2.30	9.358	9.184	0.174
2.35	9.235	9.065	0.170
2.40	9.119	8.951	0.168
2.45	9.007	8.843	0.164

The larger the trim is by the stern, the greater the decrease in KM_T.
As KM_T decreases, the ship's stability is reduced.

Table 26.8: KM_T values for the *MOANA* when trimming 0.50 m by the stern on even keel.

Ship speed (knots)	KM_T (meters)	KM_T (meters)	Difference (meters)
2.00	10.221	10.126	−0.095
2.05	10.052	9.962	−0.090
2.10	9.890	9.806	−0.084
2.15	9.737	9.658	−0.079
2.20	9.622	9.517	−0.105
2.25	9.487	9.381	−0.106
2.30	9.358	9.253	−0.105
2.35	9.235	9.156	−0.079
2.40	9.119	9.042	−0.077
2.45	9.007	8.934	−0.073

The larger the trim is by the stern, the greater the decrease in KM_T.
As KM_T decreases, the ship's stability is reduced.

Table 26.9: KM_T values for the *MOANA* when trimming 1.00 m by the bow on even keel.

Ship speed (knots)	KM_T (meters)	KM_T (meters)	Difference (meters)
2.00	10.221	10.440	0.219
2.05	10.052	10.270	0.215
2.10	9.890	10.103	0.213
2.15	9.737	9.950	0.210
2.20	9.622	9.798	0.176
2.25	9.487	9.657	0.170
2.30	9.358	9.523	0.165
2.35	9.235	9.400	0.160
2.40	9.119	9.274	0.155
2.45	9.007	9.158	0.151

The larger the trim is by the bow, the greater the increase in KM_T.
As KM_T increases, the ship's stability is increased.

Table 26.10: KM_T values for the *MOANA* when trimming 0.50 m by the bow on even keel.

Ship speed (knots)	KM_T (meters)	KM_T (meters)	Difference (meters)
2.00	10.221	10.341	0.120
2.05	10.052	10.171	0.119
2.10	9.890	10.010	0.120
2.15	9.737	9.857	0.120
2.20	9.622	9.711	0.089
2.25	9.487	9.572	0.085
2.30	9.358	9.441	0.083
2.35	9.235	9.320	0.080
2.40	9.119	9.196	0.077
2.45	9.007	9.082	0.075

The larger the trim is by the bow, the greater the increase in KM_T.
As KM_T increases, the ship's stability is increased.

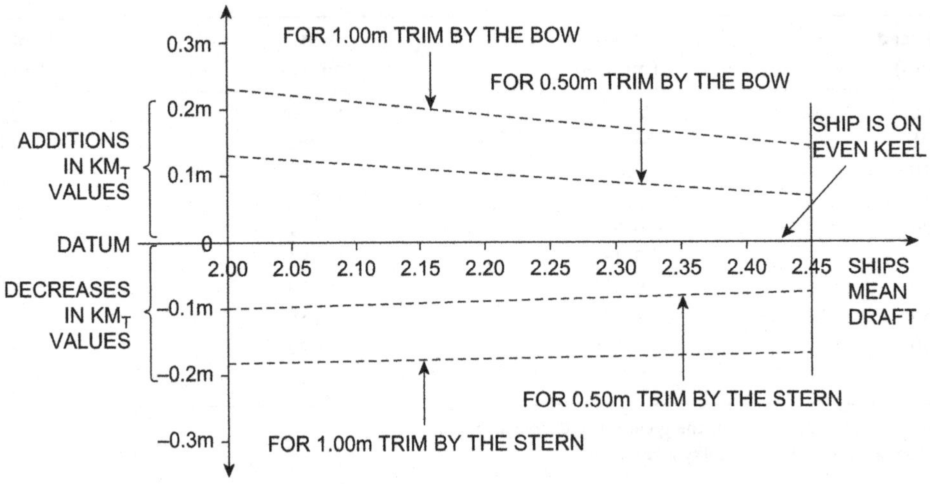

Figure 26.1: Increases and Decreases in KM_T for the *MOANA*.
Note. At lower drafts, changes in KMT are greater!

This strongly suggests that all vessels should have in their Trim and Stability Book some hydrostatic data not only for being on even keel, but also for trimmed conditions by the stern and by the bow.

CHAPTER 27

Increase in Draft Due to List

Box-Shaped Vessels

The draft of a vessel is the depth of the lowest point below the waterline. When a box-shaped vessel is floating upright on an even keel as in Figure 27.1(a) the draft is the same to port as to starboard. Let the draft be 'd' and let the vessel's beam be 'b'.

Now consider the same vessel when listed $\theta°$, as shown in Figure 27.1(b). The depth of the lowest point or draft is now increased to 'D' (xy).

In triangle OxA:

$$OA = \frac{1}{2}b \text{ and angle AxO} = 90° \qquad \therefore Ax = \frac{1}{2}b \sin \theta°$$

In triangle ABy:

$$AB = d \text{ and angle AyB} = 90° \qquad \therefore Ay = d \cos \theta°$$

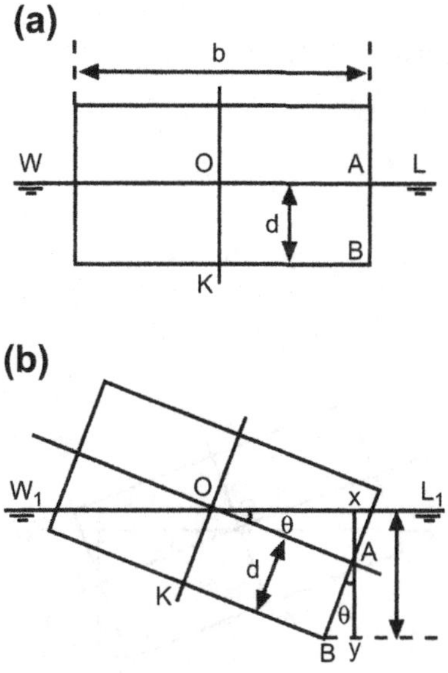

Figure 27.1

Ship Stability for Masters and Mates. http://dx.doi.org/10.1016/B978-0-08-097093-6.00027-X

$$D = xy$$
$$= Ax + Ay$$
$$D = \frac{1}{2} b \sin \theta + d \cos \theta$$

or

$$\text{New draft} = \frac{1}{2} \text{beam sin List} + \text{Old draft cos List}$$

Note. It will usually be found convenient to calculate the draft when listed by using the triangles AOx and ABy.

Example 1

A box-shaped ship with 12 m beam is floating upright at a draft of 6.7 m (see Figure 27.2(a)). Find the increase in draft if the vessel is now listed 18°.

In triangle OAx:

$$\text{Angle AOx} = 18^\circ$$

Therefore, use sin 18°:

$$\text{OA (Dist.)} = 6.0 \text{ m}$$
$$\text{Ax (Dep.)} = 1.85 \text{ m} = 6 \sin 18^\circ = 1.85 \text{ m}$$

In triangle ABy:

$$\text{Angle BAy} = 18^\circ$$

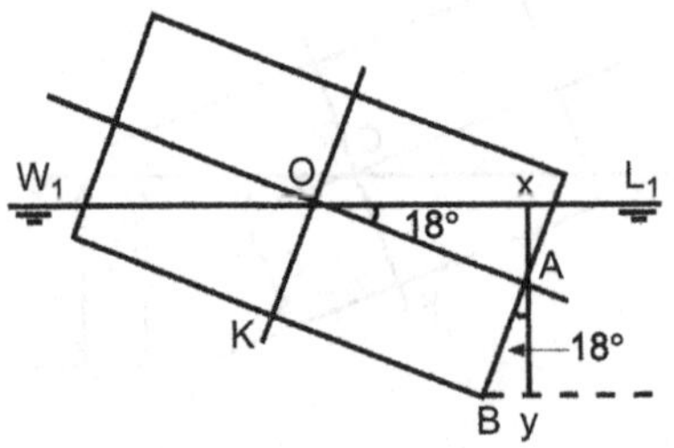

Figure 27.2a

Therefore, use cos 18°:

$$
\begin{aligned}
\text{AB (Dist.)} &= 6.70\ \text{m} \\
\text{Ay (D. Lat.)} &= 6.37\ \text{m} \\
\text{Ax} &= 1.85\ \text{m} \\
\text{Ay} &= +\underline{6.37\ \text{m}} \\
\text{New draft} &= 8.22\ \text{m} \\
\text{Old draft} &= -\underline{6.70\ \text{m}} \\
&\quad \underline{1.52\ \text{m}}
\end{aligned}
$$

Ans. Increase = 1.52 m.

Example 2

Using the dimensions of the ship given in Example 1, proceed to calculate the increase in draft at 3° intervals from 0° to 18°. Plot a graph of draft increase ∝ θ:

$$\frac{1}{2}\,\text{beam}\sin\theta = \frac{1}{2}\times 12\times\sin\theta = 6\sin\theta \qquad \text{(I)}$$

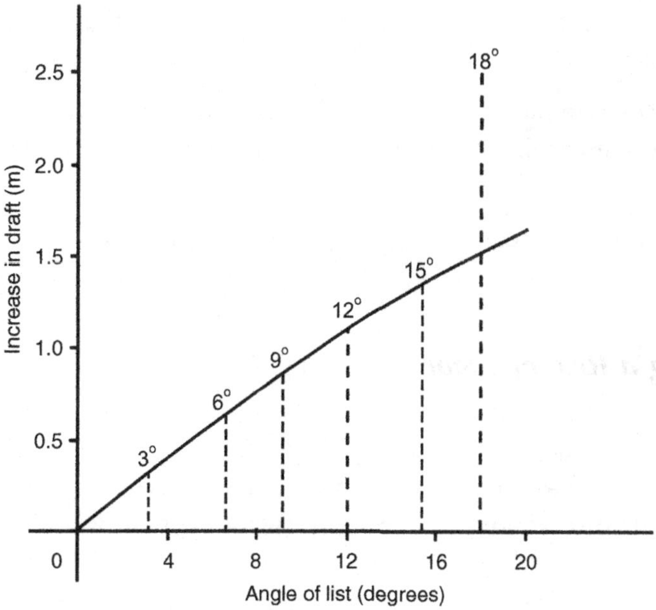

Figure 27.2b:
Draft Increase ∝ Angle of List θ for Example 2.

$$\text{Old draft} \cos\theta = 6.7\cos\theta \quad \text{(II)}$$

$$\text{Increase in draft} = \text{(I)} + \text{(II)} - \text{old draft in meter}$$

Angle of List	6 sin θ	6.7 cos θ	Old Draft	Increase in Draft (m)
0°	0	6.70	6.70	0
3°	0.31	6.69	6.70	0.30
6°	0.63	6.66	6.70	0.59
9°	0.94	6.62	6.70	0.86
12°	1.25	6.55	6.70	1.10
15°	1.55	6.47	6.70	1.32
18°	1.85	6.37	6.70	1.52

The above results clearly show the increase in draft or loss of underkeel clearance when a vessel lists.

Ships in the late 1990s and early 2000s were designed with shorter lengths and wider breadths mainly to reduce first cost and hogging/sagging characteristics. These wider ships are creating problems sometimes when initial underkeel clearance is only 10% of the ship's static draft. It only requires a small angle of list for them to go aground in way of the bilge strakes.

One such ship, for example, is the supertanker *Esso Japan*. She has an LBP of 350 m and a width of 70 m at amidships. Consequently, extra care is required should possibilities of list occur.

Figure 27.2(b) shows the requested graph of draft increase $\propto\ \theta$.

■

Vessels Having a Rise of Floor

■ Example 3

A ship has 20 m beam at the waterline and is floating upright at 6 m draft (see Figure 27.3). If the rise of floor is 0.25 m, calculate the new draft if the ship is now listed 15°.

In triangle OAx:

$$\text{Angle O} = 15^\circ$$

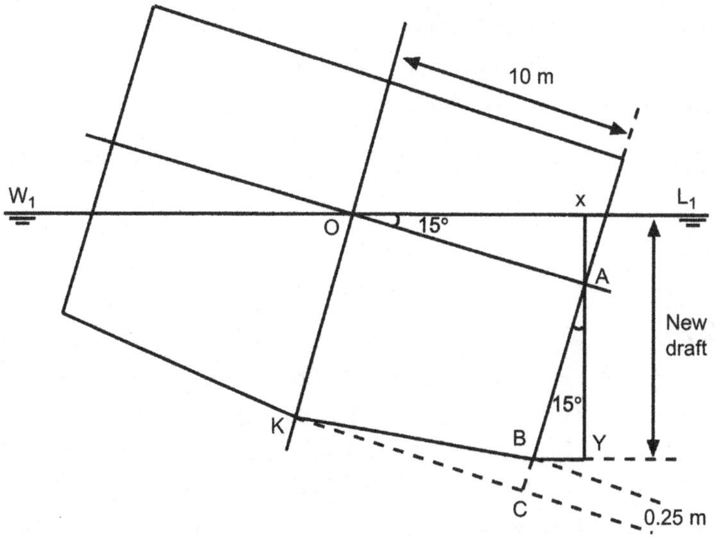

Figure 27.3

Therefore, use cos 15°:

$$AO\ (\text{Dist}) = 10\ \text{m}$$
$$Ax\ (\text{Dep.}) = 2.59\ \text{m} = OA \sin 15^\circ$$

In triangle ABY:

$$\text{Angle } A = 15^\circ$$

Therefore, use cos 15°:

$$\begin{aligned} AB\ (\text{Dist.}) &= AC - BC \\ &= \text{Old draft} - \text{Rise of floor} \\ &= 6\ \text{m} - 0.25\ \text{m} \\ AB\ (\text{Dist.}) &= 5.75\ \text{m} \\ AY\ (\text{D. Lat}) &= 5.55\ \text{m} = AB\cos 15^\circ \\ Ax &= 2.59\ \text{m} \\ AY &= +\underline{5.55\ \text{m}} \\ xY &= \underline{8.14\ \text{m}} \end{aligned}$$

Ans. New draft = 8.14 meters.

If the formula is to be used to find the new draft it must now be amended to allow for the rise of floor as follows:

$$\text{New draft} = \frac{1}{2}\text{beam} \sin \text{List} + (\text{Old draft} - \text{Rise}) \cos \text{List}$$

Note. In practice, the shipbuilder's naval architect usually calculates the increase in draft for the ship at various angles of list and supplies the information to the ship's officer by means of a table on the plans of the ship.

The rise of floor, similar to tumblehome on merchant ships, has almost become obsolete on merchant ships of today.

Exercise 27

1. A box-shaped vessel with 10 m beam and 7.5 m deep floats upright at a draft of 3 m. Find the increase of draft when listed 15° to starboard.
2. A ship 90 m long, 15 m beam at the waterline, is floating upright at a draft of 6 m. Find the increase of draft when the ship is listed 10°, allowing 0.15 m rise of floor.
3. A box-shaped vessel 60 m × 6 m × 4 m draft is floating upright. Find the increase in the draft if the vessel is now listed 15° to port.
4. A box-shaped vessel increases her draft by 0.61 m when listed 12° to starboard. Find the vessel's beam if the draft, when upright, was 5.5 m.
5. A box-shaped vessel with 10 m beam is listed 10° to starboard and the maximum draft is 5 m. A weight already on board is shifted transversely across the deck, causing the vessel to list to port. If the final list is 5° to port, find the new draft.

CHAPTER 28

Combined List and Trim

When a problem involves a change of both list and trim, the two parts must be treated quite separately. It is usually more convenient to tackle the trim part of the problem first and then the list, but no hard and fast rule can be made on this point.

Example 1

A ship of 6000 tonnes displacement has KM = 7 m, KG = 6.4 m, and MCT 1 cm = 120 tonnes m. The ship is listed 5° to starboard and trimmed 0.15 m by the head. The ship is to be brought upright and trimmed 0.3 m by the stern by transferring oil from No. 2 double-bottom tank to No. 5 double-bottom tank. Both tanks are divided at the centerline and their centers of gravity are 5.25 m out from the centerline. No. 2 holds 200 tonnes of oil on each side and is full. No. 5 holds 120 tonnes on each side and is empty. The center of gravity of No. 2 is 23.5 m forward of amidships and that of No. 5 is 21.5 m aft of amidships. Find what transfer of oil must take place and give the final distribution of the oil (neglect the effect of free surface on the GM). Assume that LCF is at amidships:

(a) *To bring the ship to the required trim*

$$\text{Present trim} = 0.15 \text{ m by the head}$$

$$\text{Required trim} = \underline{0.30} \text{ m by the stern}$$

$$\text{Change of trim} = 0.45 \text{ m by the stern}$$

$$= 45 \text{ cm by the stern}$$

$$\begin{aligned}\text{Trim moment} &= \text{Change of trim} \times \text{MCT 1 cm}\\ &= 45 \times 120\end{aligned}$$

$$\text{Trim moment} = 5400 \text{ tonnes m by the stern}$$

Let 'w' tonnes of oil be transferred aft to produce the required trim.

$$\begin{aligned}\therefore \text{Trim moment} &= w \times d\\ &= 45w \text{ tonnes m}\\ \therefore 45w &= 5400\\ w &= 120 \text{ tonnes}\end{aligned}$$

Ship Stability for Masters and Mates. http://dx.doi.org/10.1016/B978-0-08-097093-6.00028-1

From this it will be seen that, if 120 tonnes of oil are transferred aft, the ship will then be trimmed 0.30 m by the stern (see Figure 28.1(a) and (b)).

(b) *To bring the ship upright*

$$\begin{aligned} KM &= 7.0\ \text{m} \\ KG &= -\underline{6.4\ \text{m}} \\ GM &= 0.6\ \text{m} \end{aligned}$$

In triangle GG_1M:

$$\begin{aligned} GG_1 &= GM \times \tan\theta \\ &= 0.6 \times \tan 5^\circ \\ GG_1 &= 0.0525\ \text{m} \end{aligned}$$

Let 'x' tonnes of oil be transferred from starboard to port:

$$\begin{aligned} \text{Moment to port} &= x \times d \\ &= 10.5x\ \text{tonnes m} \end{aligned}$$

$$\begin{aligned} \text{Initial moment to starboard} &= W \times GG_1 \\ &= 6000 \times 0.0525 \\ &= 315\ \text{tonnes m} \end{aligned}$$

But if the ship is to complete the operation upright:

$$\text{Moment to starboard} = \text{Moment to port}$$

or

$$\begin{aligned} 315 &= 10.5x \\ x &= 30\ \text{tonnes} \end{aligned}$$

The ship will therefore be brought upright by transferring 30 tonnes from starboard to port, i.e. 15 tonnes from No. 2 starboard and 15 tonnes from No. 5 starboard.

From this it can be seen that, to bring the ship to the required trim and upright, 120 tonnes of oil must be transferred from forward to aft and 30 tonnes from starboard to

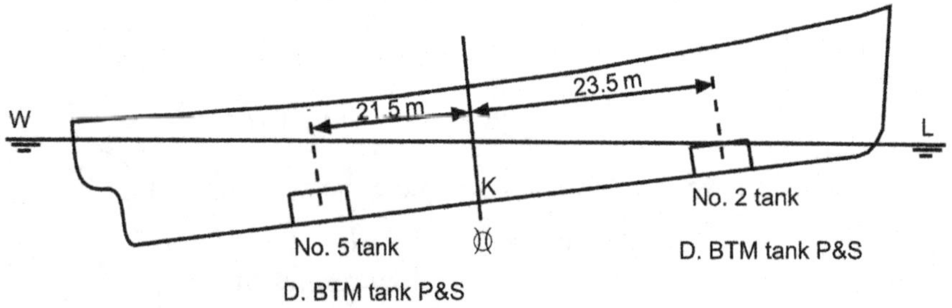

Figure 28.1a

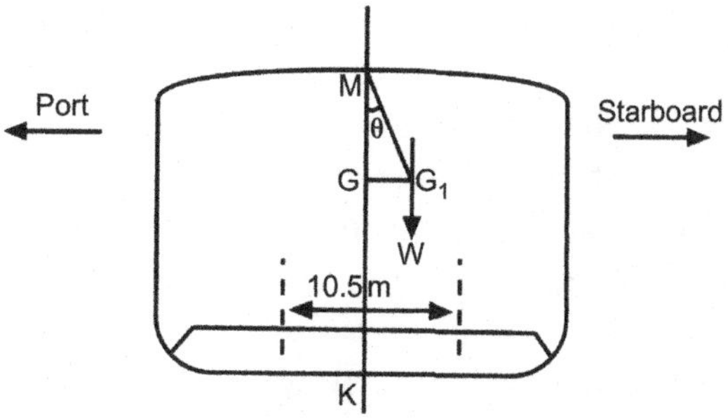

Figure 28.1b:
Looking Forward.

port. This result can be obtained by taking 60 tonnes from each of No. 2 starboard and No. 5 starboard tanks, and putting 15 tonnes in each of the No. 2 port and No. 5 port tanks.

The distributions would then be as follows:

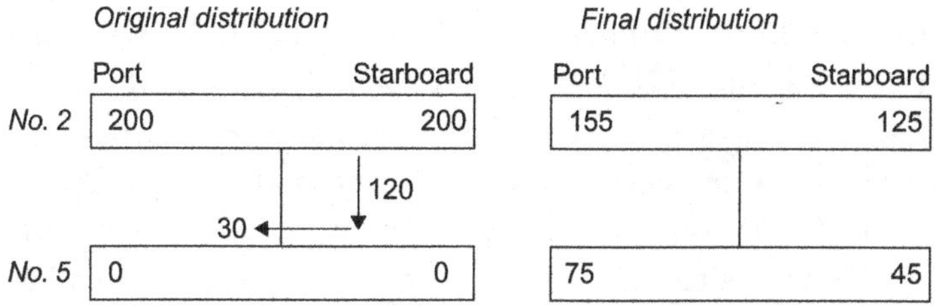

Note. There are, of course, alternative methods by which this result could have been obtained, but in each case a total of *120 tonnes* of oil must be transferred aft and *30 tonnes* must be transferred from starboard to port.

■

Exercise 28

1. A tanker has displacement of 10,000 tonnes, KM = 7 m, KG = 6.4 m, and MCT 1 cm = 150 tonnes m. There is a centerline bulkhead in both No. 3 and No. 8 tanks. The center of gravity of No. 3 tank is 20 m forward of the center of flotation and the center of gravity of No. 8 tank is 30 m aft of the center of flotation. The center of gravity of all tanks is 5 m out from the centerline. At present the ship is listed 4° to starboard and trimmed 0.15 m by the head. Find what transfer of oil must take place if the ship is to complete upright and trimmed 0.3 m by the stern.

2. A ship of 10,000 tonnes displacement is listed 5° to port and trimmed 0.2 m by the head. KM = 7.5 m, KG = 6.8 m, and MCT 1 cm = 150 tonnes m. The center of flotation is amidships. No. 1 double-bottom tank is divided at the centerline, each side holds 200 tonnes of oil and the tank is full. No. 4 double-bottom tank is similarly divided, each side having a capacity of 150 tonnes, but the tank is empty. The center of gravity of No. 1 tank is 45 m forward of amidships and the center of gravity of No. 4 tank is 15 m aft of amidships. The center of gravity of all tanks is 5 m out from the centerline. It is desired to bring the ship upright and trimmed 0.3 m by the stern by transferring oil. If the free-surface effect on GM is neglected, find what transfer of oil must take place and also the final distribution of the oil.
3. A ship of 6000 tonnes displacement, KG = 6.8 m, is floating upright in salt water, and the draft is 4 m F and 4.3 m A. KM = 7.7 m, TPC = 10 tonnes, and MCT 1 cm = 150 tonnes m. There is a locomotive to discharge from No. 2 lower hold (KG = 3 m and center of gravity 30 m forward of the center of flotation, which is amidships). If the weight of the locomotive is 60 tonnes and the height of the derrick head is 18 m above the keel and 20 m out from the centerline when plumbing overside, find the maximum list during the operation and the drafts after the locomotive has been discharged. Assume KM is constant.
4. A ship displaces 12,500 tonnes, is trimmed 0.6 m by the stern, and listed 6° to starboard. MCT 1 cm = 120 tonnes m, KG = 7.2 m, and KM = 7.3 m. No. 2 and No. 5 double-bottom tanks are divided at the centerline. The center of gravity of No. 2 is 15 m forward of the center of flotation and the center of gravity of No. 5 is 12 m aft of the center of flotation. The center of gravity of all tanks is 4 m out from the centerline. The ship is to be brought upright and on to an even keel by transferring oil from aft to forward, taking equal quantities from each side of No. 5. Find the amounts of oil to transfer.

■

CHAPTER 29

Calculating Free-Surface Effects of Slack Tanks with Divisional Bulkheads

The effect of free surface of liquids on stability was discussed in general terms in Chapter 3, but the problem will now be studied more closely and the calculations involved will be explained.

When a tank is partially filled with a liquid, the ship suffers a virtual loss in metacentric height, which can be calculated by using the formula:

$$\text{FSE} = \text{Virtual loss of GM} = \frac{i}{W} \times \rho \times \frac{1}{n^2} \text{meters} \qquad \text{(I)}$$

where i = the second moment of the free surface about the centerline (m^4), w = the ship's displacement (tonnes), ρ = the density of the liquid in the tank (tonnes/cubic meter), n = the number of longitudinal compartments into which the tank is equally subdivided, and $i \times \rho$ = free-surface moment (tonnes m). The derivation of this formula is as follows:

The ship shown in Figure 29.1(a) has an undivided tank that is partially filled with a liquid. When the ship is inclined, a wedge of the liquid in the tank will shift from the high side to the low side such that its center of gravity shifts from g to g_1. This will cause the center of gravity of the ship to shift from G to G_1, where

$$GG_1 = \frac{w \times gg_1}{W}$$

Let

$$\rho_1 = \text{Density of the liquid in the tank}$$

and

$$\rho_2 = \text{Density of the water in which the ship floats}$$

then

$$GG_1 = \frac{v \times \rho_1 \times gg_1}{V \times \rho_2}$$

Ship Stability for Masters and Mates. http://dx.doi.org/10.1016/B978-0-08-097093-6.00029-3

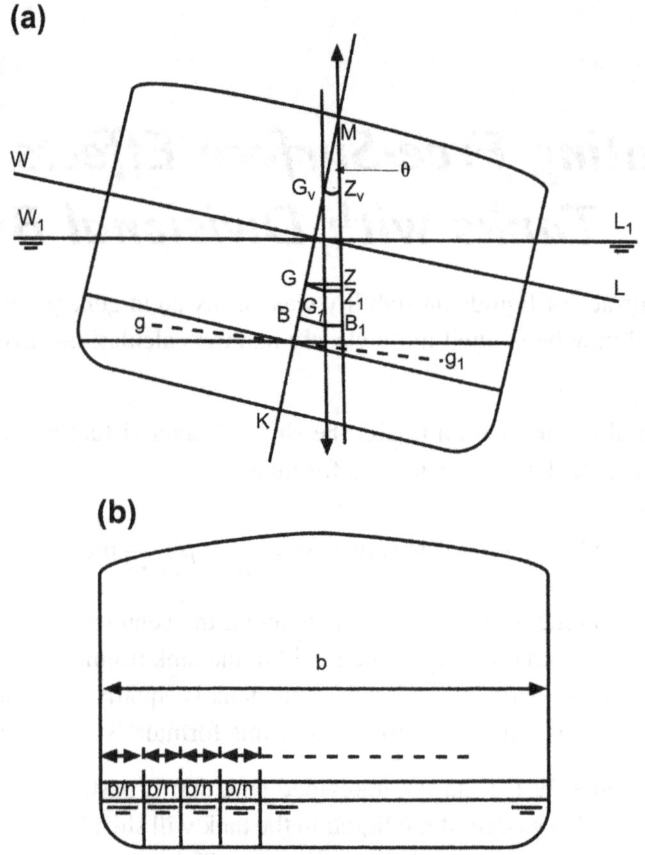

Figure 29.1

Had there been no free surface when the ship inclined, the righting lever would have been GZ. But, due to the liquid shifting, the righting lever is reduced to G_1Z_1 or G_vZ_v. The virtual reduction of GM is therefore GG_v.

For a small angle of heel:

$$GG_1 = GG_v \times \theta$$

$$\therefore GG_v \times \theta = \frac{v \times gg_1 \times \rho_1}{V \times \rho_2}$$

or

$$GG_v = \frac{v \times gg_1 \times \rho_1}{V \times \theta \times \rho_2}$$

From the proof in Chapter 10 of BM = I/V, and I = (v × gg_1) 0 we can proceed to show that $GG_v = i/V \times p_1/p_2 \times 1/n^2$.

This is the formula to find the virtual loss of GM due to the free-surface effect (FSE) in an undivided tank.

Now assume that the tank is subdivided longitudinally into 'n' compartments of equal width, as shown in Figure 29.1(b). Let

$$l = \text{Length of the tank}$$
$$b = \text{Breadth of the tank}$$

The breadth of the free surface in each compartment is thus b/n, and the second moment of each free surface is given by $[l \times (b/n)^3]/12$.

$$GG_v = \begin{array}{l}\text{Virtual loss of GM for one compartment}\\ \text{multiplied by the number of compartments}\end{array}$$
$$= \frac{i \times \rho_1}{V \times \rho_2} \times n$$

where

$$i = \begin{array}{l}\text{The second moment of the waterplane area}\\ \text{in one compartment about the centerline}\end{array}$$
$$= \frac{l \times (b/n)^3 \times \rho_1}{12 \times V \times \rho_2} \times n$$
$$= \frac{l \times b^3 \times \rho_1}{12 \times V \times n^3 \times \rho_2} \times n$$

or

$$GG_v = \frac{i}{V} \times \frac{\rho_1}{\rho_2} \times \frac{1}{n^2}$$

where

$$i = \begin{array}{l}\text{The second moment of the waterplane area of}\\ \text{the whole tank about the centerline}\end{array}$$

But

$$W = V \times \rho_2$$

so

$$GG_v = \frac{i}{W} \times \rho_1 \times \frac{1}{n^2}$$

as shown in equation (I) at the beginning of this chapter.

This is the formula to find the virtual loss of GM due to the free-surface effect in a tank that is subdivided longitudinally. From this formula it can be seen that, when a tank is subdivided longitudinally, the virtual loss of GM for the undivided tank is divided by the square of the number of compartments into which the tank is divided. Also note that the actual weight of the liquid in the tank will have no effect whatsoever on the virtual loss of GM due to the free surface.

For a rectangular area of free surface, the second moment to be used in the above formula can be found as follows:

$$i = \frac{LB^3}{12}$$

where

L = The length of the free surface
B = The total breadth of the free surface, ignoring divisions

Note. Transverse subdivisions in partially filled tanks (slack tanks) *do not* have any influence on reducing free-surface effects. However, fitting longitudinal bulkheads *does* have a very effective influence in reducing this virtual loss in GM.

Example 1

A ship of 8153.75 tonnes displacement has KM = 8 m, KG = 7. 5 m, and has a double-bottom tank 15 m × 10 m × 2 m that is full of salt water ballast. Find the new GM if this tank is now pumped out till half empty.

Note. The mass of the water pumped out will cause an actual rise in the position of the ship's center of gravity and the free surface created will cause a virtual loss in GM. There are therefore two shifts in the position of the center of gravity to consider.

In Figure 29.2 the shaded portion represents the water to be pumped out with its center of gravity at position g. The original position of the ship's center of gravity is at G. Let GG_1 represent the actual rise of G due to the mass discharged.

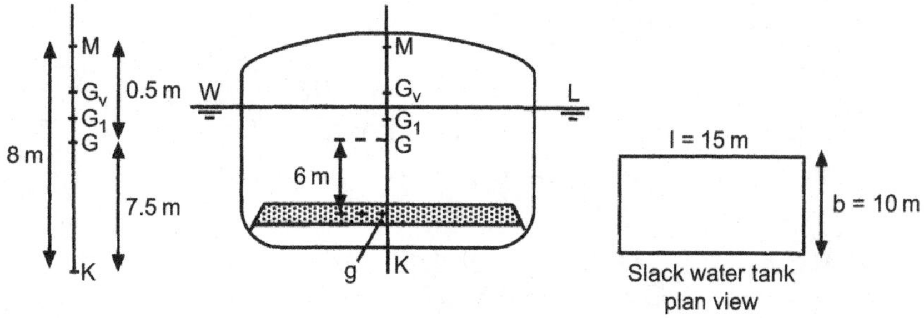

Figure 29.2

$$\text{The mass of water discharged (w)} = 15 \times 10 \times 1 \times 1.025 \text{ tonnes}$$

$$w = 153.75 \text{ tonnes}$$

$$W_2 = W_1 - w = 8153.75 - 153.75$$

$$= 8000 \text{ tonnes}$$

$$GG_1 = \frac{w \times d}{W_2}$$

$$= \frac{153.75 \times 6}{8000}$$

$$\underline{GG_1} = \underline{0.115 \text{ m}}$$

Let G_1G_v represent the virtual loss of GM due to free surface or rise in G_1. Then:

$$G_1G_v = \frac{i}{W} \times \rho_1 \times \frac{1}{n^2}$$

as per equation (I) at the beginning of this chapter:

$$n = 1$$

$$\therefore G_1G_v = \frac{i}{W_2} \times \rho_{SW}$$

or

$$G_1G_v = \frac{1b^3}{12} \times \frac{\rho_{SW}}{W_2} \times \frac{1}{n^2}$$

$$\text{Loss in GM} = \text{FSE}$$

or

$$G_1G_v = \frac{15 \times 10^3 \times 1.025}{12 \times 8000}$$

$$\underline{G_1G_v} = \underline{0.160 \text{ m}} \uparrow$$
$$\text{Old KM} = \underline{8.000 \text{ m}}$$
$$\text{Old KG} = \underline{7.500 \text{ m}}$$
$$\text{Old GM} = 0.500 \text{ m}$$
$$\text{Actual rise of G} = 0.115 \text{ m}$$
$$0.385 \text{ m} = GM_{solid}$$
$$G_1G_v = \text{Virtual rise of G} = \underline{0.160 \text{ m}} \uparrow$$
$$= \underline{0.225 \text{ m}}$$

Ans. New GM = 0.225 m = GM_{fluid}.

Hence G_1 has risen due to the discharge of the ballast water (loading change) and has also risen due to free-surface effects.

Be aware that in some cases these two rises of G do not take G above M, thereby making the ship unstable.

Example 2

A ship of 6000 tonnes displacement, floating in salt water, has a double-bottom tank 20 m × 12 m × 2 m, which is divided at the centerline and is partially filled with oil of relative density 0.82. Find the virtual loss of GM due to the free surface of the oil.

$$\text{Virtual loss of GM} = \frac{i}{W} \times \rho_{oil} \times \frac{1}{n^2}$$
$$= \frac{lb^3}{12} \times \rho_{oil} \times \frac{1}{W} \times \frac{1}{n^2}$$
$$= \frac{20 \times 12^3}{12 \times 6000} \times 0.820 \times \frac{1}{2^2}$$

Ans. Virtual loss of GM = 0.098 meters.

Example 3

A ship of 8000 tonnes displacement has KM = 7.5 m and KG = 7.0 m. A double-bottom tank is 12 m long, 15 m wide, and 1 m deep. The tank is divided longitudinally at the centerline and both sides are full of salt water (see Figure 29.3). Calculate the list if one side is pumped out until it is half empty.

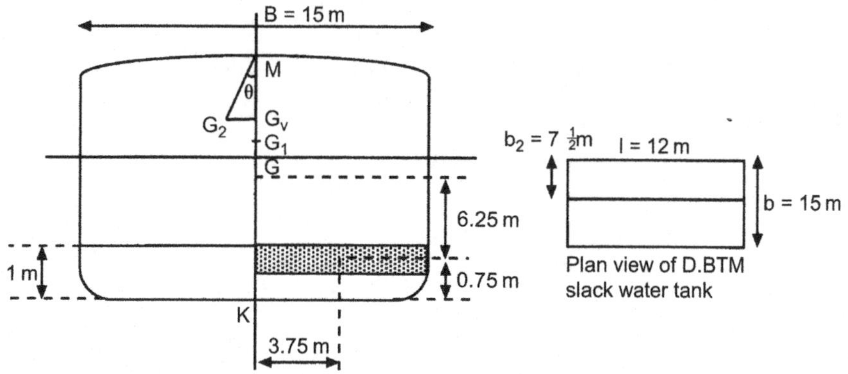

Figure 29.3

$$\text{Mass of water discharge} = l_{DB} \times b_{DB} \times \frac{d_{DB}}{2} \times \rho_{SW}$$
$$\text{Mass of water discharge} = 12 \times 7.5 \times 0.5 \times 1.025$$
$$w = 46.125 \text{ tonnes}$$

$$\text{Vertical shift of } G(GG_1) = \frac{w \times d}{W_2}$$
$$= \frac{46.125 \times 6.25}{8000 - 46.125} = \frac{46.125 \times 6.25}{7953.875}$$
$$= 0.036 \text{ meters}$$

$$\text{Horizontal shift of } G(G_vG_2) = \frac{w \times d}{W - w}$$
$$= \frac{46.125 \times 3.75}{7953.875}$$
$$= \underline{0.022 \text{ meters}}$$

$$\text{Virtual loss of } GM(G_1G_v) = \frac{l \times b_2^3}{12} \times \frac{\rho_{SW}}{W_2} \times \frac{1}{n^2}$$
$$= \frac{12 \times 7.5^3}{12} \times \frac{1.025}{7953.875} \times \frac{1}{1^2}$$
$$= 0.054 \text{ meters}$$

$$KM = 7.500 \text{ meters}$$
$$\text{Original } KG = \underline{7.000 \text{ meters}}$$
$$\text{Original } GM = 0.500 \text{ meters} = GM_{solid}$$
$$\text{Vertical rise of } G(GG_1) = \underline{0.036 \text{ meters}} \quad \uparrow$$
$$G_1M = 0.464 \text{ meters} = G_1M_{solid}$$
$$\text{Virtual loss of } GM(G_1G_v) = 0.054 \text{ meters}$$

$$\underline{\text{New GM}(G_vM) = 0.410 \text{ meters} = GM_{fluid}}$$

In triangle G_vG_2M:

$$\tan\theta = \frac{G_2G_v}{G_vM}$$

$$= \frac{0.022}{0.410} = 0.0536$$

Ans. List = 3° 04′.

So again G has risen due to discharging water ballast. It has also risen due to free-surface effects. A further movement has caused G to move off the centerline and has produced this angle of list of 3° 04′.

The following worked example shows the effect of subdivisions in slack tanks in relation to free-surface effects (FSE):

Question: A ship has a displacement of 3000 tonnes. On the vessel is a rectangular double-bottom tank 15 m long and 8 m wide. This tank is partially filled with ballast water having a density of 1.025 t/m^3.

If the GM_T without free-surface effects is 0.18 m, calculate the virtual loss in GM_T and the final GM_T when the double-bottom tank has:

(a) no divisional bulkheads fitted;
(b) one transverse bulkhead fitted at mid-length;
(c) one longitudinal bulkhead fitted on the centerline of the tank;
(d) two longitudinal bulkheads fitted giving three equal divisions.

Answer:

Part (a)

$$\text{FSE} = \text{Virtual loss in } GM_T \text{ or rise in G} = \frac{I \times \rho_{SW}}{W}$$

$$= \frac{1 \times b_1^3 \times \rho_{SW}}{12 \times W} \quad \text{(see Figure 29.4(a))}$$

$$\therefore \underline{\text{Virtual loss in } GM_T} = \frac{15 \times 8^3 \times 1.025}{3000 \times 12}$$

$$= \underline{0.2187 \text{ m}} \uparrow$$

$$\therefore GM_T \text{ finally} = 0.1800 - 0.2187$$

$$= \underline{-0.0387 \text{ m}} \uparrow$$

i.e. an unstable ship!!

Part (b)

$$\text{FSE} = \text{Virtual loss in } GM_T \text{ or rise in G} = \frac{2 @ 1_2 \times b_1^3}{12 \times W} \times \rho_{SW} \text{ (see Figure 29.4(b))}$$

$$\therefore \text{Virtual loss} = \frac{2 \times 7.5 \times 8^3 \times 1.025}{12 \times 3000}$$

$$= \underline{0.2187 \text{ m}} \uparrow$$

$$GM_T \text{ fluid} = \underline{-0.0387 \text{ m}}$$

This is same answer as for part (a). Consequently it can be concluded that fitting transverse divisional bulkheads in tanks does not reduce the free-surface effects. The ship is still unstable!!

Part (c)

$$\text{FSE} = \text{Vertical loss in } GM_T \text{ or rise in G} = \frac{2 @ l_1 b_2^3}{12 \times W} \rho_{SW}$$

$$\therefore \text{Virtual loss in } GM_T = \frac{2 \times 15 \times 4^3 \times 1.025}{12 \times 3000} \text{ (see Figure 29.4(c))}$$

$$= 0.0547 \text{ m} \uparrow \text{ i.e. } \frac{1}{4} \text{ of answer to part(a)}$$

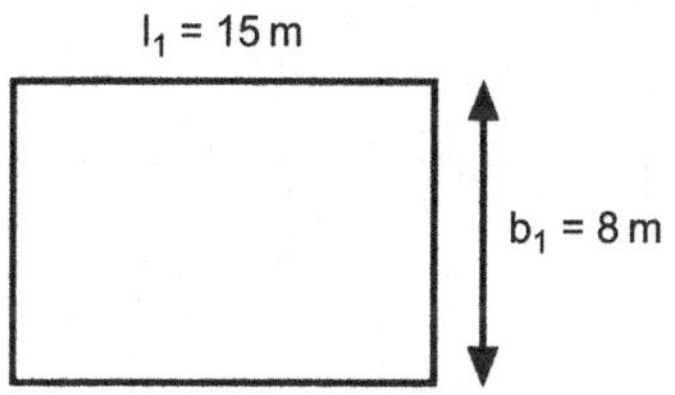

Figure 29.4a

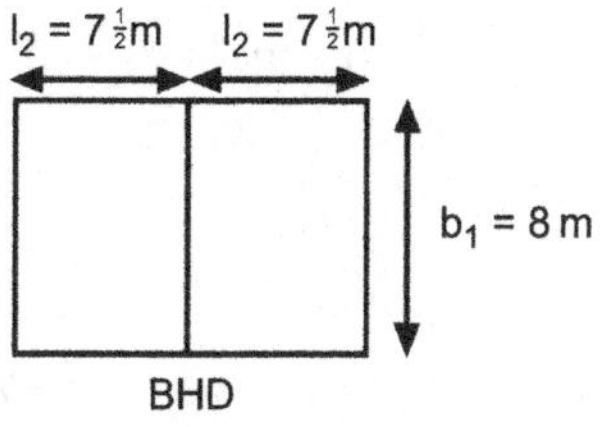

Figure 29.4b

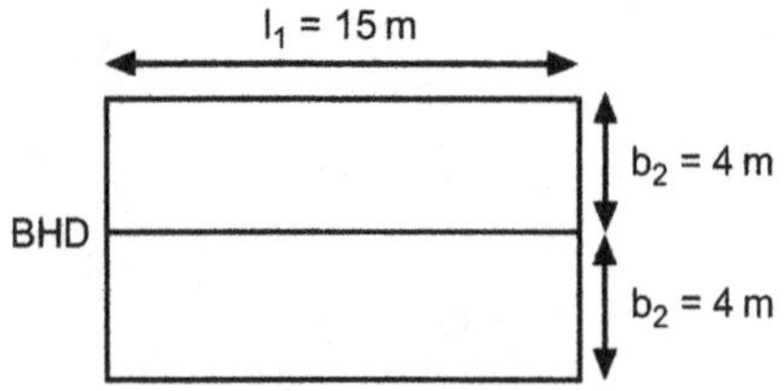

Figure 29.4c

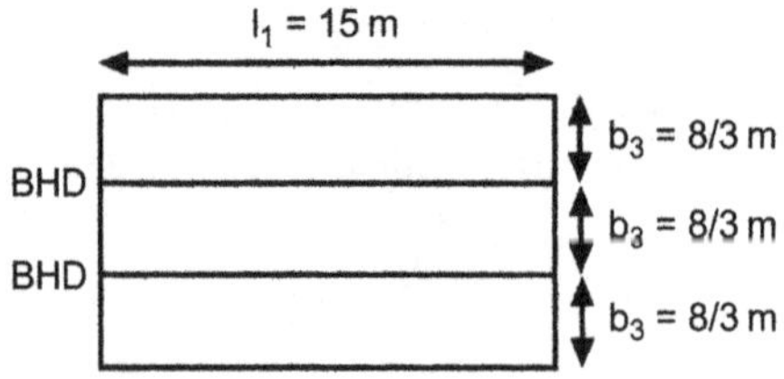

Figure 29.4d

Hence

$$\text{Final } GM_T = 0.1800 - 0.0547 \text{ m} = +0.1253 \text{ m; ship is stable}$$

GM_T is now positive, but below the minimum GM_T of 0.15 m that is allowable by DfT regulations.

Part (d)

$$\text{FSE} = \text{Virtual loss in } GM_T \text{or rise in}$$

$$G = \frac{3 \text{ @ } 1_1 \times b_3^3}{12 \times W} \times \rho_{SW} \quad \text{(see Figure 29.4(d))}$$

$$\therefore \text{Virtual loss in } GM_T = \frac{3 \times 15 \times \left(\frac{8}{3}\right)^3 \times 1.025}{12 \times 3000}$$

$$= 0.0243 \text{ m} \uparrow \text{i.e. } \frac{1}{9} \text{ of answer to part (a)}$$

Hence

$$\text{Final } GM_T = 0.1800 - 0.0243 = +0.1557 \text{ m; ship is stable}$$

The ship is stable and the above DfT minimum GM_T value of 0.15 m.

So longitudinal divisional bulkheads (watertight or wash-bulkheads) are effective. They cut down rapidly the loss in GM_T. Note the $1/n^2$ law, where n is the number of equal divisions made by the longitudinal bulkheads.

Free-surface effects therefore depend on:

1. Density of slack liquid in the tank
2. Ship's displacement in tonnes
3. Dimensions and shape of the slack tanks
4. Bulkhead subdivision within the slack tanks.

The longitudinal divisional bulkheads referred to in examples in this chapter need not be absolutely watertight; they could have openings in them. Examples on board ship are the centerline wash bulkhead in the fore peak tank and in the aft peak tank.

Exercise 29

1. A ship of 10,000 tonnes displacement is floating in dock water of density 1024 kg per cubic meter, and is carrying oil of relative density 0.84 in a double-bottom tank. The tank is 25 m long, 15 m wide, and is divided at the centerline. Find the virtual loss of GM due to this tank being slack.
2. A ship of 6000 tonnes displacement is floating in fresh water and has a deep tank (10 m × 15 m × 6 m) that is undivided and is partly filled with nut oil of relative density 0.92. Find the virtual loss of GM due to the free surface.
3. A ship of 8000 tonnes displacement has KG = 3.75 m and KM = 5.5 m. A double-bottom tank 16 m × 16 m × 1 m is subdivided at the centerline and is full of salt water ballast. Find the new GM if this tank is pumped out until it is half empty.
4. A ship of 10,000 tonnes displacement, KM = 6 m, KG = 5.5 m, is floating upright in dock water of density 1024 kg per cubic meter. She has a double-bottom tank 20 m × 15 m that is subdivided at the centerline and is partially filled with oil of relative density 0.96. Find the list if a mass of 30 tonnes is now shifted 15 m transversely across the deck.
5. A ship is at the light displacement of 3000 tonnes and has KG = 5.5 m and KM = 7.0 m. The following weights are then loaded:

   5000 tonnes of cargo KG = 5 m
   2000 tonnes of cargo KG = 10 m
   700 tonnes of fuel oil of relative density 0.96.
   The fuel oil is taken into Nos. 2, 3, and 5 double-bottom tanks, filling Nos. 3 and 5, and leaving No. 2 slack.

The ship then sails on a 20-day passage consuming 30 tonnes of fuel oil per day. On arrival at her destination Nos. 2 and 3 tanks are empty, and the remaining fuel oil is in No. 5 tank. Find the ship's GM's for the departure and arrival conditions.

Dimensions of the tanks:

No. 2 15 × 15 m × 1 m
No. 3 22 m × 15 m × 1 m
No. 5 12 m × 15 m × 1 m

Assume that the KM is constant and that the KG of the fuel oil in every case is half of the depth of the tank.

6. A ship's displacement is 5100 tonnes, KG = 4 m, and KM = 4.8 m. A double-bottom tank on the starboard side is 20 m long, 6 m wide and 1 m deep, and is full of fresh water. Calculate the list after 60 tonnes of this water has been consumed.
7. A ship of 6000 tonnes displacement has KG = 4 m and KM = 4.5 m. A double-bottom tank in the ship 20 m long and 10 m wide is partly full of salt-water ballast. Find the moment of statical stability when the ship is heeled 5°.
8. A box-shaped vessel has the following data:

   Length is 80 m, breadth is 12 m, draft even keel is 6 m, KG is 4.62 m.
   A double-bottom tank 10 m long, of full width and 2.4 m depth, is then half-filled with water ballast having a density of 1.025 t/m^3. The tank is located at amidships. Calculate the new even-keel draft and the new transverse GM after this water ballast has been put in the double-bottom tank.

CHAPTER 30

Bilging Effects of Stability – Permeability Effects

Bilging Amidships Compartments

When a vessel floats in still water it displaces its own weight of water. Figure 30.1(a) shows a box-shaped vessel floating at the waterline (WL). The weight of the vessel (W) is considered to act downwards through G, the center of gravity. The force of buoyancy is also equal to W and acts upwards through B, the center of buoyancy, $b = W$.

Now let an empty compartment amidships be holed below the waterline to such an extent that the water may flow freely into and out of the compartment. A vessel holed in this way is said to be 'bilged'.

Figure 30.1(b) shows the vessel in the bilged condition. The buoyancy provided by the bilged compartment is lost. The draft has increased and the vessel now floats at the waterline W_1L_1, where it is again displacing its own weight of water. 'X' represents the increase in draft due to bilging. The volume of lost buoyancy (v) is made good by the volumes 'y' and 'z'.

$$\therefore v = y + z$$

Let 'A' be the area of the waterplane before bilging, and let 'a' be the area of the bilged compartment. Then:

$$y + z = Ax - ax$$

or

$$v = x(A - a)$$

$$\text{Increase in draft} = x = \frac{v}{A - a}$$

i.e.

$$\text{Increase in draft} = \frac{\text{Volume of lost buoyancy}}{\text{Area of intact waterplane}}$$

Ship Stability for Masters and Mates. http://dx.doi.org/10.1016/B978-0-08-097093-6.00030-X

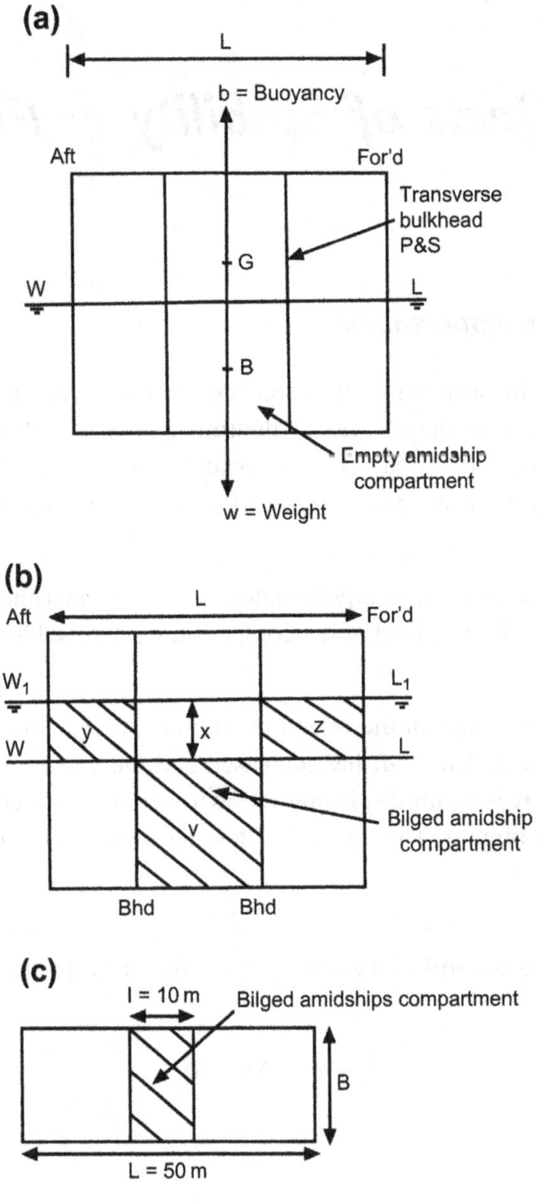
(a)
L
b = Buoyancy
Aft
For'd
Transverse
bulkhead
P&S
G
W
L
B
Empty amidship
compartment
w = Weight
(b)
L
Aft
For'd
W1
L1
y
x
z
W
L
v
Bilged amidship
compartment
Bhd
Bhd
(c)
l = 10 m
Bilged amidships compartment
B
L = 50 m

Figure 30.1

Note. Since the distribution of weight within the vessel has not been altered, the KG after bilging will be the same as the KG before bilging.

Example 1

A box-shaped vessel is 50 meters long and is floating on an even keel at 4 meters draft. An amidships compartment is 10 meters long and is empty (see Figure 30.1(c)). Find the increase in draft if this compartment is bilged.

$$x = \frac{v}{A - a} = \frac{l \times B \times d}{(L - l)B}$$

Let

$$B = \text{Breadth of the vessel}$$

then

$$x = \frac{10 \times B \times 4}{(50 \times B) - (10 \times B)}$$

$$= \frac{40B}{40B}$$

$$\underline{\text{Increase in draft} = 1 \text{ meter}}$$

Example 2

A box-shaped vessel is 150 meters long × 24 meters wide × 12 meters deep, and is floating on an even keel at 5 meters draft. GM = 0.9 meters. A compartment amidships is 20 meters long and is empty. Find the new GM if this compartment is bilged.

$$\text{Old KB} = \frac{1}{2}\text{ Old draft}$$

$$= 2.5 \text{ m}$$

$$\text{Old BM} = B^2/12d$$

$$= \frac{24 \times 24}{12 \times 5}$$

$$= 9.6 \text{ m}$$

$$\text{Old KB} = \underline{+2.5 \text{ m}}$$

$$\text{Old KM} = 12.1 \text{ m}$$

$$\text{Old GM} = \underline{-0.9 \text{ m}}$$

$$\text{KG} = \underline{11.2 \text{ m}}$$

This KG will not change after bilging has taken place:

$$\text{Increase in draft } x = \frac{v}{A - a}$$

$$= \frac{20 \times 24 \times 5}{(150 \times 24) - (20 \times 24)}$$

$$= \frac{2400}{130 \times 24} = 0.77 \text{ m}$$

$$\text{Increase in draft} = 0.77 \text{ m}$$

$$\text{Old draft} = 5.00 \text{ m}$$

$$\text{New draft} = \underline{5.77 \text{ m}} = \text{Say draft } d_2$$

$$\text{New KB} = \frac{1}{2}\text{New draft} = \frac{d_2}{2}$$

$$= 2.89 \text{ m}$$

$$\text{New BM} = B^2/12d_2$$

$$= \frac{24 \times 24}{12 \times 5.77}$$

$$= 8.32 \text{ m}$$

$$+$$

$$\text{New KB} = 2.89 \text{ m}$$

$$-$$

$$\text{New KM} = \underline{11.21 \text{ m}}$$

$$\text{As before, KG} = 11.20 \text{ m}$$

Ans. New GM = 0.01 m.

This is positive but dangerously low in value!!

Permeability, μ

Permeability is the amount of water that can enter a compartment or tank after it has been bilged. When an empty compartment is bilged, the whole of the buoyancy provided by that compartment is lost. Typical values for permeability, μ, are as follows:

Empty compartment	$\mu = 100\%$
Engine room	$\mu = 80–85\%$
Grain-filled cargo hold	$\mu = 60–65\%$
Coal-filled compartment	$\mu = 36\%$ approximately
Filled water ballast tank (when ship is in salt water)	$\mu = 0\%$

Consequently, the higher the value of the permeability for a bilged compartment, the greater will be a ship's loss of buoyancy when the ship is bilged.

The permeability of a compartment can be found from the formula:

$$\mu = \text{Permeability} = \frac{\text{Broken stowage}}{\text{Stowage factor}} \times 100\%$$

The broken stowage to be used in this formula is the broken stowage per tonne of stow.

When a bilged compartment contains cargo, the formula for finding the increase in draft must be amended to allow for the permeability. If 'μ' represents the permeability, expressed as a fraction, then the volume of lost buoyancy will be 'μv' and the area of the intact waterplane will be '$A - \mu v$' square meters. The formula then reads:

$$x = \frac{\mu v}{A - \mu a}$$

Example 1

A box-shaped vessel is 64 m long and is floating on an even keel at 3 m draft. A compartment amidships is 12 m long and contains cargo having a permeability of 25%. Calculate the increase in the draft if this compartment is bilged.

$$x = \frac{\mu v}{A - \mu a}$$

$$= \frac{\frac{1}{4} \times 12 \times B \times 3}{(64 \times B) - \left(\frac{1}{4} \times 12 \times B\right)}$$

$$= \frac{9B}{61B}$$

Ans. Increase in draft = 0.15 m.

Example 2

A box-shaped vessel 150 m × 20 m × 12 m is floating on an even keel at 5 m draft. A compartment amidships is 15 m long and contains timber of relative density 0.8 and stowage factor 1.5 cubic meters per tonne. Calculate the new draft if this compartment is now bilged.

The permeability 'μ' must first be found by using the formula given above, i.e.

$$\text{Permeability} = \frac{BS}{SF} \times 100\% = \text{'}\mu\text{'}$$

The stowage factor is given in the question. The broken stowage per tonne of stow is now found by subtracting the space which would be occupied by one tonne of solid timber from that actually occupied by one tonne of timber in the hold. One tonne of fresh water occupies one cubic meter and the relative density of the timber is 0.8.

$$\therefore \text{Space occupied by 1 tonne of solid timber} = \frac{1}{0.8}$$

$$= 1.25 \text{ cubic meters}$$

$$\text{Stowage factor} = 1.50 \text{ cubic meters}$$

$$\therefore \text{Broken stowage} = \underline{0.25} \text{ cubic meters}$$

$$\text{Permeability '}\mu\text{'} = \frac{BS}{SF} \times 100\%$$

$$= \frac{0.25}{1.50} \times 100\%$$

$$= 100/6\%$$

$$\therefore \text{'}\mu\text{'} = 1/6 \text{ or } 16.67\%$$

$$\text{Increase in draft} = x = \frac{\mu v}{A - \mu a}$$

$$= \frac{1/6 \times 15 \times 20 \times 5}{(150 \times 20) - ((1/6) \times 15 \times 20)}$$

$$= 250/2950 = 0.085 \text{ m}$$

$$\text{Increase in draft} = 0.085 \text{ meters}$$

$$\text{Old draft} = \underline{5.000 \text{ meters}} = \text{Draft } d_1$$

Ans. New draft = 5.085 meters = Draft d_2.

When the bilged compartment does not extend above the waterline, the area of the intact waterplane remains constant, as shown in Figure 30.2. In this figure:

$$\mu v = Ax$$

Let

$$d = \text{Density of the water}$$

Then

$$\mu v \times d = Ax \times d$$

But

$$\mu v \times d = \text{Mass of water entering the bilged compartment}$$

$$Ax \times d = \text{Mass of the extra layer of water displaced}$$

Therefore, when the compartment is bilged, the extra mass of water displaced is equal to the buoyancy lost in the bilged compartment. It should be carefully noted, however, that although the effect on draft is similar to that of loading a mass in the bilged compartment equal to the lost buoyancy, no mass has in fact been loaded. The *displacement* after bilging is *the same* as the displacement before bilging and there is *no alteration* in the position of the vessel's *center of gravity*. The increase in the draft is due solely to lost buoyancy.

■

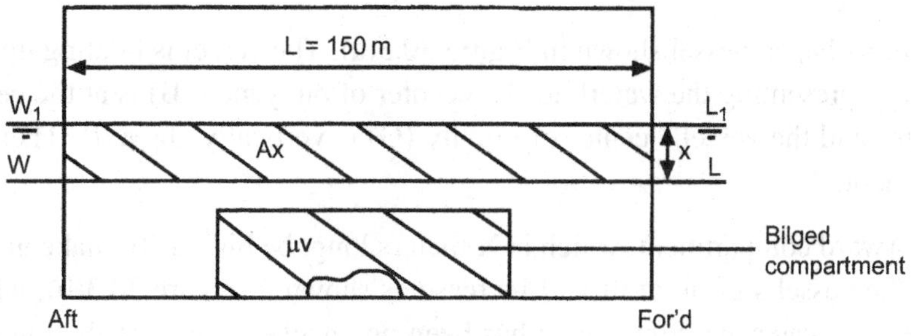

Figure 30.2

Example 3

A ship is floating in salt water on an even keel at 6 m draft. TPC is 20 tonnes. A rectangular-shaped compartment amidships is 20 m long, 10 m wide, and 4 m deep. The compartment contains cargo with permeability 25%. Find the new draft if this compartment is bilged.

$$\text{Buoyancy lost} = \frac{25}{100} \times 20 \times 10 \times 4 \times 1.025 \text{ tonnes}$$

$$= 205 \text{ tonnes}$$

$$\text{Extra mass of water displaced} = \text{TPC} \times \text{X tonnes}$$

$$\therefore X = w/\text{TPC}$$

$$= 205/20$$

$$\text{Increase in draft} = 10.25 \text{ cm}$$

$$= 0.1025 \text{ m}$$

$$\text{Plus the old draft} = \underline{6.0000 \text{ m}}$$

Ans. New draft = 6.1025 m.

Note. The *lower* the permeability, the *less* will be the changes in end drafts after bilging has taken place.

Bilging End Compartments

When the bilged compartment is situated in a position away from amidships, the vessel's mean draft will increase to make good the lost buoyancy, but the trim will also change.

Consider the box-shaped vessel shown in Figure 30.3(a). The vessel is floating upright on an even keel, WL representing the waterline. The center of buoyancy (B) is at the center of the displaced water and the vessel's center of gravity (G) is vertically above B. There is no trimming moment.

Now let the forward compartment, which is X meters long, be bilged. To make good the loss in buoyancy, the vessel's mean draft will increase as shown in Figure 30.3(b), where W_1L_1 represents the new waterline. Since there has been no change in the distribution of mass within the vessel, the center of gravity will remain at G. It has already been shown that the

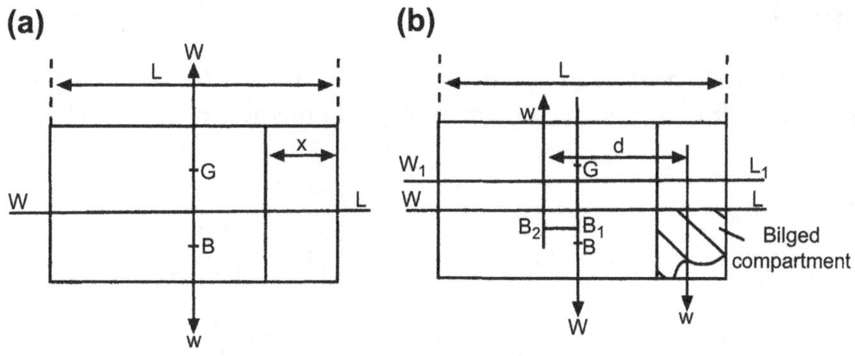

Figure 30.3

effect on mean draft will be similar to that of loading a mass in the compartment equal to the mass of water entering the bilged space to the original waterline.

The vertical component of the shift of the center of buoyancy (B to B_1) is due to the increase in the mean draft. KB_1 is equal to half of the new draft. The horizontal component of the shift of the center of buoyancy (B_1B_2) is equal to X/2.

A trimming moment of $W \times B_1B_2$ by the head is produced and the vessel will trim about the center of flotation (F), which is the center of gravity of the new waterplane area:

$$B_1B_2 = \frac{w \times d}{W}$$

or

$$W \times B_1B_2 = w \times d$$

But

$$W \times B_1B_2 = \text{Trimming moment}$$

$$\therefore w \times d = \text{Trimming moment}$$

It can therefore be seen that the effect on trim is similar to that which would be produced if a mass equal to the lost buoyancy were loaded in the bilged compartment.

Note. When calculating the TPC, MCTC, etc., it must be remembered that the information is required for the vessel in the bilged condition, using draft d_2 and intact length l_2.

Example 1

A box-shaped vessel 75 m long × 10 m wide × 6 m deep is floating in salt water on an even keel at a draft of 4.5 m (see Figure 30.4). Find the new drafts if a forward compartment 5 m long is bilged.

(a) *First let the vessel sink bodily*

$$w = x \times B \times d_1 \times 1.025 \text{ tonnes} \qquad \text{TPC} = \frac{\text{WPA}}{97.56} = \frac{L_2 \times B}{97.56}$$

$$= 5 \times 10 \times 4.5 \times 1.025 \qquad = \frac{70 \times 10}{97.56}$$

$$w = 230.63 \text{ tonnes} \qquad \text{TPC} = 7.18 \text{ tonnes}$$

$$\text{Increase in draft} = w/\text{TPC}$$

$$= 230.63/7.175$$

$$= 32.14 \text{ cm}$$

$$= 0.32 \text{ m}$$

$$+$$

$$\text{Old draft} = \underline{4.50 \text{ m}} = \text{draft } d_1$$

$$\text{New mean draft} = \underline{4.82 \text{ m}} = \text{draft } d_2$$

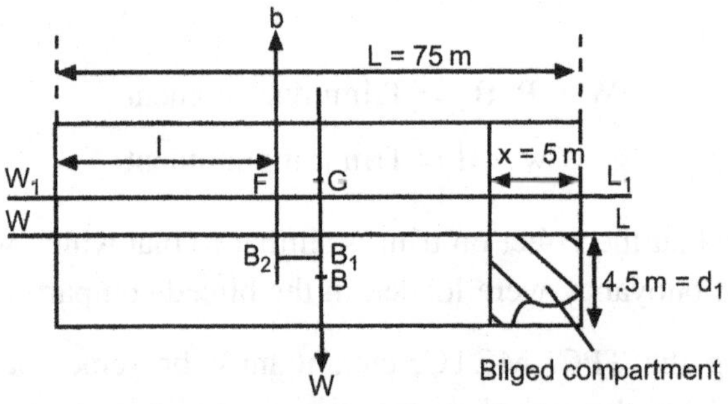

Figure 30.4

(b) *Now consider the trim*

$$W = L \times B \times d_1 \times 1.025 \text{ tonnes} \qquad BM_L = I_L/V$$

$$= 75 \times 10 \times 4.5 \times 1.025 \qquad = \frac{BL_2^3}{12V}$$

$$W = 3459 \text{ tonnes} \qquad = \frac{10 \times 70^3}{12 \times 75 \times 10 \times 4.5}$$

$$BM_L = 84.69 \text{ meters}$$

$$MCTC \simeq \frac{W \times BM_L}{100L}$$

$$= \frac{3459 \times 84.69}{100 \times 75}$$

$$= 39.06 \text{ tonnes m per cm}$$

$$\text{Change of trim} = \frac{\text{Moment changing trim}}{MCTC}$$

where

$$d = \frac{LBP}{2} = \frac{75}{2} = 37.5 \text{ m} = \text{Lever from new LCF}$$

$$\frac{w \times d}{MCTC} = \frac{230.6 \times 37.5}{39.05}$$

$$= 221 \text{ cm by the head}$$

After bilging, LCF has moved to F, i.e. $(L - x)/2$ from the stern.

$$\text{Change of draft aft} = \frac{l}{L} \times \text{Change of trim}$$

$$= \frac{35}{75} \times 221.4$$

$$= 103 \text{ cm} = 1.03 \text{ m}$$

$$\text{Change of draft forward} = \frac{40}{75} \times 221.4$$

$$= 118 \text{ cm} = 1.18 \text{ m}$$

(c) *Now find new drafts*

Drafts before trimming	A 4.82 m	F 4.82 m
Change due to trim	− 1.03 m	+ 1.18 m
Ans. New drafts	A 3.79 m	F 6.00 m

Example 2

A box-shaped vessel 100 m long × 20 m wide × 12 m deep is floating in salt water on an even keel at 6 m draft. A forward compartment is 10 m long, 12 m wide, and extends from the outer bottom to a watertight flat, 4 m above the keel. The compartment contains cargo of permeability 25% (see Figure 30.5). Find the new drafts if this compartment is bilged.

$$\left.\begin{array}{l}\text{Mass of water entering}\\ \text{the bilged compartment}\end{array}\right\} = \frac{25}{100} \times 10 \times 12 \times 4 \times 1.025$$

$$= 123 \text{ tonnes}$$

$$TPC_{sw} = \frac{WPA}{97.56} \qquad \text{Increase in draft} = w/TPC$$

$$= \frac{100 \times 20}{97.56} \qquad = 123/20.5$$

$$= 6 \text{ cm}$$

$$\underline{TPC} = 20.5 \text{ tonnes} \qquad \underline{\text{Increase in draft} = 0.06 \text{ m}}$$

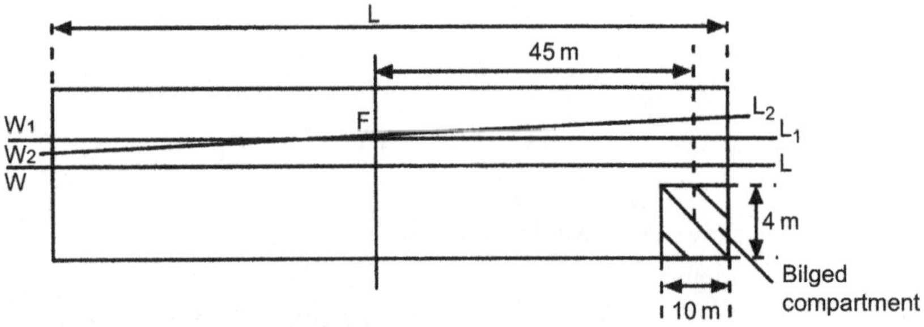

Figure 30.5

$$W = L \times B \times d_1 \times 1.025 \qquad BM_L = \frac{I_L}{V} = \frac{BL^3}{12V} = \frac{B \times L^3}{12 \times L \times B \times d} = \frac{L^2}{12 \times d_1}$$

$$= 100 \times 20 \times 6 \times 1.025 \qquad BM_L = \frac{100 \times 100}{12 \times 6}$$

$$W = \underline{12,300 \text{ tonnes}} \qquad \underline{BM_L = 139 \text{ meters}}$$

$$MCTC \simeq \frac{W \times BM_L}{100 \times L} = \frac{12,300 \times 139}{100 \times 100}$$

$$MCTC = 171 \text{ tonnes m per cm}$$

$$\text{Trimming moment} = W \times B_1B_2$$

$$= w \times d$$

$$\text{Trimming moment} = 123 \times 45 \text{ tonnes m}$$

$$\text{Change of trim} = \frac{\text{Trimming moment}}{MCTC} = \frac{123 \times 45}{171}$$

$$\text{Change of trim} = 32.4 \text{ cm by the head}$$

That is, 0.32 m by the head.

Note. The center of flotation, being the centroid of the waterplane area, remains amidships.

Old drafts	A	6.00 m	F	6.00 m
Bodily increase		+ 0.06 m		+ 0.06 m
		6.06 m		6.06 m
Change due to trim		− 0.16 m		+ 0.16 m
Ans. New drafts	A	5.90 m	F	6.22 m

■

Effect of Bilging on Stability

It has already been shown that when a compartment in a ship is bilged the mean draft is increased. The change in mean draft causes a change in the positions of the center of buoyancy and the initial metacenter. Hence KM is changed and, since KG is constant, the GM will be changed.

Example 1

A box-shaped vessel 40 m long, 8 m wide, and 6 m deep floats in salt water on an even keel at 3 m draft. GM = 1 m. Find the new GM if an empty compartment 4 m long and situated amidships is bilged.

(a) *Original condition before bilging*

Find the KG:

$$KB = \frac{d_1}{2} = 1.5 \text{ meters}$$

$$BM = \frac{I}{V} = \frac{LB^3}{12V} = \frac{B^2}{12 \times d_1} = \frac{8 \times 8}{12 \times 3}$$

$$\begin{aligned} BM &= 1.78 \text{ m} \\ KB &= +\underline{1.50 \text{ m}} \\ KM &= 3.28 \text{ m} \\ GM &= -\underline{1.00 \text{ m}} \\ KG &= \underline{2.28 \text{ m}} \end{aligned}$$

(b) *Vessel's condition after bilging*

Find the new draft:

$$\text{Lost buoyancy} = 4 \times 8 \times 3 \times 1.025 \text{ tonnes}$$

$$TPC_{SW} = \frac{WPA}{97.56} = \frac{36 \times 8}{97.56}$$

$$\text{Increase in draft} = \frac{\text{Lost buoyancy}}{TPC}$$

$$= 4 \times 8 \times 3 \times 1.025 \times \frac{100}{36 \times 8 \times 1.025} \text{ cm}$$

$$= 33.3 \text{ cm or } 0.33 \text{ m}$$

It should be noted that the increase in draft can also be found as follows:

$$\text{Increase in draft} = \frac{\text{Volume of lost buoyancy}}{\text{Area of intact waterplane}} = \frac{4 \times 8 \times 3}{36 \times 8}$$

$$= 1/3 \text{ meter}$$

$$\text{Original draft} = 3.000 \text{ m} = \text{Draft } d_1$$

$$\text{New draft} = 3.333 \text{ m} = \text{Draft } d_2$$

(c) *Find the new GM*

$$KB = \frac{d_2}{2} = 1.67 \text{ m}$$

$$BM = I/V$$ (*Note.* 'I' represents the second moment of the intact waterplane about the centerline)

$$= \frac{(L-1)B^3}{12 \times V} = \frac{36 \times 8^3}{12 \times 40 \times 8 \times 3}$$

$$BM_2 = 1.60 \text{ m}$$

$$+$$

$$KB_2 = \underline{1.67 \text{ m}}$$

$$KM_2 = 3.27 \text{ m}$$

$$-$$

$$KG = \underline{2.28 \text{ m}} \text{ as before bilging occurred}$$

$$\underline{\text{Final } GM_2} = \underline{0.99 \text{ m}}$$

GM_2 is positive so the vessel is in stable equilibrium.

■

Summary

When solving problems involving bilging and permeability the following procedure is suggested:

1. Make a sketch from the given information.
2. Calculate mean bodily sinkage using w and TPC.
3. Calculate change of trim using GM_L or BM_L.
4. Collect calculated data to evaluate the final requested end drafts.

Exercise 30

Bilging Amidships Compartments

1. (a) Define permeability, 'μ'.
   (b) A box-shaped vessel 100 m long, 15 m beam floating in salt water, at a mean draft of 5 m, has an amidships compartment 10 m long that is loaded with a general cargo. Find the new mean draft if this compartment is bilged, assuming the permeability to be 25%.
2. A box-shaped vessel 30 m long, 6 m beam, and 5 m deep has a mean draft of 2.5 m. An amidships compartment 8 m long is filled with coal stowing at 1.2 cubic meters per tonne. One cubic meter of solid coal weighs 1.2 tonnes. Find the increase in the draft if the compartment is holed below the waterline.
3. A box-shaped vessel 60 m long, 15 m beam, floats on an even keel at 3 m draft. An amidships compartment is 12 m long and contains coal (SF = 1.2 cubic meters per tonne and relative density = 1.28). Find the increase in the draft if this compartment is bilged.
4. A box-shaped vessel 40 m long, 6 m beam, is floating at a draft of 2 m F and A. She has an amidships compartment 10 m long, which is empty. If the original GM was 0.6 m, find the new GM if this compartment is bilged.
5. If the vessel in Question 4 had cargo stowed in the central compartment such that the permeability was 20%, find the new increase in the draft when the vessel is bilged.
6. A box-shaped vessel 60 m × 10 m × 6 m floats on an even keel at a draft of 5 m F and A. An amidships compartment 12 m long contains timber of relative density 0.8 and stowage factor 1.4 cubic meters per tonne. Find the increase in the draft if this compartment is holed below the waterline.
7. A box-shaped vessel 80 m × 10 m × 6 m is floating upright in salt water on even keel at 4 m draft. She has an amidships compartment 15 m long that is filled with timber (SF = 1.5 cubic meters per tonne). One tonne of solid timber would occupy 1.25 cubic meters of space. What would be the increase in the draft if this compartment is now bilged?

Bilging End Compartments

8. A box-shaped vessel 75 m × 12 m is floating upright in salt water on even keel at 2.5 m draft F and A. The forepeak tank, which is 6 m long, is empty. Find the final drafts if the vessel is now holed forward of the collision bulkhead.
9. A box-shaped vessel 150 m long, 20 m beam, is floating upright in salt water at drafts of 6 m F and A. The collision bulkhead is situated 8 m from forward. Find the new drafts if the vessel is now bilged forward of the collision bulkhead.

10. A box-shaped vessel 60 m long, 10 m beam, is floating upright in salt water on even keel at 4 m draft F and A. The collision bulkhead is 6 m from forward. Find the new drafts if she is now bilged forward of the collision bulkhead.
11. A box-shaped vessel 65 m × 10 m × 6 m is floating on even keel in salt water at 4 m draft F and A. She has a forepeak compartment 5 m long that is empty. Find the new drafts if this compartment is now bilged.
12. A box-shaped vessel 64 m × 10 m × 6 m floats in salt water on even keel at 5 m draft. A forward compartment 6 m long and 10 m wide extends from the outer bottom to a height of 3.5 m, and is full of cargo of permeability 25%. Find the new drafts if this compartment is now bilged.

■

CHAPTER 31

Effects of Side Winds on Ship Stability

When wind forces are on the side of a ship, it is possible to determine the heeling arm. Consider Figure 31.1 and assume that:

P = wind force = $2 \times 10^{-5} \times A_m \times (V_K)^2$ tonnes
V_K = the velocity in kts of the wind on side of the ship
$\breve{y}$ = lever from ship's VCB to the center of area exposed to the wind
θ = angle of heel produced by the beam wind
W = ship's displacement in tonnes
A_m = area at side of ship on which the wind acts in m^2
GZ = righting lever necessary to return ship back to being upright
$W \times GZ$ = ship's righting moment

$$\text{The wind heeling arm } (y_2) = \frac{(P \times \breve{y} \times \cos^2\theta)}{W} \text{ meters}$$

This wind heeling arm can be estimated for all angles of heel between 0° and 90°. These values can then be superimposed onto a statical stability curve (see Figure 31.2).

Figure 31.2 shows that:

θ_1 = Where the two curves intersect to give the angle of steady heel. It is the angle the ship would list to if the wind was steady and there were no waves.
θ_2 = Where the two curves intersect to give the angle beyond which the ship would capsize. Note how the ship capsizes at a slightly smaller angle of heel than if there had been no wind present.

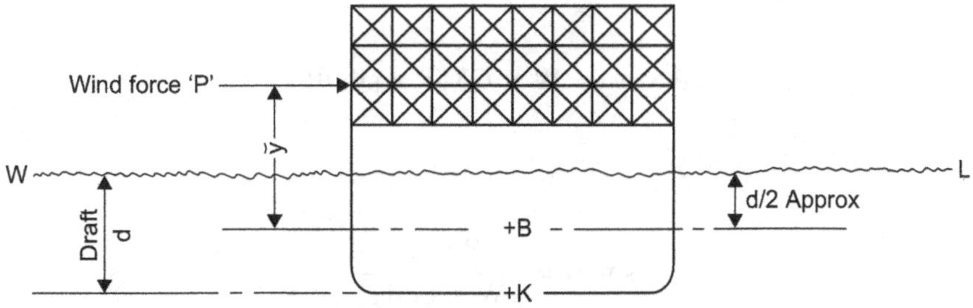

Figure 31.1:
Wind Force on Stowed Containers.

Ship Stability for Masters and Mates. http://dx.doi.org/10.1016/B978-0-08-097093-6.00031-1

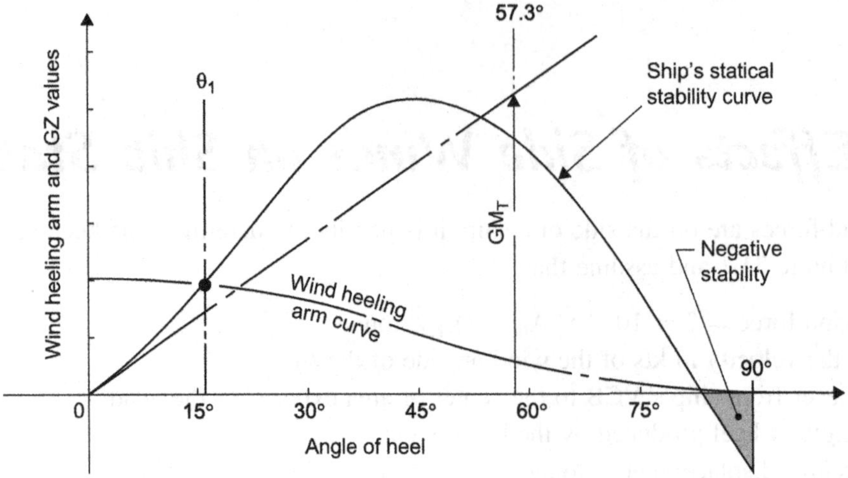

Figure 31.2:
GZ Values and Wind Heeling Arm Against Angle of Heel.

Now consider Figure 31.3. The application of the wind on the side of the ship causes an equal and opposite reaction. This causes an upsetting moment on the ship:

$$\text{The upsetting moment caused by the wind} = p \times \breve{y} \text{ tm} \qquad \text{(I)}$$

$$\begin{aligned}\text{The righting moment at small angles of heel} &= \text{W} \times \text{GZ} \\ &= \text{W} \times \text{GMT} \times \sin\theta 1\end{aligned} \qquad \text{(II)}$$

$$\text{Equation (I)} = \text{Equation (II)}$$

So

$$\text{P} \times \breve{y} = \text{W} \times \text{GMT} \times \sin\theta 1$$

Hence

$$\sin\theta_1 = \frac{(\text{P} \times \breve{y})}{(\text{W} \times \text{GM}_\text{T})}$$

where θ_1 = angle of steady heel.

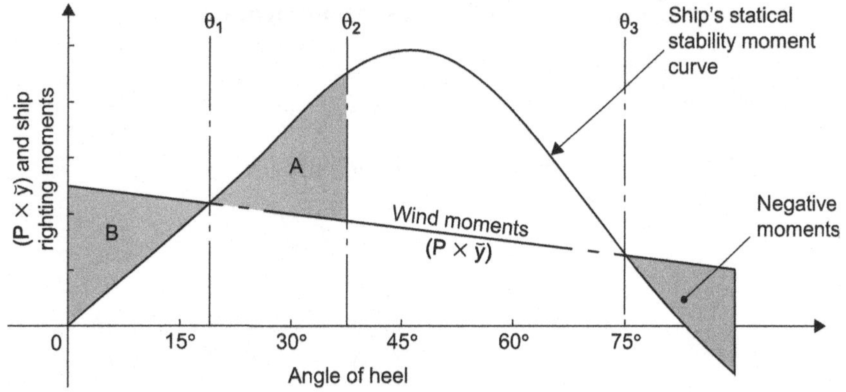

Figure 31.3:
Wind Moments and Righting Moments of a Ship.

Going back, it has been shown that $P \times \breve{y} = W \times GZ$. So

$$GZ = \frac{(P \times \breve{y})}{W}$$

where GZ = the righting lever.

Figure 31.3 shows these moments (I) and (II) superimposed on one view. From Figure 31.3 it can be seen that:

θ_1 = Angle of steady heel. This occurs with a steady beam wind.
θ_2 = Angle of lurch, where (area A) = (area B). This occurs with a gusting beam wind.
θ_3 = Angle of heel beyond which the vessel will capsize.

The problems of beam winds are exacerbated on ships with a very large side area, known as 'sail area'. These ships could be:

1. Container vessels, returning with emptied containers stowed five high on the upper deck.
2. VLCCs and ULCCs, due to their very large LBP and depth parameters.
3. Ships sailing in very light ballast condition, with consequent large freeboard values.
4. LNG and LPG ships. In their loaded departure condition, the freeboard divided by the depth molded is approximately 50%.

Exercise 31

1. (a) With the aid of a statical stability curve, clearly show how wind heeling arm effects on the side of a ship reduce a ship's stability.
   (b) List, with explanations, three types of vessels on which side winds could cause these reductions in ship stability.

2. With the aid of a sketch of a statical stability moment curve and a wind moment curve, show the following:
   (a) Angle of steady heel.
   (b) Angle of lurch.
   (c) Angle of heel beyond which the vessel will capsize.

CHAPTER 32

Icing Allowances Plus Effects on Trim and Stability

In Arctic ocean conditions, the formation of ice on the upper structures of vessels can cause several problems (see Figures 32.1 and 32.2). Ice build-up can be formed from snowfall, sleet, blizzards, freezing fog, and sea spray in sub-zero temperatures. In the Arctic, the air temperatures can be as low as −40 °C in harbor and −30 °C at sea.

Icing allowances must be made for:

- Rise in G. Loss of transverse stability.
- Increase in weight. Increased draft due to increased weight.
- Loss of freeboard due to increased weight.
- Decrease in underkeel clearance.
- Contraction of steel due to temperature.
- Increased brittleness in steel structures.
- Nonsymmetrical formation of ice.
- Angle of list. Angle of loll.
- Change of trim.
- Impairment of maneuverability.

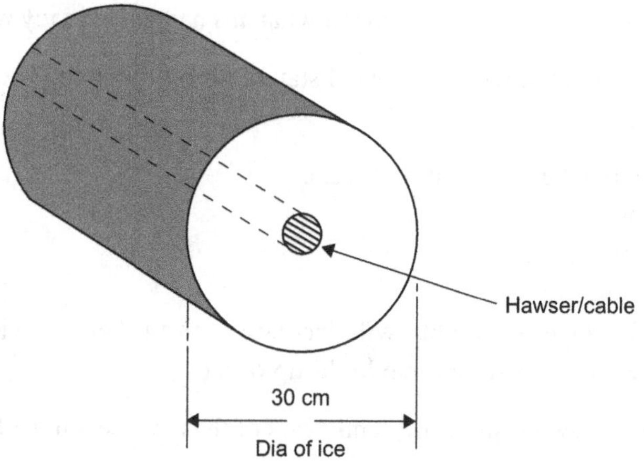

Figure 32.1:
Build-up of Ice Around a Hawser/Cable.

Ship Stability for Masters and Mates. http://dx.doi.org/10.1016/B978-0-08-097093-6.00032-3

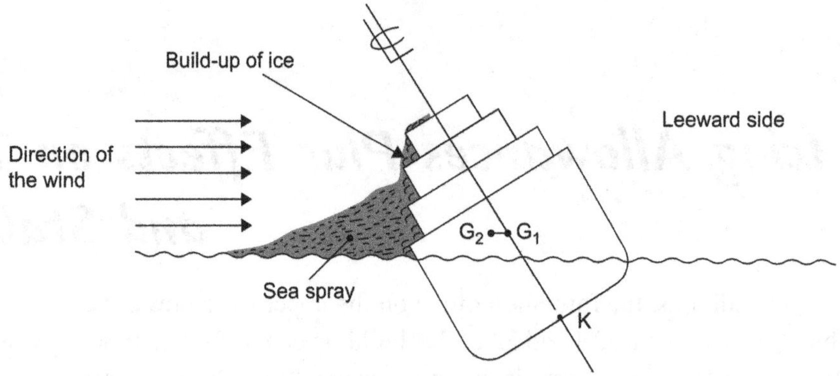

Figure 32.2: Asymmetrical Build-up of Ice, Causing an Angle of List.
G_1 moves to G_2.

- Reduction in forward speed.
- Increase in windage area on side of ship.

A 30 cm diameter of ice can form around a hawser or cable (see Figure 32.1). Blocks of ice 100 cm thick have been known to form on the poop deck of a ship in very cold weather zones. Walls of 60 cm of ice forming on the surface of a bridge front have been recorded. In ice-ridden waters the depth of ice may be up to 3 m.

In 1968, three British trawlers rolled over and sank in the North Atlantic. Icing was given as the cause of them capsizing. Only three out of 60 men survived.

Fishing vessels suffer most from the above icing effects. In 1974, the *Gaul* went down in the Barents Sea just off the northern tip of Norway. Thirty-six lives were lost.

These losses indicate not what *could* happen but what *has* happened in icy weather conditions.

Figure 32.3 shows the effect of ice on a typical statical stability curve. The changes that take place are:

- Decreased values for the GZ righting levers.
- GM_T decreases in value.
- GZ_{max} value decreases.
- Range of stability decreases.
- GZ values and the range of stability will decrease even further if a wind is present on the side of the ship with lesser transverse build-up of ice.

The IMO suggest the following structures and gears of these vessels must be kept free of ice:

- Aerials
- Running and navigational lights

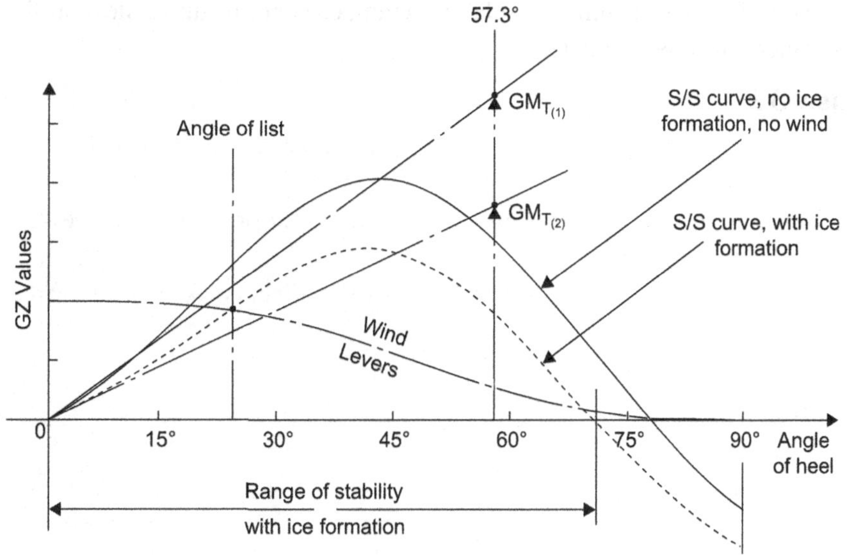

Figure 32.3:
Loss of Statical Stability Due to Wind and Formation of Ice.

- Freeing ports and scuppers
- Lifesaving craft
- Stays, shrouds, masts, and rigging
- Doors of superstructures and deckhouses
- Windlass and hawse holes.

Formation of ice on the upper works of the vessel must be removed as quickly as possible by:

- Cold water pressure.
- Hot water and steam.
- Break-up of ice with ice-crows, axes, ice-picks, metal ice-scrapers, wooden sledgehammers, and metal shovels.
- Heating of upper structures similar in effect to radiators or central heating arrangements in a house.

In making allowances for ships operating in Arctic conditions, Lloyds' Strength Regulations 'request' that transverse framing and beams forward of the collision bulkhead be decreased to about 450 mm maximum, from the more usual 610 mm spacing. In this region, the thickness of the shell plating has to be increased.

After a series of capsizes and vessel losses to fishing boats, the next generation of vessels had fewer structures and gears exposed to sub-zero temperatures. If it is under cover, it is less likely to freeze.

A final suggestion for these fishing vessels to remember is not to turn unless sufficient and safe ice allowances have been made.

Exercise 32

1. (a) List the icing allowances a master or mate must make when a vessel is in waters of sub-zero temperatures.
   (b) List four methods for the removal of ice on the upper works of a vessel operating in sub-zero temperatures.
2. On a sketch of a statical stability curve, show the effects of wind and formation of ice on the following:
   (a) GZ values
   (b) GM_T value
   (c) Range of stability
   (d) Angle of list.

CHAPTER 33

The Sectional Area Curve

The reason for having a Sectional Area curve is that for every ship there is an optimum underwater hull form to give the best ship performance with regard to ship speed and/or oil fuel consumption. For every ship there is an *optimum LCB* about amidships and an *optimum* C_b value in conjunction with the ship's *service speed and deadweight.*

The Sectional Area curve assists the naval architect and the towing tank specialist to develop the best hull form for the new ship design. As backup to statements in the introductory paragraph, see also Chapter 6.

This curve gives the sectional areas along the ship's length LBP. The areas are calculated from the base up to the draft mld (even keel). Distribution under the Sectional Area (SA) curve in a longitudinal direction determines the longitudinal center of buoyancy (LCB) about amidships.

A second factor determined from the curve is the prismatic coefficient (C_p), where C_p is defined as follows:

$$C_p = \text{Area enclosedby the SA curve}/(\text{midship area} \times \text{LBP})$$

SA ordinates in m^2 are usually expressed in terms of the midship area (A_m) taken as unity. All other cross-sectional areas are then related to this A_m value of 1.0 (see Figure 33.1 and Worked Example 1 for the 128 m LBP vessel).

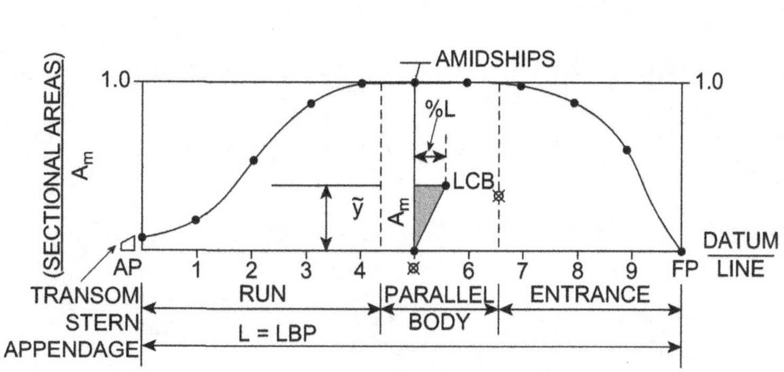

Figure 33.1:
The Sectional Area Curve.

Ship Stability for Masters and Mates. http://dx.doi.org/10.1016/B978-0-08-097093-6.00033-5

Worked Example 1

The ordinates of a curve of areas of immersed sections (SA ords) for a ship of 128 m LBP, 18.30 m br mld, and 8.25 m draft mld are shown in the table below. Midship area is 150.22 square meters. Estimate C_p, C_m, LCB about amidships, C_b, fully loaded displacement in tonnes, and $\tilde{y}$ from datum.

Station	S. Area	SM	Vol Ftn	Lever	Mmt Ftn	SA Ords	SA Squared	SM	Moment
AP 0	0.04	0.5	0.02	5.0	0.10	0.04	0.002	0.5	0.00
0.5	0.16	2.0	0.32	4.5	1.44	0.16	0.026	2.0	0.05
1	0.36	1.5	0.54	4.0	2.16	0.36	0.130	1.5	0.19
2	0.73	4.0	2.92	3.0	8.76	0.73	0.533	4.0	2.13
3	0.91	2.0	1.82	2.0	3.64	0.91	0.828	2.0	1.66
4	1.00	4.0	4.00	1.0	4.00	1.00	1.000	4.0	4.00
Amidships	1.00	2.0	2.00	0.0	20.10	1.00	1.000	2.0	2.00
6	1.00	4.0	4.00	1.0	4.00	1.00	1.000	4.0	4.00
7	0.98	2.0	1.96	2.0	3.92	0.98	0.960	2.0	1.92
8	0.86	4.0	3.44	3.0	10.32	0.86	0.740	4.0	2.96
9	0.48	1.5	0.72	4.0	2.88	0.48	0.230	1.5	0.35
9.5	0.21	2.0	0.42	4.5	1.89	0.21	0.044	2.0	0.09
FP 10	0.00	0.5	0.00	5.0	0.00	0.00	0.000	0.5	0.00
		30	22.16		23.01				19.35
			0.739			Vol Ftn = 22.16			

$$C_p = 22.16/30 = 0.739$$

$$C_m = 150.22/(18.3 \times 8.25) = 0.995$$

$$\text{LCB} = (23.01 - 20.10)/22.16 = 1.31\% \text{ L forward of amidships}$$

$$C_b = 0.995 \times 0.739 = 0.735$$

$$\text{Displacement} = 128 \times 18.30 \times 8.25 \times 0.735 \times 1.025 = 14,559 \text{ tonnes}$$

$$\tilde{y} = (19.35 \times 0.50)/22.16 = 0.437 \text{ from datum}$$

Worked Example 2

The ordinates for a curve of immersed sections (SA ords) for a ship of 91.46 m length, 14.63 m breadth mld and 3.66 m draft mld are shown in the table below. Midship area is 51.40 m^2.

Estimate the values of C_p, C_m, C_b, ship's displacement in tonnes, LCB from amidships and $\tilde{y}$ from the datum.

Station	S. Area	SM	Vol Ftn	Lever	Mmt Ftn	SA for Second Ship	SA Squared	SM	Moment
0	0.025	0.5	0.013	5	0.06	0.025	0.001	0.5	0.00
0.5	0.089	2	0.178	4.5	0.80	0.089	0.008	2	0.02
1	0.203	1.5	0.305	4	1.22	0.203	0.041	1.5	0.06
2	0.456	4	1.824	3	5.47	0.456	0.208	4	0.83
3	0.714	2	1.428	2	2.86	0.714	0.510	2	1.02
4	0.911	4	3.644	1	3.64	0.911	0.830	4	3.32
5	1.000	2	2.000	0	14.05	1.000	1.000	2	2.00
6	0.910	4	3.640	1	3.64	0.910	0.828	4	3.31
7	0.648	2	1.296	2	2.59	0.648	0.420	2	0.84
8	0.375	4	1.500	3	4.50	0.375	0.141	4	0.56
9	0.161	1.5	0.242	4	0.97	0.161	0.026	1.5	0.04
9.5	0.072	2	0.144	4.5	0.65	0.072	0.005	2	0.01
10	0.000	0.5	0.000	5	0.00	0.000	0.000	0.5	0.00
		30	16.213		12.35				12.01
			0.540			Vol Ftn = 16.213 for second ship			

$$C_p = \frac{16.213}{30} = 0.540$$

$$C_m = \frac{51.40}{14.63 \times 3.66} = 0.960$$

$$C_b = C_m \times C_p = 0.960 \times 0.540 = 0.518$$

Ship's displacement = 91.46 × 14.63 × 3.66 × 0.518 × 1.025 = 2600 tones.

$$LCB = \frac{(14.05 - 12.35)}{16.213} = 1.07\%L \text{ Aft of Amidships.}$$

$$\bar{y} = \text{from the datum.} = \frac{12.01}{16.213} \times \frac{1}{2} = 0.370$$

The first Worked Example related to a full- form vessel. The second Worked Example related to a Fine-form vessel. Note how the difference in hull-form produced changes in the C_p, C_b, LCB about Amidships and the value of $\bar{y}$ from the datum.

Definitions

Parallel body. The length of the ship where the shape of the midship section does not change. Consequently, the midship area is constant over that part of the vessel (see Figure 33.1). For ULCCs, this length can be as much as 65% LBP. For container vessels and ro-ro ships, this length will be of the order of 0% LBP.

Entrance. The length of vessel from the forward point of the parallel body to the forward perpendicular (FP). See Figure 33.1.

Run. The length of the vessel from the aft point of the parallel body to the aft perpendicular (AP). See Figure 33.1.

LCB amidships. The two-dimensional center of the SA curve measured about amidships. It is in fact the three-dimensional centroid of the underwater hull form of the ship. It is dependent upon the type of ship being considered. LCB can be as far aft as 3% LBP *aft* of amidships for very fine-form vessels. LCB can be as far forward as 3% LBP *forward* of amidships for very full-form vessels. For further details, see Chapter 6.

C_m. This is equal to (midship area)/(breadth mld × draft mld). See Chapter 5.

C_p. This is equal to (volume of displacement)/(midship area × LBP). $C_p = C_b/C_m$, as shown in Chapter 5.

$\bar{y}$. This factor is the transverse center of gravity measured from the datum line (see Figure 33.1). Using the A_m ordinate as 1, feedback shows that for very fine-form vessels $\bar{y}$ will be approximately 0.35, and for very full-form ships $\bar{y}$ will be approximately 0.45. This factor $\bar{y}$ is used where the LCB for a basis ship is being compared and modified with LCB for a new design on Sectional Area curves. Limited space within this book prevents further written discussion. For further information on procedure, see *Elements of Ship Design* by Ross Munro-Smith.

Reference

Munro-Smith, R. (1975). *Elements of Ship Design*. Marine Media Management.

CHAPTER 34

FL and PL Curves Plus Type A and Type B Vessels

Definitions

Bulkhead deck. This is the uppermost deck to which the transverse watertight bulkheads are carried.

Margin line. This is a line drawn parallel to and 76 mm below the bulkhead deck at side (see Figure 34.1).

Permeability. This is the amount of water that can enter a compartment after the compartment has been bilged. It is denoted as 'μ' and given as a percentage. If the compartment was initially empty, then 'μ' would be 100%.

Floodable length. This is the maximum allowable length of a compartment at any point along the length (with that point as center) that can be flooded without submerging the margin line. Vessel to be always upright, with no heel.

Floodable length (FL) curve. This is the curve that, at every point in the vessel's length, has an ordinate representing the length of the ship that may be flooded without the margin line being submerged. Vessel to be upright.

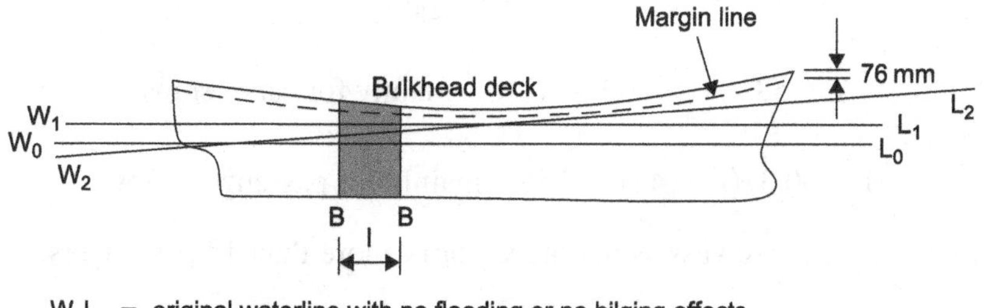

W_0L_0 = original waterline with no flooding or no bilging effects
W_1L_1 = waterline with mean bodily sinkage effects only in the bilged condition
W_2L_2 = waterline with mean bodily sinkage trimming effects in bilged condition, tangential to the margin line
l = length of flooded compartment that has produced the waterline W_2L_2
B = main watertight transverse bulkheads, giving the permissible length of the compartment

Figure 34.1

Ship Stability for Masters and Mates. http://dx.doi.org/10.1016/B978-0-08-097093-6.00034-7

Permissible length (PL) curve. This is a lower curve, obtained after the floodable length curve ordinates have been modified for contents within the compartments being considered.

Factor of subdivision (F_S). This is the factor of subdivision. It can range in value from 0.50 to a maximum of 1.00. The 1.00 value signifies that very few passengers are being carried on board. The 0.50 value signifies that a very large number of passengers are being carried on the ship.

By using the following formula, F_S is used to determine the permissible length ordinates:

$$\text{FL ordinates} \times F_S = \text{PL ordinates}$$

Subdivision load line. This is the waterline corresponding to the normal designed waterline. It is drawn parallel to the ship's keel.

Subdivision length (L). This is the length measured between the perpendiculars erected at the ends of the subdivision load line.

Subdivision beam (B). This is the greatest breadth at or below the ship's deepest subdivision load line.

Subdivision draft (d). This is the molded draft to the subdivision load waterline.

Curve of permissible lengths. In any ship, the closer the main transverse bulkheads are, the safer will be the ship. However, too many transverse bulkheads would lead to the vessel being commercially nonviable.

The Regulations Committee suggested that the PL ordinates should be some proportion of the FL ordinates. To achieve this, it was suggested that a factor of subdivision (F_S) be used, where:

$$F_S = A - \frac{\{(A - B)(C_S - 23)\}}{100}$$

$$A = 58.2/(L - 60) + 0.18 \quad \text{mainly for cargo ships}$$

$$B = 30.3/(L - 42) + 0.18 \quad \text{mainly for passenger ships}$$

Cargo–Passenger vessels are vessels that never carry more than 12 passengers.

Passenger–Cargo vessels are vessels that carry more than 12 passengers.

Changing cargo spaces into accommodation spaces will alter the factor of subdivision. It will decrease its value. This will make the permissible lengths smaller and make for a safer ship.

If some compartments are designed to carry cargo on some voyages and passengers on others, then the ship will be assigned more than one subdivisional load line.

Criterion of service numeral (C_S). If the ship's subdivision length is greater than 131 m, then C_S will have, as per regulations, a range of values of 23–123. The lower limit of 23 applies for Type 'A' ships (carrying liquid in bulk). The upper limit of 123 applies for Type 'B' ships.

The regulations state C_S is to be:

$$C_S = 72(M + 2P)/V$$

where M = total volume of machinery spaces below the margin line and P = total volume of passenger space and crew space below the margin line. This will obviously take into account the number of passengers and crew on board ship. V = total volume of ship from keel to the margin line.

Worked Example 1

Calculate the criterion of service numeral (C_S) when M is 3700 m^3, P is 2800 m^3, and V is 12,000 m^3.

$$C_S = 72(M + 2P)/V = 72(3700 + 5600)/12,000$$
$$\text{so } C_S = 55.8$$

Worked Example 2

Calculate the factor of subdivision when a ship has a subdivision length (L) of 140 m and a criterion of service numeral of 54.5.

$$A = 58.2/(L - 60) + 0.18 = 58.2/(140 - 60) + 0.18 = 0.908$$
$$B = 30.3/(L - 42) + 0.18 = 30.3/(140 - 42) + 0.18 = 0.489$$

$$F_S = A - \frac{\{(A - B)(C_S - 23)\}}{100}$$

$$F_S = 0.908 - \frac{\{(0.908 - 0.489)(54.5 - 23)\}}{100}$$

$$\text{so } F_S = 0.776$$

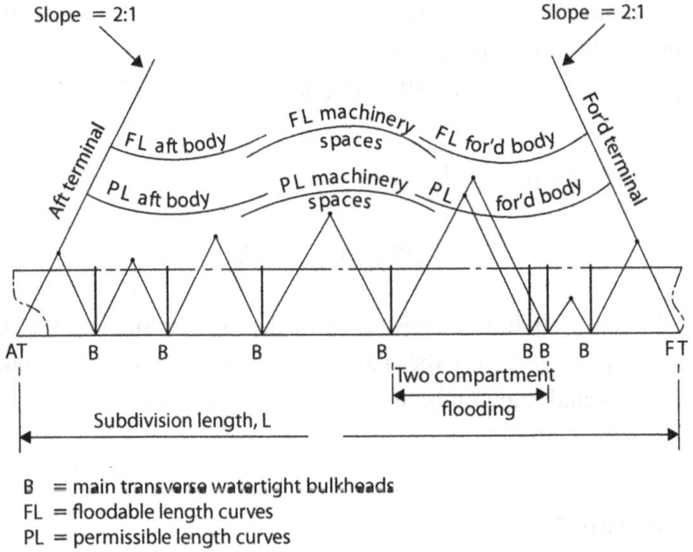

Figure 34.2:
Subdivision Curves for a Passenger Liner: Diagrammatic Sketch.

Figure 34.2 shows a set of subdivision curves. Observations can be made from these curves to give a greater understanding of why they exist.

The triangles all have a height that is equal to the base. Thus, the slope is 2:1. The base in fact is the permissible length of the newly designed compartment.

If need be, the apex of any of these triangles can go up as far as the PL curve. This would make the compartment have the maximum length within the regulations.

In most cases, the top apices of these shown triangles are not connected to the PL curves. However, in 'two-compartment flooding', the regulations do allow the PL curve to be crossed. This is when the adjacent bulkhead sloping line does not extend beyond the FL curve (see illustration of this in Figure 34.2). It is also possible to arrange for 'three-compartment flooding'. The resulting smaller length compartments may be used as baggage spaces or storerooms.

Note how the curves of aft body and forward body do not join those of the machinery space. This is because there are differences in permeability 'μ' in these localities of the ship. For example, passenger spaces have a permeability of about 95%, grain spaces have a permeability of 60–65% whilst the machinery spaces will have a permeability of 80–85%.

Floodable lengths. The basic features that affect the floodable curves for a ship are the block coefficient, sheer ratio, freeboard ratio, and permeability.

$$\text{Block coefficient} = \frac{\text{Molded displacement (excluding bossings)}}{\text{L} \times \text{B} \times \text{d}}$$

$$\text{Sheer ratio} = (\text{Sheer aft or forward})/\text{d}$$

$$\text{Freeboard ratio} = \frac{\text{Freeboard to the margin line at amidships}}{\text{d}}$$

$$\text{Permeability} = (\text{Ingress of water}/\text{Volume of compartment}) \times 100/1$$

A DfT 'standard ship' is used as a basis ship. This ship is assigned two permeability values. One is 60% and the other is 100%. Interpolation methods are used to obtain a first estimation of the FL values for the new design being considered. These values are adjusted for sectional area ratios and permeability factors (PF), where:

$$\text{PF} = 1.5(100 - \mu)/\mu$$

Summary of Procedure Steps

1. Determine the subdivision length of the new ship.
2. Calculate the block coefficient, sheer ratios, freeboard ratios, and permeability values for the new design.
3. Evaluate the values of 'A' and 'B' coefficients.
4. Determine, for the new design, volumes for 'P', 'V', and 'M'.
5. Calculate the criterion of service numeral (C_S).
6. Evaluate the factor of subdivision (F_S).
7. Multiply the FL ordinates by F_S to obtain PL ordinates for the aft body, the machinery spaces, and the forward body.
8. Plot FL and PL curves, and superimpose the main transverse bulkhead positions together with isosceles triangles having 2:1 slopes as per Figure 34.2.
9. If desired, opt for a 'two-compartment flooding' system, as previously described.
10. Adjust and decide upon the *final* positions of main transverse bulkheads for the new design.

Exercise 34

1. Sketch a set of subdivision curves for a passenger ship. Include one example of two-compartment flooding. Label the important parts on your diagram.
2. For a passenger ship, the subdivision length is 145 m and the criterion of service numeral is 56.5. Calculate her factor of subdivision (F_S).
3. Define the following floodable and permissible length terms:
   (a) Subdivision length (L)
   (b) Margin line

(c) Bulkhead deck
(d) Factor of subdivision (F_S)
(e) Criterion of service numeral (C_S).

4. Calculate the criterion of service numeral (C_S) when the total volume of machinery spaces below the margin line (M) is 3625 m^3, the total volume of passenger space and crew spaces below the margin line (P) is 2735 m^3, and the total volume of ship from keel to the margin line (V) is 12,167 m^3.

■

CHAPTER 35

Load Lines and Freeboard Marks

The Link

Freeboard and stability curves are inextricably linked. With an *increase* in the freeboard:

- Righting levers (GZ) are increased.
- GM_T increases.
- Range of stability increases.
- Deck edge immerses later at greater angle of heel.
- Dynamical stability increases.
- Displacement decreases.
- KB decreases.

Overall, both the stability and the safety of the vessel are *improved*.

Historical Note

In 1876, Samuel Plimsoll introduced a law into Parliament that meant that ships were assigned certain freeboard markings above which, in particular conditions, they were not allowed to load. Prior to this law a great many ships were lost at sea, mainly due to overloading.

In 1930 and in 1966, international conferences modified and expanded these statutory regulations dealing with the safety of ships. These regulations have been further improved over the years by conference meetings every 3 or 4 years up to the present day. One such organization was the Safety of Life at Sea organization (SOLAS). In recent years, the IMO has become another important maritime regulatory body.

Definitions

Type 'A' vessel. A ship that is designed to carry only liquid cargoes in bulk, and in which cargo tanks have only small access openings, closed by watertight gasketed covers of steel or equivalent material.

The exposed deck must be one of high integrity. It must have a high degree of safety against flooding, resulting from the low permeability of loaded cargo spaces and the degree of bulkhead subdivision usually provided.

Ship Stability for Masters and Mates. http://dx.doi.org/10.1016/B978-0-08-097093-6.00035-9

Type 'B' vessels. All ships that do not fall under the provisions for Type 'A' vessels. For these ships, careful consideration must be given to the following:

- The vertical extent of damage is equal to the depth of the ship.
- The penetration of damage is not more than one-fifth of the breadth molded (B).
- No main transverse bulkhead is damaged.
- Ship's KG is assessed for homogeneous loading of cargo holds, and for 50% of the designed capacity of consumable fluids and stores, etc.

Type 'B-60' vessels. The vessel must have an LBP of between 100 and 150 m. It must survive the flooding of any *single* compartment (excluding the machinery space). If greater than 150 m LBP, the machinery space must be considered as a floodable compartment. A typical ship type for a Type 'B-60' vessel is a bulk carrier.

Type 'B-100' vessels. The vessel must have an LBP of between 100 and 150 m. It must survive the flooding of any two adjacent fore and aft compartments (excluding the machinery space). If greater than 150 m LBP, the machinery space must be considered as a floodable compartment. Such a vessel may be classified as a Type 'A' vessel.

The minimum DfT tabular freeboard values are shown in graphical form in Figure 35.1.

Freeboards of Oil Tankers and General Cargo Ships

Oil tankers are permitted to have more Summer freeboard than general cargo ships with a similar LBP. They are considered to be safer ships for the following reasons:

1. They have much *smaller deck openings* in the main deck.
2. They have *greater subdivision*, by the additional longitudinal and transverse bulkheads.
3. Their cargo oil has *greater buoyancy* than grain cargo.
4. They have *more pumps* to quickly control ingress of water after a bilging incident.
5. Cargo oil has a permeability of about 5% whilst grain cargo has a permeability of 60–65%. The *lower permeability* will instantly allow less ingress of water following a bilging incident.
6. Oil tankers will have *greater GM values.* This is particularly true for modern double-skin tankers and wide shallow draft tankers.

Tabulated Freeboard Values: Procedure

When the freeboard for a vessel is being assigned, the procedure is to compare a basic Department for Transport (DfT) 'standard ship' with the 'new design' about to enter service. Figure 35.2 shows profiles of these two vessels. Both vessels will have the same freeboard length (L_F):

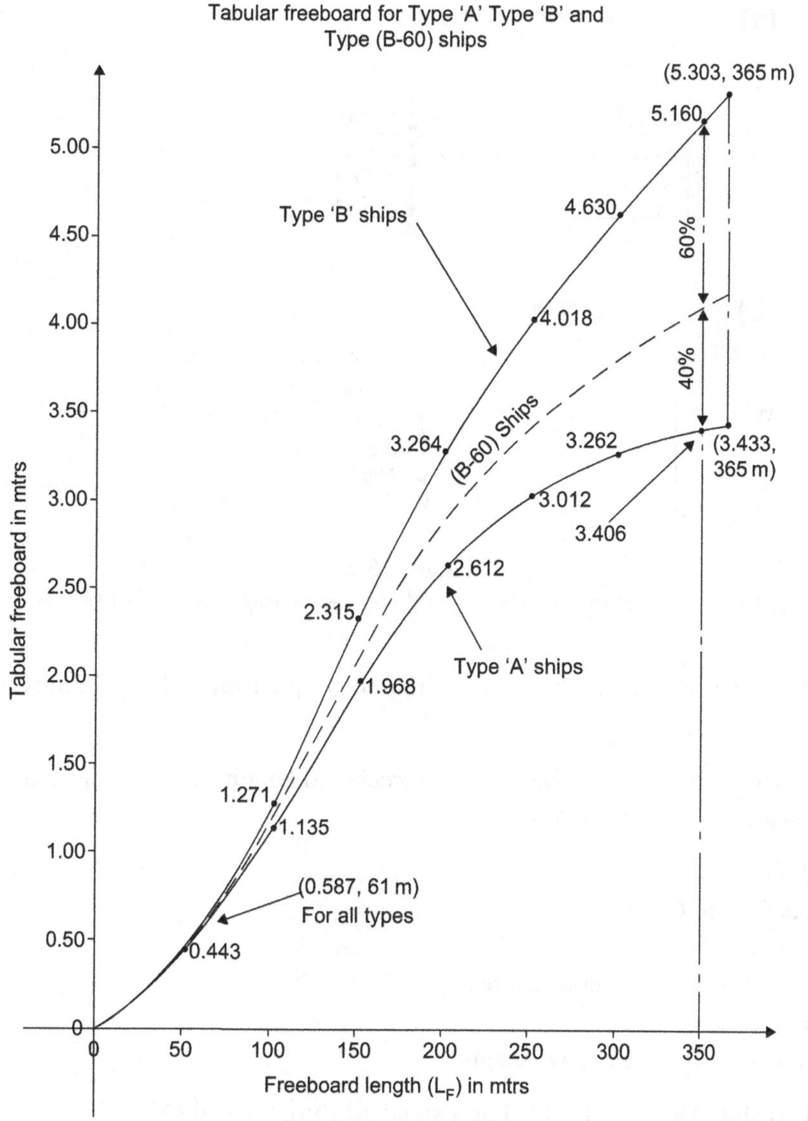

Figure 35.1:
Tabular Freeboard for Type 'A', Type 'B', and Type 'B-60' Ships.

$$f_1 = \text{The basic or tabular freeboard}$$

$$f_2 = \text{The final assigned freeboard for the new design}$$

Differences in hull form and structure are considered and compared with these two vessels. If the new design has hull form and structures that would increase the danger of

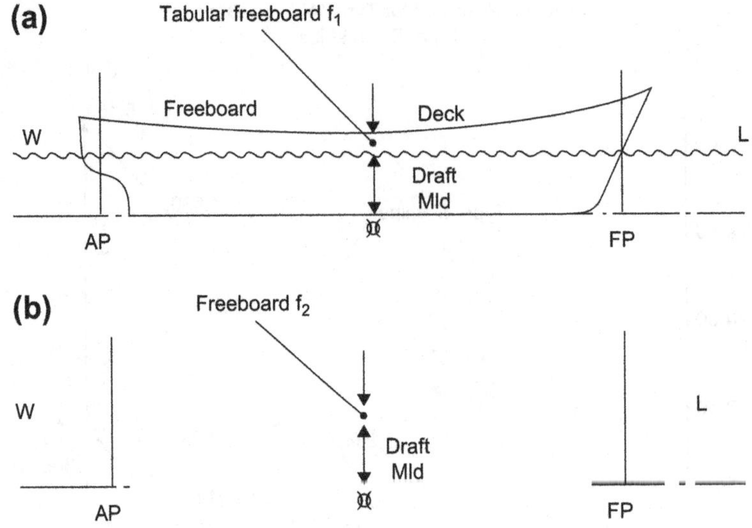

Figure 35.2:
(a) Basic DfT Standard Design and (b) New Design Being Considered.

operation, then the tabular freeboard is increased by prearranged regulations and formulae.

The DfT tabular freeboard value, based on the freeboard length value, is adjusted or modified for the following six characteristics:

1. Depth D.
2. Block coefficient C_b.
3. Bow height.
4. Length and height of superstructures.
5. Freeboard deck sheer.
6. Structural strength of the new design.

For a standard ship, Depth = $L_F/15$. For a standard ship, $C_b = 0.680$.

Standard camber is assumed to be parabolic and equal to B. Mld/50. Freeboard deck sheer is assumed to be parabolic with the sheer forward being twice the deck sheer aft.

$$\text{Standard sheer aft} = (L/3 + 10) \times 25\ \text{mm}$$
$$\text{Standard sheer forward} = (L/3 + 10) \times 50\ \text{mm}$$

Note of caution: for the sheer formulae, L is a ship's LBP in meters (not L_F).

A new vessel can be built to a structural strength of the Lloyds +100A1 standard. If the vessel is indeed built to this classification, then the modification (for strength) to the

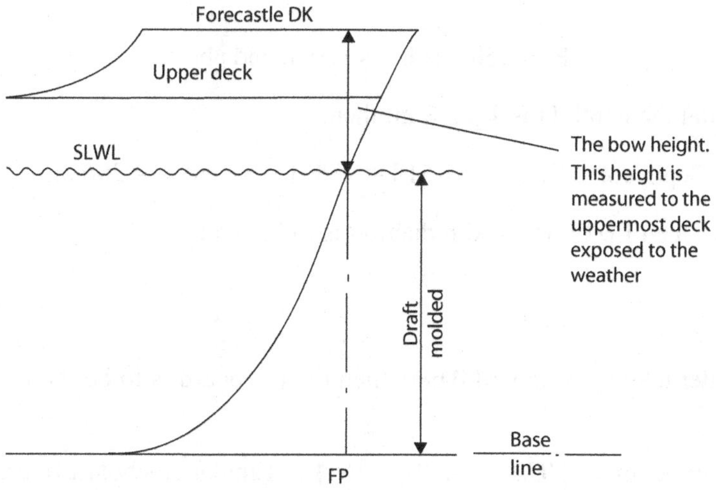

Figure 35.3:
The Bow Height, as per DfT Regulations.

tabular freeboard is zero. Whenever the new design has a characteristic that is less safe than the standard DfT standard vessel, the tabular freeboard value will be increased accordingly.

Figure 35.3 shows the bow height measurement to be considered. It is to be measured at the FP, to the uppermost deck exposed to the weather.

The final assigned statutory freeboard is always measured from the Summer load waterline (SLWL) to the *top* of the freeboard deck's stringer plate at amidships. The stringer plate is the outermost line of deck plating. It is the line of deck plating connected to the sheerstrake or gunwhale plate.

The Corrections in Detail

Depth Correction

If depth D exceeds $L_F/15$, the freeboard is to be increased. If this is so, then:

$$\text{Correction} = (D - L_F/15) \times R$$

where

$$R = L_F/0.48 \quad \text{if } L_F \text{ is less than 120 m}$$

$$R = 250 \quad \text{if } L_F \text{ is 120 m and above}$$

If L_F is 155 m and the depth D is 11.518 m, then:

$$\text{Depth correction} = (11.518 - 155/15) \times 250 = +296 \text{ mm}$$

If depth D is less then $L_F/15$, then no reduction is to be made.

C_b Correction

If the C_b is greater than the standard 0.680, then the freeboard is to be increased by the following:

$$\text{Correction} = \{(C_b + 0.680)/1.360\} \times \text{Tabular freeboard figure}$$

If the ship's C_b is 0.830 and the tabular freeboard figure is 2.048 m, then:

$$C_b \text{ correction is } \{(0.830 + 0.680)/1.360\} \times 2.048 = 2.274 \text{ m}$$

Hence, addition for actual C_b value = 2.274 – 2.048 = +0.226 m or +226 mm.

Bow Height Correction

If the bow height on the actual vessel is *less* than the standard bow height, then the freeboard must be *increased.*

If the bow height on the actual vessel is *greater* than the standard bow height, then there is *no correction* to be made to the freeboard.

The minimum bow height (mBH) for ships is as follows:

If L_F is <250 m, then mBH = $56L\{1 - L/500\} \times 1.36/(C_b + 0.680)$ mm
If L_F = 250 m or is >250 m, then mBH = $7000 \times 1.36/(C_b + 0.680)$ mm

Worked Example 1

An oil tanker has 155 m freeboard length with an actual bow height of 6.894 m and a C_b of 0.830. Does the bow height meet with the minimum statutory requirements?

$$\text{Ship is } < 250 \text{ m so mBH} = 56L\{1 - L/500\} \times 1.36/(C_b + 0.680)$$

$$\text{mBH} = 56 \times 155 \times \{1 - 155/500\} \times 1.36/(0.830 + 0.680)$$

Hence minimum bow height = 5394 mm or 5.394 m.

The actual bow height is 6.894 m so it is 1.50 m above the minimum statutory limit. No correction to the tabular freeboard is therefore necessary.

Superstructure Correction

Where the effective lengths of the superstructure and trunks is **100%** × $\mathbf{L_F}$, the freeboard can be reduced by:

350 mm	when L_F is 24 m
860 mm	when L_F is 85 m
1070 mm	when L_F is 122 m and above

These values are shown graphically in Figure 35.4.

However, if less than 100% of the vessel's length is superstructure length, then the following multiple factors should be determined.

Let the actual length of superstructure be denoted as the effective length (E). Let ratios for E/L_F range from 0 to 1.00.

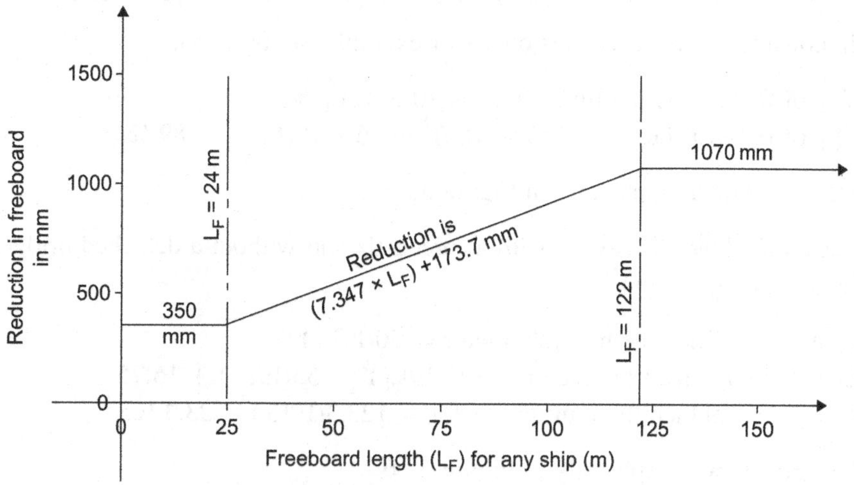

This will be later modified for actual E/L_F value, where E = E_{poop} + $E_{F'C'SLE}$

Figure 35.4:
Reduction in Freeboard for Superstructure, When E/L_F is 100%, for Any Ship.

Table 35.1: Freeboard reduction against effective length/freeboard length ratios for Type 'A' ships.

E/L_F ratio	0	0.1	0.2	0.3	0.4	0.5	0.6	0.7	0.8	0.9	1.0
Freeboard reduction	0	7%	14%	21%	31%	41%	52%	63%	75.3%	87.7%	100%

Percentages at intermediate ratios of superstructures can be obtained by linear interpolation.

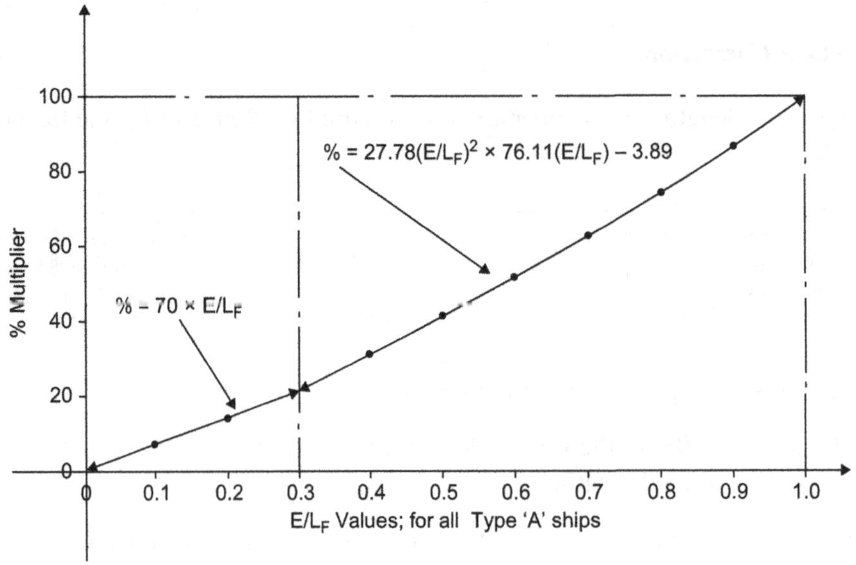

Figure 35.5:
Percent Multipliers for Superstructure Length 'E' for All Type 'A' Ships.

First of all, consider *Type 'A' vessels* only (for example, oil tankers).

For E/L_F of 0–0.3, the multiple factor is $70 \times E/L_F$ %.
For E/L_F of 0.3–1.0 factor $= 27.78(E/L_F)^2 + 76.11(E/L_F) - 3.89$ %.

These values are shown graphically in Figure 35.5.

Secondly, consider *Type 'B' vessels* with a forecastle and without a detached bridge (for example, general cargo ships).

If E/L_F is 0–0.3, then the multiple factor $= 50(E/L_F)$ %.
If E/L_F is 0.3–0.7, then the factor $= 175(E/L_F)^2 - 55(E/L_F) + 15.75$ %.
If E/L_F is 0.7–1.0 then the multiple factor $= 123.3(E/L_F) - 23.3$ %.

These values are shown graphically in Figure 35.6.

Table 35.2: Freeboard reduction against effective length/freeboard length ratios for Type 'B' vessels.

E/L_F ratio	0	0.1	0.2	0.3	0.4	0.5	0.6	0.7	0.8	0.9	1.0
Freeboard reduction	0	5%	10%	15%	23.5%	32%	46%	63%	75.3%	87.7%	100%

Percentages at intermediate ratios of superstructures can be obtained by linear interpolation.

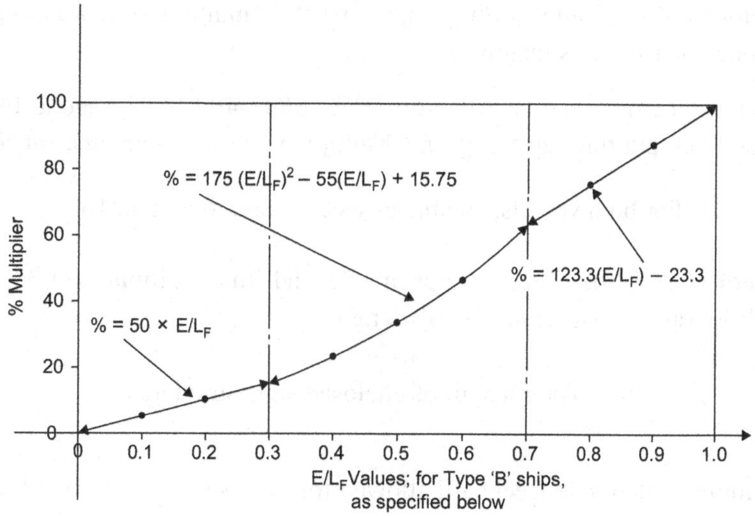

Figure 35.6:
Percent Multipliers for Superstructure Length 'E' for Type 'B' Vessels with Forecastle but no Bridge.

Worked Example 2

An oil tanker has 155 m freeboard length. Superstructure length (E) is 0.51 of the freeboard length (L_F). Estimate the superstructure negative correction to the tabular freeboard.

Oil tanker is greater than 122 m, so the first estimate is a reduction of 1070 mm.

Because E/L_F is 0.51, factor $= 27.78(E/L_F)^2 + 76.11(E/L_F) - 3.89$ %.
Therefore, factor $= 27.78(0.51)^2 + 76.11(0.51) - 3.89$ % = 42.1%.
So the superstructure correction $= -1070 \times 42.1\% = -450$ mm.

Sheer Correction

A vessel with greater than standard sheer will have a reduction in the freeboard. A vessel with less than standard sheer will have an addition in the freeboard:

$$\text{Sheer correction} = \text{Mean sheer difference} \times (0.75 - S/2) \times L_F$$

The mean sheer difference is the actual mean sheer for the ship relative to the mean sheer for the standard ship. For both cases:

$$\text{Mean sheer} = (\text{Aft sheer @ AP} + \text{Forward sheer @ FP})/6 \text{ mm}$$

The denominator of 6 is the sum of Simpson's First Rule multipliers (1,4,1) to give a mean sheer value along each vessel's length.

Deck sheer can also be measured at AP, 1/6L, 2/6L, 3/6L, 4/6L, 5/6L, and at FP. L is LBP. These sheers are then put through Simpson's Multipliers to obtain an area function.

$$\text{For both vessels, the mean sheer} = \text{area function}/16$$

The denominator of 16 is the sum of Simpson's Second Rule multipliers (1,3,3,2,3,3,1) to give a mean sheer value along each vessel's length.

$$S = \text{Total length of enclosed superstructures}$$

$$\text{Maximum reduction in freeboard allowed for excess sheer} = 1.25 \times L_F \text{ mm}$$

Worked Example 3

An oil tanker has a freeboard length of 155 m with a total length of enclosed superstructures of 71.278 m. The difference between actual mean sheer and standard ship mean sheer is −650 mm. Estimate the additional correction to the tabular freeboard figure.

$$\begin{aligned}\text{The sheer correction} &= 650 \times (0.75 - 71.278/2) \times 155\\ &= +338 \text{ mm}\end{aligned}$$

Strength Correction

A new vessel can be built to a structural strength of Lloyds +100A1 standard. If the vessel is indeed built to this classification, then the modification (for strength) to the tabular freeboard is zero.

Worked Example 4

Consider a tanker of 155 m freeboard length with a Depth Mld of 11.458 m. Calculate the Summer freeboard and the Summer loaded waterline (SLWL), i.e. draft molded.

From Figure 35.1 at a freeboard length of 155 m, the DfT tabular freeboard from the Type 'A' vessel curve is 2.048 m (recap on previous calculations).

Tabular freeboard	2.048 m
Depth correction	+0.296 m
C_b correction	+0.226 m
Bow height correction – satisfactory	0 m
Superstructure correction	−0.450 m
Freeboard deck sheer correction	+0.338 m
Strength correction – built to Lloyds +100A1 standard	0 m
Molded Summer freeboard	= 2.458 m
+ Molded Summer freeboard	= 0.018 m
Final assigned Summer freeboard	= 2.476 m

■

Summary Statement

Tabular freeboard was 2.048 m. Assigned freeboard was 2.458 m.

Depth Molded, as given	11.458 m	= D
Summer freeboard (mld)	2.458 m	= f
Draft Mld or SLWL	9.000 m	= d

So Summer load waterline = 9.00 m.

Generally, from the stability point of view, the greater the freeboard, the safer the ship will be in day-to-day operations.

The Freeboard Marks

A typical set of freeboard marks is shown in Figure 35.7.

- **S** is the Summer water mark for water of 1.025 t/m^3 density. It is determined by the Department for Transport (DfT) Tabulated Freeboard values, based on the vessel's freeboard length and various corrections. It is placed at the Summer load waterline (draft molded).
- **T** is the Tropical water mark and is 1/48 of the Summer load draft *above* the S mark.
- **F** is the Fresh water mark. The F water mark is $W/(4 \times TPC_{SW})$ *or* 1/48 of the Summer load draft *above* the S mark. W and TPC_{SW} are values applicable at the Summer load waterline.
- **TF** is the Tropical Fresh water mark and is the (T + F) marks *above* the S mark.

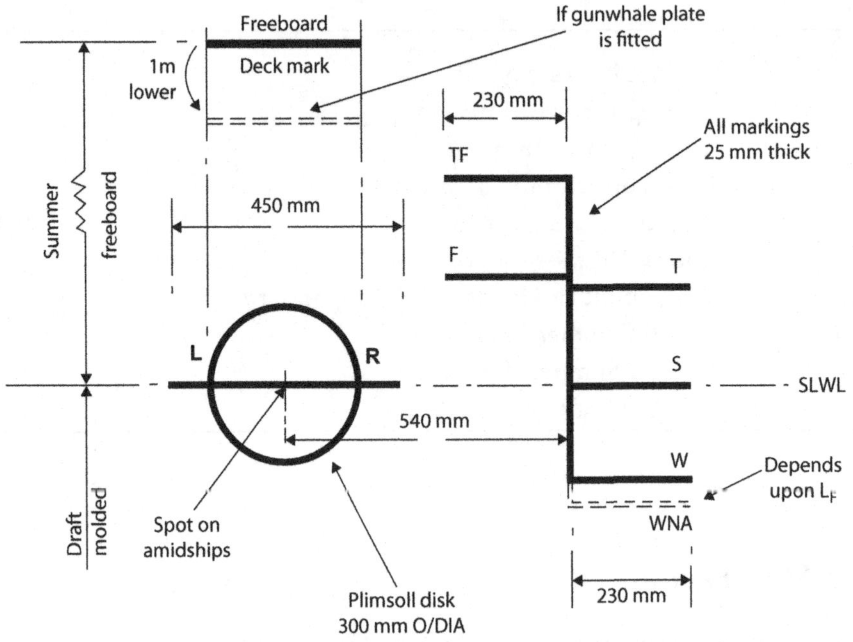

Figure 35.7: A Typical Set of Freeboard Marks.
LR stands for Lloyds Registry based in the UK with offices worldwide.
Alternative classification bodies:

AB = American Bureau of Shipping in USA
NV = Norske Veritas in Norway
BV = Bureau Veritas in France
GL = Germanischer Lloyd of Germany
CA = Commonwealth of Australia
NK = Nippon Kaiji Kyokai of Japan

No allowance to be made for the vessel being sagged. Most cargo vessels when fully loaded will be in a sagging condition.

- **W** is the Winter water mark. It is 1/48 of the Summer load draft *below* the S mark.
- **WNA** is the Winter North Atlantic water mark. It is *not* marked on the ship sides for a vessel equal to or more than 100 m freeboard length. If the vessel is less than 100 m floodable length, then the WNA is placed 50 mm *below* the W mark.

The load lines and freeboard deck line *must* be painted in white or yellow on a dark background, or in black on a light colored background. The letters on each side of the load line disk, indicating the assigning authority, should be 115 mm in height and 75 mm in width (see Figure 35.7).

Penalty warning: According to the 1998 load line regulations, if the appropriate load line on each side of the ship was submerged when it should not have been then:

> *The owner and master are liable to an additional fine that shall not exceed £1000* for each complete centimeter *overloaded.*

Worked Example

(a) Calculate the seasonal allowances and the subsequent drafts for an oil tanker having a SLWL of 9.402 m given that the LBP is 148 m, displacement W is 24,700 t, and the TPC_{SW} is 30.2.

(b) Proceed to draw the resulting freeboard marks for this ship. Label your drawing as it would appear on the side of the ship.

T = Tropical fresh water mark = SLWL/48 above SLWL
= 9042 mm/48 = 188 mm above SLWL

Hence tropical draft = 9.402 + 0.188 = 9.59 m.

F = Fresh water mark = W/(4 × TPC) above SLWL
= 24,700/(4 × 30.20) = 204 mm above SLWL

Hence fresh water draft = 9.402 + 0.204 = 9.606 m.

TF = Tropical Fresh water mark = (T + F) above SLWL
= 0.188 + 0.204 = 0.392 mm above SLWL

Hence tropical fresh water draft = 9.402 + 0.392 = 9.794 m.

W = Winter water mark = SLWL/48 below SLWL
= 9042 mm/48 = 188 mm below SLWL

Hence winter draft = 9.402 − 0.188 = 9.214 m.

WNA = Winter North Atlantic water mark. This is *not* marked on the side of this oil tanker because her LBP is greater than 100 m.

Seasonal allowances depend on a DfT World zone map (at rear of their freeboard regulations) and on three factors:

1. Time of year.
2. Geographical location of the ship.
3. LBP of the ship, relative to a demarcation value of 100 m.

Exercise 35

1. (a) Describe exactly what are Type 'A' vessels, Type 'B' vessels, and Type 'B-60' vessels.

(b) Sketch a graph to indicate how the DfT Tabular Freeboard is related to freeboard length (L_F) for each ship type listed in (a).

2. List the six characteristics that must be considered when modifying the DfT Tabular Freeboard to the ship's final assigned minimum freeboard.
3. Sketch a typical set of freeboard marks for a ship. Show how a ship's seasonal allowances are usually evaluated.
4. Discuss in detail the following corrections for freeboard:
   - (a) The freeboard deck sheer
   - (b) The ship's block coefficient
   - (c) The strength of the ship
   - (d) The bow height.
5. An oil tanker has 200 m freeboard length. Superstructure length (E) is 0.60 of the freeboard length (L_F). Estimate the superstructure negative correction to the tabular freeboard.

■

CHAPTER 36

Timber Ship Freeboard Marks

Timber deck cargo is denoted as a cargo of timber carried on an uncovered part of a freeboard or superstructure deck. Timber vessels are allowed less freeboard than other cargo vessels. Certain vessels are assigned timber freeboards but certain additional conditions have to be complied with:

1. The vessel must have a forecastle of at least 0.07 of the vessel's length in extent and of not less than standard height.
2. For a vessel 75 m or less in length the standard height is 1.80 m.
3. For a vessel 125 m in length the standard height is 2.30 m.
4. Intermediate standard heights can be evaluated for intermediate freeboard lengths ranging from 75 to 125 m.
5. A poop or raised quarter deck is also required if the vessel's freeboard length is less than 100 m.
6. The double-bottom tanks in the midship half-length must have a satisfactory watertight longitudinal subdivision.
7. Timber vessels must have either 1-m-high bulwarks with additional strengthening or 1-m-high especially strong guardrails.
8. The timber deck cargo should be compactly stowed, lashed, and secured. It should not interfere in any way with the navigation and necessary work of the ship.

Figure 36.1 shows markings for a vessel operating with two sets of freeboard marks. The markings on the left-hand side are the timber markings used when the ship is carrying timber on the upper deck. The right-hand side markings are used for conditions of loading without timber stowed on the upper deck.

Additional lumber marks are placed forward of the load line disk and indicated by the 'L' prefix. The decrease in freeboard is allowed because timber ships are considered to provide additional reserve buoyancy. This does not apply for the carriage of wood pulp or other highly absorbent cargoes.

- LS is the lumber summer mark. It is determined from the Department for Transport (DfT) tabulated freeboard values, based on the vessel's freeboard length in conjunction with the corrected Type 'B' vessel values.
- LT mark is the lumber tropical mark and is 1/48 of the summer load timber load draft above the LS mark.

Ship Stability for Masters and Mates. http://dx.doi.org/10.1016/B978-0-08-097093-6.00036-0

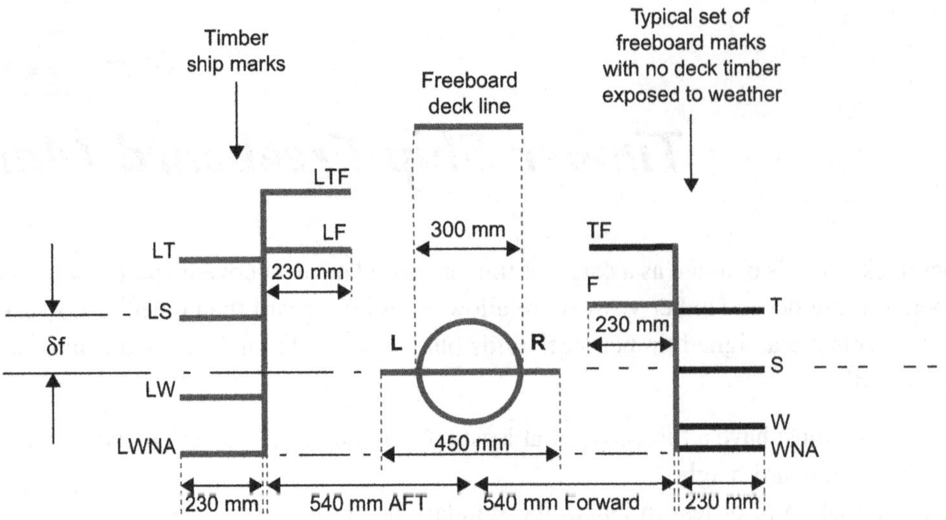

Figure 36.1: Freeboard Marks for a Vessel with Two Sets of Markings.
Notes. For the distance df, see graph in Figure 36.2, of 'LS' to 'S' against freeboard length L_F. The bow height for the timber ship must be measured *above the 'S' mark* (not the 'LS' mark). The LWNA mark is level with the WNA mark.

- LF mark is the lumber fresh water mark. It is calculated in a similar manner to the normal F mark, except that the displacement used in the formula is that of the vessel at her summer timber load draft.
- Hence LF mark is $W/(4 \times TPC_{SW})$ *or* 1/48 of the summer load timber load draft above the LS mark.
- LTF mark is the lumber tropical fresh water mark and is (LT + LF) marks above the LS mark.
- LW mark is the lumber winter water mark and is 1/36 of the summer load timber load draft below the LS mark.
- LWNA mark is the lumber Winter North Atlantic water mark. It is to be horizontally in line with the more usual WNA mark. There is no reduced freeboard allowed on either of these WNA lines.

The load lines and freeboard deck line *must* be painted in white or yellow on a dark background, or in black on a light colored background. The letters each side of the load line disk indicating the assigning authority should be 115 mm in height and 75 mm in width.

Note that in Figure 36.1 there is a vertical difference between the 'LS' on the left-hand side and the 'S' on the right-hand side. This is an allowance based on the ratio of length of superstructure used for carrying the timber.

If 100% of the vessel's length is used to carry timber above the freeboard deck, then this reduction in freeboard for carrying timber is as follows:

If a timber ship's freeboard length (L_F) is *24 m* or less, the reduction is 350 mm. If a timber ship's freeboard length (L_F) is *122 m* or more, reduction is 1070 mm. For timber ship's freeboard lengths of *24—122 m*, the reduction in freeboard is:

$$\text{LS mark to S mark is } (7.347 \times L_F) + 173.7 \text{ mm}$$

Figure 36.2 displays this information in graphical form.

However, if less than 100% of the vessel's length is used to carry timber above the freeboard deck, then Table 36.1 must be consulted. This is denoted as the effective length (E). Consider some ratios for E/L_F.

If E/L_F of 0—0.4	freeboard deduction is $(110 \times E/L_F) + 20$ %.
If E/L_F of 0.4—1.0	freeboard deduction is $(60 \times E/L_F) + 40$ %.

Figure 36.3 portrays this information in graphical form.

It is of interest to note that this vertical height 'LS' to 'S' is the approximate mean bodily rise if the timber vessel lost her deck cargo of timber overboard.

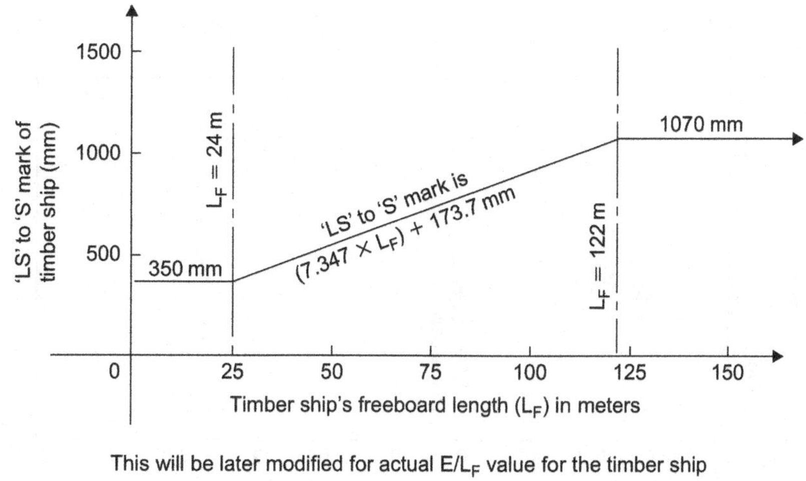

Figure 36.2:
Distance of 'LS' Mark to 'S' Mark for a Timber Ship when 100% of Weather Deck is Used to Stow Timber.

Table 36.1: Freeboard reduction against effective length/freeboard length ratios for timber ships.

E/L_F ratio	0	0.1	0.2	0.3	0.4	0.5	0.6	0.7	0.8	0.9	1.0
Freeboard reduction	20%	31%	42%	53%	64%	70%	76%	82%	88%	94%	100%

Percentages at intermediate ratios of superstructures can be obtained by linear interpolation.

Worked Example

A timber ship has a freeboard length (L_F) of 85 m and has an effective length of superstructure carrying timber of $E/L_F = 0.70$. Calculate the superstructure allowance, i.e. between the 'LS' and the 'S' marks.

$$\begin{aligned} \text{Reduction} &= (\text{allowances for } 100\%\ E) \times \{(60 \times E/L_F) + 40\} \\ &= \{(7.347 \times L_F) + 173.7\} \times \{(60 \times 0.70) + 40\} \\ &= \{(7.347 \times 85) + 173.7\} \times 82\% \\ &= 654 \text{ mm}, \quad \text{say } 0.654 \text{ m}. \end{aligned}$$

Carrying timber on the upper deck can create stability problems for those on board ship. They are:

1. The timber may become wet and saturated. This will raise the overall G of the ship, thereby possibly decreasing the ship's GM. This leads to a loss in stability. When considering the

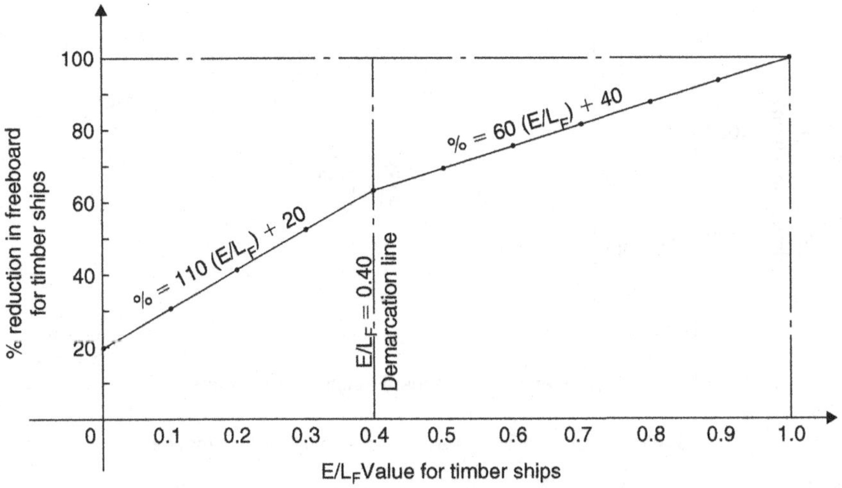

Figure 36.3:
Percent Reduction in Freeboard for the Superstructures of Timber Ships.

stability of timber ships the calculations must take into account that because of this saturation:

(a) Weight of deck timber must be increased by 15% of its dry weight.

(b) Volume available for reserve buoyancy is only 75% of the total deck timber.

2. The height of the stowed timber can produce a sailing effect, leading to an angle of list situation.
3. Icing effects in very cold weather conditions on the timber will raise the overall G of the ship. Again this will lead to a loss in stability.
4. The IMO suggest that to avoid high strain on the lashing points of the deck timber that the upright GM should not be greater than about 3% of the breadth molded.
5. As far as statical stability curves are concerned for timber ships, the MCA/UK lumber regulations are as follows:

The upright GM must not be less than 0.05 m (compared to the usual 0.15 m), with the remaining requirements being the same as for any other vessel.

As far as statical stability curves are concerned for timber ships, the IMO lumber regulations are as follows:

The upright GM must not be less than 0.10 m (compared to the usual 0.15 m). Maximum GZ must not be less than 0.25 m (compared to the usual 0.20 m). Area under the statical stability curve, 0–40°, must not be less than 0.08 meter radians (compared to the usual 0.09 meter radians).

It also must be remembered that the minimum bow height (as per DfT Regulations) for timber vessels is measured from the ordinary summer load water line. It must NOT be measured from the lumber summer mark.

Exercise 36

1. Certain vessels are assigned timber freeboards with certain conditions. List six of these conditions of assignment.
2. (a) Sketch a typical set of freeboard marks for a timber ship with two sets of markings.
   (b) Explain each of the lumber seasonal allowance markings.
3. A timber ship has a freeboard length (L_F) of 75 m and has an effective length of superstructure carrying timber of $E/L_F = 0.60$. Calculate the superstructure allowance, i.e. between the 'LS' and the 'S' marks.
4. (a) List four problems that could occur when carrying timber on a deck that is exposed to the weather.
   (b) As per DfT requirements, for timber ships, from where is the bow height measured?
   (c) List three IMO stability requirements for timber ships.

stability of timber ships the calculations must take into account [illegible]. Because of this saturation:

(a) Weight of deck timber must be increased by [illegible]% of its dry weight.
(b) Volume available for reserve buoyancy is only 75% of the total deck timber.

2. The height of the stowed timber can produce [illegible] [illegible] of the situation.
3. Icing effects in very cold weather conditions on the timber [illegible] ship. Again this will lead to a loss of stability.
4. The IMO suggest that [illegible] should not be greater than about [illegible] of the [illegible].

As [illegible] stability [illegible] the IMO [illegible] regulations are as follows:

[illegible]

[illegible]

[illegible]

Exercise [illegible]

[illegible]

CHAPTER 37

IMO Grain Rules for Safe Carriage of Grain in Bulk

The intact stability characteristics of any ship carrying bulk grain must be shown to meet, throughout the voyage, the following criteria relating to the moments due to grain shift:

1. The angle of heel due to the shift of grain shall not be greater than 12° or in the case of ships constructed on or after 1 January 1994 the angle at which the deck edge is immersed, whichever is the lesser. For most grain-carrying ships, the deck edge will not be immersed at or below 12°, so this 12° is normally the lesser of the two angles.
2. In the statical stability diagram, the net or residual area between the heeling arm curve and the righting arm curve up to the angle of heel of maximum difference between the ordinates of the two curves, or 40° or the angle of flooding (θ_1), whichever is the least, shall in all conditions of loading be not less than 0.075 meter radians.
3. The initial metacentric height, after correction for free-surface effects of liquids in tanks, shall not be less than 0.30 m.

Before loading bulk grain the Master shall, if so required by the contracting government of the country of the port of loading, demonstrate the ability of the ship at all stages of any voyage to comply with the stability criteria required by this section.

After loading, the Master shall ensure that the ship is upright before proceeding to sea.

Consider Figure 37.1. The IMO Grain Code stipulates λ_0 and λ_{40} values to graphically determine the angle of list in the event of a shift of grain, where

$$\lambda_0 = \frac{\text{Assumed volumetric heeling moment due to transverse shift}}{\text{Stowage factor} \times \text{Displacement}}$$

$$\lambda_{40} = 0.8 \times \lambda_0$$

This gives a grain heeling line.

$$\text{Stowage factor} = \text{Volume per unit weight of grain cargo}$$

It can be represented by three values: 1.25, 1.50, and 1.75 m^3/tonne.

$$\text{Displacement} = \text{Weight of ship, fuel fresh water, stores, cargo, etc.}$$

Ship Stability for Masters and Mates. http://dx.doi.org/10.1016/B978-0-08-097093-6.00037-2

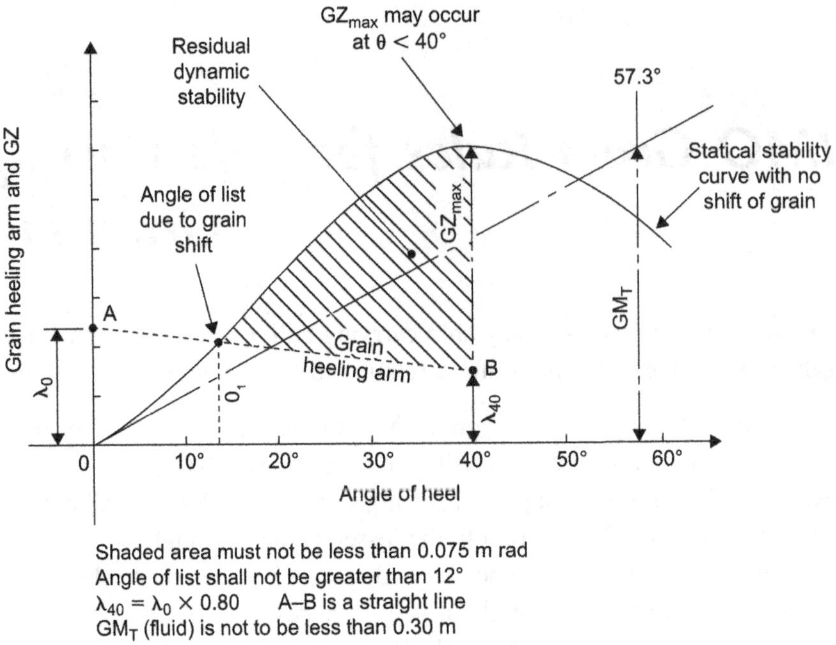

Figure 37.1:
IMO Grain Heeling Stability Requirements.

The righting arm shall be derived from cross-curves, which are sufficient in number to accurately define the curve for the purpose of these requirements and shall include cross-curves at 12° and 40°.

The angle of list occurs at the intersection of the GZ curve and the grain heeling line. The angle of list shown in Figure 37.1 may also be approximated in the following manner:

The approximate angle of heel = ahm/(mpm × SF) × 12°
ahm = actual transverse volumetric heeling moments in m^4
SF = appropriate stowage factor in m^3/tonne, say 1.50–1.70
mpm = maximum permissible grain heeling moments in tm

These are supplied to the ship by the Naval Architect in the form of a table of fluid KG against displacement W. Refer to the datasheet in the Worked Examples section.

Adjustments:

- If the holds are full with grain then the multiple shall be 1.00 × each ahm.
- A multiple of 1.06 is applied to each ahm in holds and tween decks filled to the hatchway.
- A multiple of 1.12 is applied to each ahm in holds and tween decks with partially filled full-width grain compartments.

A study of the worked examples will give an understanding of exactly how these adjustments are made.

Studies of grain cargoes after shifting has taken place have lead authorities to make the following stipulations with regard to the transverse slope 'α' of grain across the ship:

- $\alpha = 25°$ to the horizontal assumed for full-width partially filled compartments. This is at the ends of the compartment situated forward and aft of the hatchway. α also equals 25° to the horizontal for hatchway trunks (see Figure 37.2).
- $\alpha = 15°$ to the horizontal assumed for full compartments where the grain stow is situated abreast of the hatchway (see Figure 37.2).

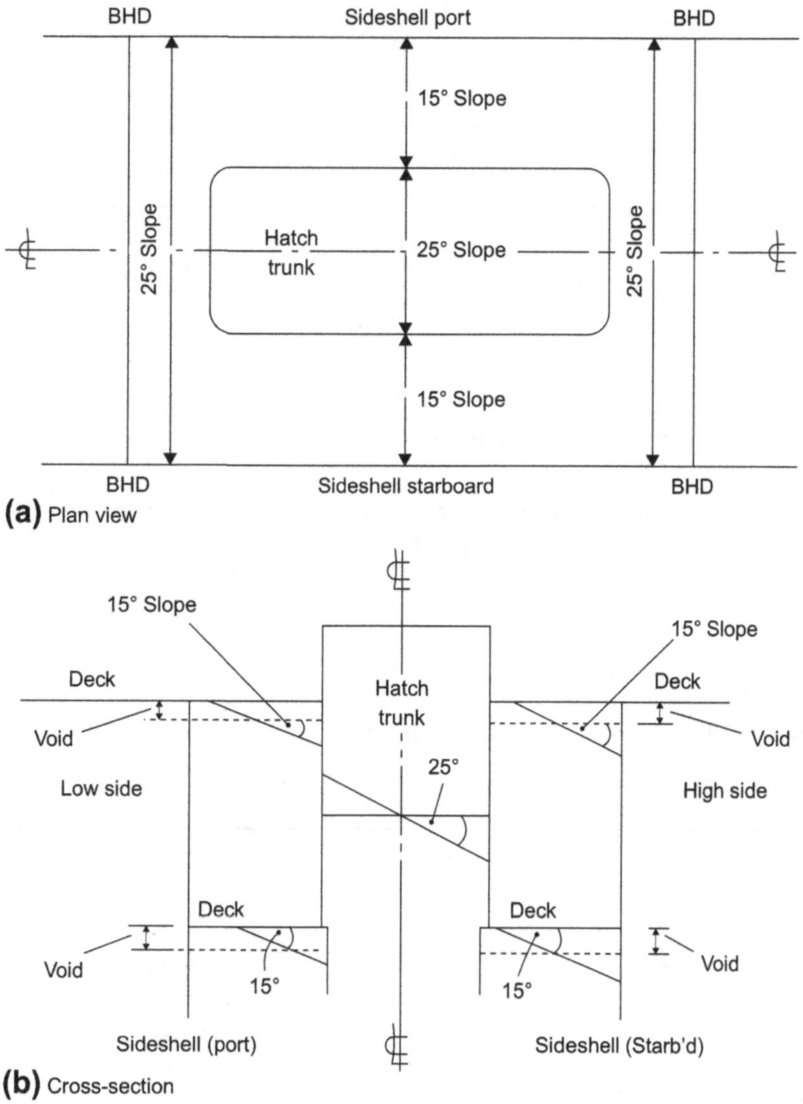

Figure 37.2:
Assumed Shifts of Grain in Degrees, as per IMO Grain Regulations for Hatch Trunks.

Worked Example 1

A vessel has loaded grain, stowage factor 1.65 m^3/tonne, to a displacement of 13,000 tonnes. In the loaded condition the effective KG is 7.18 m. All grain spaces are full, except No. 2 tween deck, which is partially full.

The tabulated transverse volumetric heeling moments are as follows:

No. 1 hold	1008 m^4
No. 2 hold	1211 m^4
No. 3 hold	1298 m^4
No. 4 hold	1332 m^4
No. 1 TD	794 m^4
No. 2 TD	784 m^4
No. 3 TD	532 m^4

The values of KG used in the calculation of the vessel's effective KG were as follows:

for lower holds, the centroid of the space;
for tween decks, the actual KG of the cargo.

(a) Using Datasheet Q.1 *Maximum Permissible Grain Heeling Moments Table*, determine the vessel's ability to comply with the statutory grain regulations. (20 marks)
(b) Calculate the vessel's approximate angle of heel in the event of a shift of grain assumed in the grain regulations. (5 marks)
(c) State the stability criteria in the current International Grain Code. (10 marks)

Using the *Maximum Permissible Grain Heeling Moments Table*, for a KG of 7.18 m and a 13,000 t displacement, using a direct interpolation procedure it can be determined that:

The maximum permissible grain heeling moment = 3979 tm, say mpm

The actual volumetric heeling moments (ahm) for this vessel are:

No. 1 hold	1008 m^4 × 1.00 = 1008
No. 2 hold	1211 m^4 × 1.00 = 1211
No. 3 hold	1298 m^4 × 1.00 = 1298
No. 4 hold	1332 m^4 × 1.00 = 1332
No. 1 TD	794 m^4 × 1.06 = 842
No. 2 TD	784 m^4 × 1.12 = 878
No. 3 TD	532 m^4 × 1.06 = 564
	Total ahm = 7133 m^4

TS002-72 STABILITY

DATASHEET Q.1

N.B. This Datasheet must be returned with your examination answer book

TABLE OF MAXIMUM PERMISSIBLE GRAIN HEELING MOMENTS (tm)										
	FLUID KG (meters)									
Displacement tonne	6.50	6.60	6.70	6.80	6.90	7.00	7.10	7.20	7.30	7.40
14 500	6141	5820	5499	5179	4858	4537	4217	3896	3575	3255
14 000	5957	5647	5338	5028	4719	4409	4099	3790	3480	3171
13 500	5924	5625	5327	5028	4730	4431	4132	3834	3535	3237
13 000	5934	5647	5359	5072	4784	4497	4209	3922	3634	3347
12 500	5891	5614	5338	5062	4785	4509	4232	3956	3679	3403
12 000	5857	5591	5326	5061	4795	4630	4265	3999	3734	3468
11 500	5893	5639	5385	5130	4876	4622	4368	4113	3859	3605
11 000	5944	5701	5457	5214	4971	4728	4484	4241	3998	3755
10 500	5948	5716	5484	5251	5019	4787	4555	4323	4090	3858
10 000	5940	5719	5498	5276	5055	4834	4613	4392	4171	3950
9500	5961	5751	5541	5331	5121	4911	4701	4491	4281	4071
9000	6027	5828	5629	5430	5231	5032	4833	4634	4435	4236
8500	6127	5939	5751	5563	5375	5187	4999	4811	4623	4435
8000	6210	6033	5856	5679	5502	5325	5148	4971	4795	4618
7500	6252	6087	5921	5755	5589	5423	5257	5091	4926	4760
7000	6343	6189	6034	5879	5724	5569	5415	5260	5105	4950
6500	6550	6406	6262	6118	5975	5831	5687	5543	5400	5256
6000	6832	6699	6566	6434	6301	6168	6035	5903	5770	5637
5500	7120	6998	6877	6755	6633	6512	6390	6268	6147	6025
5000	7320	7209	7099	6988	6877	6767	6656	6546	6435	6325

Candidate's Name ....

Examination Center ....

The factor 1.00 for the four holds is because, as stated in the question, in the full holds the centroid was at the center of the space.

The factor 1.06 is because, in the question, for the two full tween deck cargo spaces the KG was at the actual KG.

The factor of 1.12 is because, in the question, No. 2 tween deck is given as being partially full.

Let stowage factor be denoted as SF.

$$\text{The approximate angle of heel} = \text{ahm}/(\text{SF} \times \text{mpm}) \times 12^\circ$$
$$\text{Approximate angle of heel} = 7133/(1.65 \times 3979) \times 12^\circ$$
$$\text{Hence approximate angle of heel} = \underline{13.04^\circ}$$

This angle of heel is *not* acceptable since it does not comply with the IMO regulations because it is greater than 12°.

Worked Example 2

A vessel has loaded grain, stowage factor 1.65 m^3/tonne, to a displacement of 13,000 tonnes. In the loaded condition the effective KG is 7.18 m. All grain spaces are full, except No. 2 tween deck, which is partially full.

The tabulated transverse volumetric heeling moments are as follows:

No. 1 hold	851 m^4
No. 2 hold	1022 m^4
No. 3 hold	1095 m^4
No. 4 hold	1124 m^4
No. 1 TD	669 m^4
No. 2 TD	661 m^4
No. 3 TD	448 m^4

The values of KG used in the calculation of the vessel's effective KG were as follows:

for lower holds, the centroid of the space;
for tween decks, the actual KG of the cargo.

(a) Using Datasheet Q.1 *Maximum Permissible Grain Heeling Moments Table*, determine the vessel's ability to comply with the statutory grain regulations. (20 marks)
(b) Calculate the vessel's approximate angle of heel in the event of a shift of grain assumed in the grain regulations. (5 marks)
(c) State the stability criteria in the current International Grain Code. (10 marks)

Using the *Maximum Permissible Grain Heeling Moments Table*, for a KG of 7.18 m and a 13,000 tonnes displacement, by a direct interpolation procedure it can be determined that:

The maximum permissible grain heeling moment = 3979 tm, say mpm

The actual volumetric heeling moments (ahm) for this vessel are:

No. 1 hold	$851\ m^4 \times 1.00 = 851$
No. 2 hold	$1022\ m^4 \times 1.00 = 1022$
No. 3 hold	$1095\ m^4 \times 1.00 = 1095$
No. 4 hold	$1124\ m^4 \times 1.00 = 1124$
No. 1 TD	$669\ m^4 \times 1.06 = 709$
No. 2 TD	$661\ m^4 \times 1.12 = 740$
No. 3 TD	$448\ m^4 \times 1.06 = 475$
	Total ahm = $\underline{6016\ m^4}$

The factor 1.00 for the four holds is because, as stated in the question, in the full holds the centroid was at the center of the space.

The factor 1.06 is because, in the question, for the two full tween deck cargo spaces the KG was at the actual KG.

The factor of 1.12 is because, in the question, No. 2 tween deck is given as being partially full.

Let stowage factor be denoted as SF.

$$\text{The approximate angle of heel} = \text{ahm}/(\text{SF} \times \text{mpm}) \times 12^\circ$$
$$\text{Approximate angle of heel} = 6016/(1.65 \times 3979) \times 12^\circ$$
$$\text{Hence approximate angle of heel} = \underline{11^\circ}$$

This angle of heel is quite acceptable because it complies with the IMO regulations in that it is less than 12°.

■

Worked Example 3

A vessel has loaded grain, stowage factor 1.60 m^3/tonne, to a displacement of 13,674 tonnes. In this loaded condition, the fluid GM_T is 0.90 m. All grain spaces are full, except No. 2 tween deck, which is partially full.

The tabulated transverse volumetric heeling moments are as follows:

No. 1 hold	774 m^4
No. 2 hold	929 m^4
No. 3 hold	995 m^4
No. 4 hold	1022 m^4
No. 1 TD	608 m^4
No. 2 TD	601 m^4
No. 3 TD	407 m^4

The values of KG used in the calculation of the vessel's effective KG were as follows:

for lower holds, the centroid of the space;
for tween decks, the actual KG of the cargo.

The righting levers for GZ in meters at angles of heel in degrees are as shown in the table:

Angle of heel	0°	5°	10°	15°	20°	25°	30°	35°	40°	45°	50°
GZ ordinate	0	0.09	0.22	0.35	0.45	0.51	0.55	0.58	0.59	0.58	0.55

(a) Use λ_0 and λ_{40} and the tabulated GZ values to graphically determine the angle of list in the event of a shift of grain.
(b) Calculate the enclosed area between the GZ curve and the grain heeling arm line.

The actual volumetric heeling moments (ahm) for this vessel are:

No. 1 hold	774 $m^4 \times 1.00 = 774$
No. 2 hold	929 $m^4 \times 1.00 = 929$
No. 3 hold	995 $m^4 \times 1.00 = 995$
No. 4 hold	1022 $m^4 \times 1.00 = 1022$
No. 1 TD	608 $m^4 \times 1.06 = 644$
No. 2 TD	601 $m^4 \times 1.12 = 673$
No. 3 TD	407 $m^4 \times 1.06 = 431$
	Total ahm = $\underline{5468\ m^4}$

The factor 1.00 for the four holds is because, in the question, in the full holds the centroid was at the center of the space.

The factor 1.06 is because, in the question, for the two full tween deck cargo spaces the KG was at the actual KG.

The factor of 1.12 is because, in the question, No. 2 tween deck is given as being partially full.

Let stowage factor be denoted as SF and let W = vessel's displacement in tonnes.

$$\lambda_0 = \text{ahm}/(\text{W} \times \text{SF})$$
$$\lambda_0 = 5468/(13,674 \times 1.60) = 0.25 \text{ meters}$$

This value of 0.25 m is plotted in Figure 37.3 at an angle of heel of 0°.

Now according to IMO Grain Regulations:

$$\lambda_{40} = \lambda_0 \times 0.80$$

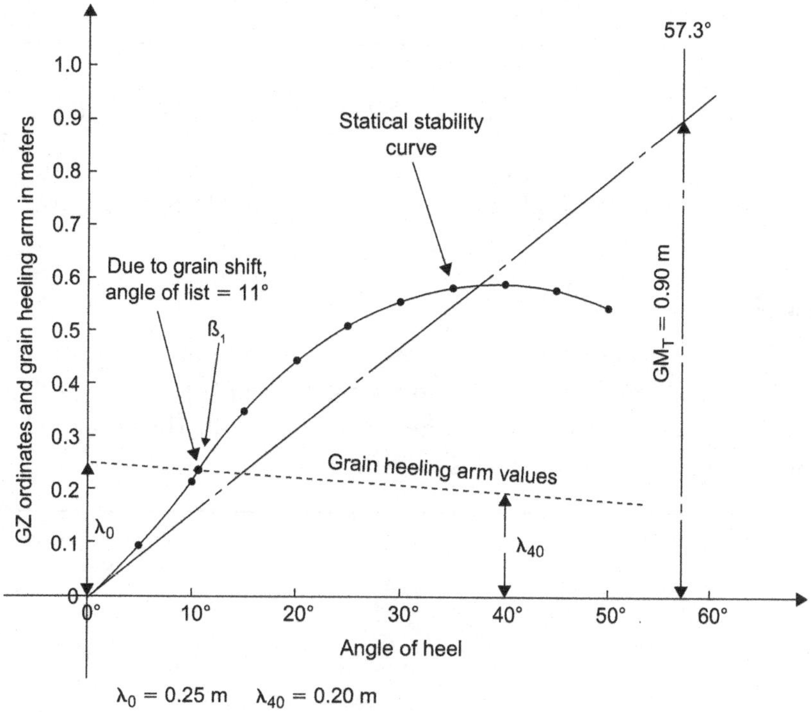

Figure 37.3:
Angle of List Due to Grain Shift.

So

$$\lambda_{40} = 0.25 \times 0.80 = 0.20 \text{ m}$$

This value of 0.20 m is then plotted in Figure 37.3 at an angle of heel of 40°.

The next step is to plot a curve of GZ against angle of heel (to give part of a statical stability curve) and then to superimpose a line connecting the λ_0 and λ_{40} values. This sloping line is a line of grain heeling arm values. Do not forget also to plot the GM_T value at 57.3°. This assists in establishing the guidance triangle that runs into the 0–0 axis.

At the intersection of these two lines the value for the angle of list may be lifted off. Figure 37.3 shows this angle of list to be 11°.

Area enclosed between 11° and 40° angles of heel:

Ordinate	Simpson's Multiplier	Area Function
0	1	0
0.31	4	1.24
0.39	1	0.39
		$1.63 = \Sigma_1$

$$\text{Area enclosed} = 1/3 \times \text{CI} \times \Sigma_1$$
$$= 1/3 \times 14.5°/57.3° \times 1.63 = 0.137 \text{ m radians}$$

where CI = common interval of (40° – 11°)/2.
This is well above the IMO stipulated minimum value of 0.075 m radians.

Exercise 37

1. A vessel has loaded grain, stowage factor 1.55 m^3/tonne, to a displacement of 13,500 tonnes. In the loaded condition the effective KG is 7.12 m. All grain spaces are full, except No. 3 tween deck, which is partially full. The tabulated transverse volumetric heeling moments are as follows:

No. 1 hold	810 m^4
No. 2 hold	1042 m^4
No. 3 hold	1075 m^4
No. 4 hold	1185 m^4
No. 1 TD	723 m^4
No. 2 TD	675 m^4
No. 3 TD	403 m^4

The value of the KG used in the calculation of the vessel's effective KG were as follows:

for lower holds, the centroid of the space;
for tween decks, the actual KG of the cargo.

(a) Using Datasheet Q.1, determine the vessel's ability to comply with the statutory grain regulations.
(b) Calculate the vessel's approximate angle of heel in the event of a shift of grain assumed in the grain regulations.

2. State the stability criteria in the current IMO International Grain Code.
3. A vessel has loaded grain, stowage factor 1.69 m^3/tonne, to a displacement of 13,540 tonnes. All grain spaces are full, except No. 2 tween deck, which is partially full. The tabulated transverse volumetric heeling moments are as follows:

No. 1 hold	762 m^4
No. 2 hold	941 m^4
No. 3 hold	965 m^4
No. 4 hold	1041 m^4
No. 1 TD	618 m^4
No. 2 TD	615 m^4
No. 3 TD	414 m^4

The value of the KG used in the calculation of the vessel's effective KG were as follows:

for lower holds, the centroid of the space;
for tween decks, the actual KG of the cargo.
Calculate the grain heeling arms λ_0, λ_{40}, and λ_{20} in meters.

4. (a) In grain-carrying ships, what is an angle of repose?
(b) For a grain-carrying ship, discuss the assumed angles of repose suggested by IMO, in way of a hatch trunk. Show these assumed angles of repose in an elevation view and in a plan view.

■

CHAPTER 38

True Mean Draft

In previous chapters it has been shown that a ship trims about the center of flotation. It will now be shown that, for this reason, a ship's true mean draft is measured at the center of flotation and may not be equal to the average of the drafts forward and aft. This is only true when LCF has an average value.

Consider the ship shown in Figure 38.1(a), which is floating on an even keel and whose center of flotation is FY aft of amidships. The true mean draft is KY, which is also equal to ZF, the draft at the center of flotation. Now let a weight be shifted aft within the ship so that she trims about F, as shown in Figure 38.1(b). The draft at the center of flotation (ZF) remains unchanged. Let the new draft forward be F and the new draft aft be A, so that the trim (A − F) is equal to 't'. Since no weights have been loaded or discharged, the ship's displacement will not have changed and the true mean draft must still be equal to KY. It can be seen from Figure 38.1(b) that the average of the drafts forward and aft is equal to KX, the draft amidships. Also,

$$ZF = KY = KX + XY$$

or

$$\text{True mean draft} = \text{Draft amidships} + \text{Correction}$$

Referring to Figure 38.1(b) and using the property of similar triangles:

$$\frac{XY}{FY} = \frac{t}{L} \quad XY = \frac{t \times FY}{L}$$

or

$$\textit{Correction FY} = \frac{\textit{Trim} \times \textit{FY}}{\textit{Length}}$$

where FY is the distance of the center of flotation from amidships.

It can also be seen from the figure that, when a ship is trimmed by the stern and the center of flotation is aft of amidships, the correction is to be added to the mean of the drafts forward and aft. Also, by substituting forward for aft and aft for forward in Figure 38.1(b), it can be seen that the correction is again to be added to the mean of the drafts forward and aft when the ship is trimmed by the head and the center of flotation is forward of amidships.

Ship Stability for Masters and Mates. http://dx.doi.org/10.1016/B978-0-08-097093-6.00038-4

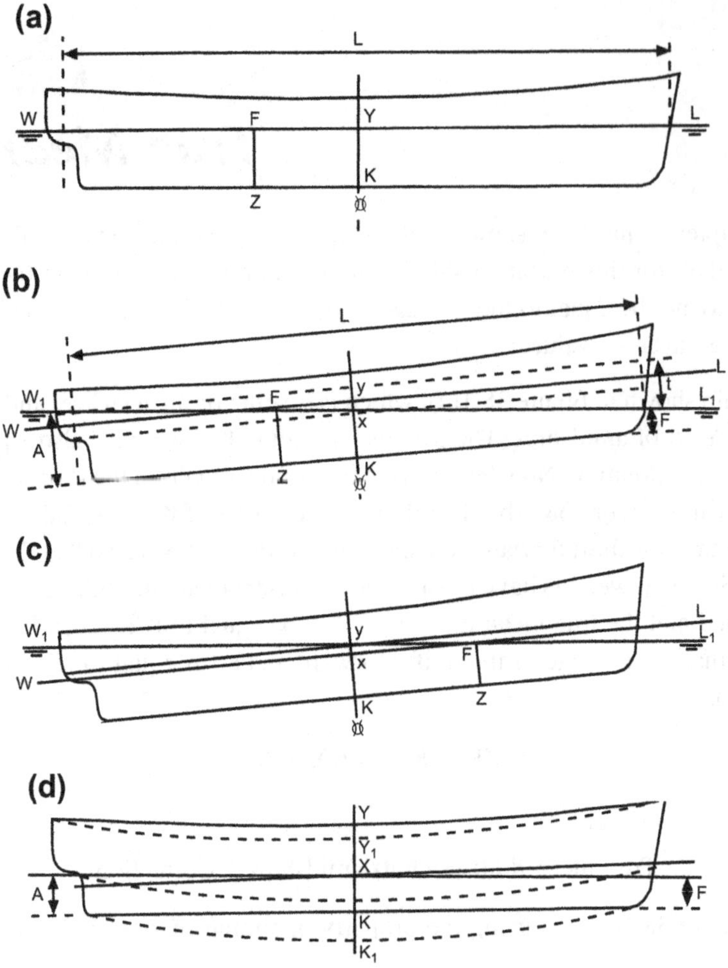

Figure 38.1

Now consider the ship shown in Figure 38.1(c), which is trimmed by the stern and has the center of flotation forward of amidships. In this case:

$$ZF = KY = KX - XY$$

or

$$\text{True mean draft} = \text{Draft amidships} - \text{Correction}$$

The actual correction itself can again be found by using the above formula, but in this case the correction is to be subtracted from the mean of the drafts forward and aft. Similarly, by substituting forward for aft and aft for forward in this figure, it can be seen that the correction is again to be subtracted from the average of the drafts forward and aft when the ship is trimmed by the head and the center of flotation is aft of amidships.

A general rule may now be derived for the application of the correction to the draft amidships in order to find the true mean draft.

Rule

When the center of flotation is in the *same direction* from amidships as the maximum draft, the correction is to be *added* to the mean of the drafts. When the center of flotation is in the *opposite direction* from amidships to the maximum draft, the correction is to be *subtracted.*

Example

A ship's minimum permissible freeboard is at a true mean draft of 8.5 m. The ship's length is 120 m, the center of flotation being 3 m aft of amidships. TPC = 50 tonnes. The present drafts are 7.36 m F and 9.00 m A. Find how much more cargo can be loaded.

$$\text{Draft forward} = 7.36\ \text{m}$$
$$\text{Draft aft} = \underline{9.00\ \text{m}}$$
$$\text{Trim} = 1.64\ \text{m by the stern}$$
$$\text{Correction} = \frac{t \times FY}{L} = \frac{1.64 \times 3}{120}$$
$$\text{Correction} = 0.04\ \text{m}$$
$$\text{Draft forward} = 7.36\ \text{m}$$
$$\text{Draft aft} = 9.00\ \text{m}$$
$$\text{Sum} = 16.36\ \text{m}$$
$$\text{Average} = \text{Draft amidships} = 8.18\ \text{m}$$
$$\text{Correction} = +0.04\ \text{m}$$
$$\text{True mean draft} = 8.22\ \text{m}$$
$$\text{Load mean draft} = \underline{8.50\ \text{m}}$$
$$\text{Increase in draft} = 0.28\ \text{m or 28 cm}$$
$$\text{Cargo to load} = \text{Increase in draft required} + \text{TPC}$$
$$= 28 \times 50$$

Ans. Cargo to load = 1400 tonnes.

Effect of Hog and Sag on Draft Amidships

When a ship is neither hogged nor sagged the draft amidships is equal to the mean of the drafts forward and aft. In Figure 38.1(d) the vessel is shown in hard outline floating without being hogged or sagged. The draft forward is F, the draft aft is A, and the draft amidships (KX) is equal to the average of the drafts forward and aft.

Now let the vessel be sagged as shown in Figure 38.1(d) by the broken outline. The draft amidships is now K_1X, which is equal to the mean of the drafts forward and aft (KX), plus the sag (KK_1). The amount of hog or sag must therefore be taken into account in calculations involving the draft amidships. The depth of the vessel amidships from the keel to the deck line (KY or K_1Y_1) is constant, being equal to the draft amidships plus the freeboard.

Example

A ship is floating in water of relative density 1.015. The present displacement is 12,000 tonnes, KG = 7.7 m, and KM = 8.6 m. The present drafts are F 8.25 m, A 8.65 m, and the present freeboard amidships is 1.06 m. The Summer draft is 8.53 m and the Summer freeboard is 1.02 m, FWA = 160 mm, and TPC = 20. Assuming that the KM is constant, find the amount of cargo (KG = 10.0 m) which can be loaded for the ship to proceed to sea at the loaded Summer draft. Also find the amount of the hog or sag and the initial GM on departure.

Summer freeboard	1.02 m	Present mean freeboard	1.06 m
Summer draft	+ 8.53 m	Depth Mld	9.55 m
Depth Mld	9.55 m	Present draft amidships	8.49 m
		Average of drafts F and A	8.45 m
		Ship is sagged by	0.04 m

$$\text{Dock water allowance (DWA)} = \frac{(1025 - \rho_{DW})}{25} \times \text{FWA} = \frac{10}{25} \times 160 = 64 \text{ mm}$$
$$= 0.064 \text{ m}$$

$$\text{TPC in dock water} = \frac{RD_{DW}}{RD_{SW}} \times TPC_{SW} = \frac{1.015}{1.025} \times 20$$
$$= 19.8 \text{ tonnes}$$

Summer freeboard = 1.020 m
DWA = 0.064 m
Min. permissible freeboard = 0.956 m
Present freeboard = 1.060 m
Mean sinkage = 0.104 m or 10.4 cm

Cargo to load = Sinkage × TPC_{DW} = 10.4 × 19.8
Cargo to load = 205.92 tonnes

$$GG_1 = \frac{w \times d}{W + w} = \frac{205.92 \times (10 - 7.7)}{12,000 + 205.92} = \frac{473.62}{12,205.92}$$

∴ Rise of G = 0.039 m

Present GM (8.6 − 7.7) = 0.900 m

GM on departure = 0.861 m

and ship has a sag of 0.04 m.

■

Exercise 38

1. The minimum permissible freeboard for a ship is at a true mean draft of 7.3 m. The present draft is 6.2 m F and 8.2 m A. TPC = 10. The center of flotation is 3 m aft of amidships. Length of the ship is 90 m. Find how much more cargo may be loaded.
2. A ship has a load salt water displacement of 12,000 tonnes, load draft in salt water 8.5 m, length 120 m, TPC = 15 tonnes, and center of flotation 2 m aft of amidships. The ship is at present floating in dock water of density 1015 kg per cubic meter at drafts of 7.2 m F and 9.2 m A. Find the cargo that must still be loaded to bring the ship to the maximum permissible draft.
3. Find the weight of cargo the ship in Question 2 could have loaded had the center of flotation been 3 m forward of amidships instead of 2 m aft.
4. A ship is floating in dock water of relative density 1.020. The present displacement is 10,000 tonnes, KG = 6.02 m, and KM = 6.92 m. Present drafts are F 12.65 m and A 13.25 m. Present freeboard is 1.05 m. Summer draft is 13.10 m and Summer freeboard is 1.01 m. FWA = 150 mm, TPC = 21. Assuming that the KM is constant, find the amount of cargo (KG = 10.0 m) that can be loaded for the ship to sail at the load Summer draft. Find also the amount of the hog or sag and the initial metacentric height on departure.

■

CHAPTER 39

Inclining Experiment (Stability Test) Plus Fluctuations in a Ship's Lightweight

It has been shown in previous chapters that, before the stability of a ship in any particular condition of loading can be determined, the initial conditions must be known. This means knowing the ship's lightweight, the VCG or KG at this lightweight, plus the LCG for this lightweight measured from amidships. For example, when dealing with the height of the center of gravity above the keel, the initial position of the center of gravity must be known before the final KG can be found. It is in order to find the KG for the light condition that the Inclining Experiment is performed.

The experiment is carried out by the builders when the ship is as near to completion as possible; that is, as near to the light condition as possible. The ship is forcibly inclined by shifting weights a fixed distance across the deck. The weights used are usually concrete blocks, and the inclination is measured by the movement of plumb lines across specially constructed battens that lie perfectly horizontal when the ship is upright. Usually two or three plumb lines are used and each is attached at the centerline of the ship at a height of about 10 m above the batten. If two lines are used then one is placed forward and the other aft. If a third line is used it is usually placed amidships. For simplicity, in the following explanation only one weight and one plumb line is considered.

The following conditions are necessary to ensure that the KG obtained is as accurate as possible:

1. There should be little or no wind, as this may influence the inclination of the ship. If there is any wind the ship should be head on or stern on to it.
2. The ship should be floating freely. This means that nothing outside the ship should prevent her from listing freely. There should be no barges or lighters alongside; mooring ropes should be slacked right down, and there should be plenty of water under the ship to ensure that at no time during the experiment will she touch the bottom.
3. Any loose weights within the ship should be removed or secured in place.
4. There must be no free surfaces within the ship. Bilges should be dry. Boilers and tanks should be completely full or empty.
5. Any persons not directly concerned with the experiment should be sent ashore.
6. The ship must be upright at the commencement of the experiment.
7. A note of 'weights on' and 'weights off' to complete the ship each with a VCG and $LCG_{amidships}$.

Ship Stability for Masters and Mates. http://dx.doi.org/10.1016/B978-0-08-097093-6.00039-6

When all is ready and the ship is upright, a weight is shifted across the deck transversely, causing the ship to list. A little time is allowed for the ship to settle and then the deflection of the plumb line along the batten is noted. If the weight is now returned to its original position the ship will return to the upright. She may now be listed in the opposite direction. From the deflections the GM is obtained as follows:

In Figure 39.1 let a mass of 'w' tonnes be shifted across the deck through a distance 'd' meters. This will cause the center of gravity of the ship to move from G to G_1 parallel to the shift of the center of gravity of the weight. The ship will then list to bring G_1 vertically under M, i.e. to $\theta°$ list. The plumb line will thus be deflected along the batten from B to C. Since AC is the new vertical, angle BAC must also be $\theta°$.

In triangle ABC,

$$\cot\theta = \frac{AB}{BC}$$

In triangle GG_1M,

$$\tan\theta = \frac{GG_1}{GM}$$

$$\therefore \frac{GM}{GG_1} = \frac{AB}{BC}$$

or

$$GM = GG_1 \times \frac{AB}{BC}$$

But

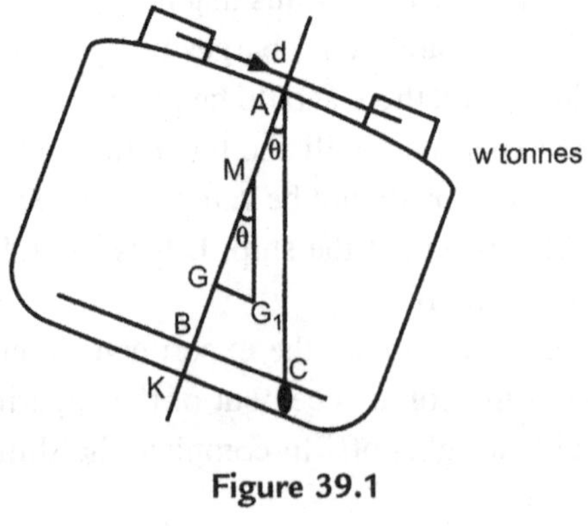

Figure 39.1

$$GG_1 = \frac{w \times d}{W}$$

$$\therefore GM = \frac{w \times d}{W} \times \frac{AB}{BC}$$

Hence

$$GM = \frac{w \times d}{W \tan \theta}$$

In this formula AB, the length of the plumb line and BC, the deflection along the batten, can be measured. 'w' the mass shifted, 'd' the distance through which it was shifted, and 'W' the ship's displacement, will all be known. The GM can therefore be calculated using the formula.

The naval architects will already have calculated the KM for this draft and hence the present KG is found. By taking moments about the keel, allowance can now be made for weights that must be loaded or discharged to bring the ship to the light condition. In this way the light KG is found.

Example 1

When a mass of 25 tonnes is shifted 15 m transversely across the deck of a ship of 8000 tonnes displacement, it causes a deflection of 20 cm in a plumb line 4 m long (see Figure 39.2). If the KM = 7 m, calculate the KG.

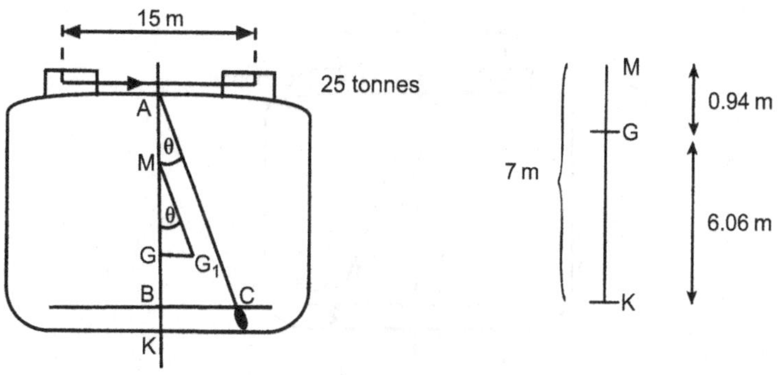

Figure 39.2

$$\frac{GM}{GG_1} = \frac{AB}{BC} = \frac{1}{\tan\theta}$$

$$\therefore \tan\theta \, GM = GG_1$$

$$GM = \frac{w \times d}{W} \times \frac{1}{\tan\theta}$$

$$= \frac{25 \times 15}{8000} \times \frac{4}{0.2}$$

$$GM = 0.94 \text{ m}$$

$$KM = 7.00 \text{ m}$$

Ans. KG = 6.06 m, as shown in the sketch.

Example 2

When a mass of 10 tonnes is shifted 12 m transversely across the deck of a ship with a GM of 0.6 m, it causes 0.25 m deflection in a 10 m plumb line (see Figure 39.3). Calculate the ship's displacement.

$$GM = \frac{w \times d}{W} \times \frac{1}{\tan\theta}$$

$$W = \frac{w \times d}{GM} \times \frac{1}{\tan\theta}$$

$$= \frac{10 \times 12 \times 10}{0.6 \times 0.25}$$

$$W = 8000$$

Ans. Displacement = 8000 tonnes.

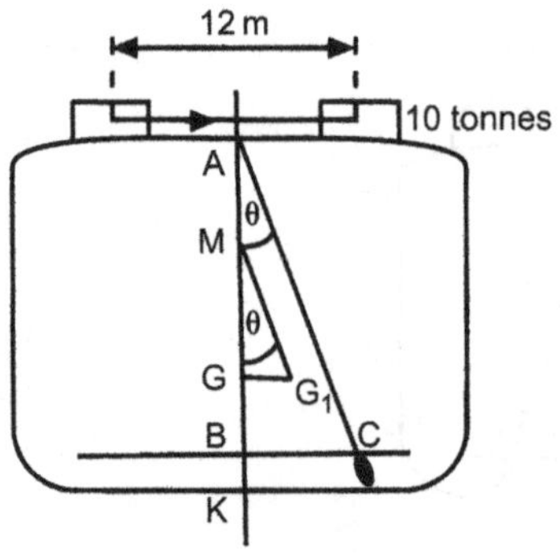

Figure 39.3

Summary

Every *new* ship should have an Inclining Experiment. However, some shipowners do not request one if their ship is a sister ship to one or more in the company's fleet.

Fluctuations in a Ship's Lightweight Over a Period of Time

Lightweight is made up of the steel weight plus the wood and outfit weight plus the machinery weight. The lightweight of a ship is the weight of the ship when completely *empty*. There will be no deadweight items on board.

Over the years in service, there will be increases in the lightweight due to:

- Accretion of paintwork
- Formation of oxidation or rust
- Build-up of cargo residue
- Sediment in bottom of oil tanks
- Mud in bottom of ballast tanks
- Dunnage
- Gradual accumulation of rubbish
- Lashing material
- Retrofits on accommodation fittings and in navigational aids
- Barnacle attachment or animal growth on the shell plating
- Vegetable growth on shell plating
- Additional engineers' spares and electricians' spares
- Changing a bulbous bow to a soft-nosed bow
- Major ship surgery such as an addition to ship's section at amidships.

Each item in the above list will change the weight of an empty ship. It can also be accumulative. One example of increase in lightweight over a period of years is the *Herald of Free Enterprise*, which capsized in 1987. At the time of capsize it was shown that the lightweight had increased by 270 t, compared to when the ship was newly built.

Regular drydocking of the ship will decrease the animal and vegetable growth on the shell plating. It has been known to form as much as 5 cm of extra weight around the hull.

Regular tank-cleaning programs will decrease the amount of oil sediment and mud in the bottom of tanks. Regular routine inspections should also decrease the accumulation of rubbish.

Over years in service, there will also be decreases in the lightweight due to:

- Oxidation or corrosion of the steel shell plating, and steel decks exposed to the sea and to the weather

- Wear and tear on moving parts
- Galvanic corrosion at localities having dissimilar metals joined together.

Corrosion and loss of weight is prevalent and vulnerable in the boot-topping area of the side-shell of a vessel, especially in way of the machinery spaces. Feedback has shown that the side-shell thickness can decrease over the years from being 18 mm thickness to being only 10 mm in thickness. This would result in an appreciable loss of weight.

Wear and tear occurs on structures such as masts and derricks, windlass, winches, hawse pipes, and capstans.

These additions and reductions will all have their own individual centers of gravity and moments of weight. The result will be an overall change in the lightweight itself, plus a new value for the KG corresponding to this new lightweight.

One point to also consider is the weight that crew and passengers bring onto ships. It is usually denoted as ‘effects’. Strictly speaking, these ‘effects’ are part of the deadweight. However, some people bring more onto the ship than average. Some leave the ship without all their belongings. It is difficult as time goes by to keep track of all these small additions. Conjecture suggests that ships with large numbers of persons will experience greater additions. Examples are passenger liners, ro-ro ships, and Royal Naval vessels.

It has been documented that the lightweight of a vessel can amount to an average addition of 0.5% of the lightweight for each year of the ship’s life. As a ship gets older it becomes heavier.

Major ship surgery can involve the insertion of a length of structure at amidships. However, it can involve raising a deck to give greater tween deck height. There is also on record a ship that was cut right down through the centerline (stern to bow) and pulled apart, and structure was added in to increase the breadth molded. Each installation of course changed each ship’s lightweight.

Returning to the line on bows, there was one notable example where the shipowner requested the removal of a bulbous bow. It was replaced with a soft-nosed bow. The modified ship actually gave an increase in service speed for similar input of engine power. This increase in ship performance was mainly due to the fact that this vessel had a low service speed (about 12 knots) and a small summer deadweight. After this retrofit, obviously there would be a change in lightweight.

These notes indicate that sometimes the lightweight will increase, for example due to plate renewal or animal and vegetable growth. Other times it will decrease, for example due to wear and tear or build-up of corrosion. There will be fluctuations. It would seem judicial to plan for an inclining experiment perhaps every 5 years. This will re-establish for the age of the ship exactly the current lightweight. Passenger liners are required to do just this.

Exercise 39

1. A ship of 8000 tonnes displacement has KM = 7.3 m and KG = 6.1 m. A mass of 25 tonnes is moved transversely across the deck through a distance of 15 m. Find the deflection of a plumb line that is 4 m long.
2. As a result of performing the inclining experiment it was found that a ship had an initial metacentric height of 1 m. A mass of 10 tonnes, when shifted 12 m transversely, had listed the ship 3° and produced a deflection of 0.25 m in the plumb line. Find the ship's displacement and the length of the plumb line.
3. A ship has KM = 6.1 m and displacement of 3150 tonnes. When a mass of 15 tonnes, already on board, is moved horizontally across the deck through a distance of 10 m it causes 0.25 m deflection in an 8-m-long plumb line. Calculate the ship's KG.
4. A ship has an initial GM = 0.5 m. When a mass of 25 tonnes is shifted transversely a distance of 10 m across the deck, it causes a deflection of 0.4 m in a 4 m plumb line. Find the ship's displacement.
5. A ship of 2304 tonnes displacement has an initial metacentric height of 1.2 m. Find the deflection in a plumb line that is suspended from a point 7.2 m above a batten when a mass of 15 tonnes, already on board, is shifted 10 m transversely across the deck.
6. During the course of an inclining experiment in a ship of 4000 tonnes displacement, it was found that, when a mass of 12 tonnes was moved transversely across the deck, it caused a deflection of 75 mm in a plumb line that was suspended from a point 7.5 m above the batten. KM = 10.2 m, KG = 7 m. Find the distance through which the mass was moved.
7. A box-shaped vessel 60 m × 10 m × 3 m is floating upright in fresh water on an even keel at 2 m draft. When a mass of 15 tonnes is moved 6 m transversely across the deck a 6 m plumb line is deflected 20 cm. Find the ship's KG.
8. The transverse section of a barge is in the form of a triangle, apex downwards. The ship's length is 65 m, breadth at the waterline 8 m, and the vessel is floating upright in salt water on an even keel at 4 m draft. When a mass of 13 tonnes is shifted 6 m transversely it causes 20 cm deflection in a 3 m plumb line. Find the vessel's KG.
9. A ship of 8000 tonnes displacement is inclined by moving 4 tonnes transversely through a distance of 19 m. The average deflection of two pendulums, each 6 m long, was 12 cm. 'Weights on' to complete this ship were 75 t centered at KG of 7.65 m. 'Weights off' amounted to 25 t centered at KG of 8.16 m.
   (a) Calculate the GM and angle of heel relating to this information for the ship as inclined.
   (b) From hydrostatic curves for this ship as inclined, the KM was 9 m. Calculate the ship's final lightweight and VCG at this weight.

Exercise 39

CHAPTER 40

The Calibration Book Plus Soundings and Ullages

The purpose of a calibration book is to give volumes, displacements, and center of gravity at preselected tank levels. The Naval Architect calculates the contents of a tank at, say, 0.01−0.20 m intervals of height within the tank. For an example of 0.10 m sounding height intervals, see Table 40.1. This information is supplied by the shipbuilder to each ship, for use by masters and mates.

A sounding is the vertical distance between the base of the tank and the surface of the liquid. A sounding pipe is a plastic pipe of about 37 mm diameter, down which a steel sounding tape is lowered (see Figures 40.1 and 40.2 and Table 40.1).

An ullage is the vertical distance between the surface of the liquid and the top of the ullage plug or top of the sounding pipe (see Figure 40.1 and Tables 40.2 and 40.3).

For tanks containing fresh water, distilled water, water ballast, etc. it is advisable to take soundings. These tanks can be filled up to 100% capacity.

For tanks containing cargo oil, oil fuel, diesel oil, lubricating oil, etc. it is advisable to take ullage readings. Oil tanks are usually filled to a maximum of 98% full capacity. This is to make allowance for expansion due to heat.

Table 40.1: Sample of a calibrated oil fuel tank.

Sounding (meters)	Volume (m^3)
0.10	0.60
0.20	0.90
0.30	1.40
0.40	1.90
0.50	2.60
0.60	3.30
0.70	4.10
0.80	5.00
0.90	6.10
1.00	7.20

Ship Stability for Masters and Mates. http://dx.doi.org/10.1016/B978-0-08-097093-6.00040-2

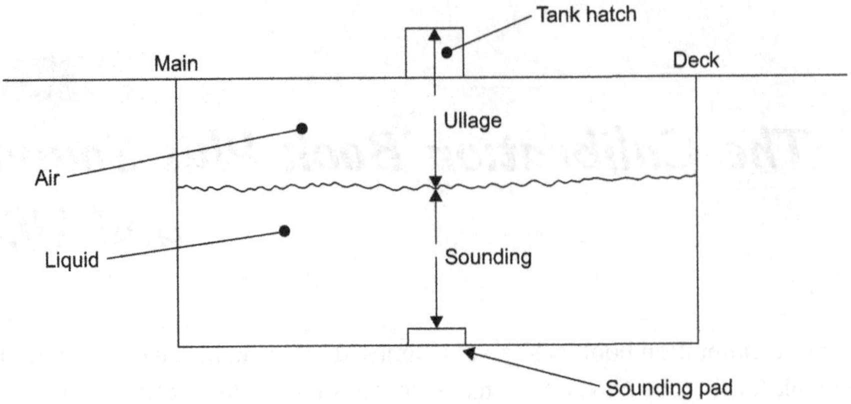

Figure 40.1:
A Partially Filled Tank.

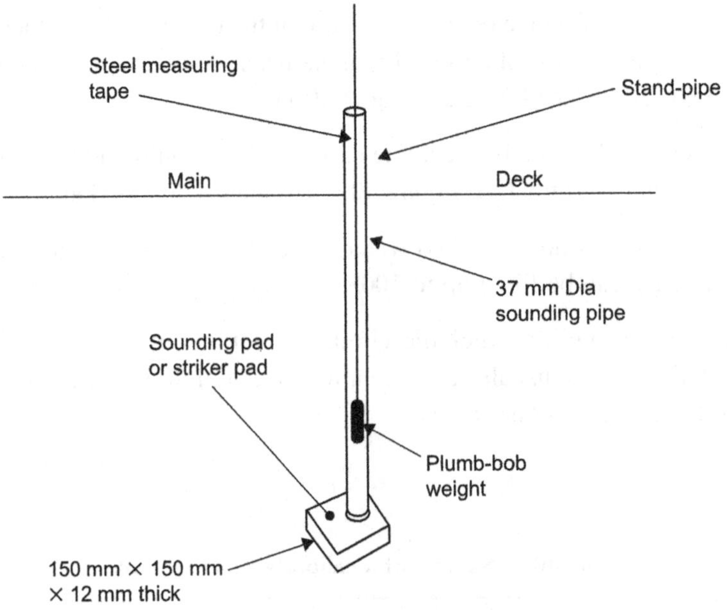

Figure 40.2:
Soundings Using a Steel Tape.

Methods for Reading Soundings and Ullages

A reading may be taken by using:

1. Steel measuring tape with a weight attached to its end (see Figure 40.2).
2. Calibrated glass tube (see Figure 40.3).

Table 40.2: Sample of a calibrated cargo oil tank: Volumes and weights.

Ullage (meters)	Volume (m^3)	Oil @ 0.860 m^3 per tonne	Oil @ 0.880 m^3 per tonne	Oil @ 0.900 m^3 per tonne
24.30	4011.73	3450	3530	3611
24.32	4012.72	3451	3531	3611
24.34	4013.61	3452	3532	3612
24.36	4014.37	3452	3532	3613
24.38	4015.01	3453	3533	3614
24.40	4015.55	3453	3534	3614
24.42	4015.97	3454	3534	3614
24.44	4016.28	3454	3534	3615
24.46	4016.47	3454	3534	3615
24.48	4016.55	3454	3535	3615
24.50	4016.58	3454	3535	3615

Weights have been rounded off to the nearest tonne.

Table 40.3: Calibrated fresh water tank: Weights, vertical and longitudinal moments.

Sight Glass Tube Ullage Reading (m)	Weight of Fresh Water (tonnes)	Vertical Moment, Weight × KG (tm)	Longitudinal Moment, Weight × LCG (tm)
0.20	6.90	23.10	68.90
0.40	5.70	18.50	57.00
0.60	4.60	14.30	45.30
0.80	3.50	10.50	33.80
1.00	2.40	6.90	22.80
1.20	1.50	4.00	13.20
1.40	0.80	2.00	6.90
1.60	0.30	0.80	2.70
1.80	0.10	0.10	0.50

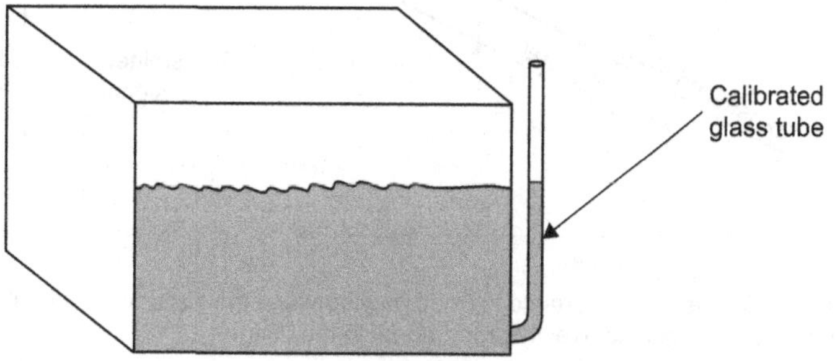

Figure 40.3:
Calibration Using a Glass Sounding Tube.

3. Whessoe gastight tank gage (see Figure 40.4).
4. Saab Radar tank gage (see Figure 40.5).

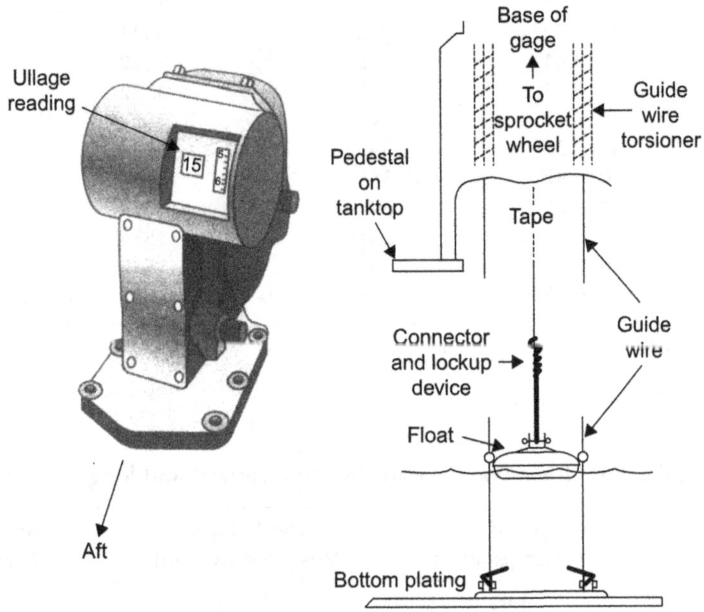

Figure 40.4:
The 'Whessoe' Gastight Tank Gage. *Source: DJ House.*

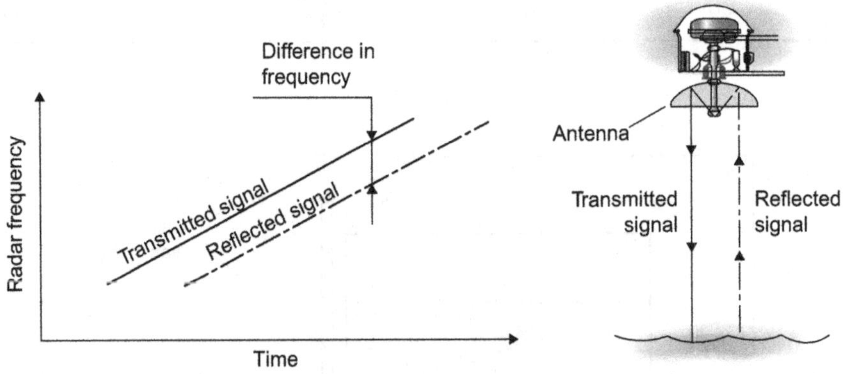

The radar principle. The difference in frequency between the transmitted signal and the reflected signal is directly proportional to the ullage.

Figure 40.5:
The Saab Gaging Method. *Source: Saab Radar Gauge Company.*

Consider an oil fuel tank (see Table 40.1). The master or mate can take a sounding in the tank on a particular day. If it happens to be 0.60 m then the contents of the tank are 3.30 cubic meters.

1. The steel measuring tape is marked with white chalk and lowered into the tank of liquid until the user hears a dull sound as the end weight makes contact with the sounding pad.

   The scale on the tape is then checked to the level where the chalk has been washed off and where the chalk has remained in dry condition. At this interface, the vertical measurement is lifted off. This is the sounding depth for the tank (see Figure 40.2 and Table 40.1).
2. A calibrated tube can be fitted on the outside of a small tank. This may be a fresh water tank, distilled water tank, lubricating oil tank, or heavy oil tank. The observer can quickly read the height on the calibrated glass tube, thus obtaining the ullage or sounding in the tank (see Figure 40.3 and Table 40.3).
3. When carrying poisons, corrosives, highly inflammable products, and cargoes that react with air, means of taking ullages and samples without releasing any vapor must be provided. Such a system is the 'Whessoe' gastight tank gage (see Figure 40.4).

   Remote reading temperature sensors and high-level alarms are also commonly fitted on modern ships. When these alarms are set off the pumps stop automatically.

   The distance of the float from the tank top is read on the 'Whessoe' ullage meter. This is moved by the measuring tape passing over a sprocket in the gage head, which in turn moves the counter drum whose rotation is transmitted to the ullage meter. The ullage for the tank is then read off the screen, as shown in Figure 40.4.
4. The Saab Radar cased cargo tank and ballast monitoring system is illustrated in Figure 40.5. The tank gage unit measures the distance to the surface of the liquid in the tank using a continuous radar signal. This signal can have a measurement range of 0–60 m. The reflected signal is received and processed via tank gage electronics. When converted, the calculated values can then be sent to a communications center, for example to officers working on the bridge.

This system has several very useful advantages. They are as follows:

- From the bridge or other central control, it is easy to check and recheck the contents in any tank containing liquid.
- Continuous minute-by-minute readings of liquid level in the tank can be made.
- Radar waves are extremely robust to any condition in the tank.
- Radar waves are generally not affected by the atmosphere above the product in the tank.
- The only part located inside the tank is the antenna without any moving parts.
- High accuracy and high reliability is assured.
- The electronics can be serviced and replaced during closed tank conditions.

The bosun on board a ship used to check all tanks at least once a day. This was to ensure that water had not entered the ship via the hull plating and that liquid had not leaked into an adjacent tank. With modern calibration systems continuous assessment is now possible.

Cofferdams must also be checked to ensure no seepage has occurred. A cofferdam is a drainage tank placed between compartments carrying dissimilar liquids; for example, it may be a water ballast tank and an oil fuel tank. Cofferdams should not be used to carry any liquids so they do not usually have filling pipes fitted.

Adjustments for Angle of Heel and Trim

If the sounding pipe or ullage plug is not transversely in line with the tank's centerline then an adjustment to the ullage reading must be made.

1. If vessel has a list to *port* and the ullage plug is a horizontal distance 'L' to the port side of the tank's VCG, then the *positive* correction to the ullage reading is:

$$+\text{L} \times \tan\theta \text{ meters} \quad \text{where } \theta = \text{Angle of list}$$

2. If vessel has a list to *port* and the ullage plug is a horizontal distance 'L' to the starboard side of the tank's VCG, then the *negative* correction to the ullage reading is:

$$-\text{L} \times \tan\theta \text{ meters} \quad \text{where } \theta = \text{Angle of list}$$

3. If vessel has a list to *starboard* and the ullage plug is a horizontal distance 'L' to the port side of the tank's VCG, then the *negative* correction to the ullage reading is:

$$-\text{L} \times \tan\theta \text{ meters} \quad \text{where } \theta = \text{Angle of list}$$

4. If vessel has a list to *starboard* and the ullage plug is a horizontal distance 'L' to the starboard side of the tank's VCG, then the *positive* correction to the ullage reading is:

$$+\text{L} \times \tan\theta \text{ meters} \quad \text{where } \theta = \text{Angle of list}$$

Worked Example 1

A ship has an ullage plug that is 1.14 m to the port side of a tank's VCG. Estimate the adjustment to the ullage readings in cm, when this ship has angles of heel to port up to 3° at intervals of 0.5°.

Angle of heel θ (°)	0	0.5	1.0	1.5	2.0	2.5	3.0
$+114 \times \tan\theta$ (cm)	0	+1	+2	+3	+4	+5	+6

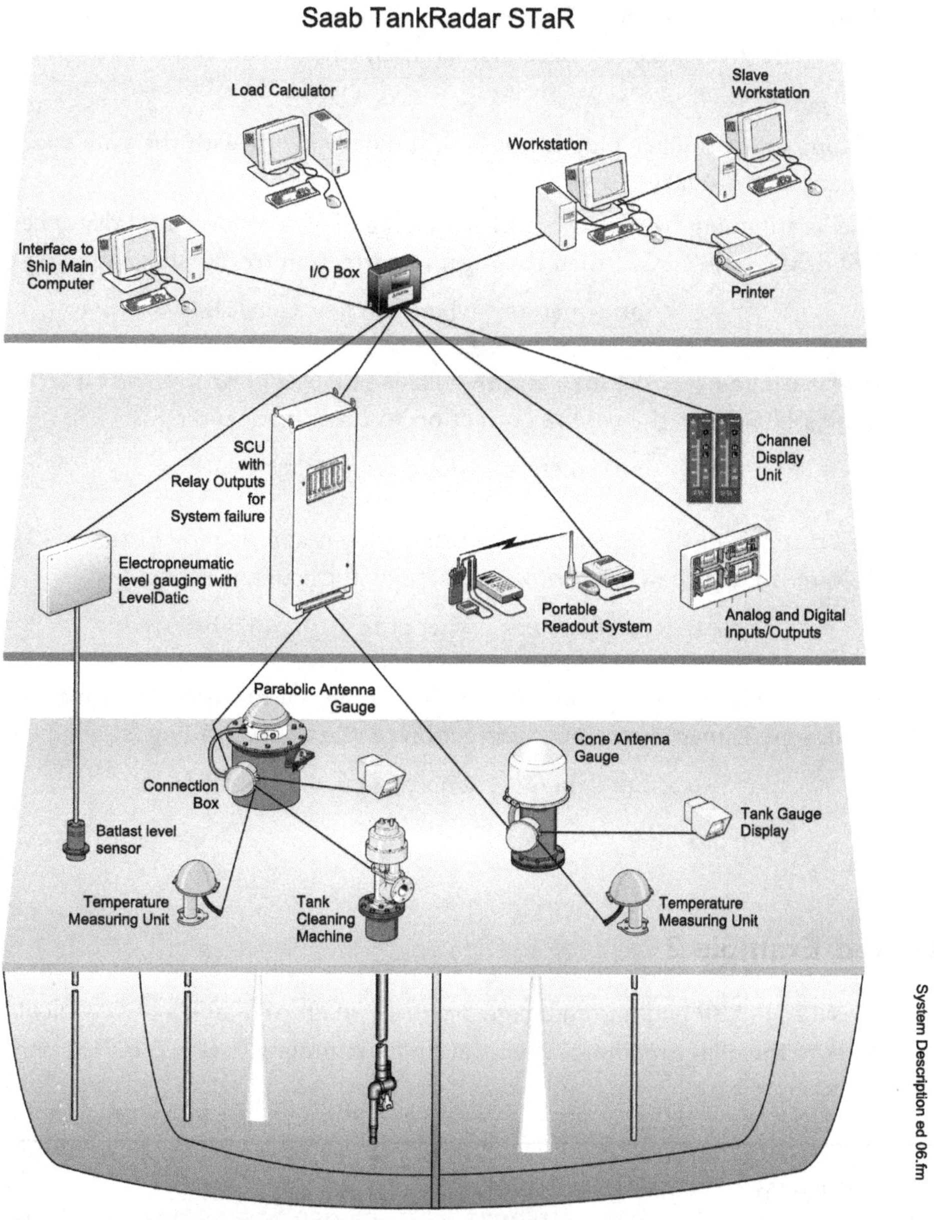

System and configuration of a Saab TankRadar STaR level gauging system.
The example shows a Parabolic Antenna Gauge and a Cone Antenna Gauge on different tanks.
LevelDatic is an Electropneumatic Level Measurement System from SF Control.

Figure 40.6

The table for Worked Example 1 shows the transverse adjustments are 0–6 cm. Consequently, they do need to be considered by the master and mate on board ship.

If the sounding pipe or ullage plug is not longitudinally in line with the tank's LCG then an adjustment to the ullage reading must be made.

1. If vessel is trimming by the *stern* and the ullage plug is a horizontal distance 'L' *forward* of the tank's LCG, then the *negative* correction to the ullage reading is:

$$-L \times \tan\theta \text{ meters} \quad \text{where } \tan\theta = \text{trim}/\text{LBP}$$

2. If vessel is trimming by the *stern* and the ullage plug is a horizontal distance 'L' *aft* of the tank's LCG, then the *positive* correction to the ullage reading is:

$$+L \times \tan\theta \text{ meters} \quad \text{where } \tan\theta = \text{trim}/\text{LBP}$$

3. If vessel is trimming by the *bow* and the ullage plug is a horizontal distance 'L' *forward* of the tank's LCG, then the *positive* correction to the ullage reading is:

$$+L \times \tan\theta \text{ meters} \quad \text{where } \tan\theta = \text{trim}/\text{LBP}$$

4. If vessel is trimming by the *bow* and the ullage plug is a horizontal distance 'L' *aft* of the tank's LCG, then the *negative* correction to the ullage reading is:

$$-L \times \tan\theta \text{ meters} \quad \text{where } \tan\theta = \text{trim}/\text{LBP}$$

Worked Example 2

A vessel has 120 m LBP and has an ullage pipe 2.07 m aft of a tank's LCG. Calculate the adjustments to the ullage readings when the ship is trimming 0, 0.3, 0.6, 0.9, and 1.2 m by the stern.

Trim (meters)	0	0.3	0.6	0.9	1.2
Trim angle (°)	0	0.0025	0.0050	0.0075	0.0100
$+207 \times \tan\theta$ (cm)	0	+0.009	+0.018	+0.027	+0.036

The table for Worked Example 2 indicates the adjustments for trim are very small and can, in practice, be ignored by the master and mate on board the ship.

Adjustments for Temperature

Oil expands when heated and consequently its density *decreases* with a *rise* in temperature. This means that the density in t/m^3 must be adjusted to give a better reading. The change of relative density due to a change of one degree in temperature is known as the relative density coefficient. For most oils, this lies between 0.00025 and 0.00050 per °C. The following worked example shows how this is done.

Worked Example

A sample of oil has a density of 0.8727 at 16 °C. Its expansion coefficient is 0.00027 per °C. Proceed to calculate its density at 26 °C.

$$\text{Difference in temperature} = 26^\circ - 16^\circ = 10\ ^\circ\text{C}$$
$$\text{Change in density} = 10 \times 0.00027 = 0.0027$$
$$\text{Density at } 26^\circ = 0.8727 - 0.0027 = 0.870\ \text{t/m}^3$$

Worked Example

A reworking of the previous example using degrees fahrenheit.

A sample oil has a density of 0.8727 at 60.8 °F. Its expansion coefficient is 0.00015 per °F. Proceed to calculate its density at 78.8 °F.

$$\text{Difference in temperature} = 78.8^\circ - 60.8^\circ = 18\ ^\circ\text{C}$$
$$\text{Change in density} = 18 \times 0.00015 = 0.0027$$
$$\text{Density at } 78.8^\circ = 0.8727 - 0.0027 = 0.870\ \text{t/m}^3\ (\text{as before})$$

A tank with a zero sounding will not always have a zero volume reading. This is because at the base of a tank there could be a 12-mm-thick sounding pad to conserve the bottom of the ship's tank.

Exercise 40

1. A ship 120 m long is trimmed 1.5 m by the stern. A double-bottom tank is 15 m long × 20 m × 1 m and has the sounding pipe situated at the after end of the tank. Find the sounding that will indicate that the tank is full.

2. A ship 120 m long is trimmed 2 m by the stern. A double-bottom tank 36 m × 15 m × 1 m is being filled with oil fuel of relative density 0.96. The sounding pipe is at the after end of the tank and the present sounding is 1.2 m. Find how many tonnes of oil are still required to fill this tank and also find the sounding when the tank is full.

CHAPTER 41

Drydocking and Stability – Procedures and Calculations

When a ship enters a dry dock she must have a positive initial GM, be upright, and trimmed slightly, usually by the stern. On entering the dry dock the ship is lined up with her centerline vertically over the centerline of the keel blocks and the shores are placed loosely in position. The dock gates are then closed and pumping out commences. The rate of pumping is reduced as the ship's stern post nears the blocks. When the stern lands on the blocks the shores are hardened up commencing from aft and gradually working forward so that all of the shores will be hardened up in position by the time the ship takes the blocks overall. The rate of pumping is then increased to quickly empty the dock.

As the water level falls in the dry dock there is no effect on the ship's stability so long as the ship is completely waterborne, but after the stern lands on the blocks the draft aft will decrease and the trim will change by the head. This will continue until the ship takes the blocks overall throughout her length, when the draft will then decrease uniformly forward and aft.

The interval of time between the stern post landing on the blocks and the ship taking the blocks overall is referred to as the *critical period.* During this period part of the weight of the ship is being borne by the blocks, and this creates an upthrust at the stern that increases as the water level falls in the dry dock. The upthrust causes a virtual loss in metacentric height and it is essential that positive effective metacentric height be maintained throughout the critical period, or the ship will heel over and perhaps slip off the blocks with disastrous results.

The purpose of this chapter is to show the methods by which the effective metacentric height may be calculated for any instant during the drydocking process.

Figure 41.1 shows the longitudinal section of a ship during the critical period. 'P' is the upthrust at the stern and 'l' is the distance of the center of flotation from aft. The trimming

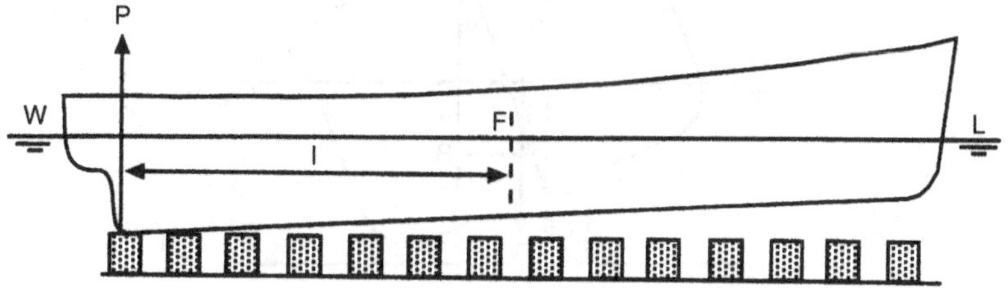

Figure 41.1

Ship Stability for Masters and Mates. http://dx.doi.org/10.1016/B978-0-08-097093-6.00041-4

moment is given by P × l. But the trimming moment is also equal to MCTC × change of trim.

Therefore,

$$P \times l = MCTC \times t$$

or

$$P = \frac{MCTC \times t}{l}$$

where P = the upthrust at the stern in tonnes, t = the change of trim since entering the drydock in centimeters, and l = the distance of the center of flotation from aft in meters.

Now consider Figure 41.2, which shows a transverse section of the ship during the critical period after she has been inclined to a small angle ($\theta°$) by a force external to the ship. For the sake of clarity the angle of heel has been magnified. The weight of the ship (W) acts downwards through the center of gravity (G). The force P acts upwards through the keel (K) and is equal to the weight being borne by the blocks. For equilibrium the force of buoyancy must now be (W − P) and will act upwards through the initial metacenter (M).

There are, thus, three parallel forces to consider when calculating the effect of the force P on the ship's stability. Two of these forces may be replaced by their resultant in order to find the effective metacentric height and the moment of statical stability.

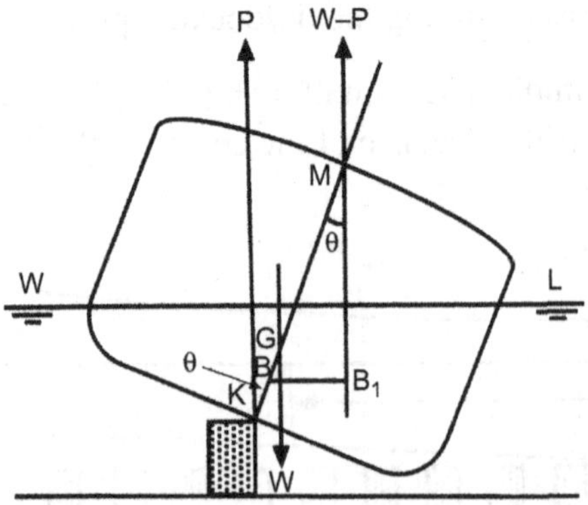

Figure 41.2

Method (a)

In Figure 41.3 consider the two parallel forces P and (W − P). Their resultant W will act upwards through M_1 such that:

$$(W - P) \times y = P \times X$$

or

$$(W - P) \times MM_1 \times \sin\theta = P \times KM_1 \times \sin\theta$$
$$(W - P) \times MM_1 = P \times KM_1$$
$$W \times MM_1 - P \times MM_1 = P \times KM_1$$
$$W \times MM_1 = P \times KM_1 + P \times MM_1$$
$$= P\ (KM_1 + MM_1)$$
$$= P \times KM$$
$$MM_1 = \frac{P \times KM}{W}$$

There are now two forces to consider: W acting upwards through M_1 and W acting downwards through G. These produce a righting moment of $W \times GM_1 \times \sin\theta$.

Note also that the original metacentric height was GM but has now been reduced to GM_1. Therefore, MM_1 is the virtual loss of metacentric height due to drydocking, or

$$\text{Virtual loss of GM}(MM_1) = \frac{P \times KM}{W}$$

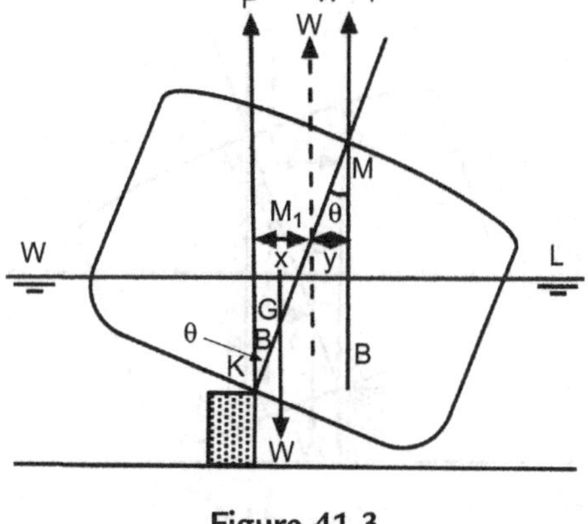

Figure 41.3

Method (b)

Now consider the two parallel forces W and P in Figure 41.4. Their resultant (W − P) acts downwards through G_1 such that:

$$W \times y = P \times X$$

or

$$\begin{aligned} W \times GG_1 \times \sin\theta &= P \times KG_1 \times \sin\theta \\ W \times GG_1 &= P \times KG_1 \\ &= P(KG + GG_1) \\ &= P \times KG + P \times GG_1 \\ W \times GG_1 - P \times GG_1 &= P \times KG \\ GG_1(W - P) &= P \times KG \\ GG_1 &= \frac{P \times KG}{W - P} \end{aligned}$$

There are now two forces to consider: (W − P) acting upwards through M and (W − P) acting downwards through G_1. These produce a righting moment of $(W - P) \times G_1M \times \sin\theta$.

The original metacentric height was GM but has now been reduced to G_1M. Therefore, GG_1 is the virtual loss of metacentric height due to drydocking, or

$$\text{Virtual loss of GM } (GG_1) = \frac{P \times KG}{W - P}$$

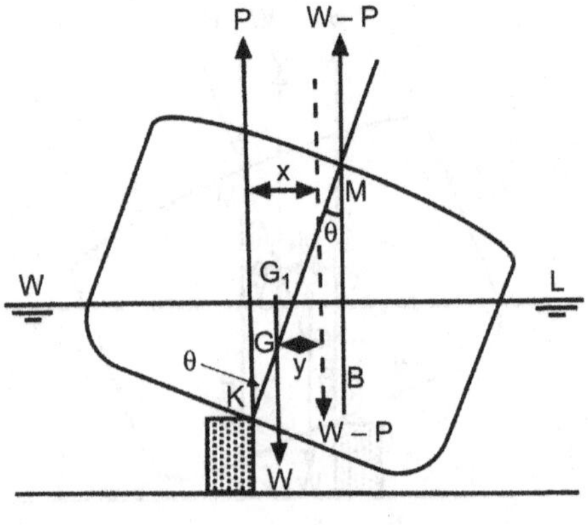

Figure 41.4

Example 1

A ship of 6000 tonnes displacement enters a dry dock trimmed 0.3 m by the stern. KM = 7.5 m, KG = 6 m, and MCTC = 90 tonnes m. The center of flotation is 45 m from aft. Find the effective metacentric height at the critical instant before the ship takes the blocks overall.

Note. Assume that the trim at the critical instant is zero.

$$P = \frac{MCTC \times t}{l}$$
$$= \frac{90 \times 30}{45}$$
$$P = 60 \text{ tonnes}$$

Method (a)

$$\text{Virtual loss of GM } (MM_1) = \frac{P \times KM}{W}$$
$$= \frac{60 \times 7.5}{6000}$$
$$= 0.075 \text{ m}$$
$$\text{Original GM} = 7.5 - 6.0 = 1.500 \text{ m}$$
$$\text{Righting moment} = 8550 \sin\theta \text{ tonnes meters}$$

Ans. New GM = 1.425 m.

Method (b)

$$\text{Virtual loss of GM} = \frac{P \times KG}{W - P}$$
$$GG_1 = \frac{60 \times 6}{5940}$$
$$= 0.061 \text{ m}$$
$$\text{Original GM} = 1.500 \text{ m}$$
$$\text{Righting moment} = 8550 \sin\theta \text{ tonnes meters}$$

Ans. New GM = 1.439 m.

From these results it would appear that there are two possible answers to the same problem, but this is not the case. The ship's ability to return to the upright is indicated by the righting moment and not by the effective metacentric height alone.

To illustrate this point, calculate the righting moments given by each method when the ship is heeled to a small angle ($\theta°$).

Method (a)

$$\begin{aligned}\text{Righting moment} &= W \times GM_1 \times \sin\theta \\ &= 6000 \times 1.425 \times \sin\theta \\ &= (8550 \times \sin\theta)\ \text{tonnes m}\end{aligned}$$

Method (b)

$$\begin{aligned}\text{Righting moment} &= (W - P) \times G_1M \times \sin\theta \\ &= 5940 \times 1.439 \times \sin\theta \\ &= (8549 \times \sin\theta)\ \text{tonnes meters}\end{aligned}$$

Thus, each of the two methods used gives a correct indication of the ship's stability during the critical period.

■

■ Example 2

A ship of 3000 tonnes displacement is 100 m long, has KM = 6 m, and KG = 5.5 m. The center of flotation is 2 m aft of amidships and MCTC = 40 tonnes m. Find the maximum trim for the ship to enter a dry dock if the metacentric height at the critical instant before the ship takes the blocks forward and aft is to be not less than 0.3 m.

$$\begin{aligned}KM &= 6.0\ \text{m} \\ KG &= \underline{5.5\ \text{m}} \\ \text{Original GM} &= 0.5\ \text{m} \\ \text{Virtual GM} &= \underline{0.3\ \text{m}} \\ \underline{\text{Virtual loss}} &= \underline{0.2\ \text{m}}\end{aligned}$$

Method (a)

$$\text{Virtual loss of GM } (MM_1) = \frac{P \times KM}{W}$$

or

$$P = \frac{\text{Virtual loss} \times W}{KM}$$
$$= \frac{0.2 \times 3000}{6}$$
$$\underline{\text{Maximum P}} = \underline{100 \text{ tonnes}}$$

But

$$P = \frac{MCTC \times t}{l}$$

or

$$\text{Maximum } t = \frac{P \times l}{MCTC}$$
$$= \frac{100 \times 48}{40}$$

Ans. Maximum trim = 120 cm by the stern.

Method (b)

$$\text{Virtual loss of GM } (GG_1) = \frac{P \times KG}{W - P}$$
$$0.2 = \frac{P \times 5.5}{3000 - P}$$
$$600 - 0.2P = 5.5P$$
$$5.7P = 600$$
$$\text{Maximum P} = \frac{600}{5.7} = 105.26 \text{ tonnes}$$

But

$$P = \frac{MCTC \times t}{l}$$

or

$$\text{Maximum } t = \frac{P \times l}{MCTC}$$
$$= \frac{105.26 \times 48}{40}$$

Ans. Maximum trim = 126.3 cm by the stern.

There are therefore two possible answers to this question, depending on the method of solution used. The reason for this is that although the effective metacentric height at the critical instant in each case will be the same, the righting moments at equal angles of heel will not be the same.

Example 3

A ship of 5000 tonnes displacement enters a dry dock trimmed 0.45 m by the stern. KM = 7.5 m, KG = 6.0 m, and MCTC = 120 tonnes m. The center of flotation is 60 m from aft. Find the effective metacentric height at the critical instant before the ship takes the blocks overall, assuming that the transverse metacenter rises 0.075 m.

$$P = \frac{MCTC \times t}{l}$$

$$= \frac{120 \times 45}{60}$$

$$P = 90 \text{ tonnes}$$

Method (a)

$$\text{Virtual loss } (MM_1) = \frac{P \times KM}{W}$$

$$= \frac{90 \times 7.575}{5000}$$

$$= 0.136 \text{ m}$$

$$\text{Original KM} = 7.500 \text{ m}$$

$$\text{Rise of M} = \underline{0.075 \text{ m}}$$

$$\text{New KM} = 7.575 \text{ m}$$

$$\text{KG} = \underline{6.000 \text{ m}}$$

$$\text{GM} = 1.575 \text{ m}$$

$$\text{Virtual loss } (MM_1) = 0.136 \text{ m}$$

Ans. New GM = 1.439 m.

Method (b)

$$\text{Virtual loss } (GG_1) = \frac{P \times KG}{W - P}$$
$$= \frac{90 \times 6.0}{4910}$$
$$= 0.110 \text{ m}$$

$$\text{Old KG} = 6.000 \text{ m}$$
$$\text{Virtual loss } (GG_1) = \underline{0.110 \text{ m}}$$
$$\text{New KG} = 6.110 \text{ m}$$
$$\text{New KM} = 7.575 \text{ m}$$

Ans. New GM = 1.465 m.

The Virtual Loss of GM After Taking the Blocks Overall

When a ship takes the blocks overall, the water level will then fall uniformly about the ship, and for each centimeter fallen by the water level P will be increased by a number of tonnes equal to the TPC. Also, the force P at any time during the operation will be equal to the difference between the weight of the ship and the weight of water she is displacing at that time.

Example

A ship of 5000 tonnes displacement enters a dry dock on an even keel. KM = 6 m, KG = 5.5 m, and TPC = 50 tonnes. Find the virtual loss of metacentric height after the ship has taken the blocks and the water has fallen another 0.24 m.

$$P = \text{TPC} \times \text{Reduction in draft in cm}$$
$$= 50 \times 24$$
$$P = 1200 \text{ tonnes}$$

Method (a)

$$\text{Virtual loss } (MM_1) = \frac{P \times KM}{W}$$
$$= \frac{1200 \times 6}{5000}$$

Ans. Virtual loss = 1.44 m.

Method (b)

$$\text{Virtual loss } (GG_1) = \frac{P \times KG}{W - P}$$

$$= \frac{1200 \times 5.5}{3800}$$

Ans. Virtual loss = 1.74 m.

Note to Students

In the DfT examinations, when sufficient information is given in a question, either method of solution may be used. It has been shown in this chapter that both are equally correct. In some questions, however, there is no choice, as the information given is sufficient for only one of the methods to be used. It is therefore advisable for students to learn both of the methods.

Example

A ship of 8000 tonnes displacement takes the ground on a sandbank on a falling tide at an even-keel draft of 5.2 m, KG = 4.0 m. The predicted depth of water over the sandbank at the following low water is 3.2 m. Calculate the GM at this time assuming that the KM will then be 5.0 m and that the mean TPC is 15 tonnes.

$$P = \text{TPC} \times \text{Fall in water level (cm)} = 15 \times (520 - 320)$$

$$= 15 \times 200$$

$$P = 3000 \text{ tonnes}$$

Method (a)

$$\text{Virtual loss of GM } (MM_1) = \frac{P \times KM}{W}$$

$$= \frac{3000 \times 5}{8000}$$

$$= 1.88 \text{ m}$$

$$\text{Actual KM} = 5.00 \text{ m}$$

$$\text{Virtual KM} = 3.12 \text{ m}$$

$$\text{KG} = 4.00 \text{ m}$$

Ans. New GM = −0.88 m.

Method (b)

$$\text{Virtual loss of GM } (GG_1) = \frac{P \times KG}{W - P}$$
$$= \frac{3000 \times 4}{5000}$$
$$= 2.40 \text{ m}$$
$$KG = 4.00 \text{ m}$$
$$\text{Virtual } KG = 6.40 \text{ m}$$
$$KM = 5.00 \text{ m}$$

Ans. New GM = −1.40 m.

Note that in this example, this vessel has developed a negative GM. Consequently she is *unstable*. She would capsize if transverse external forces such as wind or waves were to remove her from zero angle of heel. Suggest a change of loading to reduce KG and make GM a positive value greater than the DfT minimum of 0.15 m.

Exercise 41

1. A ship being drydocked has a displacement of 1500 tonnes. TPC = 5 tonnes, KM = 3.5 m, GM = 0.5 m, and the blocks have been taken fore and aft at 3 m draft. Find the GM when the water level has fallen another 0.6 m.
2. A ship of 4200 tonnes displacement has GM = 0.75 m and present drafts 2.7 m F and 3.7 m A. She is to enter a dry dock. MCTC = 120 tonnes m. The after keel block is 60 m aft of the center of flotation. At 3.2 m mean draft KM = 8 m. Find the GM on taking the blocks forward and aft.
3. A box-shaped vessel 150 m long, 10 m beam, and 5 m deep has a mean draft in salt water of 3 m and is trimmed 1 m by the stern, KG = 3.5 m. State whether it is safe to drydock this vessel in this condition or not, and give reasons for your answer.
4. A ship of 6000 tonnes displacement is 120 m long and is trimmed 1 m by the stern. KG = 5.3 m, GM = 0.7 m, and MCTC = 90 tonnes m. Is it safe to drydock the ship in this condition? (Assume that the center of flotation is amidships.)
5. A ship of 4000 tonnes displacement, 126 m long, has KM = 6.7 m and KG = 6.1 m. The center of flotation is 3 m aft of amidships, MCTC = 120 tonnes m. Find the maximum trim at which the ship may enter a dry dock if the minimum GM at the critical instant is to be 0.3 m.

CHAPTER 42

Ship Squat in Open Water and in Confined Channels

What Exactly is Ship Squat?

When a ship proceeds through water, she pushes water ahead of her. In order not to have a 'hole' in the water, this volume of water must return down the sides and under the bottom of the ship. The streamlines of return flow are speeded up under the ship. This causes a drop in pressure, resulting in the ship dropping vertically in the water.

As well as dropping vertically, the ship generally trims forward or aft. The overall decrease in the static underkeel clearance, forward or aft, is called ship squat. It is *not* the difference between the draughts when stationary and the draughts when the ship is moving ahead.

If the ship moves forward at too great a speed when she is in shallow water, say where this static even-keel underkeel clearance is 1.0−1.5 m, then grounding due to excessive squat could occur at the bow or at the stern.

For full-form ships such as supertankers or OBO vessels, grounding will occur generally at the *bow*. For fine-form vessels such as passenger liners or container ships the grounding will generally occur at the stern. This is assuming that they are on even keel when stationary. It must be *generally*, because in the last two decades several ship types have tended to be shorter in LBP and wider in breadth molded. This has led to reported groundings due to ship squat at the bilge strakes at or near to amidships when slight rolling motions have been present.

Why has ship squat become so important in the last 40 years? Ship squat has always existed on smaller and slower vessels when underway. These squats have only been a matter of centimeters and thus have been inconsequential.

However, from the mid-1960s and into the new millennium, ship size has steadily grown until we have supertankers of the order of 350,000 tonnes dwt and above. These supertankers have almost outgrown the ports they visit, resulting in small static even-keel underkeel clearances of 1.0−1.5 m. Alongside this development in ship size has been an increase in service speed on several ships, for example container ships, where speeds have gradually increased from 16 knots up to about 25 knots.

Ship Stability for Masters and Mates. http://dx.doi.org/10.1016/B978-0-08-097093-6.00042-6

Ship design has seen tremendous changes in the 1980s, 1990s, and 2000s. In oil tanker design we have the *Jahre Viking* with a dwt of 564,739 tonnes and an LBP of 440 m. This is equivalent to the length of five football pitches. In 2002, the biggest container ship to date, the *Hong Kong Express*, came into service with a dwt of 82,800 tonnes, a service speed of 25.3 knots, an LBP of 304 m, breadth molded of 42.8 m, and a draft molded of 13 m.

As the static underkeel clearances have decreased and as the service speeds have increased, ship squats have gradually increased. They can now be of the order of 1.50−1.75 m, which are of course by no means inconsequential.

Masters and mates therefore need to know about the phenomenon of ship squat.

Recent Ship Groundings

To help emphasize the dangers of excessive squat, one only has to remember the grounding of the 11 vessels listed in Table 42.1 in recent years.

I currently have a database of 117 vessels that have recently gone aground attributable to excessive ship squat and other reasons. There may well be others not officially reported.

In the United Kingdom, over the last 40 years the D.Tp. have shown their concern by issuing eight 'M' notices concerning the problems of ship squat and accompanying problems in shallow water. These alert all mariners to the associated dangers.

Table 42.1: Vessels that have grounded in recent years.

Date	Name of Ship	Type of Ship	Location
16 Jan 2009	*Mirabelle*	Cargo ship	Svendborg, Denmark
20 Jan 2009	*CSL Argosy*	Bulk carrier	Chesapeake Bay, Baltimore
21 Jan 2009	*Gunay 2*	Cargo ship	Marseilles
17 Feb 2009	*Ocean Nova*	Cruise ship	San Martin, Antarctica
22 March 2009	*Karin Schepers*	Container ship	Drogden, Baltic Sound
10 June 2009	*Akti N*	Oil tanker	Vlissingen, Netherlands
3 April 2010	*Shen Neng 1*	Bulk coal carrier	Great Barrier Reef Australia
13 Aug 2010	*Flinterforest*	Cargo ship	Oresund Strait, Sweden
31 Jan 2011	*Jack Alby 11*	Trawler	Isle of Rum, Shetlands
16 Feb 2011	*K-Wave*	Container ship	Malaga coastline, Spain
3 Aug 2011	*Karin Schepers*	Container ship	St Just, near West Cornwall

Signs that a ship has entered shallow water conditions can be one or more of the following:

1. Maximum ship squat increases.
2. Mean bodily sinkage increases.
3. Ship will generally develop extra trim by the bow or by the stern.
4. Wave-making increases, especially at the forward end of the ship.
5. Ship becomes more sluggish to maneuver. A pilot's quote, 'almost like being in porridge,' is one that many masters and mates will surely agree with.
6. Draught indicators on the bridge or echo-sounders will indicate changes in the end draughts.
7. Propeller r.p.m. indicator will show a decrease. If the ship is in 'open water' conditions, i.e. without any breadth restrictions, this decrease may be up to 15% of the service r.p.m. in deep water. If the ship is in a confined channel, this decrease in r.p.m. can be up to 20% of the service r.p.m.
8. There will be a drop in speed. If the ship is in open water conditions, this decrease may be up to 35%. If the ship is in a confined channel such as a river or a canal, then this decrease can be up to 75%.
9. The ship may start to vibrate suddenly. This is because of the entrained water effects causing the natural hull frequency to become resonant with another frequency associated with the vessel.
10. Any rolling, pitching, and heaving motions will all be reduced as the ship moves from deep water to shallow water conditions. This is because of the cushioning effects produced by the narrow layer of water under the bottom shell of the vessel.
11. The appearance of mud could suddenly show in the water around the ship's hull, say in the event of passing over a raised shelf or a submerged wreck.
12. Turning circle diameter (TCD) increases. TCD in very shallow water could increase 100%.
13. Stopping distances and stopping time increase, compared to when a vessel is in deep waters.
14. Effectiveness of the rudder helm decreases when a ship is in shallow waters.
15. Width of the wake increases considerably.

What Are the Factors Governing Ship Squat?

The main factor is ship speed V_k. Squat varies approximately with the speed squared. In other words, we can take as an example that if we halve the speed we quarter the squat. In this context, speed V_k is the ship's speed relative to the water; in other words, effect of current/tide speed with or against the ship must be taken into account.

Another important factor is the block coefficient C_b. Squat varies directly with C_b. Oil tankers will therefore have comparatively more squat than passenger liners.

The blockage factor 'S' is another factor to consider. This is the immersed cross-section of the ship's midship section divided by the cross-section of water within the canal or river (see Figure 42.1). If the ship is in open water the width of influence of water can be calculated. This ranges from about 8.25b for supertankers to about 9.50b for general cargo ships, to about 11.25 ship breadths for container ships.

The presence of another ship in a narrow river will also affect squat, so much so that squats can *double in value* as they pass/cross the other vessel.

Formulae have been developed that will be satisfactory for estimating maximum ship squats for vessels operating in confined channels and in open water conditions (see Figure 42.2). These formulae are the results of analyzing about 600 results, some measured on ships and some on ship models. Some of the empirical formulae developed are as follows:

Let

b = breadth of ship
B = breadth of river or canal
H = depth of water
T = ship's even-keel static draft
C_b = block coefficient
V_k = ship speed relative to the water or current
CSA = cross-sectional area (see Figure 42.1)

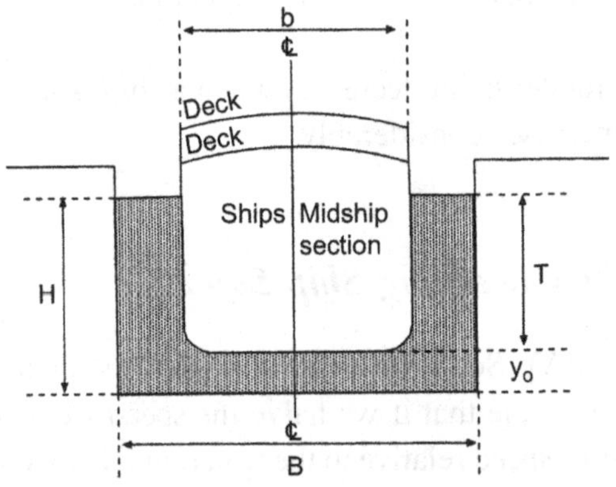

Figure 42.1:
Ship in a Canal in Static Condition.

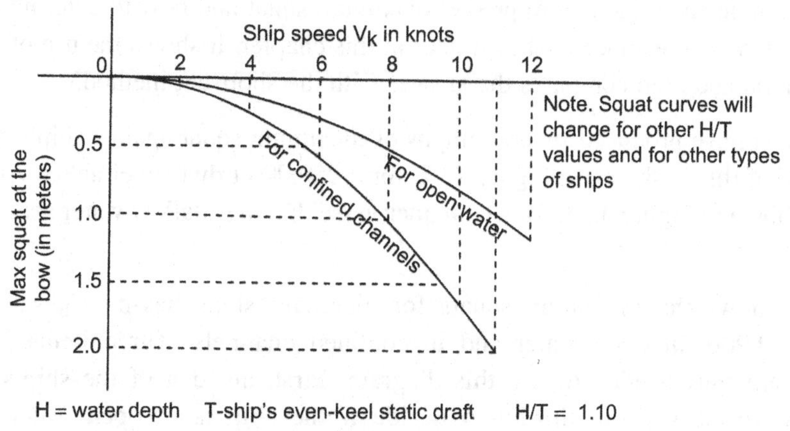

Figure 42.2:
Maximum Squats Against Ship Speed for a 250,000 t dwt Supertanker.

Let

$$S = \text{blockage factor} = \text{CSA of ship/CSA of river or canal}$$

If ship is in open water conditions, then the formula for B becomes

$$B = \{7.04/(C_b{}^{0.85})\} \times b, \text{ known as the 'width of influence'}$$

$$\text{Blockage factor} = S = \frac{b \times T}{B \times H}$$

$$\text{Maximum squat} = \delta_{max} = \frac{C_b \times S^{0.81} \times V_k^{2.08}}{20} \text{ meters,}$$

for *open water* and *confined channels*

Two short-cut formulae relative to the previous equation are:

$$\text{Maximum squat} = \frac{C_b \times V_k^2}{100} \text{ meters for } \textit{open water} \text{ conditions } \textit{only}, \text{ with H/T of 1.1 to 1.4}$$

$$\text{Maximum squat} = \frac{C_b \times V_k^2}{50} \text{ meters for } \textit{confined channels}, \text{ where S} = 0.100 \text{ to } 0.250$$

A worked example showing how to predict maximum squat and how to determine the remaining underkeel clearance is shown later in this chapter. It shows the use of the more detailed formula and then compares the answer with the short-cut method.

These formulae have produced several graphs of maximum squat against ship's speed V_K. One example of this is shown in Figure 42.2, for a 250,000 t dwt supertanker. Another example is shown in Figure 42.3, for a container vessel having shallow water speeds up to 18 knots.

Figure 42.4 shows the maximum squats for merchant ships having C_b values from 0.500 up to 0.900, in open water and in confined channels. Three items of information are thus needed to use this diagram. First, an idea of the ship's C_b value, secondly the speed V_K, and thirdly to decide if the ship is in open water or in confined river/canal conditions. A quick graphical prediction of the maximum squat can then be made.

In conclusion, it can be stated that if we can predict the maximum ship squat for a given situation then the following advantages can be gained:

1. The ship operator will know which speed to reduce to in order to ensure the safety of his/her vessel. This could save the cost of a very large repair bill. It has been reported in the technical press that the repair bill for the *QE2* was *$13 million*, plus an estimation for lost passenger booking of *$50 million*!!

   In Lloyd's lists, the repair bill for the *Sea Empress* had been estimated to be in the region of *$28 million*. In May 1997, the repairs to the *Sea Empress* were completed at Harland &

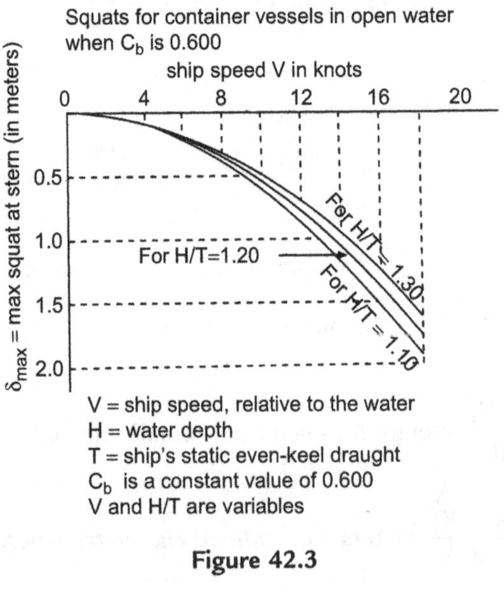

Figure 42.3

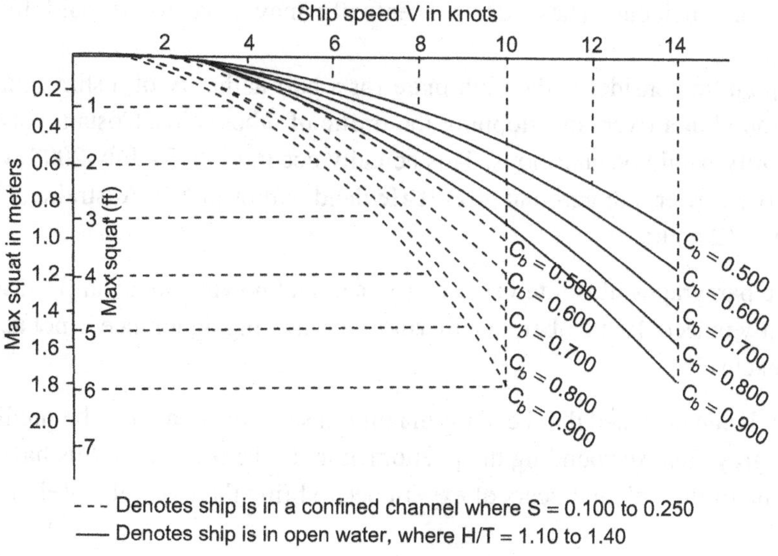

Figure 42.4:
Maximum Ship Squats in Confined Channels and in Open Water Conditions.

Ship Type	Typical C_b, Fully Loaded	Ship Type	Typical C_b, Fully Loaded
ULCC	0.850	General cargo	0.700
Supertanker	0.825	Passenger liner	0.575–0.625
Oil tanker	0.800	Container ship	0.575
Bulk carrier	0.775–0.825	Coastal tug	0.500

Wolff Ltd of Belfast, for a reported cost of £*20 million.* Rate of exchange in May 1997 was of the order of £1 = $1.55. She was then renamed the *Sea Spirit.*

2. The ship officers could load the ship up an extra few centimeters (except of course where load-line limits would be exceeded). If a 100,000-tonne dwt tanker is loaded by an extra 30 cm or an SD14 general cargo ship is loaded by an extra 20 cm, the effect is an extra 3% onto their dwt. This gives these ships extra earning capacity.
3. If the ship grounds due to excessive squatting in shallow water, then apart from the large repair bill, there is the time the ship is 'out of service'. Being 'out of service' is indeed very costly because loss of earnings can be greater than £*100,000* per *day.*
4. When a vessel goes aground there is always a possibility of leakage of oil, resulting in compensation claims for oil pollution and fees for cleanup operations

following the incident. These costs eventually may have to be paid for by the shipowner.

5. Another point to consider is the nightmare for a port authority of a ship squatting and going aground in a river, and stopping the traffic of other ships. Costs involved for the port authority would be enormous. This actually occurred on 31 July 2008 when the *Iron King*, an ore carrier, ran aground at Port Hedland harbor in NW Australia. The port was blocked for 12 hours.

These last five paragraphs illustrate very clearly that not knowing about ship squat can prove to be very costly indeed. Remember, in a marine court hearing, ignorance is not acceptable as a legitimate excuse!!

Summarizing, it can be stated that because maximum ship squat can now be predicted, it has removed the 'gray area' surrounding the phenomenon. In the past ship pilots have used 'trial and error', 'rule of thumb', and years of experience to bring their vessels safely in and out of port.

Empirical formulae quoted in this study, modified and refined by the author over a period of 40 years' research on the topic, give *firm guidelines*. By maintaining the ship's trading availability a shipowner's profit margins are not decreased. More important still, this research can help prevent loss of life as occurred with the *Herald of Free Enterprise* grounding.

It should be remembered that the quickest method for reducing the danger of grounding due to ship squat is to *reduce the ship's speed*. 'Prevention is better than cure' and *much cheaper.*

$$A_S = \text{cross-section of ship at amidships} = b \times T$$

$$A_C = \text{cross-section of canal} = B \times H$$

$$\text{Blockage factor} = S = \frac{A_S}{A_C} = \frac{b \times T}{B \times H}$$

$$y_o = \text{Static underkeel clearance}$$

$$\frac{H}{T} \text{ range is 1.10 to 1.40}$$

Blockage factor range is 0.100 to .0250

$$\text{Width of influence} = F_B = \frac{\text{Equivalent 'B'}}{b} \text{ in open water}$$

$$V_k = \text{speed of ship relative to the water, in knots}$$

$$\text{'B'} = \frac{7.04}{C_b^{0.85}} \text{ ship breadths}$$

Ship Squat for Ships with Static Trim

Each ship to this point has been assumed to be on *even keel* when static. For a given forward speed the maximum ship squat has been predicted. Based on the C_b, the ship will have this maximum squat at the bow, at the stern or right away along the length of the ship (see Figure 42.5(a)).

However, some ships will have *trim by the bow* or *trim by the stern* when they are stationary. This static trim will decide where the maximum squat will be located when the ship is underway.

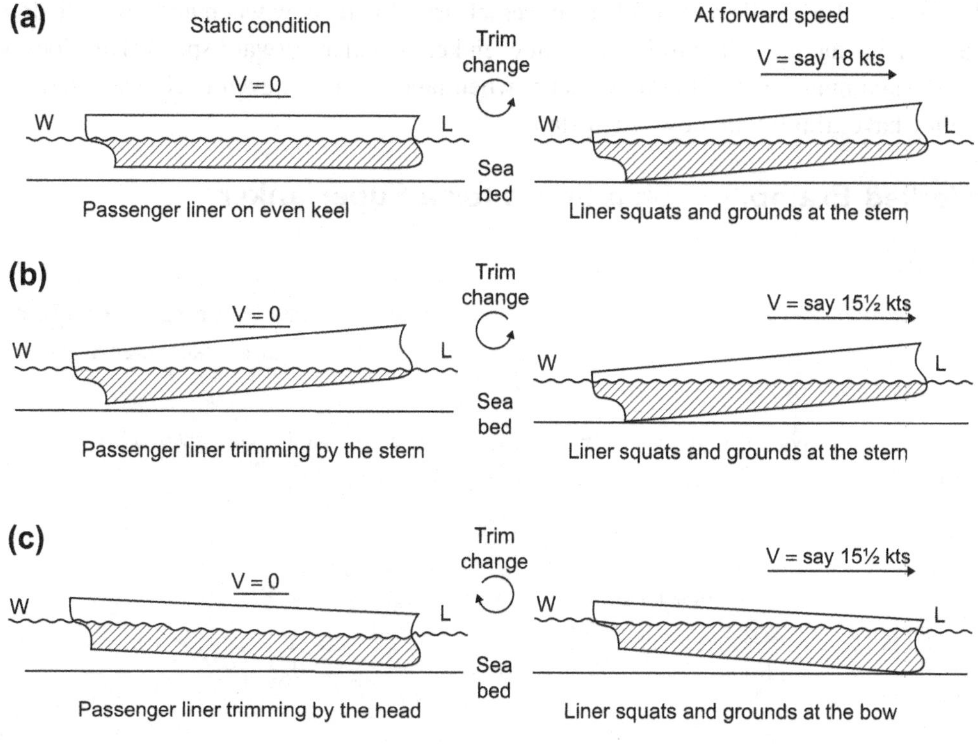

Figure 42.5:
The Link Between Static Trim and Location of Maximum Squat.

Tests on ship models and from full-size squat measurements have shown that:

1. If a ship has static trim by the stern when static, then when underway she will have a maximum squat (as previously estimated) at the stern. The ship will have dynamic trim in the same direction as the static trim. In other words, when underway she will have increased trim and could possibly go aground at the stern (see Figure 42.5(b)).

   This is due to streamlines under the vessel at the stern moving faster than under the vessel at the bow. Cross-sectional area is less at the stern than under the bow and this causes a greater suction at the stern. The vessel trims by the stern. In hydraulics, this is known as the Venturi effect.
2. If a ship has static trim by the bow when static, then when underway she will have a maximum squat (as previously estimated) at the bow. The ship will have dynamic trim in the same direction as the static trim (see Figure 42.5(c)). In other words, when underway she will have increased trim and could possibly go aground at the bow. The *Herald of Free Enterprise* grounding is a prime example of this trimming by the bow when at forward speed.

Note of caution. Some masters on oil tankers trim their vessels by the stern before going into shallow waters. They believe that full-form vessels trim by the bow when underway. In doing so they believe that the ship will level out at even keel when at forward speed. This does not happen!! Maximum squat at the bow occurs when tankers are on even keel when static, or when they have trim by the bow when static.

Worked Example – Ship Squat for a Supertanker

Question:

A supertanker operating in open water conditions is proceeding at a speed of 11 knots. Her $C_b = 0.830$; static even-keel draft = 13.5 m with a static underkeel clearance of 2.5 m. Her breadth molded is 55 m with LBP of 320 m.

Calculate the maximum squat for this vessel at the given speed via *two* methods, and her remaining ukc (underkeel clearance) at V_k of 11 knots.

Answer:

$$\text{Width of influence} = \frac{7.04}{C_b^{0.85}} \times b = \text{'B'}$$

$$\therefore \text{'B'} = \frac{7.04}{0.830^{0.85}} \times 55$$

$$\therefore \text{'B'} = 454 \text{ m}$$

i.e. artificial boundaries in open water or wide rivers

$$\text{Blockage factor, } S = \frac{b \times T}{\text{'B'} \times H} = \frac{55 \times 13.5}{454 \times (13.5 + 2.5)} = \underline{0.102}$$
(water depth)

C_b	V	Static Draft Even Keel	Br. Mld	Water Depth
0.830	11.00	13.50	55.00	16.00

Width of influence = 453.65 meters

Blockage factor = 0.102

Method 1

$$\text{Maximum squat} = \frac{C_b \times S^{0.81} \times V_k^{2.08}}{20} = \delta_{max}$$

$$\therefore \delta_{max} = 0.830 \times 0.102^{0.81} \times 11^{2.08} \times \frac{1}{20}$$

$$\therefore \underline{\delta_{max} = 0.96 \text{ m}}$$

at the bow, because $C_b > 0.700$ and because, when static, the ship was on even keel.

Method 2

Simplified approximate formula:

$$\delta_{max} = \frac{C_b \times V_k^2}{100} \text{ meters}$$

$$\therefore \delta_{max} = \frac{0.830 \times 11^2}{100} = 1.00 \text{ m}$$

Maximum squat = 1.00 meter at the bow

Average squat = 0.98 meters at the bow

Remaining ukc under the bow = 1.52 meters

i.e. slightly above the previous answer, so overpredicting on the *safe side*.

■

Fitting a Trench

What if there is a trench at the base of a navigation channel? If one is fitted then it will allow larger vessels to enter and leave port. There is an '*economy of scale*' because larger vessels earn more per deadweight tonne than smaller ones. Because of having greater draft, shipowners are able to transport larger cargoes and make greater profits.

Predicting Ship Squat with a Vessel in a Trench

In Melbourne there is a navigational channel that has an 180-m-wide trench at its base (see Tables 42.2 and 42.3). Using the values given in these tables we can proceed to evaluate the maximum squat at the stern and the remaining ukc under the stern. When static, the container ship was on even keel. Assume her forward speed is 10 knots.

$$\text{Width of influence} = \{7.04/(C_b{}^{0.85})\} \times b = \{7.04/(0.575^{0.85})\} \times 32.26$$

$$= 364 \text{ (see Table 42.2)}$$

$$A_{SHIP} = b \times T = 32.26 \times 12.10 = 390 \text{ m}^2 \text{ (see Table 42.2)}$$

$$A_{CHANNEL} = (H \times \text{Trench width}) + (F_b - \text{Trench width})(H - 3.6)$$

$$= (13.1 \times 180) + (364 - 180)(13.1 - 3.6)$$

$$= 2358 + 1748 = 4101 \text{ m}^2 \text{ (see Table 42.2)}$$

Table 42.2: Port of Melbourne data – 1.

Ship Type	b (m)	T (m)	V (knots)	C_b	F_b (m)	Wedge Breadth (m)	A_{ship} (m^2)	A_{canal} (m^2)	'S' Factor	$S^{0.81}$	Max. Squat (m)	Dynamic ukc (m)
Container ship	32.26	12.10	10.00	0.575	364	180	390	4101	0.095	0.149	0.51	0.49
Bulker	35.00	11.50	10.00	0.775	306	180	403	3555	0.113	0.171	0.80	0.80
VLCC	42.00	11.50	10.00	0.825	348	180	483	3956	0.122	0.182	0.90	0.70
Products tanker	32.26	11.50	10.00	0.800	275	180	371	3256	0.114	0.172	0.83	0.77

Water depth = 13.10 meters.
Mean water 9.5 m for >180 m breadth.

Table 42.3: Port of Melbourne data – 2.

	Ship Type							
	Container Ship		Bulker Ship		VLCC		Products' Tanker	
Squat at 10 knots Ship Speed	0.51	meters y2	0.80	meters y2	0.90	meters y2	0.83	Meters y2
0	0.00	1.00	0.00	1.60	0.00	1.60	0.00	1.60
1	0.00	1.00	0.01	1.59	0.01	1.59	0.01	1.59
2	0.02	0.98	0.03	1.57	0.03	1.57	0.03	1.57
3	0.04	0.96	0.07	1.53	0.07	1.53	0.07	1.53
4	0.08	0.92	0.12	1.48	0.13	1.47	0.12	1.48
5	0.12	0.88	0.19	1.41	0.21	1.39	0.20	1.40
6	0.18	0.82	0.28	1.32	0.31	1.29	0.29	1.31
7	0.24	0.76	0.38	1.22	0.43	1.17	0.39	1.21
8	0.32	0.68	0.50	1.10	0.57	1.03	0.52	1.08
9	0.41	0.59	0.64	0.96	0.73	0.87	0.66	0.94
10	0.51	0.49	0.80	0.80	0.90	0.70	0.83	0.77
11	0.63	0.37	0.97	0.63	1.10	0.50	1.01	0.59
12	0.75	0.25	1.17	0.43	1.32	0.28	1.21	0.39
13	0.89	0.11	1.38	0.22	1.56	0.04	1.43	0.17
14	1.04	−0.04	1.61	−0.01	1.82	−0.22	1.67	−0.07

Water depth = 13.10 meters.
Mean water 9.5 m for >180 m breadth.
y2 = dynamical ukc (m) at forward speed.
Ship speed in knots, relative to the water.
Container ship: maximum squats occur at the stern and the y2 values will occur under the stern in meters.
For the other three vessels maximum squats occur at the bow and the y2 values will occur under the bow in meters.

$$\text{Blockage factor (S)} = A_{SHIP}/A_{CHANNEL} = 390/401$$

$$\text{Thus } S = 0.095 \text{ (see Table 42.2)}$$

$$S^{0.81} = 0.095^{0.81} = 0.149 \text{ (see Table 42.2)}$$

$$\delta\text{max} = (C_b \times S^{0.81} \times V^{2.08})/20 \text{ meters} \quad V^{2.08} = 10^{2.08} = 120.23$$

$$\delta\text{max} = (0.575 \times 0.149 \times 120.23)/20 \text{ meters}$$

$$\delta\text{max} = 0.51 \text{ m at the stern}$$

This occurs at the stern because $C_b < 0.700$ and the ship when static was on even keel. This value of 0.51 m is shown in column 12 of Table 42.2.

$$\text{Remaining ukc at the stern} = \text{Static ukc} - \delta\text{max}$$

$$= 1.00 - 0.51 = 0.49 \text{ m at V} = 10 \text{ knots}$$

This value of 0.49 m is shown in the far right-hand column of Table 42.2.

This example considered the procedure for predicting maximum squats for a container ship. Tables 42.2 and 42.3 go on to consider squat predictions for a bulker, a VLCC, and a products tanker. Figures 42.6–42.9 show these maximum squats graphically together with the remaining ukc at the bow or at the stern.

In Tables 42.2 and 42.3, it is possible to have *input* statements for the following variables:

- Any conventional type of ship
- Any breadth mold of ship
- Any ship draft even keel
- Any ship speed relative to the water
- Any C_b value, from ballast up to loaded departure condition
- Any water depth
- Any width of trench.

In Tables 42.2 and 42.3, it IS possible to have *output* statements for the following information:

- Blockage factor (S)
- Cross-sectional area of ship at amidships
- Cross-sectional area of channel
- Maximum squat and dynamical ukc initially for a speed of 10 knots
- Maximum squats and dynamical ukc for speeds up to 14 knots at 1-meter intervals.

Finally, graphs of the information contained in Tables 42.2 and 42.3 are shown in Figures 42.6–42.9.

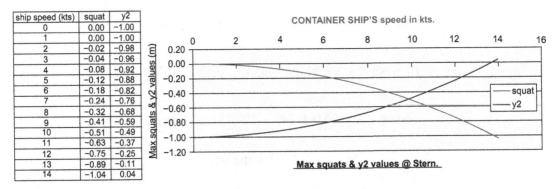

ship speed (kts)	squat	y2
0	0.00	−1.00
1	0.00	−1.00
2	−0.02	−0.98
3	−0.04	−0.96
4	−0.08	−0.92
5	−0.12	−0.88
6	−0.18	−0.82
7	−0.24	−0.76
8	−0.32	−0.68
9	−0.41	−0.59
10	−0.51	−0.49
11	−0.63	−0.37
12	−0.75	−0.25
13	−0.89	−0.11
14	−1.04	0.04

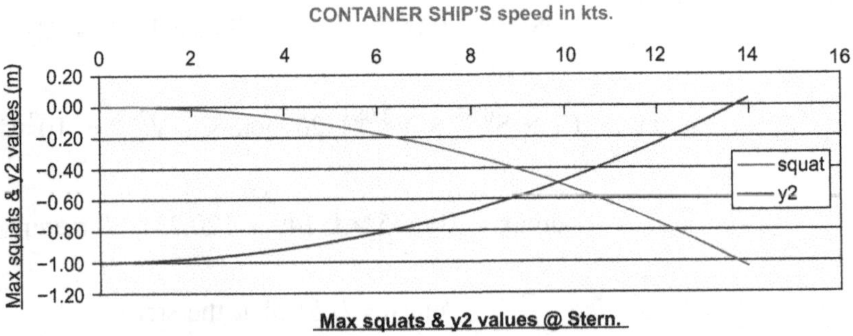

Figure 42.6:
Maximum Squats and y2 Values at the Stern for a Container Ship.

ship speed (kts)	squat	y2
0	0.00	−1.60
1	−0.01	−1.59
2	−0.03	−1.57
3	−0.07	−1.53
4	−0.12	−1.48
5	−0.19	−1.41
6	−0.28	−1.32
7	−0.38	−1.22
8	−0.50	−1.10
9	−0.64	−0.96
10	−0.80	−0.80
11	−0.97	−0.63
12	−1.17	−0.43
13	−1.38	−0.22
14	−1.61	0.01

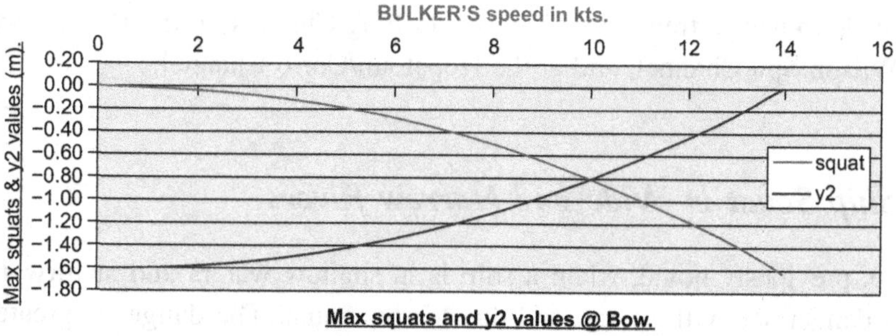

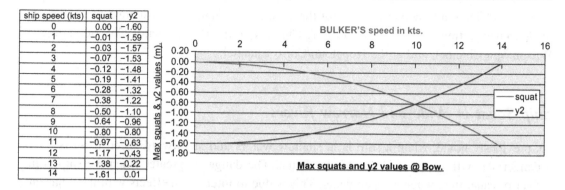

Figure 42.7:
Maximum Squats and y2 Values at the Bow for a Bulker.

ship speed (kts)	squat	y2
0	0.00	−1.60
1	−0.01	−1.59
2	−0.03	−1.57
3	−0.07	−1.53
4	−0.13	−1.47
5	−0.21	−1.39
6	−0.31	−1.29
7	−0.43	−1.17
8	−0.57	−1.03
9	−0.73	−0.87
10	−0.90	−0.70
11	−1.10	−0.50
12	−1.32	−0.28
13	−1.56	−0.04
14	−1.82	0.22

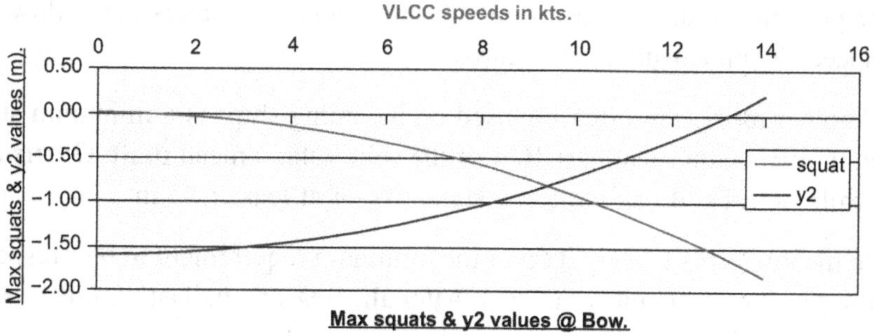

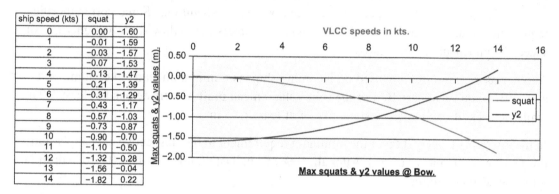

Figure 42.8:
Maximum Squats and y2 Values at the Bow for a VLCC.

ship speed (kts)	squat	y2
0	0.00	−1.60
1	−0.01	−1.59
2	−0.03	−1.57
3	−0.07	−1.53
4	−0.12	−1.48
5	−0.20	−1.40
6	−0.29	−1.31
7	−0.39	−1.21
8	−0.52	−1.08
9	−0.66	−0.94
10	−0.83	−0.77
11	−1.01	−0.59
12	−1.21	−0.39
13	−1.43	−0.17
14	−1.67	0.07

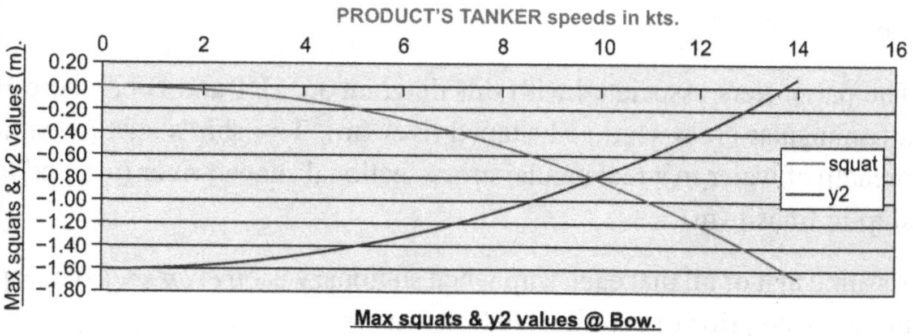

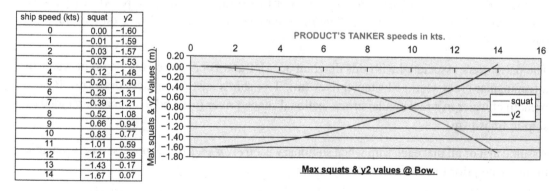

Figure 42.9:
Maximum Squats and y2 Values at the Bow for a Products Tanker.

As well as having a trench at the base of the Melbourne Channel, considerations have been made to have a trench fitted at the Geelong Channel, at the Port Richards Channel, at the Wilson Spit Channel, and at the Hopetoun/Corio Channel.

Ship Squat in Wide and Narrow Rivers

As previously noted, when a ship is in shallow waters and at forward speed there is a danger she will go aground due to ship squat. The danger is greater when the ship is in a river or canal than when in open water. This is due to interaction effects with the adjacent banks and the sides of the moving vessel. The narrower the river, the greater will be the ship squats.

Because of erosion of riverbanks and interaction with moored ships in a river, the forward speeds of ships are a lot less than at sea or in open water conditions. Some port authorities request maximum speeds of only 4 knots in rivers. Others may allow slightly higher transit speeds, commensurate with ship size.

Another restriction often imposed on incoming ships is a minimum static ukc. Some port authorities require at least 10% of the ship's static mean draft. Others demand at least 1.0 m while others still can request a static ukc of at least 1.25 m.

In the interests of safety, I favor the minimum requirement to be irrespective of a ship's static draft and to be in units of meters. After all, 10% of a ballast draft is less than 10% of a draft in a fully loaded condition.

Note of caution. Whilst Figure 42.4 shows C_b values for fully loaded conditions, it must be realized that ships do go aground at drafts less than their draft molded.

To predict maximum ship squat in river conditions, I have produced a diagram that involves 'K' coefficients (see Figure 42.10). For this study, the value of 'K' will range only from 1.0 to 2.0. If 'K' < 1 on the diagram, then use 'K' = 1. If 'K' > 2 on the diagram, then use 'K' = 2.

The parameters associated with this diagram are H/T and B/b, where H = depth of water in a rectangular cross-sectional shaped river (m), T = ship's static even-keel draft (m), B = breadth of water in a rectangular cross-sectional shaped river (m), and b = breadth molded of ship in transit (m).

Assume first of all that each ship when stationary is on *even keel.* This appears to be a sensible prerequisite prior to entering shallow waters.

In this section I have concentrated on shallow waters where the static H/T ranges from 1.10 to 1.30. It is in this range of 10–30% ukc that there is greater chance of

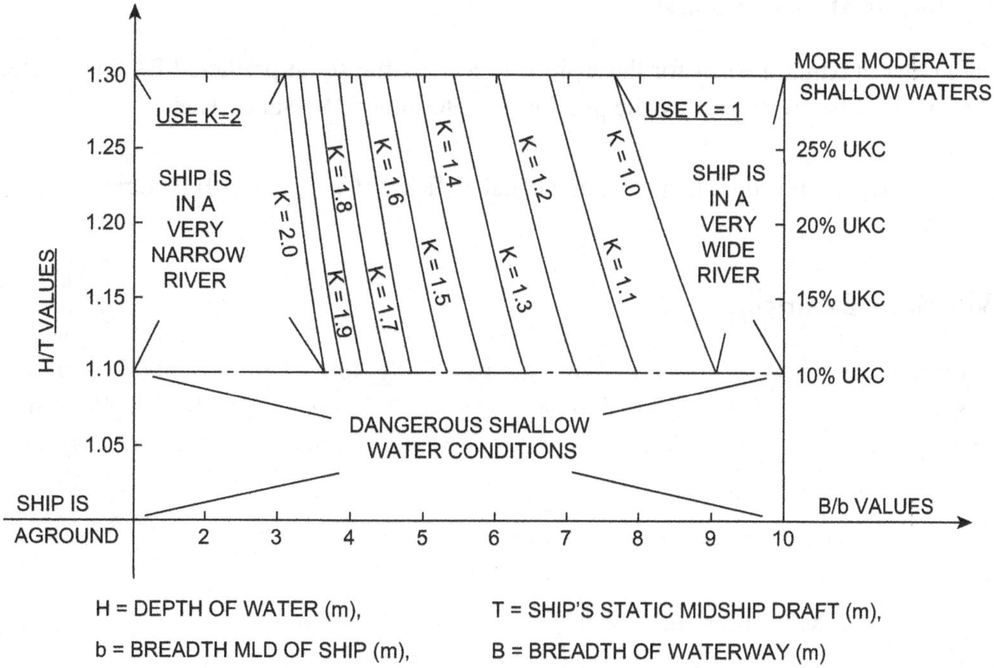

Figure 42.10:
'K' Coefficients for Rectangular Section Rivers, Where $K = 5.74 \times S^{0.76}$.

a ship going aground. At greater than 30% ukc, likelihood of touchdown is greatly decreased.

Procedure for Using Figure 42.10

The following procedure should be used:

1. Determine and record the values of H, T, B, and b.
2. Calculate the values H/T and B/b.
3. Enter Figure 42.10 and plot the intersection point of H/T with B/b. See the results of the next three worked examples.
4. Using this intersection point and the 'K' contour lines, determine the value of 'K' appropriate to the river condition under consideration.

Having now obtained a value for 'K', the next step is to link it with two more variables. They are the block coefficient C_b and the ship speed V measured in knots. C_b will depend on the ship being considered and her condition of loading. As shown in Chapter 5, C_b values will decrease with a decrease in draft.

V is the ship speed relative to the water. The tidal speed and the direction of current must always be taken into account on the bridge by the master, mate, or ship pilot.

Calculation of Maximum Squat

To obtain the maximum squat for these ships in a river, the squat predicted for open water conditions must be multiplied by the previously determined 'K' value. Hence:

$$\text{Maximum squat in a river or a canal} = \text{'K'} \times (C_b \times V^2/100) \text{ meters}$$

Worked Example 1

A container ship has a C_b of 0.575 and is proceeding upriver at a speed (V) of 6 knots. This river of rectangular cross-section has an H/T of 1.25 and a B/b of 3.55. When static, the ship was on even keel. Calculate the maximum squat at this speed and where it will occur.

Using Figure 42.10, at the point of intersection of 1.25 and 3.55, the 'K' value lifted off is 1.847. So:

$$\begin{aligned}\text{Maximum squat} &= \text{'K'} \times (C_b \times V^2/100) \text{ meters}\\ &= 1.847 \times (0.575 \times 6^2/100) = 0.38 \text{ m}\end{aligned}$$

This 0.38 m will be located at the stern, because $C_b < 0.700$ and the ship when static was on even keel.

■

Worked Example 2

A general cargo ship has a C_b of 0.700 and is proceeding upriver at a speed (V) of 5 knots. This river of rectangular cross-section has an H/T of 1.10 and a B/b of 5.60. When static, the ship was on even keel. Calculate the maximum squat at this speed and where it will occur.

Using Figure 42.10, at the point of intersection of 1.10 and 5.60, the 'K' value lifted off is 1.439. So:

$$\begin{aligned}\text{Maximum squat} &= \text{'K'} \times (C_b \times V^2/100) \text{ meters}\\ &= 1.439 \times (0.700 \times 5^2/100) = 0.25 \text{ m}\end{aligned}$$

This 0.25 m will be located from the stern to amidships to the bow, because C_b is 0.700 and the ship when static was on even keel.

■

Worked Example 3

A supertanker has a C_b of 0.825 and is proceeding upriver at a speed (V) of 7 knots. This river of rectangular cross-section has an H/T of 1.15 and a B/b of 7.25. When static, the ship had trim by the stern. Calculate the maximum squat at this speed and where it will occur.

Using Figure 42.10, at the point of intersection of 1.15 and 7.25, the 'K' value lifted off is 1.146. So:

$$\text{Maximum squat} = \text{'K'} \times (C_b \times V^2/100) \text{ meters}$$

$$= 1.146 \times (0.825 \times 7^2/100) = 0.46 \text{ m}$$

This 0.46 m will be located at the stern, because when static this ship had trim by the stern.

Worked Example 4

Table 42.4 shows the maximum squats for a modern cruise ship. The graphs in Figure 42.11 illustrate these squats at different ship speeds and in different depths of water for a specific H/T of 1.12.

Any water breadth of greater than 324 m indicates that this ship is in open water conditions. Equal to or less than 334 m means the ship is in confined channel conditions.

In Table 42.4, it must be realized that the bold/italic/roman text alluded to below only applies when the ship's draft is 8.30 m even keel:

- Bold entries in Table 42.4 indicate that the ship has gone aground!!
- Italic entries in Table 42.4 indicate that for each ship speed there is greater than 0.60 m remaining under the stern. Consequently, the ship is reasonably safe.
- Roman entries in Table 42.4 indicate the remaining ukc under the stern is critical. The master, mate, or pilot must take extra care over this period of transit of the ship.

Numerically, the maximum squats shown are applicable for drafts of 7.30 m with 0.88 m static ukc, to drafts of 8.70 m with 1.04 m static ukc. For each pair of values across the table, H/T is 1.12.

Figure 42.11 shows graphical plots of the maximum squats in Table 42.4. Notice how the width of water ranges from 100 m up to 300 m, and then up to 324 m for open

Table 42.4: Squats for a modern cruise ship H/T = 1.12.

Water Width B (m) Max. Squat at 10 knots Ship Speed (knots)	100 **−1.44**	150 **−1.04**	200 **−0.82**	250 **−0.68**	300 **−0.59**			>324 **−0.55**
0	0.00	0.00	0.00	0.00	0.00			0.00
1	−0.01	−0.01	−0.01	−0.01	0.00			0.00
2	−0.05	−0.04	−0.03	−0.02	−0.02			−0.02
3	−0.12	−0.09	−0.07	−0.06	−0.05			−0.04
4	−0.21	−0.15	−0.12	−0.10	−0.09			−0.08
5	−0.34	−0.25	−0.19	−0.16	−0.14			−0.13
6	−0.50	−0.36	−0.28	−0.23	−0.20			−0.19
7	−0.69	−0.50	−0.39	−0.32	−0.28			−0.26
8	−0.91	−0.65	−0.52	−0.43	−0.37			−0.35
9	**−1.16**	−0.84	−0.66	−0.55	−0.47			−0.44
10	**−1.44**	**−1.04**	−0.82	−0.68	−0.59			−0.55
11	**−1.76**	**−1.27**	**−1.00**	−0.83	−0.72			−0.67
12	**−2.10**	**−1.52**	**−1.20**	−0.99	−0.86			−0.80
Vessel has grounded!!			if Dr Mld was 8.30 m		>0.60 m water under stern			
Squat values applicable for:						Dr Mld		
Draft (m) even keel	7.30	7.50	7.70	7.90	8.10	8.30	8.50	8.70
Water depth (m)	8.18	8.40	8.62	8.85	9.07	9.30	9.52	9.74
Static ukc (m)	0.88	0.90	0.92	0.95	0.97	1.00	1.02	1.04

water. As expected, greater squats occur when at lower widths of water and/or at greater forward ship speeds.

If a master, mate, or pilot feels his/her ship may go aground due to ship squat, then simply reduce ship speed. After all is said and done, **prevention is better then cure — and a lot cheaper!!**

■

Conclusions

By using this procedure, it becomes possible to quickly and accurately predict the maximum ship squat and its location. This procedure caters for all types of merchant ships operating in confined channels such as a river or in a canal. It will be suitable for typical speeds of transits of merchant ships operating in rivers of any size.

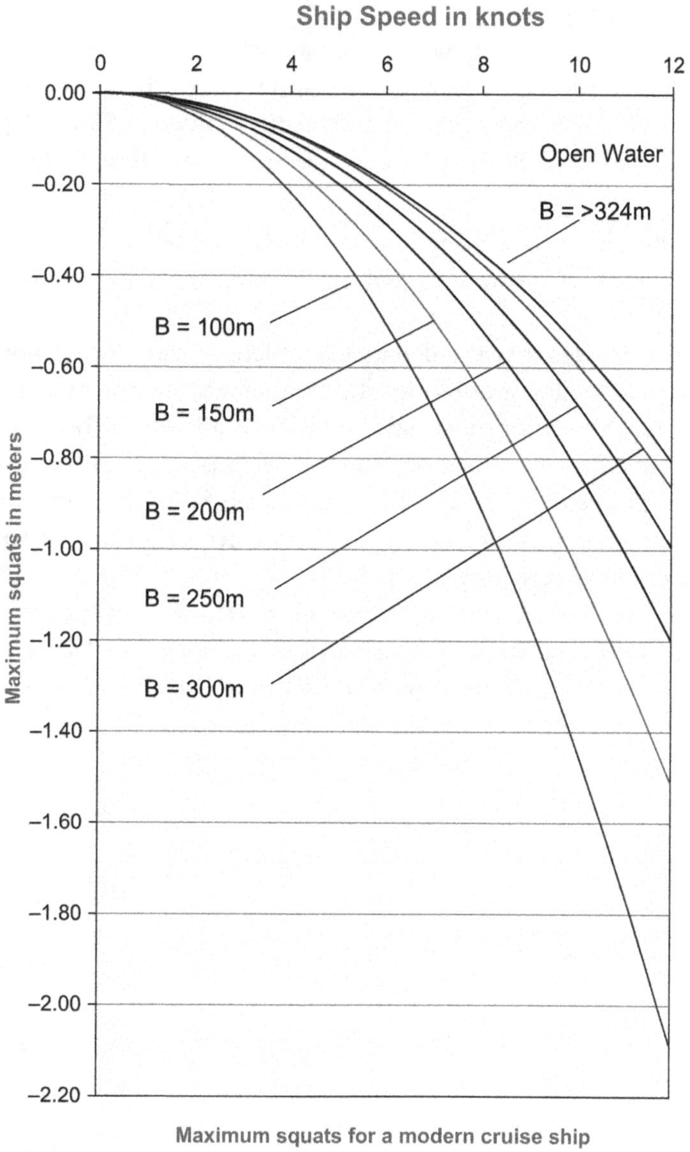

Figure 42.11:
Maximum Squats for a Modern Cruise Ship, H/T = 1.12.

Exercise 42

1. A container ship is operating in open water river conditions at a forward speed of 12.07 knots. Her C_b is 0.572 with a static even-keel draft of 13 m. Breadth molded is 32.25 m. If the depth of water is 14.5 m, calculate the following:
   (a) Width of influence for this wide river

   (b) Blockage factor
   (c) Maximum squat, stating with reasoning where it occurs
   (d) Dynamical underkeel clearance corresponding to this maximum squat.
2. A vessel has the following particulars: Breadth molded is 50 m, depth of water in river is 15.50 m, C_b is 0.817, width of water in river is 350 m, static even-keel draft is 13.75 m.
   (a) Prove this ship is operating in a 'confined channel' situation.
   (b) Draw a graph of maximum squat against ship speed for a range of speeds up to 10 knots.
   (c) The pilot decides that the dynamical underkeel clearance is not to be less than 1.00 m. Determine graphically and mathematically the maximum speed of transit that this pilot must have in order to adhere to this prerequisite.
3. A 75,000-tonne dwt oil tanker has the following particulars: Breadth molded is 37.25 m, static even-keel draft is 13.5 m, C_b is 0.800, depth of water is 14.85 m, width of river is 186 m. Calculate the forward speed at which, due to ship squat, this vessel would just go aground at the bow.
4. A passenger liner has a C_b of 0.599. She is proceeding upriver at a speed (V) of 7.46 knots. This river of rectangular cross-section has an H/T of 1.25 and a B/b of 4.36. When static, the ship was on even keel. Calculate her maximum squat at this speed and state the whereabouts it will occur. Use Figure 42.10.

■

CHAPTER 43

Turning Circle Diameter (TCD) Values for Vessels in Shallow Waters

Introduction

First of all, what exactly is a turning circle diameter (TCD)? It is measured on ship trials that are carried out in deep-water conditions. H/T is usually greater than 6.00. The vessel is turned completely through 360° with, say, starboard helm and then with port helm. This will produce two TCDs of different diameters.

This hypothesis is for:

1. Single-screw merchant ships only
2. Each ship fitted with a horizontal propeller shaft rotating within a shaft tunnel
3. TCD maneuver carried out at full speed
4. Turning maneuver conducted with a rudder helm held at 35°.

It should be observed in Figure 43.1 that at the beginning of the starboard turning maneuver, the ship turns initially to port. There are two reasons for this. Forces acting on the rudder itself will cause this move at first to port. Larger centrifugal forces acting on the ship's hull will then cause the vessel to move the ship on a course to starboard (see Figure 43.1).

The maximum angle of heel must be recorded. If the ship has port rudder helm this final angle of heel will be to starboard and vice versa. Again, this is due to centrifugal forces acting on the ship's hull.

During the TCD maneuver the ship will experience transfer, advance, drift angle, and angle of heel. With regard to grounding at the bilge plating, in *deep water* the angle of heel will present few or no problems. However, if the trials were undertaken in *shallow waters* there is an increased possibility of grounding at the bilge plating.

Updates of 1987 and 2002

In November 1987, the IMO regulation was that H/T should be greater than 4.00, where H is the water depth and T is the mean static draft of the vessel. In December 2002, further resolutions were made by the IMO (Ref A 601 (15).

Ship Stability for Masters and Mates. http://dx.doi.org/10.1016/B978-0-08-097093-6.00043-8

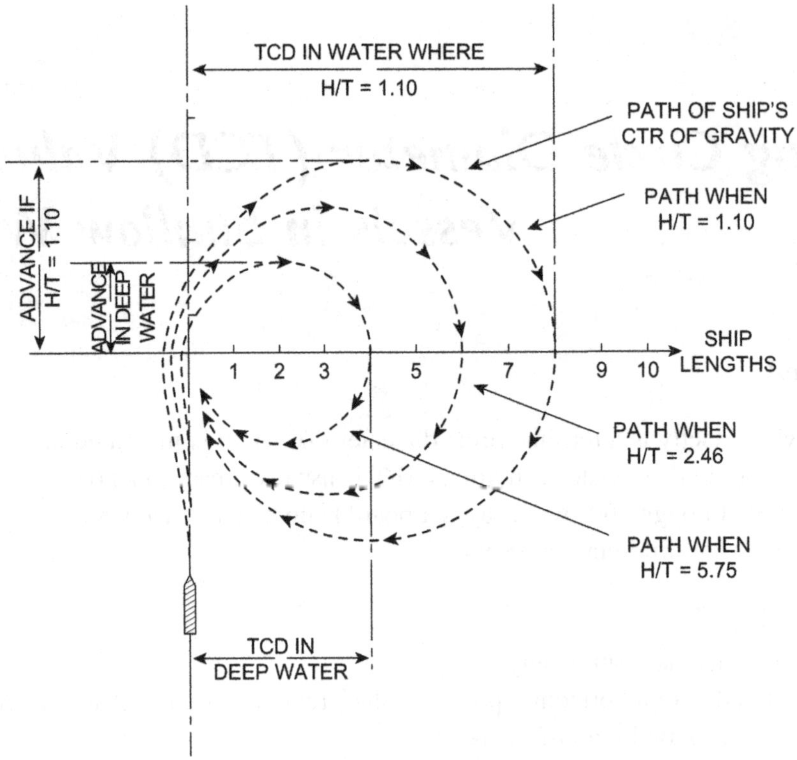

Figure 43.1: Variation of TCD with H/T Values for a Supertanker.
Note how TCD in very shallow waters is approximately *twice* that when ship is in deep waters.

In their Section 4 on values of TCD in shallow water, they stated the following:

For all ships of 100 m in length and over:

4.1.1 TCD is to be measured with the ship model and ship as for the fully loaded condition.

4.1.2 The initial speed of the ship should be at half-ahead speed.

4.1.3 Times and speeds at 90°, 180°, 270°, and 360° turning should be specifically shown, together with an outline of the ship.

4.1.4 Rudder angle should be at a maximum and H/T ratio to be 1.20.

The obtained information for the above should be presented (a) on a pilot card, (b) on a wheelhouse posting, and (c) in a maneuvering booklet.

Angles of heel should be presented on a spreadsheet (see Tables 43.1–43.4).

Table 43.1:

C_b	Draft T (m)	F_d	(F_d − 1.10)	$(H/T)_{actual}$	F_d − $(H/T)_{actual}$	Y2	K = 1.00 −2.00	Ship Speed (knots)	Max. Squat (m)	Br. Mld. (m)	LBP (m)
0.800	10.00	5.93	4.83	4.63	1.30	0.07	1.07	10.00	0.80	32.26	200

Table 43.2:

Angle of heel (°)	0	2	4	6	8	10
½bsin α (m)	0.00	0.56	1.13	1.69	2.24	2.80

Table 43.3: Meters at station 7.5 or 2.5 with ship at zero angle of heel and with trim or even keel.

K_{mbs}	mbs	Trim Change
0.800	0.64	0.72
At 0° angle of heel, squat at station 2.5 or 7.5 is 0.72 m at the bilge strakes P and S		
At 2° angle of heel, squat at station 2.5 or 7.5 is 1.28 m at the bilge strakes P and S		
At 4° angle of heel, squat at station 2.5 or 7.5 is 1.85 m at the bilge strakes P and S		
At 6° angle of heel, squat at station 2.5 or 7.5 is 2.41 m at the bilge strakes P and S		
At 8° angle of heel, squat at station 2.5 or 7.5 is 2.96 m at the bilge strakes P and S		
At 10° angle of heel, squat at station 2.5 or 7.5 is 3.52 m at the bilge strakes P and S		

Table 43.4:

Heel Angle (°)	Squats (m)
0	−0.72
2	−1.28
4	−1.85
6	−2.41
8	−2.96
10	−3.52

Tanker's speed is 10 knots.

Some Observations

- A 1% increase in trim by the stern in deep water increases the TCD by 10%.
- Changes in a ship's displacement do not alter changes in TCD values.
- Fine-form ships have greater TCD values than full-form ships. Hence container ships and passenger liners will have greater TCD values than supertankers.
- In deep water TCD for merchant ships is approximately three to four times ship's LBP (see details in Figure 43.1).
- Ships with comparatively larger rudders have smaller TCD values.
- Any change in ship speed for another test run will not cause a change in TCD value.
- When H/T is 1.10, TCD can be *double* that for when ship is in deep waters (see details in Figure 43.1 and Table 43.5).

Calculations

When a ship has a water depth greater than depth of influence, the ship is in deep water and the coefficient (K_{TCD}) for TCD is 1.00. If the ship has a water depth of 1.10 times the ship's mean draft, then the coefficient (K_{TCD}) for TCD is about 2.00. So K_{TCD} is 1.00–2.00.

$$F_D = 4.44/(C_b)^{1.3} \text{ times each ship's mean static even-keel draft}$$

This will alter with each ship type. For a specific ship it will change with each condition of loading.

A detailed study of this by the author produced the following equation:

$$K_{TCD} = 1 + \{(F_D - H/T_{actual})/(F_D - 1.10)\}^2$$

Figure 43.2 shows the values for K_{TCD} against H/T, for several C_b values ranging from full-form to fine-form ships.

Table 43.5:

H/T	C_b	Water Depth	K_{TCD}
5.75	0.820	Deep water	1.00
2.46	0.820	Shallow	1.50
1.10	0.820	Very shallow water	2.00

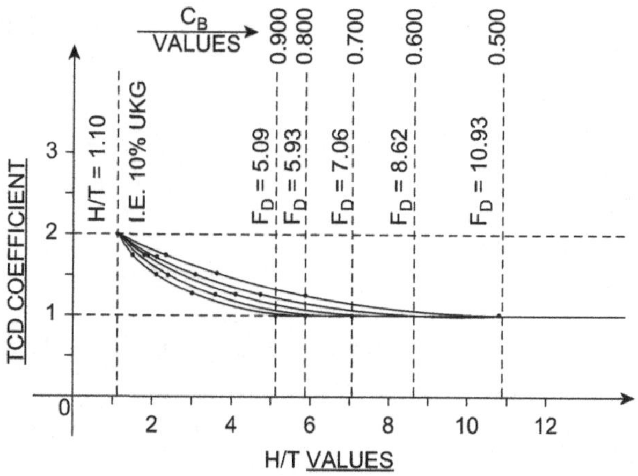

Figure 43.2: TCD Coefficient Plotted Against H/T.
At $H/T = 1.10$, $K_{TCD} = 2.0$ in very shallow water at each FD value; $K_{TCD} = 1.0$ in deep water. Hence, TCD is *twice* the the TCD in deep water. When $H/T = 1.00$, the vessel has grounded. Each $F_D = 4.44/C_b^{1.3}$, varying with ship type and C_b. $H/T > F_D$ means ship is in deep water; $H/T \leq F_D$ means that the ship is in shallow waters, and K_{TCD} will increase above 1.00. $K_{TCD} = 1 + \{(F_D - H/T_{actual})/(F_D - 1.10)\}^2$. Range of $K_{TCD} = 1.00-2.00$.

Worked Example

Let us now consider one hull form, that of an oil tanker with LBP of 200 m and a C_b of 0.800. Let breadth molded be 32.26 m, the static mean draft be 10.00 m, the water depth be 46.30m, and the ship speed be 10 knots.

$$\text{TCD in deep waters} \approx 3.25 \times \text{LBP} = 650 \text{ m}$$

$$(F_D - 1.10) = 5.93 - 1.10 = 4.83 \text{ m} \quad \text{(column 4 in Table 43.1)}$$

$$(F_D - H/T_{actual}) = 5.93 - 4.63 = 1.30 \text{ m} \quad \text{(column 6 in Table 43.1)}$$

$$\{(F_D - H/T_{actual})/(F_d - 1.10)\}^2 = 0.07 \quad \text{(column 7 in Table 43.1)}$$

$$K_{TCD} = 1 + 0.07 = 1.07 \quad \text{(column 8 in Table 43.1)}$$

Consequently, at H/T of 4.63, the TCD is $1.07 \times 650 = 760.5$ m instead of 650 m.

Continuing:

Maximum squat at the bow $= (C_b \times V^2)/100 = 0.800 \times 10 \times 10/100 = 0.80$ m

$$\text{Loss of ukc at bilge plating} = 1/_2\,\text{bsin}\,\alpha = 16.13\text{sin}\,\alpha \quad \text{(see Table 43.2)}$$

where α is the angle of heel, ranging from 0° to 10°.

$$\text{Mean bodily sinkage} = K_{mbs} \times \text{max. squat} = \{1 - 20(0.700 - C_b)^2\} \times 0.80$$

$$= \{1 - 20(0.700 - 0.80)^2\} \times 0.80$$

$$= 0.80 \times 0.80 = 0.64\text{ m from bow to stern} \quad \text{(see Table 43.3)}$$

$$\text{Trim at station 7.5} = (\text{mbs} + \text{max. squat})/2$$

$$= 0.64 + 0.80 = 0.72\text{m} \quad \text{(see column 3 in Table 43.3)}$$

Transverse squat at the bilge plating at station 7.5

$$= 0.72 + 1/_2\,\text{bsin}\,\alpha \quad \text{(Table 43.4)}$$

where α is angle of heel ranging from 0° to 10°.

■

Summary

This oil tanker will go aground at the bilge plating port or starboard at an angle of heel of 4.82°. This is shown in graphical form in Figure 43.3, where the transverse squat due to mean bodily sinkage plus angle of heel is equal to the static ukc even keel of 2.00 m.

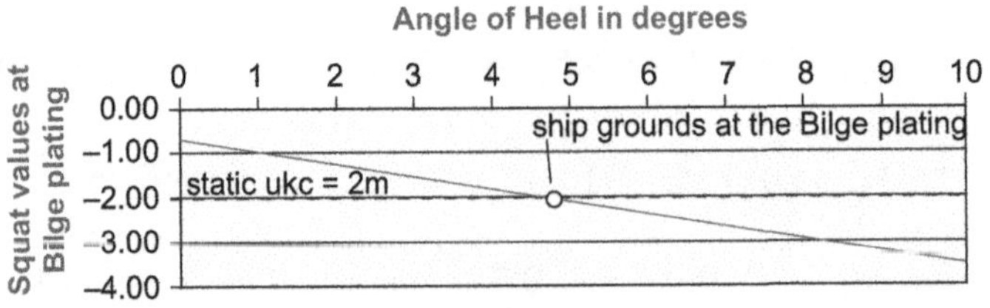

Figure 43.3:
Squats at Bilge Plating.

Exercise 43

1. Estimate the turning circle diameters and TCD coefficients with respect to the LBP for the following vessels:

Ship Type	Br Mld (m)	LBP (m)	Water Depth H (m)	Draft T (m)	C_b Value
(a) Container vessel	32.26	250	30.00	7.50	0.581
(b) General cargo ship	20.00	140	20.00	8.20	0.705
(c) Supertanker	55.00	320	14.85	13.50	0.820

2. For the supertanker in Question 1, predict the angle of heel at which the bilge plating would just go aground at station 7.5 were her forward speed to be 9.00 knots.

Reference

Captain Eduardo O. Gilardoni (2006). Manejo del buque en aguas restringidas (2nd edn), p. 160. *Instituto Iberoamericano De Derecho Maritimo*, Lavalle, Argentine.

CHAPTER 44

Interaction Effects, Including Two Case Studies

What Exactly is Interaction?

Interaction occurs when a ship comes too close to another ship or too close to, say, a river or canal bank. As ships have increased in size (especially in breadth molded), interaction has become very important to consider. In February 1998, the Marine Safety Agency (MSA) issued a Marine Guidance note 'Dangers of Interaction', alerting owners, masters, pilots, and tug-masters on this topic.

Interaction can result in one or more of the following characteristics:

1. If two ships are on a passing or overtaking situation in a river the squats of both vessels could be doubled when their amidships are directly in line.
2. When they are directly in line each ship will develop an angle of heel and the smaller ship will be drawn bodily towards the larger vessel.
3. Both ships could lose steerage efficiency and alter course without change in rudder helm.
4. The smaller ship may suddenly veer off course and head into the adjacent riverbank.
5. The smaller ship could veer into the side of the larger ship or, worse still, be drawn across the bows of the larger vessel, bowled over, and capsized.

In other words there is:

- a ship-to-ground interaction
- a ship-to-ship interaction
- a ship-to-shore interaction.

What causes these effects of interaction? The answer lies in the pressure bulbs that exist around the hull form of a moving ship model or a moving ship (see Figure 44.1). As soon as a vessel moves from rest, hydrodynamics produce the shown positive and negative pressure bulbs. For ships with a greater parallel body such as tankers, these negative bulbs will be comparatively longer in length. When a ship is stationary in water of zero current speed these bulbs disappear.

Note the elliptical domain that encloses the vessel and these pressure bulbs. This domain is very important. When the domain of one vessel interfaces with the domain of another vessel,

Ship Stability for Masters and Mates. http://dx.doi.org/10.1016/B978-0-08-097093-6.00044-X

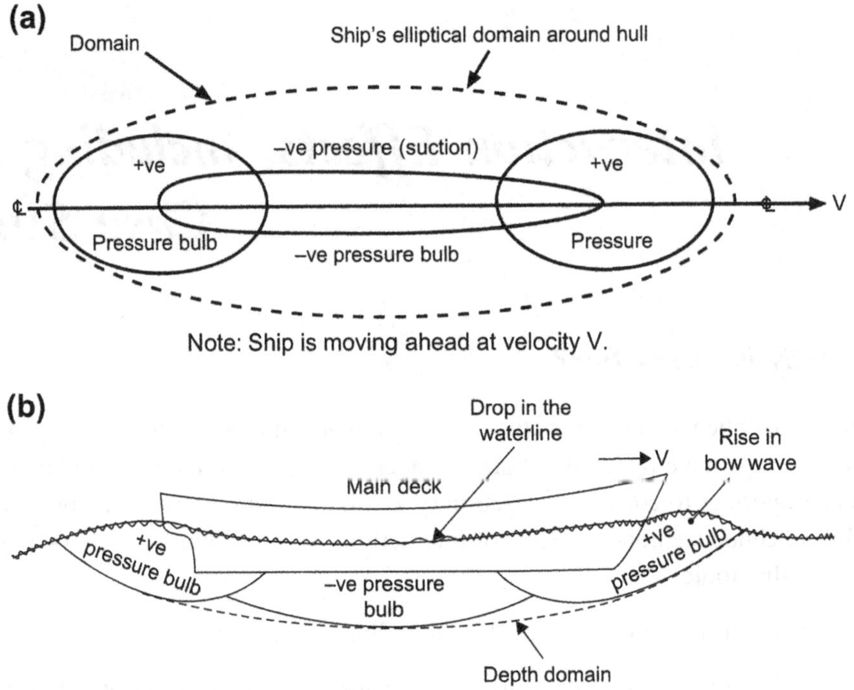

Figure 44.1:
(a) Pressure Distribution Around Ship's Hull (not drawn to scale). (b) Pressure Bulbs Around a Ship's Profile when at Forward Speed.

interaction effects will occur. Effects of interaction are increased when ships are operating in shallow waters.

Masters, mates, and pilots need to know about the phenomenon of interaction.

Ship-to-Ground (Squat) Interaction

In a report on measured ship squats in the St Lawrence seaway, A.D. Watt stated: 'meeting and passing in a channel also has an effect on squat. It was found that when two ships were moving at the low speed of 5 knots that squat increased up to double the normal value. At higher speeds the squat when passing was in the region of one and a half times the normal value.' Unfortunately, no data relating to ship types, gaps between ships, blockage factors, etc. accompanied this statement.

Thus, at speeds of the order of 5 knots the squat increase is +100%, while at higher speeds, say 10 knots, this increase is +50%. Figure 44.2 illustrates this passing maneuver. Figure 44.3 interprets the percentages given in the previous paragraph.

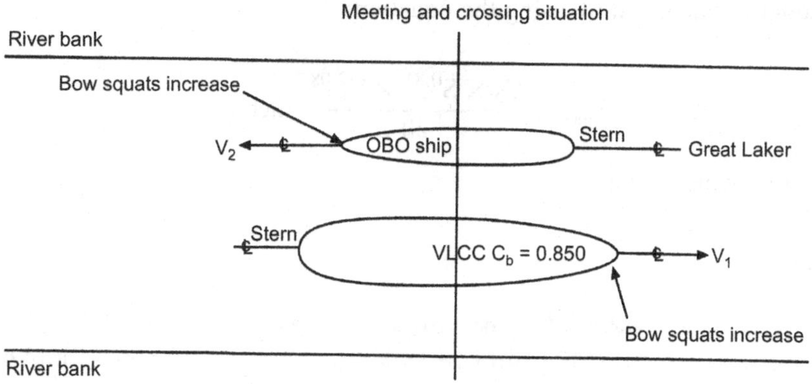

Figure 44.2:
Amidships of VLCC Directly in Line with Amidships of OBO Ship in St Lawrence Seaway.

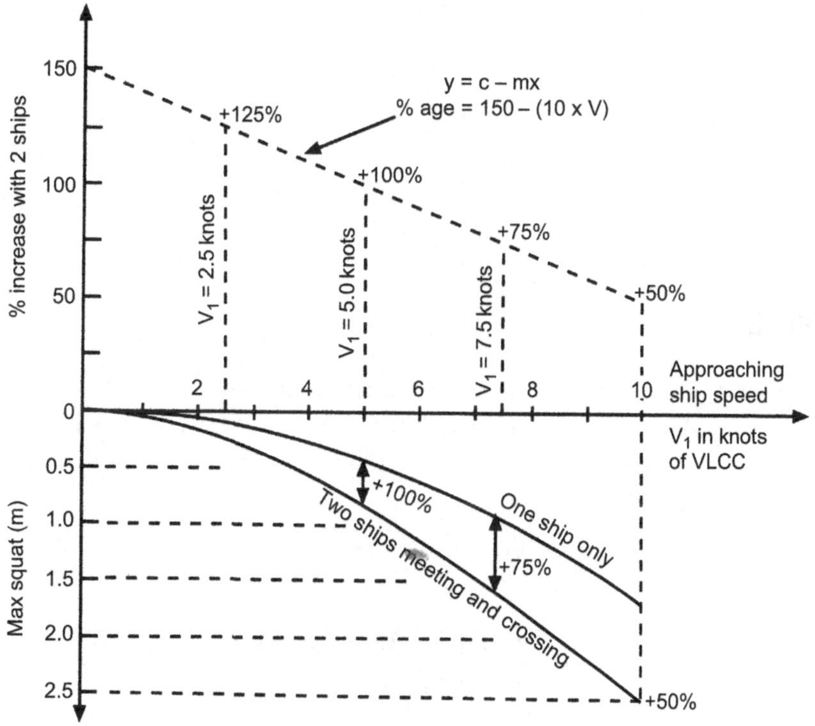

Figure 44.3:
Maximum Squats for One Ship, and for the Same Ship with Another Ship Present.

How may these squat increases be explained? It has been shown in the chapter on ship squat that its value depends on the ratio of the ship's cross-section to the cross-section of the river. This is the blockage factor 'S'. The presence of a second ship meeting and crossing will of course increase the blockage factor. Consequently the squat on each ship will increase.

Maximum squat is calculated by using the equation:

$$\delta_{max} = \frac{C_b \times S^{0.81} \times V_k^{2.08}}{20} \text{ meters}$$

Consider the following example.

Example 1

A supertanker has a breadth of 50 m with a static even-keel draft of 12.75 m. She is proceeding along a river of 250 m and 16 m depth rectangular cross-section. If her speed is 5 knots and her C_b is 0.825, calculate her maximum squat when she is on the centerline of this river.

$$S = \frac{b \times T}{B \times H} = \frac{50 \times 12.75}{250 \times 16} = 0.159$$

$$\delta_{max} = \frac{0.825 \times 0.159^{0.81} \times 5^{2.08}}{20} = \underline{0.26 \text{ m}}$$

Example 2

Assume now that this supertanker meets an oncoming container ship also traveling at 5 knots (see Figure 44.4). If this container ship has a breadth of 32 m, a C_b of 0.580, and a static even-keel draft of 11.58 m, calculate the maximum squats of both vessels when they are transversely in line as shown.

$$S = \frac{(b_1 \times T_1) + (b_2 \times T_2)}{B \times H}$$

$$S = \frac{(50 \times 12.75) + (32 \times 11.58)}{250 \times 16} = 0.252$$

Supertanker:

$$\delta_{max} = \frac{0.825 \times 0.252^{0.81} \times 5^{2.08}}{20}$$

$$= \underline{0.38 \text{ m at the bow}}$$

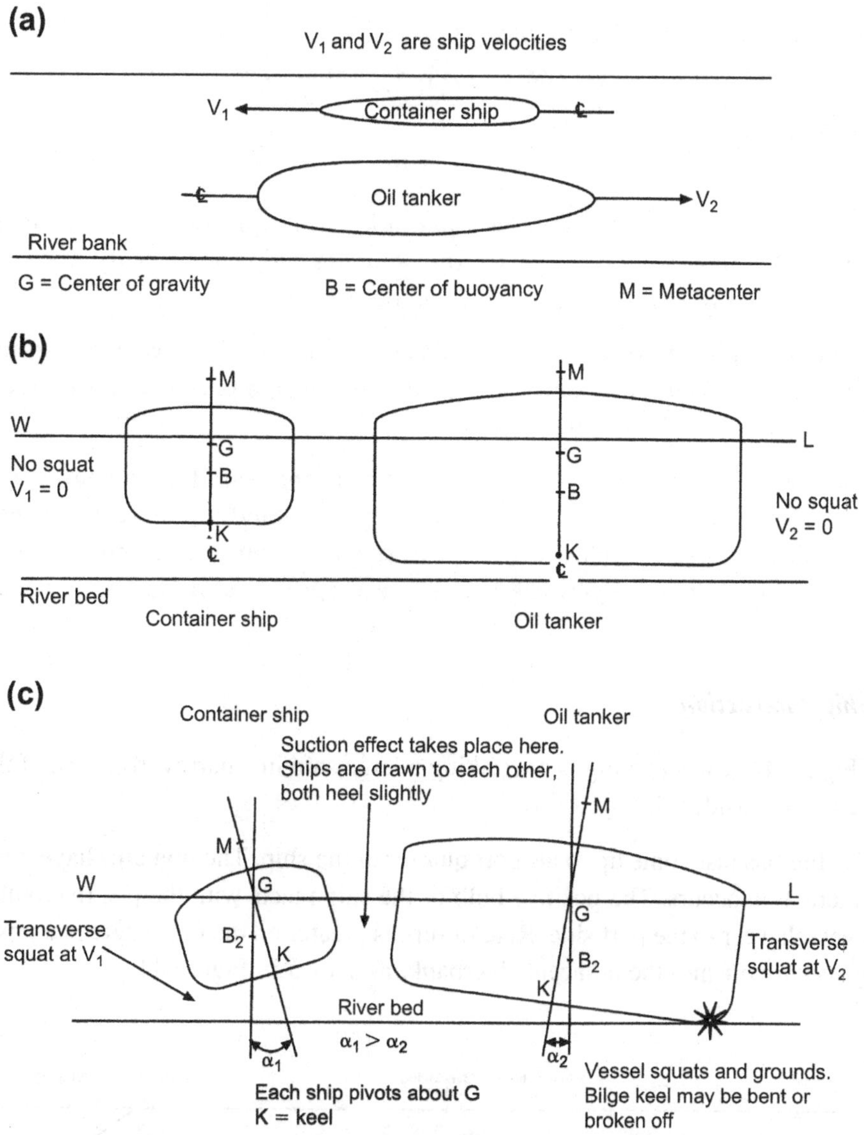

Figure 44.4:
Transverse Squat Caused by Ships Crossing in a Confined Channel.

Container ship:

$$\delta_{max} = \frac{0.580 \times 0.252^{0.81} \times 5^{2.08}}{20}$$
$$= 0.27 \text{ m at the stern}$$

The maximum squat of 0.38 m for the supertanker will be at the bow because her C_b is greater than 0.700. Maximum squat for the container ship will be at the stern, because her C_b is less than 0.700. As shown, this will be 0.27 m.

If this container ship had traveled alone on the centerline of the river then her maximum squat at the stern would have only been 0.12 m. Thus, the presence of the other vessel has more than doubled her squat.

Clearly, these results show that the presence of a second ship does increase ship squat. Passing a moored vessel would also make blockage effect and squat greater. These values are not qualitative but only illustrative of this phenomenon of interaction in a ship-to-ground (squat) situation. Nevertheless, they are supportive of A.D. Watt's statement.

■

Ship-to-Ship Interaction

Consider Figure 44.5, where a tug is overtaking a large ship in a narrow river. The following cases have been considered:

Case 1. The tug has just come up to aft port quarter of the ship. The domains have come into contact. Interaction occurs. The positive bulb of the ship reacts with the positive bulb of the tug. Both vessels veer to the port side. Rate of turn is greater on the tug. There is a possibility of the tug veering off into the adjacent riverbank, as shown in Figure 44.5.

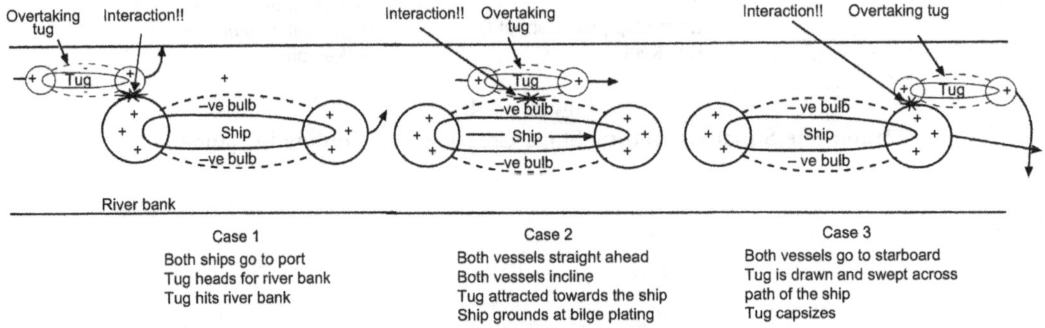

Figure 44.5:
Ship-to-Ship Interaction in a Narrow River During an Overtaking Maneuver.

Case 2. The tug is in danger of being drawn bodily towards the ship because the negative pressure (suction) bulbs have interfaced. The bigger the differences between the two deadweights of these ships, the greater will be this transverse attraction. Each ship develops an angle of heel as shown. There is a danger of the ship losing a bilge keel or indeed fracture of the bilge strakes occurring. This is 'transverse squat', the loss of underkeel clearance at forward speed. Figure 44.4 shows this happening with the tanker and the container ship.

Case 3. The tug is positioned at the ship's forward port quarter. The domains have come into contact via the positive pressure bulbs (see Figure 44.5). Both vessels veer to the starboard side. Rate of turn is greater on the tug. There is great danger of the tug being drawn across the path of the ship's heading and bowled over. This has actually occurred with resulting loss of life.

Note how in these three cases it is the smaller vessel, be it a tug, a pleasure craft, or a local ferry involved, that ends up being the casualty!!

Figures 44.6 and 44.7 give further examples of ship-to-ship interaction effects in a river.

Methods for Reducing the Effects of Interaction in Cases 1–5

Reduce speed of both ships and then if safe increase speeds after the meeting crossing maneuver time slot has passed. Resist the temptation to go for the order 'increase revs'. This is because the forces involved with interaction vary as the *speed squared*. However, too much of a reduction in speed produces a loss of steerage because rudder effectiveness is decreased. This is even more so in shallow waters, where the propeller r.p.m. decreases for similar input of deep water power. Care and vigilance are required.

Keep the distance between the vessels as large as practicable, bearing in mind the remaining gaps between each ship side and nearby riverbank.

Keep the vessels from entering another ship's domain, for example crossing in wider parts of the river.

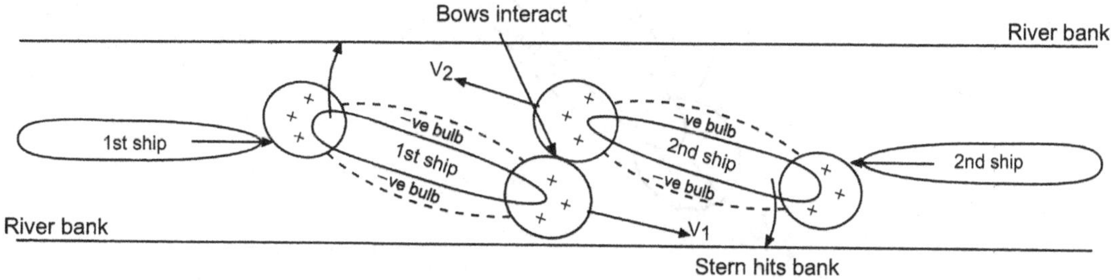

Figure 44.6:
Case 4. Ship-to-Ship Interaction, Both Sterns Swing Towards Riverbanks (the Approach Situation).

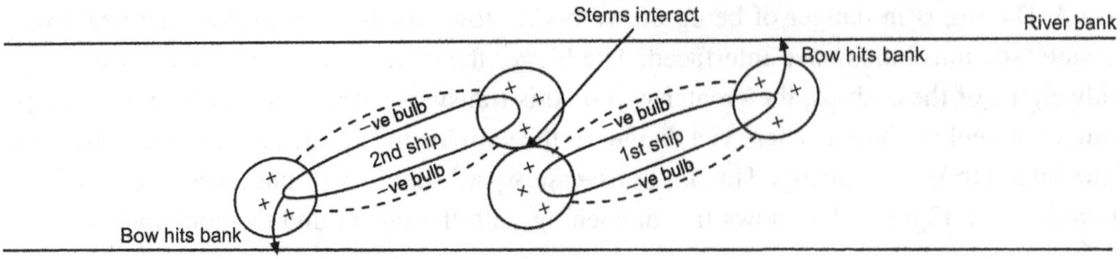

Figure 44.7:
Case 5. Ship-to-Ship Interaction, Both Bows Swing Towards Riverbanks (the Leaving Situation).

Cross in deeper parts of the river rather than in shallow waters, bearing in mind those increases in squat.

Make use of rudder helm. In Case 1, starboard rudder helm could be requested to counteract loss of steerage. In Case 3, port rudder helm would counteract loss of steerage.

Ship-to-Shore Interaction

Figures 44.8 and 44.9 show the ship-to-shore interaction effects. Figure 44.8 shows the forward positive pressure bulb being used as a pivot to bring a ship alongside a riverbank.

Figure 44.9 shows how the positive and negative pressure bulbs have caused the ship to come alongside and then to veer away from the jetty. Interaction could in this case cause the stern to swing and collide with the wall of this jetty.

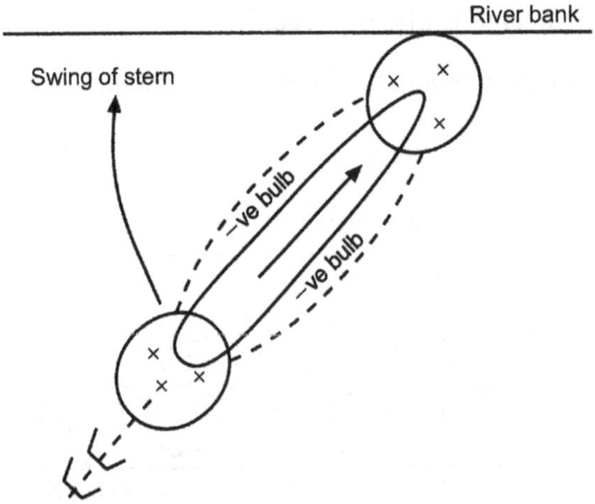

Figure 44.8: Ship-to-Bank Interaction.
Ship approaches slowly and pivots on forward positive pressure bulb.

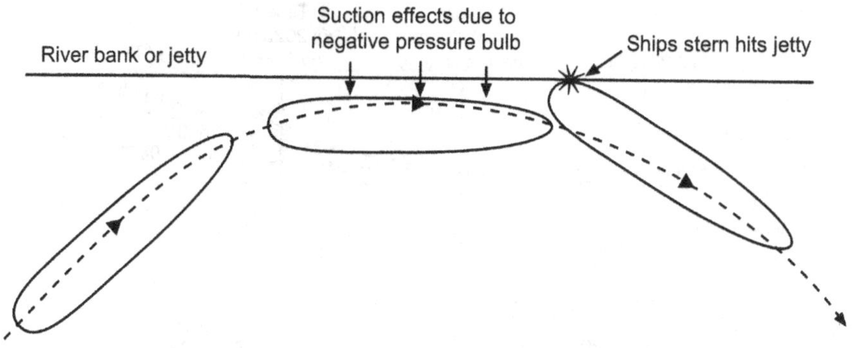

Figure 44.9: Ship-to-Bank Interaction.
Ship comes in at too fast a speed. Interaction causes stern to swing towards riverbank and then hits it.

Summary

An understanding of the phenomenon of interaction can avert a possible marine accident. For masters, mates, and pilots generally, a reduction in speed is the best preventive procedure. This could prevent an incident leading to loss of seaworthiness, loss of income for the shipowner, cost of repairs, compensation claims, and maybe loss of life.

A Collision Due to Interaction? An Example of Directional Stability

Case Study 1

On 23 October 1970, the *Pacific Glory* (a 77,648-tonne dwt oil tanker) collided with the *Allegro* (a 95,455-tonne dwt oil tanker). The collision took place just south of the Isle of Wight.

Both vessels were traveling from west to east up the English Channel. It was approximately 2000 hours as the vessels approached each other. It was a clear night with good visibility of about 10 nautical miles. Both oil tankers had full loads of cargo oil.

The *Pacific Glory,* on transit from Nigeria to Rotterdam, had been to Brixham to pick up a North Sea pilot. This vessel had a service speed of 15 knots and was fitted with diesel machinery. It was on a course of 087° that changed at 2020 hours to a heading of 080°.

The *Allegro* had been on route from Libya to the Fawley Oil Terminal at Southampton. It had a service speed of 15.5 knots, was fitted with steam turbine machinery, and was proceeding on a heading of 060°. The *Allegro* was larger, longer, and faster than the *Pacific Glory.*

At 2018 hours, the vessels were 915 m apart. Figure 44.10 shows the converging courses of these two ships prior to their collision at 2023 hours. The top diagram shows the movements in the 5 minutes before the collision. The middle diagram provides details 1–2 minutes before the collision. The lower diagram shows the *Pacific Glory* being drawn towards the larger *Allegro.*

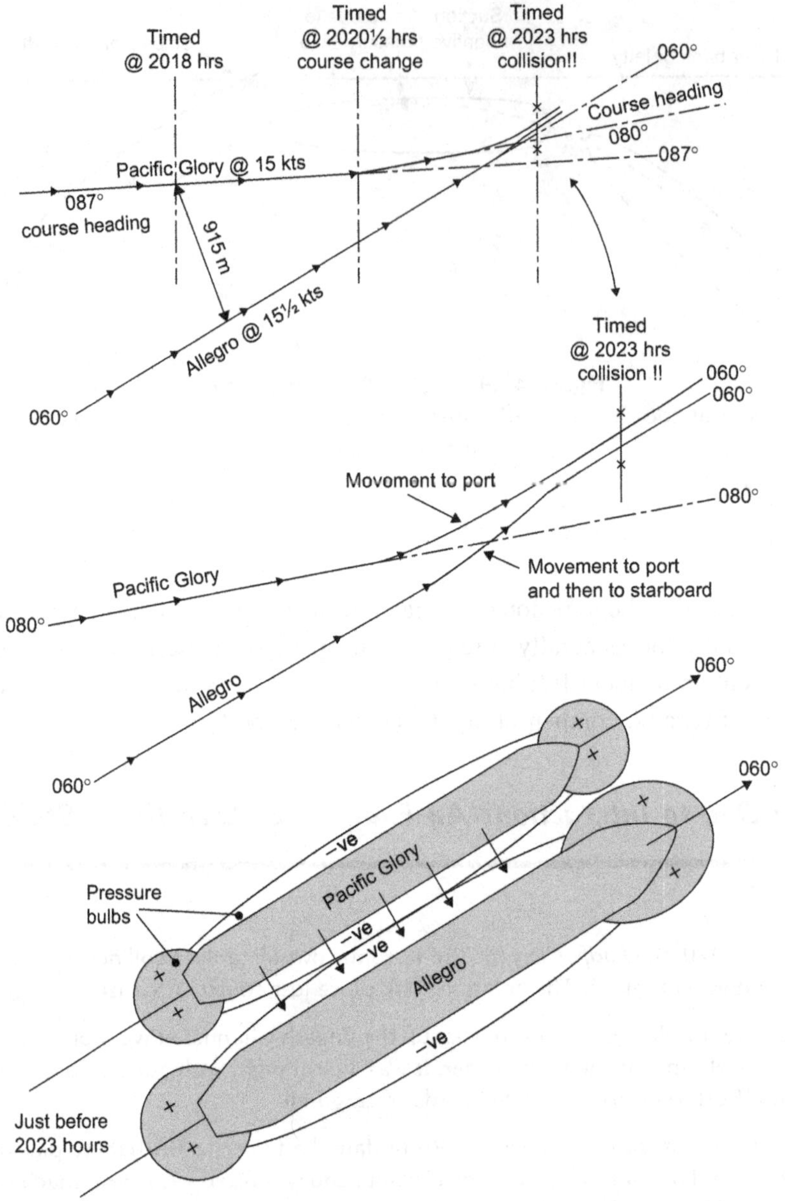

Figure 44.10:
Interaction Effects of *Pacific Glory* with the *Allegro*. *Information source: 'Collisions and their causes' by Captain R.A. Cahill.*

Just moments before the collision, the vessels must have been very nearly on parallel headings, each one close to 060°.

The three diagrams in Figure 44.10 clearly show belated attempts to steer clear of one another. Interaction between the parallel bodies of the vessels caused them to attract and then to collide.

Case Study 2

On 11 May 1972 at 0500 hours, the *Royston Grange* (a 10,262-tonne dwt reefer) collided with the *Tien Chee* (a 19,700-tonne dwt Liberian oil tanker). The collision took place in the Indio Channel of Montevideo in the River Plate.

The *Royston Grange* was outward bound, on route to Montevideo and then on to London. It had a static mean draft of 7 m and a trim by the stern of 0.30 m. On board were grains, butter, and refrigerated meats (see Figure 44.11(a)).

The *Tien Chee* was inward bound, on route to Buenos Aires, and had a mean draft of 9 m with a trim of 0.30 m by the stern. It was almost fully loaded with cargo oil and forward speed just prior to the collision was 12 knots.

Both vessels were traveling at forward speed and each had their pilot aboard. Both vessels were also on an approaching maneuver, passing port to port, and were in shallow waters. They were both in the southern part of the channel and it was at low tide.

Just after their bows crossed, the *Royston Grange* suddenly veered to port and crashed into the amidships port side of the *Tien Chee*. The photograph, showing extensive bow damage to the reefer (see Figure 44.11(b)), suggests that the vessels must have been traveling towards each other at a fast speed.

Analysis and investigation into the collision has drawn the following conclusions:

1. The *Royston Grange* hit the *Tien Chee* at an angle of 40° (see Figure 44.12). The first point of contact was at No. 7 cargo wing tank port. Damage then occurred at Nos. 8, 9, and 10 wing tanks of the *Tien Chee.* As shown in the photograph, substantial structural damage to the bow structure of the *Royston Grange* occurred and 800 tonnes of oil were spilled.
2. The pilot on the bridge of the *Tien Chee* made a claim to the master on 11 May 1972 that there would be an extra 0.60 m of water than was shown for low water. He convinced the ship's master that this would be sufficient to proceed upriver without the need to wait and the master was persuaded.
3. The *Tien Chee* had been delayed and was 3 hours late in arriving at the river. The master was in a hurry to take the ship upriver because of this delay, resulting in human error of judgment.
4. The approach speed of the *Tien Chee* was too fast. This can be considered another human error of judgment.
5. The Tien Chee was ploughing through silt and mud on the riverbed.
6. At 12 knots speed the *Tien Chee* had squatted about 0.90 m at the stern and would have increased the trim by the stern. The total trim would then be about 0.66 m, thereby drawing the stern further into the mud.
7. The bow of the *Royston Grange* was drawn into the port side of the *Tien Chee* (see Figure 44.12).
8. Just before the collision both vessels were drawn as close as 50 m on a parallel course. Much too close!!
9. The negative pressure bulb of the *Tien Chee's* side-shell was much larger than the negative pressure bulb of the *Royston Grange's* side-shell.

Figure 44.11

10. If the forward speed (not given) of the *Royston Grange* had been 6–10 knots (average 8 knots) then the closing speed would have been a very dangerous speed of 20 knots (see Figure 44.12).
11. A northerly wind present at the time of the collision would have decreased the water depth. This would have increased squat and interaction.
12. The approach maneuver took place in the southern part of this channel in a confined channel condition (see Figure 44.13).
13. The build-up of silt, due to poor local dredging arrangements, meant there was an appreciable build-up of mud and silt on the riverbed. The bottom saucers of silt had been

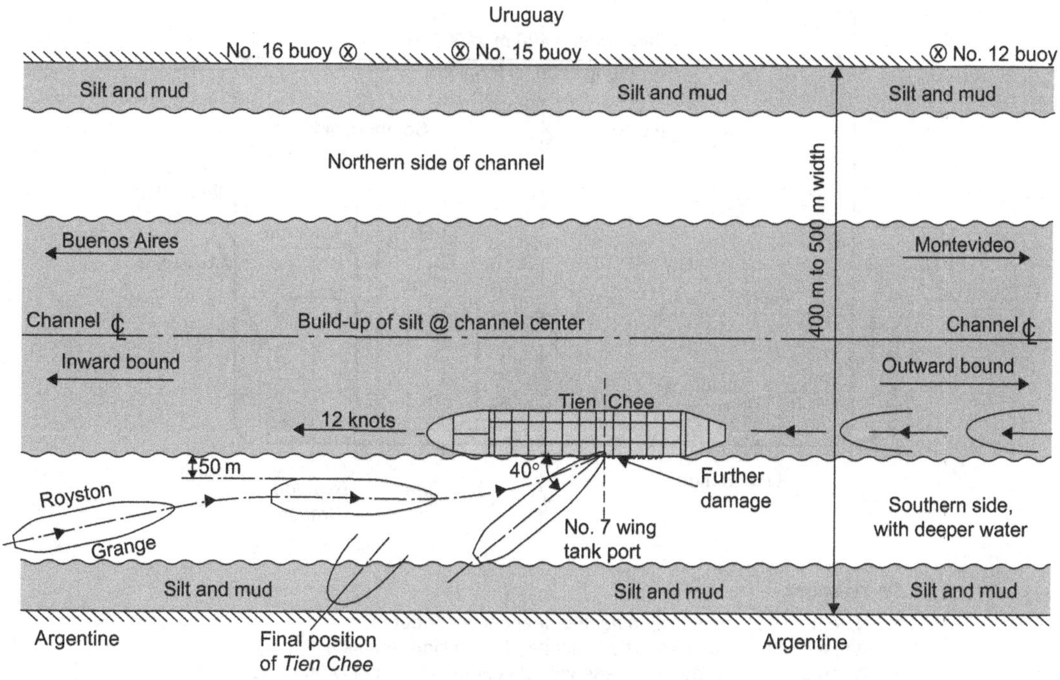

Figure 44.12

built up by the transit of many vessels going up and downriver. This silt and mud would increase the blockage factor and would increase squat and interaction effects.

14. When the vessel collided it was at low tide and was in a narrow part of the river. In these conditions, rudder helm is less effective.
15. The deadweight of the *Tien Chee* was 19,700 t. The deadweight of the *Royston Grange* was only 10,262 t. This gives a ratio of almost 2:1. The greater the difference of deadweight between ships in this situation, the greater will be the effects of interaction.
16. The mean drafts of the *Tien Chee* and the *Royston Grange* were in the ratio of about 1.25:1.00. Interaction increases with greater ratios of drafts, so this would definitely contribute to the collision.
17. The maintenance (dredging arrangements) by the Argentine and Uruguay port authorities left a lot to be desired. To say the least, it allegedly was very suspect (see Figure 44.13).
18. The *Tien Chee* was loaded, so much so that there was evidence of transitting through mud when at 12 knots forward speed. The ship would have experienced the greater squat effects with its H/T of 1.10 (see Figure 44.13).
19. In this river the greatest water depth is not on the centerline of the channel but rather in the southern part of the channel (see Figure 44.13).
20. The *Royston Grange* was turned to port because the large negative pressure bulb of the *Tien Chee* was moving upriver whilst the smaller negative pressure bulb was moving downriver and at the same time towards it. This set up an anticlockwise couple that moved the *Royston Grange* from being parallel, to being at a 40° angle to the *Tien Chee* (see Figure 44.14).

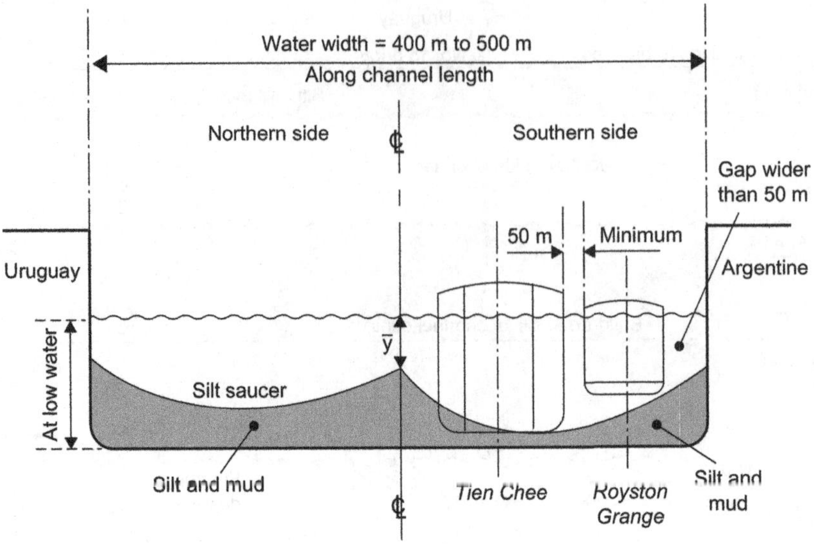

Observations

1. Note how ȳ at channel ℄ is *not* the deepest water depth.
2. Water depth in Southern side is deeper than in Northern side.
3. When on parallel courses, these vessels came to being only 50 m apart.
4. Silt 'saucers' were scoured out by many vessels in transit along this channel.
5. *Tien Chee* was traveling through silt and mud.
6. *Royston Grange* was very much in a confined channel situation.

Figure 44.13:
Cross-Section Through the Indio Channel of Montevideo.

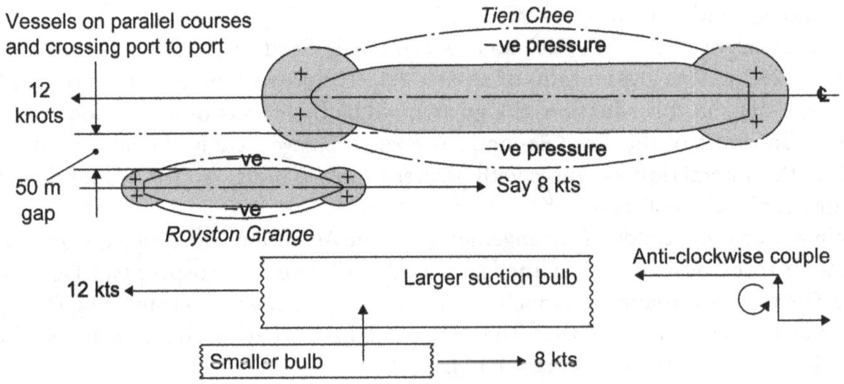

Observations

1. Interaction effects would exist between the port negative bulbs of the vessels.
2. The *Royston Grange* would be drawn bodily towards the *Tien Chee*, because she was much smaller than the *Tien Chee*.
3. The ACW couple would also cause the *Royston Grange* to veer ACW and collide with the *Tien Chee* at the reported angle of 40°.

Figure 44.14:
Interaction Effects Between the Negative Pressure (Suction) Bulbs.

21. The parallel body of the *Tien Chee* would have been 50–60% of its LBP. This represents a long wall of moving water going upriver at 12 knots, almost like a moving riverbank. The greater the percentage, the greater the effect of interaction.

Due to the butter cargo the *Royston Grange* was carrying, the oil vapors from the *Tien Chee*, and the sparking from the collision contact, the reefer was incinerated. Both vessels were later scrapped.

The total complement (74) of people aboard the *Royston Grange* lost their lives. Eight people aboard the *Tien Chee* also died.

Summary and Conclusions

These case studies on interaction effects have clearly shown the perils of allowing the domain of one ship to enter the domain of another ship.

If one ship is *overtaking* another ship in a narrow river and the amidships are coming in line, then it is possible that the smaller vessel will be drawn bodily towards the other vessel.

However, if the vessels are on an *approaching* maneuver and transversely close together, then as well as this suction effect the smaller ship may suddenly veer, turn, and then run into the side of the other vessel.

Exercise 44

1. A river is 150 m wide and has a 12 m depth of water. A passenger liner having a breadth of 30 m, a static even-keel draft of 10 m, and a C_b of 0.625 is proceeding along this river at 8 knots. It meets an approaching general cargo vessel having a breadth of 20 m, a static even-keel draft of 8 m, and a C_b of 0.700, moving at 7 knots. Estimate the maximum squats for each vessel when their amidships are transversely in line.
2. With the aid of sketches, define interaction and describe how its effects may be reduced. Show clearly how interaction and transverse squat are interrelated.
3. With the aid of sketches, show the domains of a moving ship in plan view and in profile view.

CHAPTER 45

Rolling, Pitching, and Heaving Motions

A ship will not normally roll in still water but if a study is made of such rolling some important conclusions may be reached. For this study it is assumed that the amplitude of the roll is small and that the ship has positive initial metacentric height. Under these conditions rolling is considered to be simple harmonic motion so it will be necessary to consider briefly the principle of such motion.

Let XOY in Figure 45.1 be a diameter of the circle whose radius is 'r' and let OA be a radius vector that rotates about O from position OY at a constant angular velocity of 'w' radians per second. Let P be the projection of the point A on to the diameter XOY. Then, as the radius vector rotates, the point P will oscillate backwards and forwards between Y and X. The motion of the point P is called 'simple harmonic'.

Let the radius vector rotate from OY to OA in 't' seconds; then angle AOY is equal to 'wt'. Let the time taken for the radius vector to rotate through one complete revolution (2π radians) be equal to 'T' seconds; then

$$2\pi = wT$$

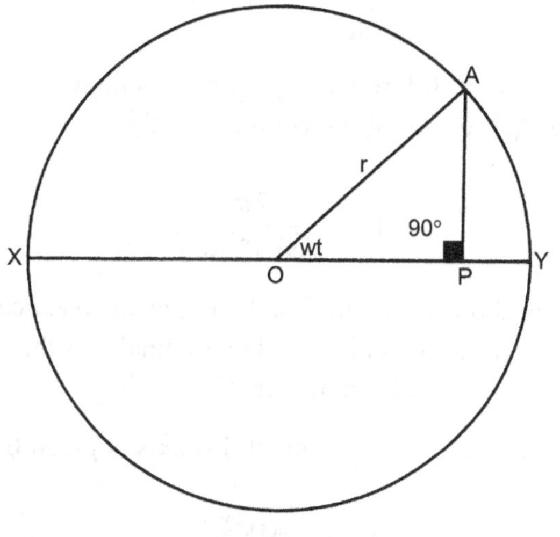

Figure 45.1

Ship Stability for Masters and Mates. http://dx.doi.org/10.1016/B978-0-08-097093-6.00046-3

or

$$T = 2\pi/w$$

Let

$$OP = x$$

then

$$x = r\cos wt$$

$$\frac{dx}{dt} = -rw\sin wt$$

$$\frac{d^2x}{dt^2} = -rw^2\cos wt$$

but

$$r\cos wt = x$$

$$\therefore \frac{d^2x}{dt^2} = -w^2x$$

or

$$\frac{d^2x}{dt^2} + w^2x = 0$$

The latter equation is the type of differential equation for simple harmonic motion and since $T = 2\pi/w$ and 'w' is the square root of the coefficient of x in the above equation, then

$$T = \frac{2\pi}{\sqrt{\text{coeff. of } x}}$$

When a ship rolls, the axis about which thc oscillation takes place cannot be accurately determined but it would appear to be near to the longitudinal axis through the ship's center of gravity. Hence the ship rotates or rolls about her 'G'.

The mass moment of inertia (I) of the ship about this axis is given by

$$I = MK^2$$

where

$$M = \text{the ship's mass}$$
$$K = \text{the radius of gyration about this axis}$$

But

$$M = \frac{W}{g}$$

where

$$W = \text{the ship's weight}$$
$$g = \text{the acceleration due to gravity}$$

$$\therefore I = \frac{W}{g}K^2$$

When a ship is inclined to a small angle (θ) the righting moment is given by

$$\text{Righting moment} = W \times GZ$$

where

$$W = \text{the ship's weight}$$
$$GZ = \text{the righting lever}$$

But

$$GZ = GM \sin\theta$$
$$\therefore \text{Righting moment} = W \times GM \times \sin\theta$$

And since θ is a small angle, then

$$\text{Righting moment} = W \times GM \times \theta$$

$$\text{The angular acceleration} = \frac{d^2\theta}{dt^2}$$

$$\therefore I \times \frac{d^2\theta}{dt^2} = -W \times GZ$$

or

$$\frac{W}{g}K^2 \times \frac{d^2\theta}{dt^2} = -W \times GM \times \theta$$

$$\frac{W}{g}K^2 \times \frac{d^2\theta}{dt^2} + W \times GM \times \theta = 0$$

$$\frac{d^2\theta}{dt^2} + \frac{gGM\theta}{K^2} = 0$$

This is the equation for a simple harmonic motion having a period 'T' given by the equation:

$$T = \frac{2\pi}{\sqrt{\text{Coeff. of } \theta}}$$

or

$$T = \frac{2\pi K}{\sqrt{gGM}} = \frac{2\pi}{\sqrt{g}} \times \frac{K}{\sqrt{GM}} = \frac{2K}{\sqrt{GM}} \text{ approx.}$$

$$T_R = 2 \times \pi \times (K^2/(g \times GM_T)^{0.5}) \text{ seconds}$$
$$= 2(K^2/GM_T)^{0.5}$$
$$= 2 \times K/(GM_T)^{0.5} \text{ sec}$$

The transverse K value can be approximated with acceptable accuracy to be $K = 0.35 \times B$, where $B =$ breadth mold of the ship.

$$T_R = 2 \times \pi \times (K^2/(g \times GM_T)^{0.5})$$
$$T_R = (2 \times 3.142 \times 0.35 \times B)/3.131 \times (GM_T)^{0.5}$$
$$T_R = (0.7 \times B)/(GM_T)^{0.5} \text{ sec approx.}$$

Figure 45.2 shows the resulting rolling periods based on the above formula, with the variables of GM_T up to 5 m and breadth B up to 60 m incorporated.

From the above it can be seen that:

1. The time period of roll is completely independent of the actual amplitude of the roll so long as it is a small angle.
2. The time period of roll varies directly as K, the radius of gyration. Hence if the radius of gyration is increased, then the time period is also increased. K may be increased by moving weights away from the axis of oscillation. Average K value is about 0.35 × Br. Mld.

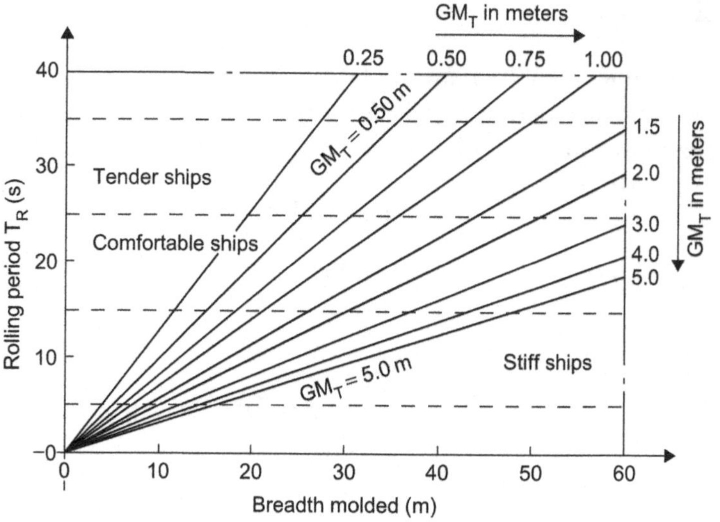

Figure 45.2

3. The time period of roll varies inversely as the square root of the initial metacentric height. Therefore, ships with a large GM will have a short period and those with a small GM will have a long period.
4. The time period of roll will change when weights are loaded, discharged, or shifted within a ship, as this usually affects both the radius of gyration and the initial metacentric height.

Example 1

Find the still water period of roll for a ship when the radius of gyration is 6 m and the metacentric height is 0.5 m.

$$T = \frac{2\pi K}{\sqrt{gGM}} = \frac{2K}{\sqrt{GM_T}} \text{ approx.}$$

$$T = \frac{2\pi K}{\sqrt{9.81 \times 0.5}} = \frac{2 \times 6}{\sqrt{0.5}} \text{ apprrox.}$$

$$= 16.97 \text{ s} \quad \text{Average} = 17\text{s}$$

(99.71 % correect giving only 0.29% error !!)

Ans. $T = 17.02$ s

Note. In the SI system of units the value of g to be used in problems is 9.81 m per second per second, unless another specific value is given.

Example 2

A ship of 10,000 tonnes displacement has GM = 0.5 m. The period of roll in still water is 20 seconds. Find the new period of roll if a mass of 50 tonnes is discharged from a position 14 m above the center of gravity. Assume

$$g = 9.81\ m/s^2$$

$$W_2 = W_0 - w = 10,000 - 50 = 9950 \text{ tonnes}$$

$$T = \frac{2\pi K}{\sqrt{gGM}} \qquad GG_1 = \frac{w \times d}{W_2}$$

$$20 = \frac{2\pi K}{\sqrt{9.81 \times 0.5}} \qquad = \frac{50 \times 14}{9950}$$

$$400 = \frac{4\pi^2 K^2}{9.81 \times 0.5} \qquad GG_1 = 0.07 \text{ m}$$

$$GM = \underline{0.50 \text{ m}}$$

$$\text{New GM} = 0.57 \text{ m}$$

or

$$K^2 = \frac{400 \times 9.81 \times 0.5}{4 \times \pi^2}$$

$$= 49.69$$

$$\therefore K = 7.05$$

$$\text{I (originally)} = MK^2$$

$$I_o = 10,000 \times 49.69$$

$$I_o = 496,900 \text{ tonnes } m^2$$

$$\text{I of discharged mass about G} = 50 \times 14^2$$

$$= 9800 \text{ tonnes } m^2$$

$$\text{New I of ship about the original C of G} = \text{Original I} - \text{I of discharged mass}$$

$$= 496,900 - 9800$$

$$= 487,100 \text{ tonnes } m^2$$

By the theorem of parallel axes, let

$$I_2 = 487{,}100 - 9950 \times 0.07^2$$
$$I_2 = 487{,}100 - 49$$
$$I_2 = 487{,}051 \text{ tonnes m}^2$$
$$I_2 = M_2K_2^2$$
$$\therefore \text{New } K^2 = \frac{I_2}{M_2}$$

$$\therefore K_2^2 = \frac{487{,}051}{9950}$$

Let

$$K_2 = \text{New } K = \sqrt{\frac{487{,}051}{9950}}$$
$$K_2 = 7 \text{ m}$$

Let

$$T_2 = \text{New } T = \frac{2\pi K_2}{\sqrt{gGM_2}}$$
$$T_2 = \frac{2\pi 7}{\sqrt{9.81 \times 0.57}}$$

Ans. $T_2 = 18.6$ s.

Procedure Steps for Example 2

1. Calculate the new displacement in tonnes (W_2).
2. Estimate the original radius of gyration (K).
3. Evaluate the new displacement and new GM (W_2 and GM_2).
4. Calculate the new mass moment of inertia (I_2).
5. Calculate the new radius of gyration (K_2).
6. Finally, evaluate the new period of roll (T_2).

For 'stiff ships' the period of roll could be as low as 8 seconds due to a large GM. For 'tender ships' the period of roll will be, say, 25–35 seconds, due to a small GM. A good comfortable period of roll for those on board ship will be 15–25 seconds, averaging out at 20 seconds (see Figure 45.3).

Pitching Motions

Pitching is the movement of the ship's bow, from the lowest position to the highest position and back down to its lowest position. Pitching can also be assumed to be longitudinal rolling

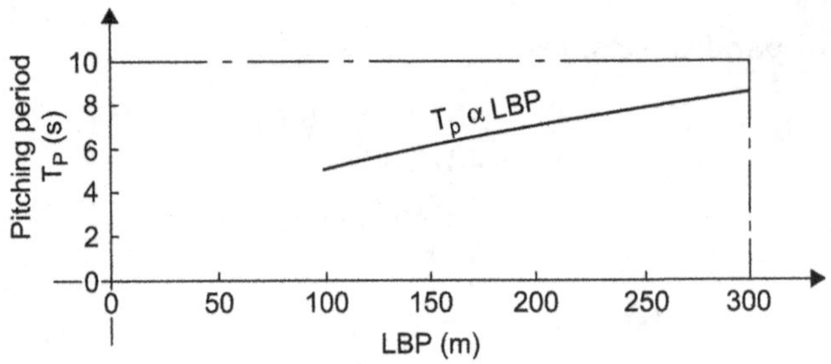

Figure 45.3

motion, where the ship is oscillating around a point at or very near to amidships. This point is known as the 'quiescent point' and is very close to the longitudinal center of gravity (LCG).

$$
\begin{aligned}
T_P &= 2 \times \pi \times (K^2/(g \times GM_L)^{0.5}) \text{ seconds} \\
&= 2(K^2/GM_L)^{0.5} \\
&= 2 \times K/(GM_L)^{0.5} \text{ sec}
\end{aligned}
$$

The longitudinal K value can be approximated with acceptable accuracy to be $K = 0.25 \times L$, where $L = \text{LBP}$ of the ship.

The longitudinal GM value can be approximated to $GM_L = 1.1 \times L$.

$$
\begin{aligned}
T_P &= 2 \times \pi \times (K^2/(g \times GM_L)^{0.5}) \\
T_P &= (2 \times 3.142 \times 0.25 \times L)/3.131 \times (1.1 \times L)^{0.5} \\
T_P &= 1/2 \times (L)^{0.5} \text{sec approx.}
\end{aligned}
$$

Figure 45.3 shows the resulting pitching periods, based on the above formula, with the variable of LBP up to 300 m incorporated.

Heaving Motions

This is the vertical upward or downward movement in the water of the ship's center of gravity 'G'.

Let the heaving period $= T_H$ seconds.

$$T_H = 2 \times \pi \times (W/(TPC \times 100 \times g)))^{0.5} \text{ seconds} \qquad (1)$$

where

$$W = \text{ship's displacement in tonnes} = L \times B \times d \times C_b \times p$$

$$TPC = \text{tonnes per centimeter immersion} = WPA \times p/100$$

$$WPA = \text{waterplane area in } m^2 = L \times B \times C_W$$

$$g = \text{gravity} = 9.81\ m/s^2$$

$$p = \text{density of salt water} = 1.025\ t/m^3$$

Substituting all of these values back into equation (1):

$$T_H = 2 \times 3.142 \times (L \times B \times d \times C_b \times 1.025 \times 100)/(L \times B \times C_w \times 1.025 \times 100 \times 9.81)^{0.5}$$

Hence

$$T_H = 2 \times (d \times C_b/C_W)^{0.5}\ s$$

Hence the heaving period depends on the draft 'd' and the ratio C_b/C_W. This ratio varies with each type of ship, as shown in Table 45.1.

Figure 45.4 shows the resulting heaving periods from Table 45.1, with the type of hull form and the variable of SLWL (up to 22.00 m) incorporated.

Table 45.1: Approximate heaving periods for several merchant ships.

Ship Type	Typically Fully Loaded C_W	Corresponding Loaded C_b	Approximate C_b/C_W	Approximate Heaving Period (seconds)
ULCC	0.850	0.900	0.944	$1.94 \times (d)^{0.5}$
VLCC	0.825	0.883	0.934	$1.93 \times (d)^{0.5}$
Bulk carrier	0.775	0.850	0.912	$1.91 \times (d)^{0.5}$
General cargo	0.700	0.800	0.875	$1.87 \times (d)^{0.5}$
Passenger liner	0.600	0.733	0.819	$1.81 \times (d)^{0.5}$
Container ship	0.575	0.717	0.802	$1.79 \times (d)^{0.5}$

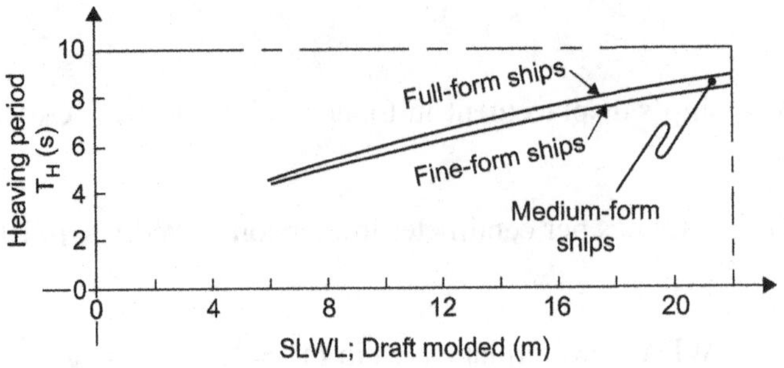

Figure 45.4

Worked Example

For a general cargo ship in a particular loaded condition, LBP = 140 m, B = 19.17 m, $C_b = 0.709$, $C_W = 0.806$, draft = 8.22 m, and $GM_T = 0.45$ m. Estimate the rolling period, the pitching period, and the heaving period in seconds.

From the notes

$$T_R = (0.7 \times B)/(GM_T)^{0.5} \text{ seconds approx.}$$

So

$$T_R = (0.7 \times 19.17)/(0.45)^{0.5}$$

$$\text{Rolling period } T_R = 20 \text{ seconds approx.}$$

From the notes

$$T_P = \frac{1}{2} \times (L)^{0.5} \text{ seconds approx.}$$

So

$$T_P = \frac{1}{2} \times (L)^{0.5}$$

$$T_P = \frac{1}{2} \times (140)^{0.5}$$

$$\text{Pitching period } T_P = 5.92 \text{ s}$$

From the notes

$$T_H = 2 \times (d \times C_b/C_W)^{0.5} \text{ seconds}$$

$$T_H = 2 \times (8.22 \times 0.709/0.806)^{0.5} \text{ seconds}$$

$$\text{Heaving period } T_H = 5.38 \text{ s}$$

Note how the rolling period is much greater in value than the pitching period and the heaving period. Figures 45.2–45.4 graphically confirm the mathematics of this statement.

In examinations, only use the approximate formulae in the absence of being given more detailed information. Calculations are to be based wherever possible on real stability information.

Exercise 45

1. Find the still water period of roll for a ship when the radius of gyration is 5 m and the initial metacentric height is 0.25 m.
2. A ship of 5000 tonnes displacement has GM = 0.5 m. The still water period of roll is 20 seconds. Find the new period of roll when a mass of 100 tonnes is discharged from a position 14 m above the center of gravity.
3. A ship of 9900 tonnes displacement has GM = 1 m and a still water rolling period of 15 seconds. Calculate the new rolling period when a mass of 100 tonnes is loaded at a position 10 m above the ship's center of gravity.
4. A vessel has the following particulars: Displacement is 9000 tonnes, natural rolling period is T_R of 15 seconds, GM is 1.20 m. Determine the new natural rolling period after the following changes in loading have taken place:

   2000 tonnes added at 4.0 m above ship's VCG
   500 tonnes discharged at 3.0 m below ship's VCG

   Assume that KM remains at the same value before and after changes of loading have been completed. Discuss if this final condition results in a 'stiff ship' or a 'tender ship'.
5. For a bulk carrier, in a particular loaded condition, the LBP is 217 m, B is 32.26 m, C_b is 0.795, C_W is 0.863, draft is 12.20 m, and the GM_T is 0.885 m. Estimate the rolling period, the pitching period, and the heaving period in seconds.

CHAPTER 46

Synchronous Rolling and Parametric Rolling of Ships

Synchronous Rolling of Ships

Synchronous rolling is caused by the ship's rolling period T_R becoming synchronous or resonant with the wave period. When this occurs, the ship will heel over and, in exceptional circumstances, be rolled further over by the action of the wave. Consequently, there is a serious danger that the vessel will heel beyond a point or angle of heel from which it cannot return to an upright condition. The ship ends up having negative stability, and will capsize. Figure 46.1 shows a ship with synchronous rolling problems.

To reduce synchronous rolling:

1. Use water ballast changes to alter the KG of the vessel. This should alter the GM_T and hence the natural rolling period T_R to a nonsynchronous value.
2. Change the course heading of the ship so that there will be a change in the approaching wave frequencies. In other words, introduce a yawing effect.
3. Alter the ship's speed until synchronism or resonance no longer exists with the wave frequency.

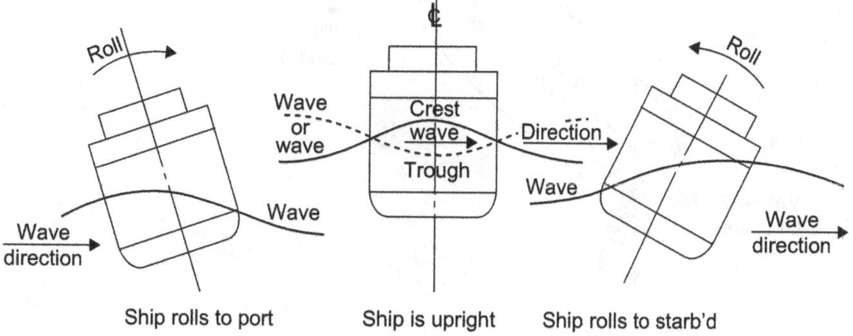

Figure 46.1:
Synchronous Rolling in Waves.

Ship Stability for Masters and Mates. http://dx.doi.org/10.1016/B978-0-08-097093-6.00047-5

Parametric Rolling of Ships

Parametric rolling is produced by pitching motions on vessels that have very fine bowlines together with very wide and full stern contours. One such ship type is the container ship. Figure 46.2 shows a ship with parametric rolling problems.

The cause depends very much on the parameters of the vessel, hence the name 'parametric rolling'. It is most marked when the pitching period T_P is either equal to or half that of the vessel's rolling period T_R.

As the stern dips into the waves it produces a rolling action. This remains unchecked as the bow next dips into the waves due to pitching forces. It is worst when $T_P = T_R$ or when $T_P = \frac{1}{2} \times T_R$.

In effect, the rolling characteristics are different at the stern to those at the bow. This causes a twisting or torsioning along the ship, leading to extra rolling motions.

If $T_P = T_R$ or $T_P = \frac{1}{2} \times T_R$, then interaction exists and the rolling of the ship is increased. A more dangerous situation develops because of the interplay between the pitching and rolling motions.

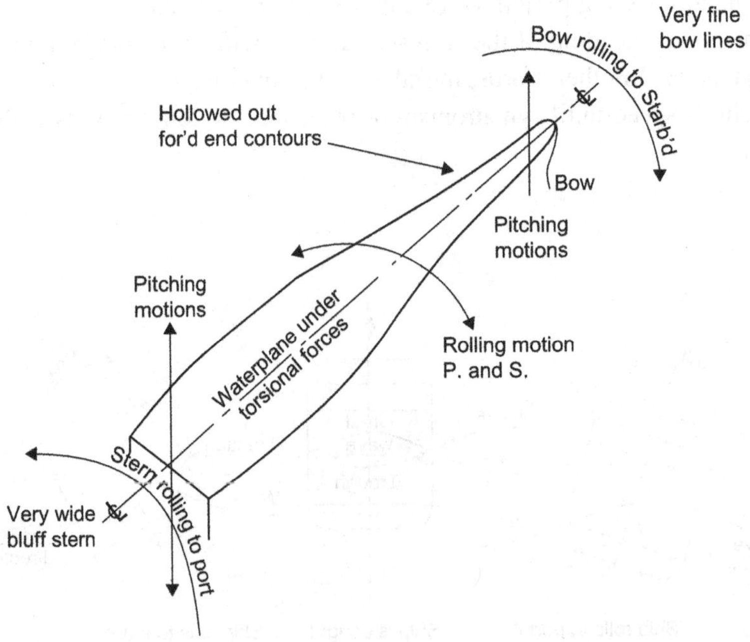

Figure 46.2:
Pitch Induced or Parametric Rolling on a Container Vessel.

Parametric rolling is worse when a ship is operating at reduced speed in heavy sea conditions. Such condition can cause containers to be lost overboard due to broken deck lashings. The IMO suggests that parametric rolling is particularly dangerous when the wavelength is 1.0−1.5 times the ship's length. Parametric rolling problems are least on box-shaped vessels or full-form barges where the aft and forward contours are not too dissimilar. Very little transverse and longitudinal interplay occurs.

To reduce parametric rolling:

1. A water ballast could be used to alter the GM_T, and hence the natural rolling period T_R, to a nonsynchronous value.
2. The ship needs to have an anti-rolling acting stabilizing system. Anti-rolling stability tanks that transfer water across the ship or vertically between two tanks are effective for all ship speeds. A quick response time is vital to counteract this type of rolling.
3. Hydraulic fin stabilizers would also help to reduce parametric rolling. They may be telescopic or hinged into the sides of the vessel at or near to amidships.
4. Alter the ship's forward speed.
5. Alter the ship's course.

Exercise 46

1. With the aid of a sketch, describe *synchronous rolling*. Suggest three methods a master or mate can consider for reducing the effects of *synchronous rolling*.
2. With the aid of sketches, describe *parametric rolling*. Suggest five methods a master or mate can consider for reducing the effects of *parametric rolling*.

CHAPTER 47

Effects of Change of Density on a Ship's Draft and Trim

When a ship passes from water of one density to water of another density the mean draft is changed and, if the ship is heavily trimmed, the change in the position of the center of buoyancy will cause the trim to change.

Let the ship in Figure 47.1 float in salt water at the waterline WL. B represents the position of the center of buoyancy and G the center of gravity. For equilibrium, B and G must lie in the same vertical line.

If the ship now passes into fresh water, the mean draft will increase. Let W_1L_1 represent the new waterline and b the center of gravity of the extra volume of the water displaced. The center of buoyancy of the ship, being the center of gravity of the displaced water, will move from B to B_1 in a direction directly towards b. The force of buoyancy now acts vertically upwards through B_1 and the ship's weight acts vertically downwards through G, giving a trimming moment equal to the product of the displacement and the longitudinal distance between the centers of gravity and buoyancy. The ship will then change trim to bring the centers of gravity and buoyancy back into the same vertical line.

Example

A box-shaped pontoon is 36 m long, 4 m wide, and floats in salt water at drafts F 2.00 m, A 4.00 m (see Figure 47.2). Find the new drafts if the pontoon now passes into fresh water. Assume salt water density is 1.025 t/m^3. Assume fresh water density = 1.000 t/m^3.

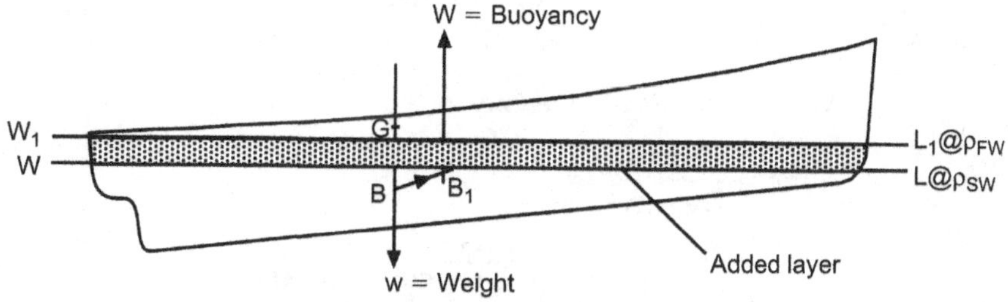

Figure 47.1

Ship Stability for Masters and Mates. http://dx.doi.org/10.1016/B978-0-08-097093-6.00048-7

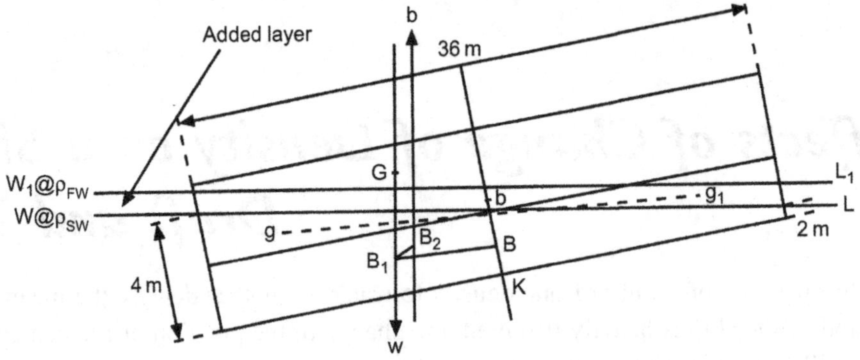

Figure 47.2

(a) *To find the position of B_1*

$$BB_1 = \frac{v \times gg_1}{V}$$

$$v = \frac{1}{2} \times 1 \times \frac{36}{2} \times 4 \qquad gg_1 = \frac{2}{3} \times 36 \qquad V = 36 \times 4 \times 3$$

$$= 36 \text{ cubic meters} \qquad gg_1 = 24 \text{ m} \qquad = 432 \text{ cubic meters}$$

$$\therefore BB_1 = \frac{36 \times 24}{432}$$

$$= 2 \text{ m}$$

Because the angle of trim is small, BB_1 is considered to be the horizontal component of the shift of the center of buoyancy.

Now let the pontoon enter fresh water, i.e. from ρ_{SW} into ρ_{FW}. The pontoon will develop mean bodily sinkage.

(b) *To find the new draft*

In salt water:

$$\text{Mass} = \text{Volume} \times \text{Density}$$

$$= 36 \times 4 \times 3 \times 1.025$$

$$= 442.8 \text{ tonnes}$$

In fresh water:

$$\text{Mass} = \text{Volume} \times \text{Density}$$

$$\therefore \text{Volume} = \frac{\text{Mass}}{\text{Density}}$$

$$= \frac{36 \times 4 \times 3 \times 1.025}{1.000} \text{ cubic meters}$$

(Mass in salt water = Mass in fresh water)

Let

$$\text{MBS} = \text{Mean bodily sinkage} \qquad \rho_{SW} = \text{Higher density}$$
$$\rho_{FW} = \text{Lower density}$$

$$\text{MBS} = \frac{W}{\text{TPC}_{SW}} \times \frac{(\rho_{SW} - \rho_{FW})}{\rho_{FW}}$$

$$\text{MBS} = \frac{\cancel{L}\,\cancel{B}\,d\,\cancel{\rho_{SW}}}{\frac{\cancel{L} \times \cancel{B}}{100} \times \cancel{\rho_{SW}}}\left\{\frac{\rho_{SW} - \rho_{FW}}{\rho_{FW}}\right\}$$

$$\therefore \text{MBS} = \frac{d(\rho_{SW} - \rho_{FW})}{\rho_{FW}} \times 100$$

$$\text{MBS} = \frac{3 \times 0.025}{1.000} \times 100 = \underline{0.075\text{ m}}$$

$$\therefore \text{MBS} = 0.075\text{ m}$$

$$\text{Original mean draft} = 3.000\text{ m}$$

$$\text{New mean draft} = 3.075\text{ m}, \text{ say draft } d_2$$

(c) *Find the change of trim*

Let B_1B_2 be the horizontal component of the shift of the center of buoyancy. Then

$$B_1B_2 = \frac{w \times d}{W} \qquad W = LBd_{SW} \times \rho_{SW}$$

$$= \frac{(442.8 - 432.0) \times 2}{442.8} \qquad = 36 \times 4 \times 3 \times 1.025$$

$$\therefore B_1B_2 = 0.0487\text{ m} \qquad \therefore W = 442.8\text{ tonnes}$$

$$\text{Trimming moment} = W \times B_1B_2$$

$$= 36 \times 4 \times 3 \times \frac{1.025}{1.000}\text{ t} \times 0.0487\text{ m} = 21.56\text{ t m}$$

$$BM_{L(2)} = \frac{L^2}{12d_{(2)}}$$

$$= \frac{36^2}{12 \times 3.075}$$

$$= \frac{36}{1.025}\text{ m} = 35.12\text{ m}$$

$$\text{MCTC} \simeq \frac{W \times BM_L}{100 \times L}$$

$$= \frac{442.8 \times 35.12}{100 \times 36}$$

$$= 4.32\text{ tonnes meters}$$

$$\text{Changes of trim} = \frac{\text{Trimming moment}}{\text{MCTC}}$$

$$= \frac{21.56}{4.32} = 5 \text{ cm}$$

$$\text{Change of trim} = 5 \text{ cm by the stern}$$

$$= 0.05 \text{ m by the stern}$$

Drafts before trimming	A	4.075 m	F	2.075 m
Change due to trim	A	+0.025 m	F	−0.025 m
New drafts	A	4.100 m	F	2.050 m

In practice the trimming effects are so small that they are often ignored by shipboard personnel. Note in the above example the trim ratio forward and aft was only 2.5 cm.

However, for DfT examinations, they must be studied and fully understood.

Exercise 47

1. A box-shaped vessel is 72 m long, 8 m wide, and floats in salt water at drafts F 4.00 m, A 8.00 m. Find the new drafts if the vessel now passes into fresh water.
2. A box-shaped vessel is 36 m long, 5 m wide, and floats in fresh water at drafts F 2.50 m, A 4.50 m. Find the new drafts if the vessel now passes into salt water.
3. A ship has a displacement of 9100 tonnes, LBP of 120 m, and even-keel draft of 7 m in fresh water of density of 1.000 t/m^3. From her hydrostatic curves it was found that:

   $MCTC_{SW}$ is 130 t m/cm
   TPC_{SW} is 17.3 t
   LCB is 2 m forward of amidships and
   LCF is 1.0 m aft of amidships

   Calculate the new end drafts when this vessel moves into water having a density of 1.02 t/m^3 without any change in the ship's displacement of 9100 tonnes.

CHAPTER 48

The Deadweight Scale

The deadweight scale provides a method for estimating the additional draft or for determining the extra load that could be taken onboard when a vessel is being loaded in water of density less than that of salt water. For example, the vessel may be loading in a port where the water density is that of fresh water at 1.000 t/cubic meter.

This deadweight scale (see Figure 48.1) displays columns of scale readings for:

- Freeboard (f)
- Dwt in salt water and in fresh water
- Draft of ship (mean)
- Displacement in tonnes in salt water and in fresh water
- Tonnes per cm (TPC) in salt water and in fresh water
- Moment to change trim 1 cm (MCTC).

On every dwt scale the following constants must exist:

$$\text{Any freeboard (f)} + \text{Any draft (d)} = \text{Depth of ship (D)}$$

hence $f + d = C1$.

$$\text{Any displacement (W)} - \text{Any Dwt} = \text{Lightweight(Lwt)}$$

hence $W - Dwt = C2$.

The main use of the Dwt scale is to observe Dwt against draft. Weight in tonnes remains the same but the volume of displacement will change with a change in density of the water in which the ship floats. The salt water and fresh water scales relate to these changes.

On many ships this Dwt scale has been replaced by the data being presented in tabular form. The officer onboard only needs to interpolate to obtain the information that is required. Also, the Dwt scale can be part of a computer package supplied to the ship. In this case the officer only needs to key in the variables and the printout supplies the required data.

The following worked example shows the use of the Dwt scale.

Ship Stability for Masters and Mates. http://dx.doi.org/10.1016/B978-0-08-097093-6.00049-9

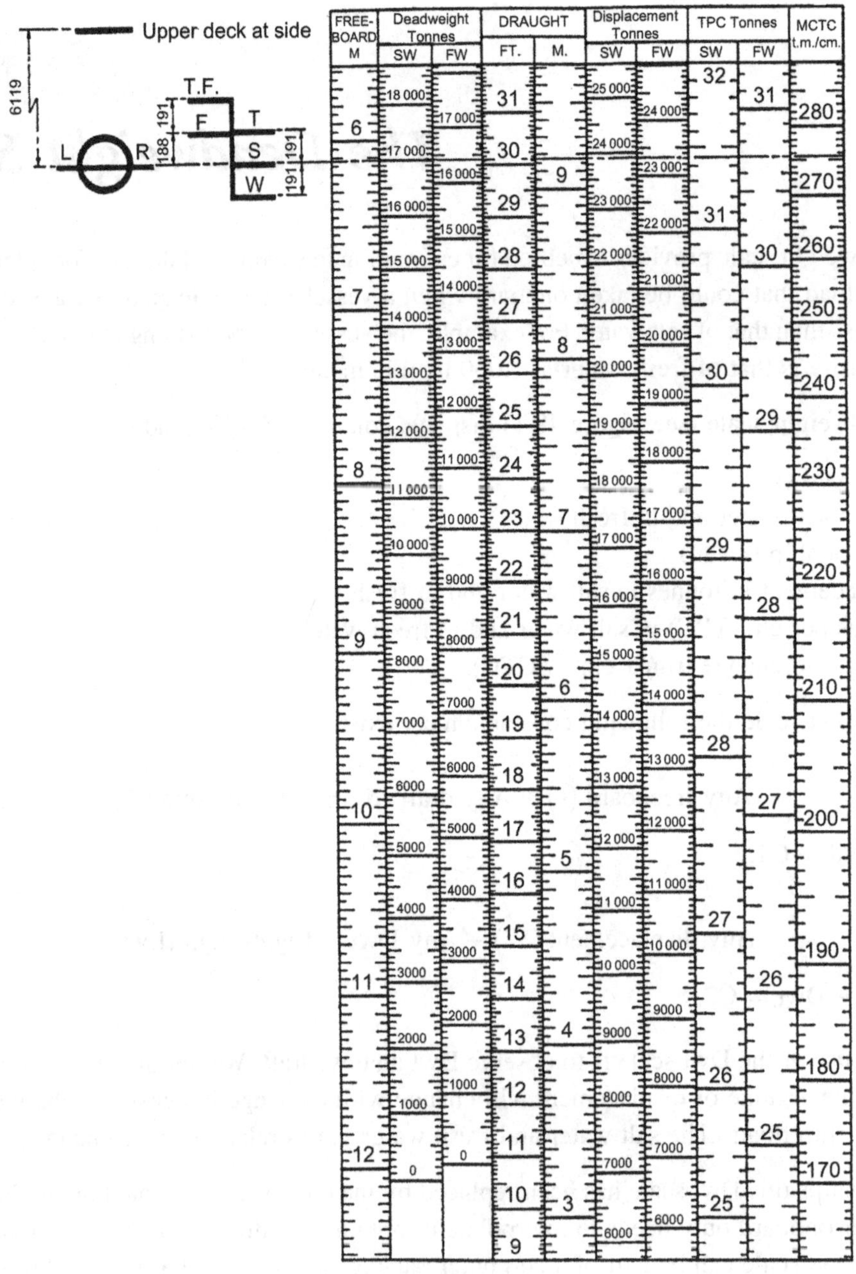

Figure 48.1:
Deadweight Scale.

Worked Example

Question:

Determine the TPC at the fully loaded draft from the Dwt scale shown in Figure 48.1 and show the final displacement in tonnes remains similar for fresh and salt water.

From Figure 49.1 TPC is 31.44 and the permitted fresh water sinkage as shown on the freeboard marks is 19 cm with displacement in salt water being almost 23,900 t. Consequently, the approximate load displacement in fresh water is given by:

$$\text{FW sinkage} = \text{W/TPC} \times 40 \text{ cm}$$

So

$$\text{W} = \text{TPC} \times \text{FW sinkage} \times 40 = 31.44 \times 19 \times 40 = 23,894 \text{ tonnes}$$

Hence this vessel has loaded up an extra 19 cm of draft in fresh water whilst keeping her displacement at 23,894 t (equivalent to salt water draft of 9.17 m).

■

CHAPTER 49

The Trim and Stability Book

When a new ship is nearing completion, a Trim and Stability book is produced by the shipbuilder and presented to the shipowner. Masters and mates will use this for the day-to-day operation of the vessel. The Trim and Stability book contains the following technical data:

1. General particulars of the ship and General Arrangement Plan.
2. Inclining experiment report and its results.
3. Capacity, VCG, and LCG particulars for all holds, compartments, tanks, etc.
4. Cross curves of stability. These may be GZ curves or KN curves.
5. Deadweight scale data. May be in diagrammatic form or in tabular form.
6. Hydrostatic curves. May be in graphical form or in tabular form.
7. Example conditions of loading such as:
   - Lightweight (empty vessel) condition.
   - Full-loaded departure and arrival conditions.
   - Heavy-ballast departure and arrival conditions.
   - Medium-ballast departure and arrival conditions.
   - Light-ballast departure and arrival conditions.

For the arrival conditions, a ship should arrive at the end of the voyage (with cargo and/or passengers as per loaded departure conditions) with at least '10% stores and fuel remaining'.

A mass of 75 kg should be assured for each passenger, but may be reduced to not less than 60 kg where this can be justified.

On each condition of loading there is a profile and plan view (at upper deck level usually). A color scheme is adopted for each item of deadweight. Examples could be red for cargo, blue for fresh water, green for water ballast, brown for oil. Hatched lines for this Dwt distribution signify wing tanks port and starboard.

For each loaded condition, in the interests of safety, it is necessary to show:

- Deadweight
- End draughts, thereby signifying a satisfactory and safe trim situation
- KG with no free-surface effects (FSE) and KG with FSE taken into account
- Final transverse metacentric height (GM). This informs the officer if the ship is in stable, unstable, or neutral equilibrium. It can also indicate if the ship's stability is approaching a dangerous state
- Total free-surface effects of all slack tanks in this condition of loading

Ship Stability for Masters and Mates. http://dx.doi.org/10.1016/B978-0-08-097093-6.00050-5

A statical stability curve relevant to the actual loaded condition should be drawn with the important characteristics clearly indicated. For each S/S curve it is important to observe the following:

- Maximum GZ and the angle of heel at which it occurs.
- Range of stability.
- Area enclosed from 0° to 30° (A1) and the area enclosed from 30° to 40° (A2), as shown in Figure 49.1.
- Shear force and bending moment curves, with upper limit lines clearly superimposed, as shown in Figure 49.2.

8. Metric equivalents. For example, TPI″ to TPC, MCTI″ to MCTC, or tons to tonnes.

All of this information is sent to a DfT surveyor for appraisal. On many ships today this data is part of a computer package. The deadweight items are keyed in by the officer prior to any weights actually being moved. The computer screen will then indicate if the prescribed condition of loading is indeed safe from the point of view of stability and strength.

Summary

The results and graphs can also be obtained using computer packages supplied to the ships. Each time a ship is loaded or discharged, the computer will give a printout of all the stability information required by the master or mate.

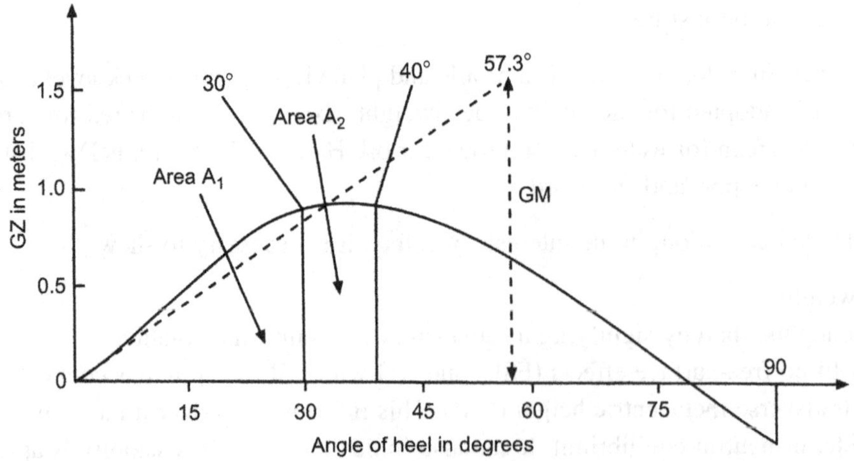

Figure 49.1:
Enclosed Areas on a Statical Stability Curve.

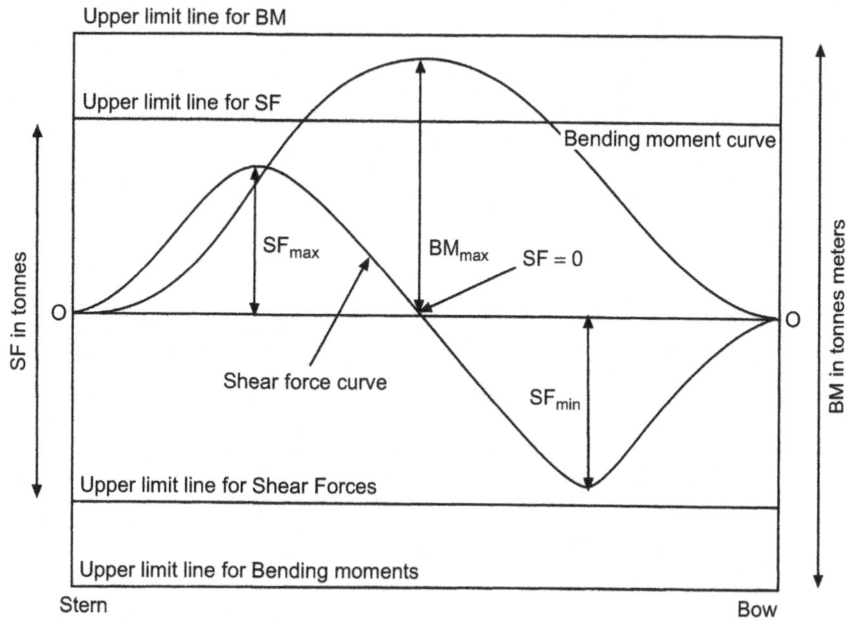

Figure 49.2: SF and BM Curves with Upper Limit Lines.
BM = bending moments; SF = shear forces.

The computer can be programmed to illustrate the information shown in Figures 49.1 and 49.2. Furthermore, if the limiting lines are exceeded for SF and/or BM, a flashing light or audio alarm will signal to warn the officer. Corrective loading arrangements can then be undertaken.

CHAPTER 50

Simplified Stability Information

SIMPLIFIED STABILITY INFORMATION

Notice to Shipowners, Masters, and Shipbuilders: Department for Transport Merchant Shipping Notice No. 1122

1. It has become evident that the master's task of ensuring that his ship complies with the minimum statutory standards of stability is in many instances not being adequately carried out. A feature of this is that undue traditional reliance is being placed on the value of GM alone, while other important criteria that govern the righting lever GZ curve are not being assessed as they should be. For this reason the Department, appreciating that the process of deriving and evaluating GZ curves is often difficult and time-consuming, strongly recommends that in future simplified stability information be incorporated into ships' stability booklets. In this way masters can more readily assure themselves that safe standards of stability are met.
2. Following the loss of the *Lairdsfield*, referred to in Notice M.627, the Court of Inquiry recommended that simplified stability information be provided. This simplified presentation of stability information has been adopted in a large number of small ships and is considered suitable for wider application in order to overcome the difficulties referred to in Paragraph 1.
3. Simplified stability information eliminates the need to use cross curves of stability and develop righting lever GZ curves for varying loading conditions by enabling a ship's stability to be quickly assessed, to show whether or not all statutory criteria are complied with, by means of a single diagram or table. Considerable experience has now been gained and three methods of presentation are in common use. These are:
   (a) The Maximum Deadweight Moment Diagram or Table.
   (b) The Minimum Permissible GM Diagram or Table.
   (c) The Maximum Permissible KG Diagram or Table.

   In all three methods the limiting values are related to salt water displacement or draft. Free surface allowances for slack tanks are, however, applied slightly differently.
4. Consultation with the industry has revealed a general preference for the Maximum Permissible KG approach, and graphical presentation also appears to be preferred rather than a tabular format. The Department's view is that any of the methods may be adopted subject to:
   (a) clear guidance notes for their use being provided and
   (b) submission for approval being made in association with all other basic data and sample loading conditions.

   In company fleets it is, however, recommended that a single method be utilized throughout.
5. It is further recommended that the use of a *Simplified Stability Diagram* as an adjunct to the *Deadweight Scale* be adopted to provide a direct means of comparing stability relative to

Ship Stability for Masters and Mates. http://dx.doi.org/10.1016/B978-0-08-097093-6.00051-7

other loading characteristics. Standard work forms for calculating loading conditions should also be provided.

6. It is essential for masters to be aware that the standards of stability obtainable in a vessel are wholly dependent on exposed openings such as hatches, doorways, air pipes and ventilators being securely closed weathertight, or in the case of automatic closing appliances such as airpipe ball valves that these are properly maintained in order to function as designed.
7. Shipowners bear the responsibility to ensure that adequate, accurate, and up-to-date stability information for the master's use is provided. It follows that it should be in a form that should enable it to be readily used in the trade in which the vessel is engaged.

Maximum Permissible Deadweight Moment Diagram

This is one form of simplified stability data diagram in which a curve of maximum permissible deadweight moments is plotted against displacement in tonnes on the vertical axis and deadweight moment in tonnes meters on the horizontal axis, the deadweight moment being the moment of the deadweight about the keel.

The total deadweight moment at any displacement must not, under any circumstances, exceed the maximum permissible deadweight moment at that displacement.

Figure 50.1 illustrates this type of diagram. The ship's displacement in tonnes is plotted on the vertical axis from 1500 to 4000 tonnes while the deadweight moments in tonnes meters are plotted on the horizontal axis. From this diagram it can be seen that, for example, the maximum deadweight moment for this ship at a displacement of 3000 tonnes is 10,260 tonnes meters (Point 1). If the light displacement for this ship is 1000 tonnes then the deadweight at this displacement is 2000 tonnes. The maximum KG for the deadweight tonnage is given by:

$$\begin{aligned}\text{Maximum KG} &= \frac{\text{Deadweight moment}}{\text{Deadweight}}\\ &= \frac{10,260}{2000}\\ &= 5.13\text{ m}\end{aligned}$$

Example 1

Using the simplified stability data shown in Figure 50.1, estimate the amount of cargo (KG = 3 m) that can be loaded so that after completion of loading the ship does not

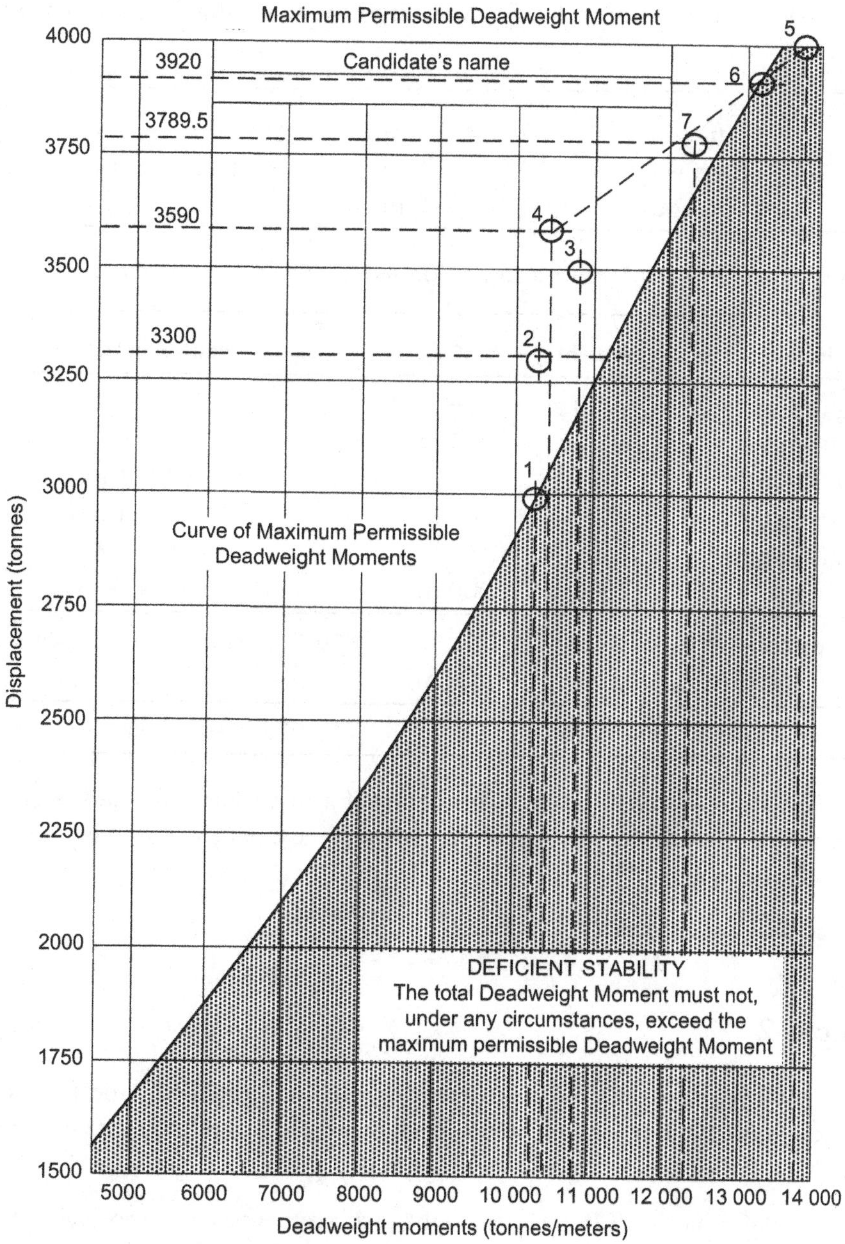
Maximum Permissible Deadweight Moment
Candidate's name
3920
3789.5
3590
3300
1
2
3
4
5
6
7
4000
3750
3500
3250
3000
2750
2500
2250
2000
1750
1500
Displacement (tonnes)
Curve of Maximum Permissible Deadweight Moments
DEFICIENT STABILITY
The total Deadweight Moment must not, under any circumstances, exceed the maximum permissible Deadweight Moment
5000
6000
7000
8000
9000
10 000
11 000
12 000
13 000
14 000
Deadweight moments (tonnes/meters)

Figure 50.1

have deficient stability. Prior to loading the cargo the following weights were already on board:

250 t	fuel oil	KG = 0.5 m	Free surface moment 1400 tm
50 t	fresh water	KG = 5.0 m	Free surface moment 500 tm
2000 t	cargo	KG = 4.0 m	

The light displacement is 1000 t and the loaded Summer displacement is 3500 t.

Item	Weight	KG	Deadweight Moment
Light disp.	1000 t	—	—
Fuel oil	250 t	0.5 m	125 t m
Free surface	—	—	400 t m
Fresh water	50 t	5.0 m	250 t m
Fresh surface	—	—	500 t m
Cargo	2000 t	4.0 m	8000 t m
Present cond.	3300 t		10,275 t m – Point 2 (satisfactory)
Maximum balance	200 t	3.0 m	600 t m
Summer displ.	3500		10,875 t m – Point 3 (satisfactory)

Since 10,875 tonnes meters is less than the maximum permissible deadweight moment at a displacement of 3500 tonnes, the ship will not have deficient stability and may load 200 tonnes of cargo.

Ans. Go ahead and load the 200 tonnes.

■

■ Example 2

Using the maximum permissible deadweight moment (Figure 50.1) and the information given below, find the quantity of timber deck cargo (KG = 8.0 m) that can be loaded, allowing 15% for water absorption during the voyage.

Summer displacement is 4000 tonnes. Light displacement is 1000 tonnes. Weights already on board:

Fuel oil 200 tonnes	KG = 0.5 m, Free surface moment 1400 t m
Fresh water 40 tonnes	KG = 5.0 m, Free surface moment 600 t m
Cargo 2000 tonnes	KG = 4.0 m
Ballast 350 tonnes	KG = 0.5 m

The following weights will be consumed during the voyage:

Fuel oil 150 tonnes	KG = 0.5 m, Free surface moment will be reduced by 800 t m
Fresh water 30 tonnes	KG = 5.0 m, Free surface moment will be by 200 t m

Departure Condition

Item	Weight	KG	Deadweight Moment
Light ship	1000	–	–
Fuel oil	200	0.5	100
Free surface			1400
Fresh water	40	5.0	200
Free surface			600
Cargo	2000	4.0	8000
Ballast	350	0.5	175
Departure disp. (without deck cargo)	3590		10,475 – Point 4 (satisfactory)
Maximum deck cargo	410	8.0	3280
Summer disp.	4000		13,755 – Point 5 (deficient stability)

From Figure 50.1, where the line joining Points 4 and 5 cuts the curve of maximum permissible deadweight moments (Point 6), the displacement is 3920 tonnes.

Total departure displacement	3920 tonnes
Departure displacement without deck cargo	3590 tonnes
∴ Max deck cargo to load	330 tonnes

$$\text{Absorption during voyage} = \frac{15}{100} \times 330 = 49.5 \text{ tonnes}$$

Arrival Condition

Item	Weight	KG	Deadweight Moment
Departure disp. (without deck cargo)	3590		10,475
Fuel oil	−150	0.5	−75
Free surface			−800

(Continued)

Item	Weight	KG	Deadweight Moment
Fresh water	−30	5.0	−150
Free surface			−200
Arrival disp. (without deck cargo)	3410		9250
Deck cargo	330	8.0	2640
Absorption	49.5	8.0	396
Total arrival disp.	3789.5		12,286 − Point 7 (satisfactory stability)

Ans. Go ahead and load the 330 tonnes of deck cargo.

Exercise 50

1. Using the maximum permissible deadweight moment (Figure 50.1), find the amount of deck cargo (KG = 8.0 m) that can be loaded allowing 15% for water absorption during the voyage given the following data:

   Light displacement 1000 tonnes. Loaded displacement 4000 tonnes. Weights already on board:

Item	Weight	KG	Free Surface Moment
Cargo	1800	4.0	–
Fuel oil	350	0.5	1200
Fresh water	50	5.0	600
Ballast	250	0.5	–

During the voyage the following will be consumed (tonnes and KG):

Fuel oil	250	0.5	Reduction in free surface moment 850 t m
Fresh water	40	5.0	Reduction in free surface moment 400 t m

CHAPTER 51

The Stability Pro-Forma

There is now a trend to supply a blank pro-forma for the master or mate to fill in that covers the loading condition for a ship. As the trim and stability calculations are made this information is transferred or filled in on the pro-forma. The following example shows the steps in this procedure.

Worked Example

Use the supplied blank 'Trim and Stability' pro-forma and the supplied 'Hydrostatic Particulars Table' to fill in the pro-forma and determine each of the following:

(a) The effective metacentric height GM_T
(b) The final drafts, forward and aft.

The first steps are to estimate the total displacement, the VCG, and the LCG for this displacement. They came to 14,184 t, 7.07 m, and 66.80 m forward of aft perpendicular (foap).

The true mean draft is then required. This is calculated by direct interpolation of draft for displacements between 14,115 t and 14,345 t in the 'Hydrostatic Particulars Table'.

Based on the ship's displacement of 14,184 t, this direct interpolation must be between 6.80 and 6.90 m, a difference of 0.10 m.

$$\text{True mean draft} = 6.80 + \{(14,184 - 14,115)/(14,354 - 14,115)\} \times 0.10$$

$$= 6.80 + \{0.3 \times 0.10\} = 6.83 \text{ meters}$$

$$MCTC_{sw} = 181.4 + \{0.3 \times 1.6\} = 181.88, \text{ say } 181.9 \text{ t m/cm}$$

$$KM_T = 8.36 - \{0.3 \times 0.01\} = 8.36 \text{ m}$$

$$LCB = 70.12 - \{0.3 \times 0.04\} = 70.11 \text{ m foap}$$

$$LCF = 67.57 - \{0.3 \times 0.11\} = 67.54 \text{ m foap}$$

Ship Stability for Masters and Mates. http://dx.doi.org/10.1016/B978-0-08-097093-6.00052-9

TS002-72 SHIP STABILITY

WORKSHEET Q.1

(This Worksheet must be returned with your examination answer book)

TRIM & STABILITY

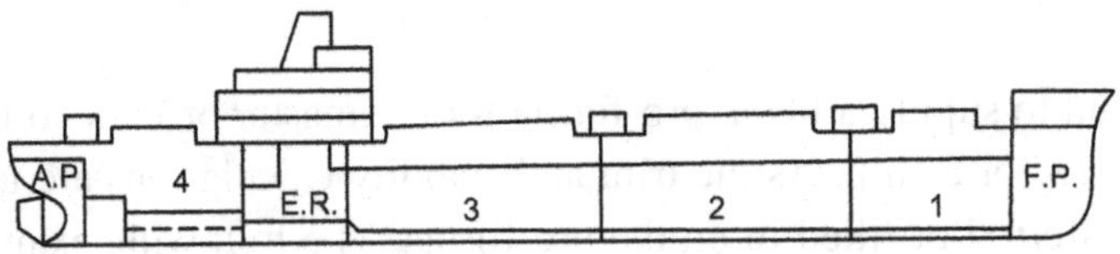

CONDITION: FULLY LOADED – GENERAL CARGO

Compartment	Capacity	Stowage Factor	Weight	KG	Vertical Moment	Free Surface Moment	LCG foap	Longitudinal Moment
	m^3	m^3/t	t	m	tm	tm	m	tm
All Holds	13,507	2.01		6.76			74.90	
1 TD	936	2.38		10.89			114.60	
2 TD	1297	2.65		10.71			92.80	
3 TD	1579	2.13		10.51			63.80	
Consumables			1812	–	4784	2324	–	58,736
Deadweight								
Lightship			4029	8.13			61.50	
DISPLACEMENT								

HYDROSTATICS	True Mean Draught	LCB foap	LCF foap
LENGTH B.P. 133.0m	MCTC		
TRIM			KM_T
			KG
DRAUGHTS: F.	A.		GM

Candidate's Name ....

Examination Center ....

TS002-72 SHIP STABILITY

DATASHEET Q.1

(This Worksheet must be returned with your answer book)

HYDROSTATIC PARTICULARS 'A'

DRAUGHT	DISPLACEMENT t		TPC t		MCTC tm		KM_T	KB	LCB foap	LCF foap
	SW RD 1.025	FW RD 1.000	SW RD 1.025	FW RD 1.000	SW RD 1.025	FW RD 1.000				
m							m	m	m	m
7.00	14576	14220	23.13	22.57	184.6	180.1	8.34	3.64	70.03	67.35
6.90	14345	13996	23.06	22.50	183.0	178.5	8.35	3.58	70.08	67.46
6.80	14115	13771	22.99	22.43	181.4	177.0	8.36	3.53	70.12	67.57
6.70	13886	13548	22.92	22.36	179.9	175.5	8.37	3.48	70.16	67.68
6.60	13657	13324	22.85	22.29	178.3	174.0	8.38	3.43	70.20	67.79
6.50	13429	13102	22.78	22.23	176.8	172.5	8.39	3.38	70.24	67.90
6.40	13201	12879	22.72	22.17	175.3	171.0	8.41	3.33	70.28	68.00
6.30	12975	12658	22.66	22.11	173.9	169.6	8.43	3.28	70.32	68.10
6.20	12748	12437	22.60	22.05	172.5	168.3	8.46	3.22	70.35	68.20
6.10	12523	12217	22.54	21.99	171.1	167.0	8.49	3.17	70.38	68.30
6.00	12297	11997	22.48	21.93	169.8	165.7	8.52	3.11	70.42	68.39
5.90	12073	11778	22.43	21.87	168.5	164.4	8.55	3.06	70.46	68.43
5.80	11848	11559	22.37	21.82	167.3	163.2	8.59	3.01	70.50	68.57
5.70	11625	11342	22.32	21.77	166.1	162.1	8.63	2.95	70.53	68.65
5.60	11402	11124	22.26	21.72	165.0	161.0	8.67	2.90	70.57	68.73
5.50	11180	10908	22.21	21.66	163.9	160.0	8.71	2.85	70.60	68.80
5.40	10958	10691	22.15	21.61	162.9	158.9	8.76	2.80	70.64	68.88
5.30	10737	10476	22.10	21.56	161.8	157.9	8.81	2.74	70.68	68.95
5.20	10516	10260	22.05	21.51	160.8	156.9	8.86	2.69	70.72	69.02
5.10	10296	10045	22.00	21.46	159.8	155.9	8.92	2.63	70.75	69.09
5.00	10076	9830	21.95	21.41	158.8	154.9	8.98	2.58	70.79	69.16
4.90	9857	9616	21.90	21.36	157.9	154.0	9.06	2.53	70.82	69.23
4.80	9638	9403	21.85	21.32	156.9	153.1	9.13	2.48	70.86	69.29
4.70	9420	9190	21.80	21.27	156.0	152.2	9.22	2.43	70.90	69.35
4.60	9202	8978	21.75	21.22	155.1	151.3	9.30	2.38	70.93	69.42
4.50	8985	8766	21.70	21.17	154.2	150.5	9.40	2.32	70.96	69.48
4.40	8768	8554	21.65	21.12	153.3	149.6	9.49	2.27	71.00	69.55
4.30	8552	8344	21.60	21.07	152.4	148.7	9.60	2.22	71.04	69.62
4.20	8336	8133	21.55	21.02	151.5	147.8	9.71	2.17	71.08	69.68
4.10	8121	7923	21.50	20.97	150.6	146.9	9.83	2.12	71.12	69.74
4.00	7906	7713	21.45	20.93	149.7	146.0	9.96	2.07	71.15	69.81
3.90	7692	7505	21.40	20.88	148.7	145.1	10.11	2.01	71.18	69.88
3.80	7478	7296	21.35	20.83	147.8	144.2	10.25	1.96	71.22	69.94
3.70	7265	7088	21.30	20.78	146.8	143.3	10.41	1.91	71.25	70.00
3.60	7052	6880	21.24	20.72	145.9	142.3	10.57	1.86	71.29	70.07
3.50	6840	6673	21.19	20.67	144.9	141.3	10.76	1.81	71.33	70.14
THESE HYDROSTATIC PARTICULARS HAVE BEEN DEVELOPED WITH THE VESSEL FLOATING ON EVEN KEEL										

Candidate's Name ....

Examination Center ....

TS002-72 SHIP STABILITY

WORKSHEET Q.1

(This Worksheet must be returned with your examination answer book)

TRIM & STABILITY

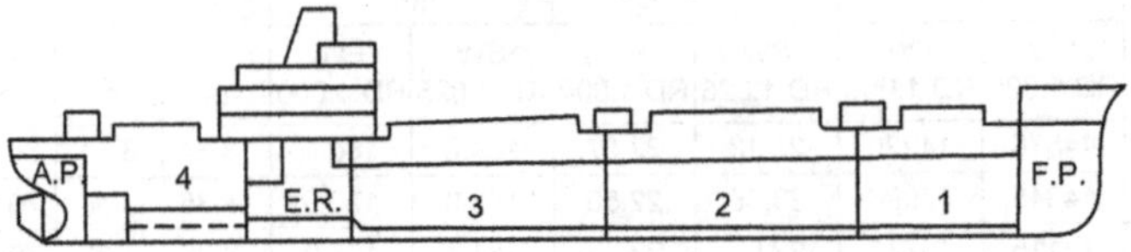

CONDITION: FULLY LOADED – GENERAL CARGO

Compartment	Capacity m^3	Stowage Factor m^3/t	Weight t	KG m	Vertical Moment tm	Free Surface Moment tm	LCG foap m	Longitudinal Moment tm
All Holds	13,507	2.01	**6720**	6.76	**45,427**		74.90	503,328
1 TD	936	2.38	**393**	10.89	**4280**		114.60	45,038
2 TD	1297	2.65	**489**	10.71	**5237**		92.80	45,379
3 TD	1579	2.13	**741**	10.51	**7788**		63.80	47,276
Consumables			1812	–	4784	2324	–	58,736
Deadweight			**10,155**					
Lightship			4029	8.13	**32,756**		61.50	247,784
DISPLACEMENT			**14,184**	**7.07**	**100,272**	**2324**	**66.80**	**947,541**

HYDROSTATICS — True Mean Draught **6.83 m** — LCB foap **70.11 m** — LCF foap **67.54 m**

LENGTH B.P. 133.0 m — MCTC = **181.9 tm**

TRIM = **2.58 m by the stern** — KM_T = **8.36 m**

KG_{FL} = **7.23 m**

DRAUGHTS: F. = 5.56 m — A. = 8.14 m — GM_T = **1.13 m** (fluid)

Also see Figure 52.1 and Figure 52.2

Candidate's Name ....

Examination Center ....

These values can now be inserted into the pro-forma sheet.

(a) *To determine the effective metacentric height GM_T*

$$KG_{solid} = \text{Sum of the vertical moments}/W$$
$$= 100,272/14,184$$
$$KG_{solid} = 7.07 \text{ m}$$

plus

$$FSM/W = 2324/14,184 = \underline{0.16} \text{ m}$$
$$KG_{fluid} = \underline{7.23} \text{ m}$$
$$\text{Effective metacentric height } GM_T = KM_T - KG_{fluid}$$
$$= 8.36 - 7.23$$
$$\text{Effective metacentric height } GM_T = 1.13 \text{ m} = GM_{fluid}$$

These values can now be inserted into the pro-forma sheet.

(b) *To determine the final end drafts*

$$\text{Change of trim} = \{W \times (LCB - LCG)\}/MCTC_{sw}$$

Note that

$$LCG = \text{Sum of the longitudinal moments}/W = 947,541/14,184$$

So

$$LCG = 66.80 \text{ m foap}$$
$$\text{Change of trim} = \{14,184\,(70.11 - 66.80)\}/181.9$$
$$= 258 \text{ cm} = 2.58 \text{ m by the stern}$$
$$\text{Final draft aft} = \text{True mean draft} + (LCF/LBP) \times \text{Change of trim}$$

LBP of 133 m for this ship is given on the blank pro-forma.

$$\text{Final draft aft} = 6.83 + (67.54/133) \times 2.58$$
$$\text{Final draft aft} = 8.14 \text{ m}$$

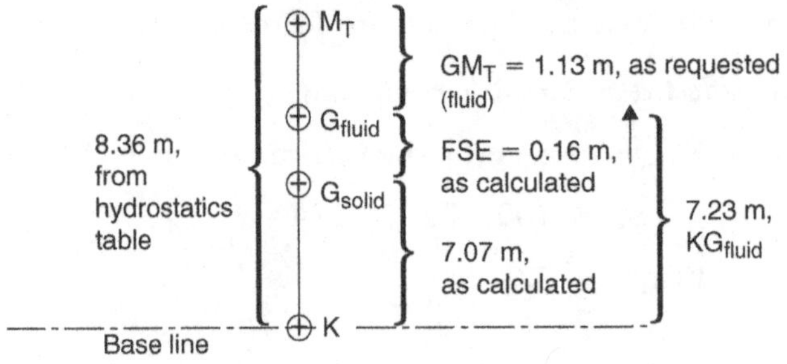

Figure 51.1:
Final Transverse Stability Details.

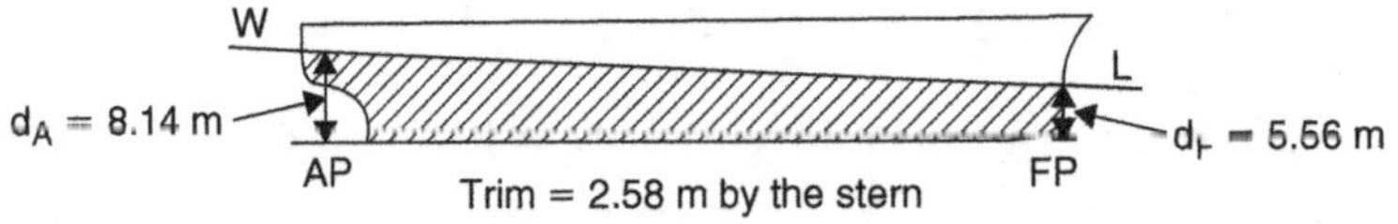

Figure 51.2:
Final Trim Details.

$$\text{Final draft forward} = \text{True mean draft} - (\text{LBP} - \text{LCF}/\text{LBP}) \times \text{Change of trim}$$
$$= 6.83 - (65.46/133) \times 2.58$$
$$\text{Final draft forward} = 5.56 \text{ m}$$

Check:

- Aft draft − forward draft = 8.14 − 5.56 = 2.58 m change of trim (as calculated earlier).
- These values can be inserted into the pro-forma sheet.
- Figure 51.1 shows the positions of all the transverse stability values.
- Figure 51.2 shows the ship's final end drafts and final trim by the stern.

Exercise 51

1. Use the supplied blank 'Trim and Stability' pro-forma and the supplied 'Hydrostatic Particulars Table' to fill in the pro-forma and determine each of the following:
   (a) Fully loaded deadweight in tonnes.
   (b) Fully loaded displacement in tonnes.
   (c) Final aft draft in meters.
   (d) Final forward draft in meters.
   (e) Final effective metacentric height GM_T in meters.

TS032-72 SHIP STABILITY

WORKSHEET Q.1

(This Worksheet must be returned with your answer book)

TRIM & STABILITY

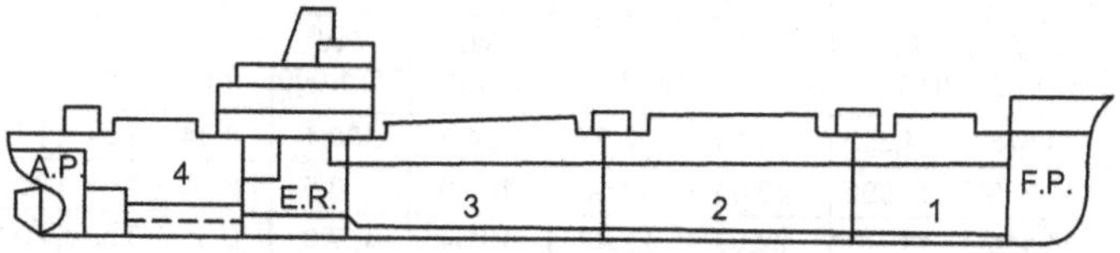

CONDITION: FULLY LOADED – GENERAL CARGO

Compartment	Capacity m^3	Stowage Factor m^3/t	Weight t	KG m	Vertical Moment tm	Free Surface Moment tm	LCG foap m	Longitudinal Moment tm
All Holds	14,562	1.86		6.78			73.15	
1 TD	264	2.48		10.71			114.33	
2 TD	1688	2.74		10.60			93.57	
3 TD	1986	2.72		10.51			63.92	
Consumables			1464	–	4112	2560	–	58,675
Deadweight								
Lightship			3831	8.21			61.67	
DISPLACEMENT								

HYDROSTATICS	True Mean Draught	LCB foap	LCF foap
LENGTH B.P. 130.00 m	MCTC		
TRIM			KM_T
			KG
DRAUGHTS: F.	A.		GM_{fluid}

Candidate's Name ....

Examination Center ....

TS032-72 SHIP STABILITY

DATASHEET Q.1

(This Worksheet must be returned with your answer book)

HYDROSTATIC PARTICULARS

DRAUGHT	DISPLACEMENT t		TPC t		MCTC tm		KM_T	KB	LCB foap	LCF foap
m	SW RD 1.025	FW RD 1.000	SW RD 1.025	FW RD 1.000	SW RD 1.025	FW RD 1.000	m	m	m	m
7.00	14,576	14,220	23.13	22.57	184.6	180.1	8.34	3.64	70.03	67.35
6.90	14,345	13,996	23.06	22.50	183.0	178.5	8.35	3.58	70.08	67.46
6.80	14,115	13,771	22.99	22.43	181.4	177.0	8.36	3.53	70.12	67.57
6.70	13,886	13,548	22.92	22.36	179.9	175.5	8.37	3.48	70.16	67.68
6.60	13,657	13,324	22.85	22.29	178.3	174.0	8.38	3.43	70.20	67.79
6.50	13,429	13,102	22.78	22.23	176.8	172.5	8.39	3.38	70.24	67.90
6.40	13,201	12,879	22.72	22.17	175.3	171.0	8.41	3.33	70.28	68.00
6.30	12,975	12,658	22.66	22.11	173.9	169.6	8.43	3.28	70.32	68.10
6.20	12,748	12,437	22.60	22.05	172.5	168.3	8.46	3.22	70.35	68.20
6.10	12,523	12,217	22.54	21.99	171.1	167.0	8.49	3.17	70.38	68.30
6.00	12,297	11,997	22.48	21.93	169.8	165.7	8.52	3.11	70.42	68.39
5.90	12,073	11,778	22.43	21.87	168.5	164.4	8.55	3.06	70.46	68.43
5.80	11,848	11,559	22.37	21.82	167.3	163.2	8.59	3.01	70.50	68.57
5.70	11,625	11,342	22.32	21.77	166.1	162.1	8.63	2.95	70.53	68.65
5.60	11,402	11,124	22.26	21.72	165.0	161.0	8.67	2.90	70.57	68.73
5.50	11,180	10,908	22.21	21.66	163.9	160.0	8.71	2.85	70.60	68.80
5.40	10,958	10,691	22.15	21.61	162.9	158.9	8.76	2.80	70.64	68.88
5.30	10,737	10,476	22.10	21.56	161.8	157.9	8.81	2.74	70.68	68.95
5.20	10,516	10,260	22.05	21.51	160.8	156.9	8.86	2.69	70.72	69.02
5.10	10,296	10,045	22.00	21.46	159.8	155.9	8.92	2.63	70.75	69.09
5.00	10,076	9,830	21.95	21.41	158.8	154.9	8.98	2.58	70.79	69.16
4.90	9,857	9,616	21.90	21.36	157.9	154.0	9.06	2.53	70.82	69.23
4.80	9,638	9,403	21.85	21.32	156.9	153.1	9.13	2.48	70.86	69.29
4.70	9,420	9,190	21.80	21.27	156.0	152.2	9.22	2.43	70.90	69.35
4.60	9,202	8,978	21.75	21.22	155.1	151.3	9.30	2.38	70.93	69.42
4.50	8,985	8,766	21.70	21.17	154.2	150.5	9.40	2.32	70.96	69.48
4.40	8,768	8,554	21.65	21.12	153.3	149.6	9.49	2.27	71.00	69.55
4.30	8,552	8,344	21.60	21.07	152.4	148.7	9.60	2.22	71.04	69.62
4.20	8,336	8,133	21.55	21.02	151.5	147.8	9.71	2.17	71.08	69.68
4.10	8,121	7,923	21.50	20.97	150.6	146.9	9.83	2.12	71.12	69.74
4.00	7,906	7,713	21.45	20.93	149.7	146.0	9.96	2.07	71.15	69.81
3.90	7,692	7,505	21.40	20.88	148.7	145.1	10.11	2.01	71.18	69.88
3.80	7,478	7,296	21.35	20.83	147.8	144.2	10.25	1.96	71.22	69.94
3.70	7,265	7,088	21.30	20.78	146.8	143.3	10.41	1.91	71.25	70.00
3.60	7,052	6,880	21.24	20.72	145.9	142.3	10.57	1.86	71.29	70.07
3.50	6,840	6,673	21.19	20.67	144.9	141.3	10.76	1.81	71.33	70.14

THESE HYDROSTATIC PARTICULARS HAVE BEEN DEVELOPED WITH THE VESSEL FLOATING ON EVEN KEEL

Candidate's Name.. Examination Center..

TS032-72 SHIP STABILITY

WORKSHEET Q.1

(This Worksheet must be returned with your examination answer book)

TRIM & STABILITY

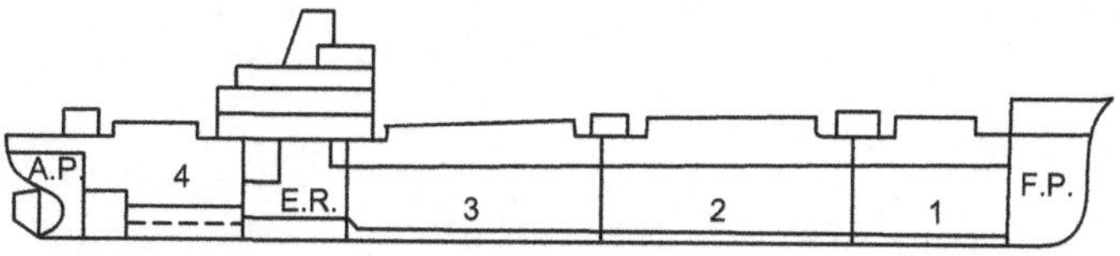

CONDITION: FULLY LOADED – GENERAL CARGO

Compartment	Capacity m^3	Stowage Factor m^3/t	Weight t	KG m	Vertical Moment tm	Free Surface Moment tm	LCG foap m	Longitudinal Moment tm
All Holds	14,562	1.86	**7829**	6.78	**53,081**		73.15	**572,691**
1 TD	264	2.48	**106**	10.71	**1135**		114.33	**12,119**
2 TD	1688	2.74	**616**	10.60	**6530**		93.57	**57,639**
3 TD	1986	2.72	**730**	10.51	**7672**		63.92	**46,662**
Consumables			1464	–	4112	2560	–	58,675
Deadweight			**10,745**					
Lightship			3831	8.21	**31,453**		61.67	**236,258**
DISPLACEMENT			**14,576**	7.13	**103,983**	**2560**	**67.51**	**984,044**

HYDROSTATICS — True Mean Draught **7.00 m** — LCB foap **70.03** — LCF foap **67.35 m**

LENGTH B.P. 130.00 m — MCTC = **184.6**

$$\text{TRIM} = \frac{\mathbf{14{,}576 \times (70.03 - 67.51)}}{\mathbf{184.6}} = \mathbf{199\ cm\ by\ the\ STERN}$$

KM_T = **8.34 m**

$$KG_{fluid} = \mathbf{7.13 + \frac{2560}{14{,}576} = 7.13 + 0.18}$$

KG = **7.31 m**

DRAUGHTS: F. = **6.04 m** A. = **8.03 m**

GM_{fluid} = **1.03 m**

Candidate's Name ....

Examination Center ....

CHAPTER 52

Looking Forward into the Next Decade

In the last decade, new designs for much larger vessels have been appearing. Allow me to select three of those designed and make several observations about them. The three I have selected are:

1. 'Q' LNG ships, namely Qflex and Qmax vessels.
2. Triple 'E' container ships.
3. FLNG 'facility' ship.

Regarding ship stability and ship handling, these much larger vessels are going to present newer and greater challenges to masters and pilots.

Qflex and Qmax LNG Vessels

This section uses data researched by C.B. Barrass (October 2011) from *Significant Ships* (RINA yearly publications 2007 and 2008).

In earlier years, LNG carriers could be divided mainly into two distinct groups, 75,000 m^3 and 150,000 m^3. Their size was denoted not by the deadweight but by the maximum amount of gas they could carry in cubic meters.

Nowadays, because of advances in ship technology, Naval Architects have obtained 'economy of scale' advantages by designing LNG vessels of capacities 200,000–266,000 m^3.

The 200,000–220,000 m^3 LNG ships are known as Qflex vessels. By October 2011, 31 orders for this Qflex design had been placed with shipbuilders. All three Qflex vessels featured in Table 52.1 have 19 officers and 18 crew. Each is chartered to operate for 25 years.

The 250,000–266,000 m^3 LNG ships are known as Qmax vessels. By October 2011, 14 orders for this Qmax design had been placed with shipbuilders. They will load their LNG cargoes eventually at the brand new Ras Laffan export terminal at Qatar.

Table 52.1 gives general particulars of five of these vessels that have already been delivered and are commercially in operation today (October 2011). All are over 300 m in length with service speeds of 19.50–20.50 knots.

Figure 52.1 shows a good example of one of the largest Qmax ships, the giant sized MOZAH operating at a service speed of 19.50 knots.

Ship Stability for Masters and Mates. http://dx.doi.org/10.1016/B978-0-08-097093-6.00053-0

Table 52.1: Recently Delivered Vessels.

Delivery Date	Name of Vessel	Cargo Gas (m^3)	LBP (L) (m)	Br. Mld (B) (m)	L/B Value (m)	Depth (D) (m)	Draft (H)	H/D Value	Speed (Knots)
			Gas Carriers – Latest Designs						
Oct 2008	Al Ghuwariya, Qmax design	263,250	330.0	55.00	6.00	27.00	12.00	0.44	19.50
Sept 2008	Mozah, Qmax design	266,000	332.0	53.80	6.17	27.00	12.00	0.44	19.50
Nov 2007	Al Gattara, Qflex design	216,250	303.0	50.00	6.06	27.00	12.00	0.44	19.50
Oct 2007	Al Ruwais, Qflex design	210,000	303.0	50.00	6.06	27.00	12.00	0.44	20.50
Oct 2007	Tembek, Qflex design	216,200	303.0	50.00	6.06	27.00	12.00	0.44	19.50

Qmax design = 14 orders, approx. 265,000 cubic meters cargo gas.
Qflex design = 31 designs, say 210,000–216,000 cubic meters cargo gas. Qatargas 4 project/Qatargas 2 project.

Figure 52.1:
Example of a Q_{max} ship, the MOZAH.

Each of the vessels in Table 52.1 has:

- An all-aft engine room
- Five-membrane tank construction
- A double-skin hull
- Power from twin low-speed diesel engines (no steam turbine machinery)

- A twin-screw propulsion system
- A bulbous bow Qflex dwt about 100,000 tonnes, Qmax dwt about 125,000 tonnes.

All cargo boil-off gas, a normal by-product of LNG transport, is reliquefied and returned to the cargo tanks instead of being used as boiler fuel in a steam turbine plant.

Renaming of ULCC Sister Ships

The Ultra Large Crude Carriers, *Hellespont Alhambra, Hellespont Fairfax, Hellespont Metropolis*, and *Hellespont Tara* (built in 2002/2003), have (in 2004) been renamed *TI Asia, TI Oceania, TI Africa*, and *TI Europe* respectively.

Triple 'E' Container Ships

This section draws upon *Size Matters for Maersk* by Mike Gerber (*Shipping Telegraph*, April 2011) and a précis of this by C.B. Barrass (October 2011).

According to the Maersk Line, big is beautiful. The company have revealed details of a massive £1.17bn order for the world's largest box-ships. The company claim, for these container ships, equally large economic and environment savings. They will be the world's largest vessels of any kind (see Figures 52.2 and 52.3).

For these massive ships, the Triple 'E' stands for:

- Economy of scale
- Energy efficiency
- Environmentally improved.

General particulars for each of these ships are as follows:

LOA = 400 m, Breadth Mld = 59 m, Draft Mld = 14.50 m, TEU = 18,000
Dwt = 165,000 t, Service speed = 23 knots, twin-screw propulsion
Power requirement = 65–70 MW, Officers/crew complement = 13
Containers can be stowed 23 across the ship; for the *Emma Maersk* it is only 22
Delivery date to be 2013–2015.

Figure 52.2:
A Triple E 18,000 TEU Container Vessel.

Figure 52.3:
Containers aboard a Triple 'E' vessel.

The main route will be Rotterdam to five ports in China via the Suez Canal. It will take only 20 days. These vessels will visit Shanghai, Ningbo, Xiamen, Yantian, and Hong Kong, They will also visit Felixstowe and Bremerhaven. Looking to the future, in September 2011 Felixstowe has already extended two of its berths to accommodate vessels of these dimensions.

Overall, they will consume around 35% less fuel per container than a standard 13,000 TEU container ship. These fuel savings will also mean operational cost savings. Maersk believe that the company will be able to cut the cost of carrying a container from China to Europe by 26%.

In July 2011, Daewoo are building 10 Triple 'E' container ships for Maersk.

The Floating Liquefied Natural Gas (FLNG) Project in NW Australia

This section uses as reference 'The gas platform that will be the world's biggest "ship",' by Chris Summers, BBC News (last updated 15 July 2011), www.bbc.co.uk/news/science-environment-13709293, and a précis of this by C.B. Barrass (October 2011).

Shell has unveiled plans to build the world's first floating liquefied natural gas (FLNG) platform. Shell refers to its FLNG platform as a 'facility' rather than a ship – fully loaded it

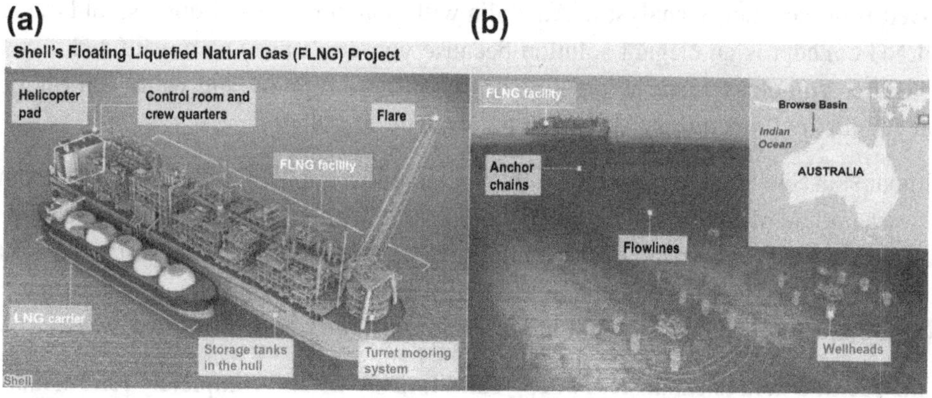

Figure 52.4:
The FLNG concept (a) vessel, (b) location

will weigh 600,000 tonnes. The double-hulled vessel is designed to last 50 years (see Figure 52.4(a)).

The FLNG facility is 488 m long and 75 m wide, with a crew of 100–120 per shift out of a total of 220–240. The 600,000-tonne behemoth – the world's biggest 'ship' – will be sited off the coast of Australia. But how will it work?

Deep beneath the world's oceans are huge reservoirs of natural gas. Some are hundreds or thousands of miles from land, or from the nearest pipeline. Tapping into these 'stranded gas' resources has been impossible – until now.

At Samsung Heavy Industries' shipyard on Geoje Island in South Korea, work is about to start on a 'ship' that, when finished and fully loaded, will weigh 600,000 tonnes.

By 2017 the vessel should be anchored off the north coast of Australia, where it will be used to harvest natural gas from Shell's Prelude field (see Figure 52.4(b)).

Once the gas is on board, it will be cooled until it liquefies and stored in vast tanks at −161 °C. Every six or seven days a huge tanker will dock beside the platform and load up enough fuel to heat a city the size of London for a week.

The tankers will then sail to Japan, China, Korea, or Thailand to offload their cargo.

Scotsman Neil Gilmour, Shell's general manager for FLNG, states: 'The traditional way of producing gas offshore was through pipelines. You brought gas up to a platform and piped it to the "beach". That is the way it's done in the North Sea.'

'Cyclone Alley'

But the Prelude gas field is 200 km (124 miles) from Western Australia's Kimberley Coast and there are no pipelines there to be used.

Johan Hedstrom, an energy analyst in Australia with Southern Cross Equities, told the BBC: 'The FLNG concept is an elegant solution because you don't need so much fixed infrastructure. You don't need the pipeline or the onshore refinery and when you run out of gas you can just pull up stumps and go to the next field.'

Mr Gilmour said Shell had to overcome a 'raft of technical challenges', ensuring for example that the vast amount of equipment on board would work in choppy seas. The Prelude field is in the middle of what is known as 'cyclone alley', an area prone to extremely stormy weather. Mr Gilmour said the vessel had been built to withstand category-5 cyclones and even a 'one-in-10,000-years storm' producing 300 km/h (185 mph) gusts and 20-m-high waves.

When the Prelude field is exhausted, in 25 years' time, it will be completely refurbished and packed off to start work on another field off the coast of Australia, Angola, Venezuela, or wherever.

Mr Hedstrom said: 'FLNG is a neat way of going forward. The way that energy prices are going it does look like a good industry to be in and I think they could make a lot of money out of it.'

The price of LNG has risen markedly as demand has increased. LNG currently sells for $14 per one million British thermal units in Japan, where the price was boosted by the tsunami, which cut the production of nuclear power.

The project, estimated to cost between $8bn (£5bn) and $15bn (£9.5bn), could provide 3.6 million tonnes of gas a year.

Flaring Off

Nick Campbell, an energy analyst with Inenco, said Shell's move into FLNG was a 'smart move'. 'Shell are positioning themselves in an emerging market, not just in China – where gas usage has increased by 20% – but in India, which is also increasing its demand,' he said.

Australia's Minister for Resources and Energy has welcomed the Prelude project, drawing attention to the reduced environmental footprint as compared with a land-based scheme.

But there has been opposition from environmentalists. Martin Pritchard from Environs Kimberley says he is concerned about the potential for 'oil leaks and spills'.

WWF Western Australia, meanwhile, argues that the underwater wellheads and pipelines will harm the tropical marine environment, and estimates the project will emit more than two million tonnes of greenhouse gases per year.

Mr Gilmour says the Prelude project could be the first of several. Shell has already identified the Sunrise gas field in the Timor Sea as having potential for FLNG.

'There are only four or five dry docks globally which could have built this facility and there are certainly no yards in the UK large enough,' says Mr Gilmour. He has been to Geoje Island. He said of Samsung's yard, 'It's an extraordinary place. It's just a phenomenal yard. Samsung is very hi-tech, world class. There are going to be some very spectacular images coming out of there during the building process.'

The ship, whose first section will be laid in 2012, has no name. Shell normally refers to it merely as a 'facility'.

PART II

Endnotes

CHAPTER 53

Draft Surveys

When a ship loads up at a port, departs from that port and travels to another port, a Draft Survey is carried out. This is to check that the cargo deadweight or 'constant' is satisfactory for the shipowner at the port of arrival.

It is virtually a check on the amount of cargo that left the first port against that arriving at the second port. This Draft Survey may be carried out by a Master, a Chief Engineer, or a Naval Architect.

Prior to starting on a Draft Survey the vessel should be in upright condition and on even keel if possible. If not on even keel then certainly within 1% of her LBP would be advantageous.

When the ship arrives in port ready for this Draft Survey, there are several items of information that have to be known by, say, the Naval Architect. They include:

- LBP and C_b relative to the ship's waterline or actual loaded condition
- Lightweight
- Density of the water in which the vessel is floating
- Draft readings port and starboard at the stern, at midships, and at the bow
- Distance from aft perp to aft draft marks
- Distance from amidships to midship draft marks
- Distance from forward perp to forward draft marks
- Distance of LCF from amidships
- Cargo deadweight or 'constant', for example 10,766 t.

Using the above data the Naval Architect will modify the actual draft readings to what they would be at AP, amidships, and FP. These values are then used to determine the mean draft at the position of the ship's LCF (see Chapter 49).

To take into account any hog or sag a 'mean of means' formula is used.

Ship Stability for Masters and Mates. http://dx.doi.org/10.1016/B978-0-08-097093-6.00054-2

Worked Example 1

Suppose the drafts at the AP, amidships, and FP were 8.994, 8.797, and 8.517 m, and LCF was 0.37 m forward of amidships with an LBP of 143.5 m. Then the mean of means draft is:

$$\text{Means of means draft} = (d_{AP} + (6 \times d_{AM}) + d_{FP})/8$$
$$= (8.994 + (6 \times 8.797) + 8.517)/8$$
$$\text{Thus, mean of means draft} = 8.787 \text{ m}$$

When corrected for LCF position, the mean draft is 8.786 m.

The Naval Architect uses this draft of 8.786 m on the ship's hydrostatic curves or tabulated data to obtain a first estimate of the displacement and TPC. These are of course for salt water density of 1.025 t/m^3. Assume this displacement was 17,622 t and the TPC was 25.86 t. The Naval Architect can now evaluate the fresh water allowance and proceed to make a correction for density.

$$\text{FWA} = W/(4 \times \text{TPC}) = 17,662/(4 \times 25.86)$$
$$\text{FWA} = 171 \text{ mm}$$

The density correction is then made (see Chapter 47). Assume the water was of 1.020 t/m^3. Consequently, this correction would be −0.043 m.

Thus, the final true draft would be 8.786 − 0.043 = 8.743 m

Interpolating once more the hydrostatic data the final displacement at the time of the Draft Survey is obtained. Presume this final displacement is 17,537 tonnes.

From this figure must be deducted the lightweight and all other deadweight items (except the 'constant' or cargo deadweight). Assume this lightweight was 5675 t and the residual dwt of oil, fresh water, stores, crew and effects, water ballast, etc. was 1112 t. This residual dwt would have been estimated after all tanks had been sounded and all compartments been checked for contents. Then the final cargo dwt or 'constant' is:

$$\text{Cargo deadweight} = 17,537 - 5675 - 1112 = 10,750 \text{ tonnes}$$

This compares favorably with the cargo dwt given at the port of departure, i.e. 10,766 tonnes, a difference of 0.15%. In a perfect world the cargo dwt at the arrival port would have been 'constant' at 10,766 t. However, ships are not built to this degree of accuracy and measurement to this standard is very hard to achieve.

Summary

Corrections have been made for:

- Draft mark positions
- Trim and LCF position
- Hog or sag
- Density of water.

  True mean draft is 8.743 m

  True displacement is 17,537 t

  Cargo deadweight or 'constant' is 10,750 t (−0.15%).

A well-conducted survey is capable of achieving an absolute accuracy of within ±0.50% of the cargo dwt. This is as good as, if not better than, other systems of direct weighing.

Error can creep in over the years because of the ship's reported lightweight. This is due to the ship herself gaining about 0.50% lightweight each year in service. Hence, over a period of 10 years this ship would have gained about 280 tonnes since her maiden voyage.

Lightweight will also alter slightly depending on the position of the anchors and cables. Obviously, with anchors fully housed the lightweight will be greater. Adjustment may have to be made but it is better if at both ports the anchors are fully housed.

Error can also be made if the draft readings were to be taken in a tidal current. The speed of the tide would cause the ship to sink slightly in the water (squat effects) and so give draft readings that were too high in value. One reported instance of this occurring resulted in a cargo reduction of over 300 t. Initially this was put down to excessive pilfering until further checks discovered that drafts had been taken in moving water. At departure and arrival ports draft readings must be read when water speed is zero.

One suggestion for improving accuracy of measurement is to have draft marks also at 1/4 L and 3/4 L from aft, in other words at stations 2.5 and 7.5. They would reduce errors where there is an appreciable hog or sag at the time of the Draft Survey.

Worked Example 2

Question

A ship has the following characteristics:

- LBP = 165 meters.
- LCF = 0.45 meters forward of amidships.
- TPC = 30.5 tonnes/cm.

- Displacement (loaded departure) = 22,000 tonnes.
- Displacement (lightship) = 9500 tonnes.
- Stores, tankage, etc. (departure) = 1200 tonnes.
- Cargo = Containers.

The following draft marks were read at departure:

- FP = 8.160 meters.
- Amidships = 8.730 meters.
- AP = 9.280 meters.

Find the following:

1. The mass of the loaded cargo.
2. The average mean draft.
3. The draft using the 'mean of means' method.
4. State the condition of the ship that results in the difference between parts 2 and 3 above.
5. Draft at the LCF.
6. Using the draft calculated in part 5, you enter the hydrostatic curves (in the Stability Book) and find the displacement at this draft is 22,000 tonnes. Calculate the Fresh Water Allowance (FWA).
7. On arrival you sound the dock water and find the relative density (ρ) to be 1.020. Calculate the Dock Water Allowance (DWA).
8. Find the final draft at the dock side.
9. Using the final draft calculated in part 8, you enter the hydrostatic curves (in the Stability Book) and find the displacement at this draft is given as 21,700 tonnes. You also estimate that the mass of the remaining stores, tankage, etc. is 950 tonnes. Calculate the cargo deadweight on arrival.
10. Calculate the percentage error in the cargo at arrival.

Answers

1. *The mass of the loaded cargo*

$$\text{Cargo loaded} = \text{Displacement (loaded departure)} - \text{Displacement (lightship)}$$
$$-\text{Stores, tankage, etc.(departure)}$$

$$\text{Cargo loaded} = 22,000 - 9500 - 1200$$

$$\text{Cargo loaded} = 11,300 \text{ tonnes}$$

2. *The average mean draft*

$$\text{Average draft} = \frac{\text{Draft(FP)} + \text{Draft(AP)}}{2}$$

$$\text{Average draft} = \frac{8.160 + 9.280}{2}$$

$$\text{Average draft} = 8.720 \text{ meters}$$

3. *The draft using the 'mean of means' method*

$$\text{Mean of means draft} = \frac{h_{FP} + (6\ h_{AM}) + h_{AP}}{8}$$

$$\text{Mean of means draft} = \frac{8.16 + (6 \times 8.73) + 9.28}{8}$$

$$\text{Meanof means draft} = 8.728 \text{ meters}$$

4. *State the condition of the ship that results in the difference between parts 2 and 3 above*

$$\text{Amidships draft} = 8.730 \text{ meters}$$

$$\text{Average draft} = 8.72 \text{ meters}$$

$$\text{Mean of means draft} = 8.728 \text{ meters}$$

$$\text{Difference} = \text{Average draft} - \text{Amidships draft}$$

$$\text{Difference} = 8.72 - 8.73$$

$$\text{Difference} = -0.010 \text{ meters}$$

The vessel is sagging by 0.010 meters.

Note. The average mean draft assumes the vessel bottom is perfectly straight between the FP and the AP; hence the difference between the average mean draft and the amidships draft will tell you if the vessel is hogging or sagging.

5. *Draft at the LCF*

$$\text{Trim} = \text{Draft}_{AP}(h_{AP}) - \text{Draft}_{FP}(h_{FP})$$

$$\text{Trim correction} = \frac{\text{Trim} \times \text{Dist. of LCF from amidships}}{\text{LPB}}$$

$$\text{Trim correction} = \frac{(h_{AP} - h_{FP}) \times \text{Dist.of LCF from amidships}}{\text{LPB}}$$

$$\text{Trim correction} = \frac{(9.28 - 8.16) \times 0.45}{165}$$

$$\text{Trim correction} = 0.003$$

$$\text{Draft at LCF} = \text{Mean of mean draft} - \text{Trim correction}$$

Note. You should always check if the correction is added or subtracted. It is subtracted if the vessel is trimming by the stern and the LCF is forward of amidships.

$$\text{Draft at LCF} = 8.728 - 0.003$$

$$\text{Draft at LCF} = 8.725 \text{ meters}$$

6. *Calculate the Fresh Water Allowance (FWA)*

$$\text{Displacement at the LCF} = 22{,}000 \text{ tonnes}$$

$$\text{FWA} = \frac{\text{Disp}(\Delta)}{(4 \times \text{TPC})}$$

$$\text{FWA} = \frac{22{,}000}{(4 \times 30.5)}$$

$$\text{FWA} = 180.328 \text{ m}$$

7. *Calculate the Dock Water Allowance (DWA)*

$$\text{DWA} = \frac{\text{FWA} \times (\rho_{SW} - \rho_{DW})}{25}$$

$$\text{DWA} = \frac{180.328 \times (1.025 - 1.020)}{25}$$

$$\text{DWA} = 0.036 \text{ m}$$

8. *Find the final draft at the dock side*

$$\text{Final draft} = \text{Draft at LCF} + \text{DWA}$$

$$\text{Final draft} = 8.725 + 0.036$$

$$\text{Final draft} = 8.761 \text{ meters}$$

9. *Calculate the cargo deadweight on arrival*

$$\text{Cargo deadweight (arrival)} = \text{Displacement (arrival)} - (\text{Displacement (lightship)} + \text{Stores, tankage, etc.})$$

$$\text{Cargo deadweight (arrival)} = 21{,}700 - (9500 + 950)$$

$$\text{Cargo deadweight (arrival)} = 11{,}250 \text{ tonnes}$$

10. *Calculate the percentage error in the cargo at arrival*

$$\text{Percentage error} = \frac{\text{Cargo(depart)} - \text{Cargo(arrival)}}{\text{Cargo(depart)}} \times 100$$

$$\text{Percentage error} = \frac{11{,}300 - 11{,}250}{11{,}300} \times 100 = +0.44\%$$

$$\text{Percentage error} = 0.44\%, \text{ i.e. within the } \pm 0.50\% \text{ of cargo dwt}$$

■

CHAPTER 54

Quality Control — Plus the Work of Ship Surveyors

Ship surveyors ensure that stability of motions and stability of ship structures are of the right standards. In the UK, *Lloyd's* surveyors look after the *strength* of the ship whilst the *Maritime Department for Transport* surveyors look after the *safety* of ships. Another classification society ensuring the ship is up to acceptable safety standards is the *International Maritime Organization (IMO).*

With regard to Lloyd's surveyors, their responsibilities include the following:

- Inspection of the design drawings with respect to the scantlings. Sometimes plate thickness has to be increased. Sometimes the sectional modulus of the plating and stiffener has to be modified. When considered satisfactory, the drawing is given the 'Lloyd's Approved' stamp.
- Inspection and testing of materials used in the construction of the vessel. Sample tests may be asked for on wood, steel, aluminum, or glass reinforced material.
- Examination of shearing forces and bending moments with respect to longitudinal bending of the ship. Racking and buckling stresses will be considered and analyzed.
- Examination of jointing within the structure of the vessel. This could mean tests carried out on specimen welds. These may be destructive tests or nondestructive tests that use X-rays and ultrasonics.
- To arrange special surveys to ensure that a vessel has retained its standard of seaworthiness perhaps after maintenance repairs or after a major accident. Special surveys are carried out on the ship every 4 or 5 years and are more severe with each successive special survey. If in doubt, it is a case of take it out and renew.
- Examining to see if corrosion or wear and tear over years of service have decreased scantlings to a state where the structure is no longer of adequate strength. There have been cases where the shell plating thickness (over a period of 10 years in service) has decreased from 18 mm at time of build to only 9 mm!!
- Each shipyard will usually provide an office for use of the Lloyd's surveyor, who will ensure that the ship is built as per drawing and that inferior material is not being substituted. The surveyor will also check the standard and quality of workmanship. For obvious reasons, the surveyor often undertakes random testing.
- Final positioning of the load line and freeboard marks together with the official ship number and the official port of registry. This responsibility is also shared with the Maritime Department for Transport surveyor.

Ship Stability for Masters and Mates. http://dx.doi.org/10.1016/B978-0-08-097093-6.00055-4

With regard to the Maritime Department for Transport surveyors their responsibilities include the following:

- Inspection of drawings. Care is taken with accommodation plans that the floor area for all cabins and communal spaces are above a certain minimum requisite area.
- Care and maintenance of all life-saving appliances. Inspection of inventory of flares, life-jackets, life-rafts, etc.
- Inspection of fire-fighting appliances. Examining arrangements for prevention, detection, and extinguishing arrangements for the vessel.
- Consideration of navigation lights. Provision of emergency lighting.
- Consideration of sound-signaling apparatus. The arrangements for radio contact in the event of incidents such as fire or flooding. Emails are now used for quick communication, especially when crew or passengers have to leave the ship for emergency medical assistancc.
- Inspection of the stability information supplied by the shipbuilder to the ship.
- Inspection of lighting, heating, and ventilation in accommodation, navigation spaces, holds, and machinery spaces.
- Attendance at the inclining experiment or stability test. This is to verify that the test had been carried out in a correct and fitting manner. The final stability calculations will also be checked out at the local DfT office.
- Written and oral examinations for masters and mates for Certificates of Competency.
- Final positioning of the load line and freeboard marks together with the official ship number and the official port of registry. This responsibility is also shared with the Lloyd's surveyor.
- If the ship has undergone major repair, or conversion to another ship type, or the ship has had major update/retrofit, then the surveyors check for strength and safety. If deemed of satisfactory quality, then new certificates will be signed, dated, and issued to the shipowner.

The International Maritime Organization (IMO) is a technical organization, established in 1958. The organization's chief task was to develop a comprehensive body of international conventions, codes, and recommendations that could be implemented by all member governments. The effectiveness of IMO measures depends on how widely they are accepted and how they are implemented.

Apart from Safety Of Life At Sea (SOLAS) regulations, the IMO Maritime Safety Committee deal with the following:

- Stability and load lines and fishing vessel safety.
- Standards of training and watchkeeping.
- Carriage of dangerous goods.
- Safety of navigation.
- Radio communications, search and rescue.

- Fire protection.
- Bulk liquids and gases.
- Solid cargoes and containers.
- Ship design and equipment.
- Flag state implementation.

One major piece of work recently produced by the IMO is the International Code for the safe carriage of grain in bulk, abbreviated to *International Grain Code* (see Chapter 37). The IMO deals with environmental protection and pollution prevention. It also has a legal committee, which was formed following the *Torrey Canyon* pollution disaster in 1967.

For further information regarding the work of these three classification societies, contact can be made at:

Lloyd's Registry of Shipping, London	www.lr.org
Department for Transport/MCA, Southampton	www.mcga.gov.uk
International Maritime Organization, London	www.imo.org

CHAPTER 55

Extracts from the 1998 Merchant Shipping (Load Line) Regulations Number MSN 1752(M)

Part VI: Information for the Master

Information as to Stability of Ships

32. (1) The owner of every ship to which these Regulations apply shall provide, for the guidance of the master, information relating to the stability of the ship in accordance with this regulation. The information shall be in the form of a book which shall be kept on the ship at all times in the custody of the master.

(2) In the case of a United Kingdom ship this information shall include all matters specified in Schedule 6 in Merchant Shipping Notice MSN 1701(M), and be in the form required by that Schedule. This information shall also be in accordance with the requirements of paragraphs (3), (4), and (5).

(3) Subject to paragraph (4), this information shall be based on the determination of stability taken from an inclining test carried out in the presence of a surveyor appointed by the Secretary of State or, for ships listed in paragraph (5), by the Assigning Authority. This information shall be amended whenever any alterations are made to the ship or changes occur to it that will materially affect this information and, if necessary, the ship shall be re-inclined.

(4) The inclining test may be dispensed with if:

(a) in the case of any ship basic stability data is available from the inclining test of a sister ship and it is known that reliable stability information can be obtained from such data; and

(b) in the case of:

(i) a ship specially designed for the carriage of liquids or ore in bulk, or

(ii) of any class of such ships,

the information available in respect of similar ships shows that the ship's proportions and arrangements will ensure more than sufficient stability in all probable loading conditions.

(5) Before this information is issued to the master:

(a) if it relates to a ship that is

(i) an oil tanker over 100 meters in length;

(ii) a bulk carrier, or an ore carrier, over 150 meters in length;

(iii) a single-deck bulk carrier over 100 meters in length but not exceeding 150 meters in length;

Ship Stability for Masters and Mates. http://dx.doi.org/10.1016/B978-0-08-097093-6.00056-6

(iv) a single-deck dry cargo ship over 100 meters in length;
(v) a purpose-built container ship over 125 meters in length; or
(vi) a column-stabilized mobile offshore drilling unit; or
(vii) a column-stabilized mobile offshore support unit,
it shall be approved either by the Secretary of State or the Assigning Authority which assigned freeboards to the ship; and

(b) if it relates to any other ship, it shall be approved by the Secretary of State.

Information as to Loading and Ballasting of Ships

33. (1) The owner of any ship of more than 150 meters in length specially designed for the carriage of liquids or ore in bulk shall provide, for the guidance of the master, information relating to the loading and ballasting of the ship.
(2) This information shall indicate the maximum stresses permissible for the ship and specify the manner in which the ship is to be loaded and ballasted to avoid the creation of unacceptable stresses in its structure.
(3) In the case of a United Kingdom ship the provisions of Regulation 32(5) shall have effect in respect of information required under this regulation, and the information so approved shall be included in the book referred to in Regulation 32(1).

Part I: Ships in General

Structural Strength and Stability

2. (1) The construction of the ship shall be such that its general structural strength is sufficient for the freeboards assigned.
(2) The design and construction of the ship shall be such as to ensure that its stability in all probable loading conditions shall be sufficient for the freeboards assigned, and for this purpose due consideration shall be given to the intended service of the ship and to the following criteria:
(a) The area under the curve of righting levers (GZ curve) shall not be less than:
(i) 0.055 meter-radians up to an angle of 30°;
(ii) 0.09 meter-radians up to an angle of 40° or the angle at which the lower edge of any openings in the hull, superstructures, or deckhouses which cannot be closed weather-tight, are immersed if that angle is less; and
(iii) 0.03 meter-radians between the angles of heel of 30° and 40° or such lesser angle as is referred to in subparagraph (ii) above.
(b) The righting lever (GZ) shall be at least 0.20 meters at an angle of heel equal to or greater than 30°.
(c) The maximum righting lever shall occur at an angle of heel not less than 30°.
(d) The initial transverse metacentric height shall not be less than 0.15 meters. In the case of a ship carrying a timber deck cargo that complies with subparagraph (a) above by taking into account the volume of timber deck cargo, the initial transverse metacentric height shall not be less than 0.05 meters.
(3) To determine whether the ship complies with the requirements of subparagraph (2) the ship shall, unless otherwise permitted, be subject to an inclining test that

shall be carried out in the presence of a surveyor appointed by the Secretary of State or, for the ships listed in Regulation 32(5), a surveyor appointed by the Assigning Authority.

Regulation 32. Schedule 6. Stability

Part I: Information as to Stability

The information relating to the stability of a ship to be provided for the master shall include the particulars specified below.

1. The ship's name, official number, port of registry, gross and register tonnages, principal dimensions, displacement, deadweight, and draught to the Summer load line.
2. A profile view and, if necessary, plan views of the ship drawn to scale showing all compartments, tanks, storerooms, and crew and passenger accommodation spaces, with their position relative to mid-ship.
3. (1) The capacity and the longitudinal and vertical center of gravity of every compartment available for the carriage of cargo, fuel, stores, feed-water, domestic or water ballast.
   (2) In the case of a vehicle ferry, the vertical center of gravity of compartments designated for the carriage of vehicles shall be based on the estimated centers of gravity of the vehicles and not on the volumetric centers of the compartments.
4. (1) The estimated total weight and the longitudinal and vertical center of gravity of each such total weight of:
   (a) the passengers and their effects; and
   (b) the crew and their effects.
   (2) In estimating such centers of gravity, passengers and crew shall be assumed to be distributed about the ship in the spaces they will normally occupy, including the highest decks to which either or both have access.
5. (1) The estimated weight and the disposition and center of gravity of the maximum amount of deck cargo which the ship may reasonably be expected to carry on an exposed deck.
   (2) In the case of deck cargo, the arrival condition shall include the weight of water likely to be absorbed by the cargo. (For timber deck cargo the weight of water absorbed shall be taken as 15% of the weight when loaded.)
6. A diagram or scale showing:
   (a) the load-line mark and load lines with particulars of the corresponding freeboards; and
   (b) the displacement, tonnes per centimeter immersion, and deadweight corresponding to a range of mean draughts extending between the waterline representing the deepest load line and the waterline of the ship in light condition.
7. (1) A diagram or tabular statement showing the hydrostatic particulars of the ship, including the heights of the transverse metacenter and the values of the moment to change trim 1 cm. These particulars shall be provided for a range of mean draughts extending at least between the waterline representing the deepest load line and the waterline of the ship in light condition.
   (2) Where a tabular statement is used to comply with subparagraph (1), the intervals between such draughts shall be sufficiently close to permit accurate interpolation.

(3) In the case of ships having raked keels, the same datum for the heights of centers of buoyancy and metacenters shall be used as for the centers of gravity referred to in paragraphs (3), (4), and (5).

8. The effect on stability of free surface in each tank in the ship in which liquids may be carried, including an example to show how the metacentric height is to be corrected.

9. (1) A diagram or table showing cross curves of stability, covering the range of draughts referred to in paragraph 7(1).

(2) The information shall indicate the height of the assumed axis from which the righting levers are measured and the trim which has been assumed.

(3) In the case of ships having raked keels and where a datum other than the top of keel has been used, the position of the assumed axis shall be clearly defined.

(4) Subject to subparagraph (5), only enclosed superstructures and efficient trunks as defined in paragraph 10 of Schedule 4 shall be taken into account in deriving such curves.

(5) The following structures may be taken into account in deriving such curves if the Secretary of State is satisfied that their location, integrity, and means of closure will contribute to the ship's stability:

(a) superstructures located above the superstructure deck;

(b) deckhouses on or above the freeboard deck whether wholly or in part only;

(c) hatchway structures on or above the freeboard deck.

(6) Subject to the approval of the Secretary of State in the case of a ship carrying timber deck cargo, the volume of the timber deck cargo, or a part thereof, may be taken into account in deriving a supplementary curve of stability appropriate to the ship when carrying such cargo.

(7) An example shall be included to show how a curve of righting levers (GZ) may be obtained from the cross curves of stability.

(8) In the case of a vehicle ferry or a similar ship having bow doors, ship-side doors, or stern doors where the buoyancy of a superstructure is taken into account in the calculation of stability information, and the cross curves of stability are based upon the assumption that such doors are secured weather-tight, there shall be a specific warning that such doors must be secured weather-tight before the ship proceeds to sea.

10. (1) The diagram and statements referred to in subparagraph (2) shall be provided separately for each of the following conditions of the ship:

(a) *Light condition.* If the ship has permanent ballast, such diagram and statements shall be provided for the ship in light condition both with and without such ballast.

(b) *Ballast condition both on departure and on arrival.* It is to be assumed that on arrival oil fuel, fresh water, consumable stores, and the like are reduced to 10% of their capacity.

(c) *Condition on departure and on arrival when loaded to the Summer load line with cargo filling all spaces available for cargo.* Cargo shall be taken to be homogeneous except where this is clearly inappropriate, for example in cargo spaces that are intended to be used exclusively for the carriage of vehicles or of containers.

(d) *Service loaded conditions both on departure and on arrival.*

(2) (a) A profile diagram of the ship drawn to a suitable small scale showing the disposition of all components of the deadweight.

(b) A statement showing the lightweight, the disposition, and the total weights of all components of the deadweight, the displacement, the corresponding positions of the center of gravity, the metacenter, and also the metacentric height (GM).

(c) A diagram showing the curve of righting levers (GZ). Where credit is given for the buoyancy of a timber deck cargo the curve of righting levers (GZ) must be drawn both with and without this credit.

(d) A statement showing the elements of stability in the condition compared to the criteria laid down in Schedule 2, paragraph 2(2).

(3) The metacentric height (GM) and the curve of righting levers (GZ) shall be corrected for liquid free surface.

(4) Where there is a significant amount of trim in any of the conditions referred to in subparagraph (1), the metacentric height and the curve of righting levers (GZ) may be required to be determined from the trimmed waterline.

(5) If in the view of the Assigning Authority the stability characteristics in either or both of the conditions referred to in subparagraph (1)(c) are not satisfactory, such conditions shall be marked accordingly and an appropriate warning to the master shall be inserted.

11. A statement of instructions on appropriate procedures to maintain adequate stability in each case where special procedures are applied such as partial or complete filling of spaces designated for cargo, fuel, fresh water, or other purposes.

12. The report on the inclining test and of the calculation derived from it to obtain information of the light condition of the ship.

Part II: Ships in Relation to Which the Secretary of State's Or the Assigning Authority's Approval of the Stability Information is Required

13. The ships referred to in Regulation 32(3), (4)(a), and (5)(a) of the Regulations are as follows:

(a) an oil tanker over 100 meters in length;

(b) a bulk carrier, or an ore carrier, over 150 meters in length;

(c) a single-deck bulk carrier over 100 meters in length but not exceeding 150 meters in length;

(d) a single-deck dry cargo ship over 100 meters in length;

(e) a purpose-built container ship over 125 meters in length.

Penalties

35. (1) to (8). Where Regulations are contravened, the owner and master of the ship shall be guilty of an offence and liable:

(a) on summary conviction, by a fine to a fine not exceeding the statutory maximum; and

(b) on conviction on indictment, by a fine or to a fine.

If the appropriate load line on each side of the ship was submerged when it should not have been according to these 1998 Load Line Regulations, then: The owner and master are liable to an additional fine that shall not exceed *one thousand pounds for each complete centimeter* overloaded.

Exercise 55

1. The *Load Line Regulations* 1998 require the master to be provided with stability particulars for various conditions. Detail the information to be provided for a given service condition, describing how this information may be presented.

CHAPTER 56

Keeping Up to Date

This book covers many facets of ship stability, ship handling, and ship strength. It is very important to keep up to date with developments within these specialist topics.

To help with this, the reader should become a member of the Institute of Nautical Studies, the Royal Institution of Naval Architects, or the Honourable Company of Master Mariners, all based in London, UK. These establishments produce monthly journals, and frequently organize conferences and one-day seminars that deal with many topical maritime subjects. Contact can be made at:

Institute of Nautical Studies	www.nautinst.org
Royal Institution of Naval Architects	www.rina.org.uk/tna
Honourable Co. of Master Mariners	www.hcmm.org.uk

Three maritime papers that are of interest and very helpful in the field of Maritime Studies are:

NuMast Telegraph	www.numast.org
Lloyd's List	www.lloydslist.com
Journal of Commerce	www.joc.com

Enrolling on short courses is an ideal way to keep up to date with developments within the shipping industry. Signing onto refresher courses is another way of re-treading the mind.

Ship handling awareness can be greatly enhanced on ship-simulator courses such as the one at Birkenhead (www.lairdside@livjmc.ac.uk). Another very useful contact is The Marine Society & Sea Cadets (bthomas@ms-sc.org).

There are many computer packages used for calculations and graphics, in designing offices, and on ships today. Many books have been written giving details of these marine packages.

Knowledge and skills have to be acquired by all involved in the stability and strength of ships. All procedures have to be planned, monitored, and adjusted. The handling of ships at sea and in port is very much about teamwork.

All members of that team must share a common understanding of the principles involved. Communication is essential, particularly on board ship. To improve techniques, it is essential

Ship Stability for Masters and Mates. http://dx.doi.org/10.1016/B978-0-08-097093-6.00057-8

to discuss evolutions and their consequences. Experiences of past problems and their solutions are always advantageous.

I hope that you have found this book of interest and of benefit to you in your work.

C.B. Barrass

PART III

Appendices

APPENDIX I

Summary of Stability Formulae[1]

Form Coefficients

$$\text{Area of waterplane} = L \times B \times C_w$$
$$\text{Area of amidships} = B \times d \times C_m$$
$$\text{Volume of displacement} = L \times B \times d \times C_b$$
$$C_b = C_m \times C_p$$

Drafts

When displacement is constant (for box shapes):

$$\frac{\text{New draft}}{\text{Old draft}} = \frac{\text{Old density}}{\text{New density}}$$

When draft is constant:

$$\frac{\text{New displacement}}{\text{Old displacement}} = \frac{\text{New density}}{\text{Old density}}$$

$$TPC_{SW} = \frac{WPA}{97.56}$$

$$FWA = \frac{W}{4 \times TPC_{SW}}$$

$$\text{Change of draft or dock water allowance} = \frac{FWA\ (1025 - \rho_{DW})}{25}$$

[1] See Note at end of the Appendix.

Variable immersion hydrometer:

$$\text{Density} = \frac{M_y}{M_y - x\left(\frac{M_y - M_x}{L}\right)}$$

Trim

$$\text{MCTC} = \frac{W \times GM_L}{100 \times L}$$

$$\text{Change of trim} = \frac{\text{Trimming moment}}{\text{MCTC}} \text{ or}$$

$$\frac{W \times (LCB_{foap} - LCG_{foap})}{\text{MCTC}}$$

$$\text{Change of draft aft} = \frac{1}{L} \times \text{Change of trim}$$

$$\text{Change of draft forward} = \text{Change of trim} - \text{Change of draft aft}$$

Effect of trim on tank soundings:

$$\frac{\text{Head when full}}{\text{Length of tank}} = \frac{\text{Trim}}{\text{Length of ship}}$$

True mean draft:

$$\frac{\text{Correction}}{\text{FY}} = \frac{\text{Trim}}{\text{Length}}$$

To keep the draft aft constant:

$$d = \frac{\text{MCTC} \times L}{\text{TPC} \times 1}$$

To find GM_L:

$$\frac{GM_L}{GG_1} = \frac{L}{t}$$

Simpson's Rules

First rule:

$$\text{Area} = h/3(a + 4b + 2c + 4d + e) \quad \text{or} \quad \frac{1}{3} \times CI \times \Sigma_1$$

Second rule:

$$\text{Area} = 3h/8(a + 3b + 3c + 2d + 3e + 3f + g) \quad \text{or} \quad \frac{3}{8} \times CI \times \Sigma_2$$

Third rule:

$$\text{Area} = h/12(5a + 8b - c) \quad \text{or} \quad \frac{1}{12} \times CI \times \Sigma_3$$

KB and BM

Transverse Stability

For rectangular waterplanes:

$$I = \frac{LB^3}{12}$$

$$BM = I/V$$

For box shapes:

$$BM = B^2/12d$$

$$KB = d/2$$

$$KM_{min} = B/\sqrt{6}$$

For triangular prisms:

$$BM = B^2/6d$$

$$KB = 2d/3$$

$$\text{Depth of center of buoyancy below the waterline} = \frac{1}{3}\left(\frac{d}{2} + \frac{V}{A}\right)$$

Longitudinal Stability

For rectangular waterplanes:

$$I_L = \frac{BL^3}{12}$$

$$BM_L = \frac{I_L}{V}$$

For box shapes:

$$BM_L = \frac{L^2}{12d}$$

For triangular prisms:

$$BM_L = \frac{L^2}{6d}$$

Transverse Statical Stability

$$\text{Moment of statical stability} = W \times GZ$$

At small angles of heel:

$$GZ = GM \times \sin\theta$$

By wall-sided formula:

$$GZ = \left(GM + \frac{1}{2} BM \tan^2\theta\right) \sin\theta$$

By Attwood's formula:

$$GZ = \frac{v \times hh_1}{V} - BG \ \sin\theta$$

Stability curves:

$$\text{New } GZ = \text{Old } GZ \pm GG_1 \sin \text{heel or}$$

$$\text{New } GZ = KN - KG \sin \text{heel}$$

$$\text{Dynamical stability} = W \times \text{Area under stability curve}$$

$$= W\left[\frac{v(gh + g_1h_1)}{V} - BG(1 - \cos\theta)\right]$$

$$\lambda_o = \frac{\text{Total VHM}}{\text{SF} \times \text{W}}$$

$$\lambda_{40} = \lambda_o \times 0.8 \quad \text{Actual HM} = \frac{\text{Total VHM}}{\text{SF}}$$

$$\text{Approx. angle of heel} = \frac{\text{Actual HM}}{\text{Maximum permissible HM}} \times 12^\circ$$

Approx angle of θ due to grain shift

$$\text{Reduction in GZ} = (\text{GG}_\text{H} \times \cos\theta) + (\text{GG}_\text{V} \times \sin\theta)$$

W = ship displacement in tonnes
GG_H = horizontal movement of 'G'
GG_V = vertical movement of 'G'
λ_o = righting arm @ $\theta = 0^\circ$ (if upright ship)
λ_{40} = righting arm @ $\theta = 40^\circ$
HM = heeling moment
SF = stowage factor
VHM = volumetric heeling moment

List

$$\text{Final KG} = \frac{\text{Final moment}}{\text{Final displacement}}$$

$$\text{GG}_1 = \frac{\text{w} \times \text{d}}{\text{Final W}}$$

$$\tan \text{list} = \frac{\text{GG}_1}{\text{GM}}$$

Increase in draft due to list:

$$\text{New draft} = \frac{1}{2} \times \text{b} \times \sin\theta + (\text{d}) \cos\theta$$ ... Rise of floor is zero; put 'r' in if rise of floor exists (measured at full Br. Mld)

Inclining experiment:

$$\frac{\text{GM}}{\text{GG}_1} = \frac{\text{Length of plumbline}}{\text{Deflection}}$$

Effect of Free Surface

$$\text{Virtual loss of GM} = \frac{lb^3}{12} \times \frac{\rho}{W} \times \frac{1}{n^2}$$

Drydocking and Grounding

Upthrust at stern:

$$P = \frac{MCTC \times t}{l}$$

or

$$P = \text{Old} - \text{New displacement}$$

$$\text{Virtual loss of GM} = \frac{P \times KM}{W}$$

$$\text{or} = \frac{P \times KG}{W - P}$$

Bilging and Permeability

$$\text{Permeability} = \frac{BS}{SF} \times 100\,\%$$

$$\text{Increase in draft} = \frac{\mu v}{A - \mu a}$$

Freeboard Marks

$$\text{Distance Summer LL to Winter LL} = \frac{1}{48} \times \text{Summer draft}$$

$$\text{Distance Summer LL to Tropical LL} = \frac{1}{48} \times \text{Summer draft}$$

Ship Squat

$$\text{Blockage factor} = \frac{b \times T}{B \times H}$$

$$\delta_{max} = \frac{C_b \times S^{0.81} \times V_k^{2.08}}{20}$$

$$y_o = H - T$$
$$y_2 = y_o - \delta_{max}$$
$$A_s = b \times T$$
$$A_c = B \times H$$

In open water:

$$\%\ \text{Vel.} = 40 + 25 \times \frac{H}{T}$$

$$\delta_{max} = \frac{C_b \times V_k^2}{100}$$

In confined channel:

$$\delta_{max} = \frac{C_b \times V_k^2}{50}$$

$$\text{Width of influence} = \frac{7.04}{C_b^{\ 0.85}} \times b$$

In moderately wide channel:

$$\delta_{max} = K \times \frac{C_b \times V_k^2}{100} \quad \text{where } K = 5.74 \times S^{0.76}$$

Miscellaneous

Angle of loll:

$$\tan \text{loll} = \sqrt{\frac{2 \times GM}{BM_T}}$$

$$GM = \frac{2 \times \text{initial GM}}{\cos \text{loll}}$$

Heel due to turning:

$$\tan \text{heel} = \frac{v^2 \times BG}{g \times r \times GM}$$

Rolling period:

$$T = 2\pi \frac{k}{\sqrt{g \times GM}} = \frac{2k}{\sqrt{GM}} \text{ approx.}$$

Theorem of parallel axes:

$$I_{CG} = I_{OZ} - Ay^2$$

or

$$I_{NA} = I_{xx} - Ay^2$$

Permeability (μ)

$$\text{Permeability} = \frac{BS}{SF} \times 100\ \% \qquad \text{Increase in draft} = \frac{\mu\theta}{A - \mu a}$$

$$\text{Permeability } (\mu) = \frac{\text{Volume available for water}}{\text{Volume available for cargo}} \times 100$$

$$\text{Permeability } (\mu) = \frac{\text{SF of cargo} - \text{Solid factor}}{\text{SF of cargo}} \times 100$$

$$\text{Solid factor} = \frac{1}{RD} \qquad \text{Effective length} = l \times \mu$$

$$\text{Sinkage} = \frac{\text{Volume of bilged compartment} \times \mu}{\text{Intact waterplane area}}$$

$$\text{Tan } \theta = \frac{BB_H}{GM_{bilged}}$$

Drafts and Trim Considerations

$$\text{Correction to observed drafts} = \frac{l_1}{L_1} \times \text{Trim}$$

$$\text{Midships draft corrected for deflection} = \frac{d_{FP} + (6 \times d_m) + d_{AP}}{8}$$

Correction of midships draft to true mean draft when LCF is not at amidships

$$= \frac{\text{Distance of LCF from midships} \times \text{Trim}}{LBP}$$

Second trim correction for position of LCF, if trimmed hydrostatics are not supplied a (form correction)

$$= \frac{\text{True trim} \times (\text{MCTC}_2 - \text{MCTC}_1)}{2 \times \text{TPC} \times \text{LBP}}$$

$$\text{Alternative form correction} = \frac{50 \times (\text{True trim})^2 \times \left(\text{MCTC}_2 - \text{MCTC}_1\right)}{\text{LBP}}$$

Note

These formulae and symbols are for guidance only and other formulae that give equally valid results are acceptable.

$$\rho = \frac{\text{Mass}}{\text{Volume}}$$

$$\text{RD} = \frac{\rho_{\text{substance}}}{\rho_{\text{FW}}}$$

$$\nabla = (\text{L} \times \text{B} \times \text{d}) \times \text{C}_\text{b}$$

$$\Delta = \nabla \times \rho$$

$$\text{DWT} = \Delta - \Delta_{\text{light}}$$

$$\text{A}_\text{W} = (\text{L} \times \text{B}) \times \text{C}_\text{W}$$

$$\text{TPC} = \frac{\text{A}_\text{W}}{100} \times \rho$$

$$\text{Sinkage/rise} = \frac{\text{W}}{\text{TPC}}$$

$$\text{FWA} = \frac{\Delta_{\text{Summer}}}{4 \times \text{TPC}_{\text{SW}}}$$

$$\text{DWA} = \frac{(1025 - \rho_{\text{dock}})}{25} \times \text{FWA}$$

$$\Delta \times \text{GZ} = \text{Moment of statical stability}$$

$$\text{GZ} = \text{GM} \times \sin\theta$$

$$\text{GZ} = \left[\text{GM} + \frac{1}{2}\,\text{BMtan}^2\theta\right] \sin\theta$$

$$\text{GZ} = \text{KN} - (\text{KG} \times \sin\theta)$$

$$\text{Dynamic stability} = \text{Area under GZ curve} \times \Delta$$

$$\text{Area under curve (SR1)} = \frac{1}{3} \times \text{h} \times (\text{y}_1 + 4\text{y}_2 + \text{y}_3)$$

$$\text{Area under curve (SR2)} = \frac{3}{8} \times \text{h} \times (\text{y}_1 + 3\text{y}_2 + 3\text{y}_3 + \text{y}_4)$$

$$\lambda_o = \frac{\text{Total VHM}}{\text{SF} \times \Delta}$$

$$\lambda_{40} = \lambda_o \times 0.8$$

$$\text{Approx. angle of heel} = \left\{\frac{\text{Actual HM}}{\text{Max. permissible HM} \times \text{SF}}\right\} \times 12^\circ$$

$$\text{Reduction in GZ} = (\text{GG}_\text{H} \times \cos\theta) + (\text{GG}_\text{v} \times \sin\theta)$$

$$\text{Rolling period T (s)} = \frac{2 \times \pi \times \text{K}}{\sqrt{\text{g} \times \text{GM}}} \quad \text{or} \quad 2\pi\sqrt{\frac{\text{K}^2}{\text{gGM}_\text{T}}}$$

$$\text{or} \quad 2\sqrt{\frac{\text{K}^2}{\text{GM}_\text{T}}} \text{ seconds}$$

$$GG_{H/V} = \frac{w \times s}{\Delta}$$

$$FSC = \frac{i}{\Delta} \times \rho_T$$

$$FSC = \frac{l \times b^3}{12 \times \Delta} \times \rho_T$$

$$FSC = \frac{FSM}{\Delta}$$

$$\tan\theta = \frac{GG_H}{GM}$$

$$KG = \frac{\Sigma \text{ Moments}}{\Sigma \text{ Weights}}$$

$$GG_H = \frac{\Sigma \text{ Moments}}{\Sigma \text{ Weights}}$$

$$GM = \frac{w \times s \times \text{length}}{\Delta \times \text{deflection}}$$

$$\tan \text{ angle of loll} = \sqrt{\frac{-2 \times GM}{BM_T}}$$

$$GM \text{ at angle of loll} = \frac{2 \times \text{initial } GM}{\cos\theta}$$

$$\tan\theta = \sqrt[3]{\frac{2 \times w \times s}{\Delta \times BM_T}}$$

$$\text{Draft when heeled} = (\text{Upright draft} \times \cos\theta) + \left(\frac{1}{2} \times \text{beam} \times \sin\theta\right)$$

Position of the metacenter

$$KM_T = KB + BM_T$$

$$BM_T = \frac{I_T}{\nabla}$$

$$BM_T(\text{box}) = \frac{L \times B^3}{12 \times \nabla}$$

$$\text{Distance Summer LL to Winter LL} = \frac{1}{48} \times \text{Summer draft}$$

$$\text{Distance Summer LL to Tropical LL} = \frac{1}{48} \times \text{Summer draft}$$

$$KM_L = KB + BM_L$$

$$BM_L = \frac{I_L}{\nabla}$$

$$BM_L(\text{box}) = \frac{L^3 \times B}{12 \times \nabla}$$

$$MCTC = \frac{\Delta \times GM_L}{100 \times LBP}$$

$$CoT = \frac{\Sigma \text{ Trimming moment}}{MCTC} = \frac{\Sigma(w \times d)}{MCTC}$$

$$\text{Change of trim aft} = \text{Change of trim} \times \frac{LCF}{LBP}$$

$$\text{Change of trim fwd} = \text{Change of trim} \times \frac{LBP - LCF}{LBP}$$

$$\text{True mean draft} = \text{Draft aft} \pm \left(\text{Trim} \times \frac{LCF}{LBP}\right)$$

$$\text{Trim} = \frac{\Delta \times (LCG - LCB)}{MCTC}$$

$$P = \frac{\text{Trim} \times \text{MCTC}}{\text{LCF}}$$

$$P = \text{Reduction in TMD} \times \text{TPC}$$

$$\text{Loss of GM} = \frac{P \times KM_T}{\Delta} \quad \text{or} \quad \frac{P \times KG}{\Delta - P}$$

$$\tan\theta = \frac{v^2 \times BG}{g \times R \times GM}$$

$$\text{Permeability } (\mu) = \frac{\text{Volume available for water}}{\text{Volume available for cargo}} \times 100$$

$$\text{Solid factor} = \frac{1}{RD}$$

$$\text{Permeability } (\mu) = \frac{\text{SF of cargo} - \text{Solid factor}}{\text{SF of cargo}} \times 100$$

$$\text{Effective length} = l \times \mu$$

$$\text{Sinkage} = \frac{\text{Volume of bilged compartment} \times \text{Permeability } (\mu)}{\text{Intact waterplane area}}$$

$$I_{\text{parallel axis}} = I_{\text{centroidal axis}} + As^2$$

$$\tan\theta = \frac{BB_H}{GM_{\text{bilged}}}$$

$$\text{Correction to observed drafts} = \frac{l_1}{L_1} \times \text{Trim}$$

$$\text{Midships draft corrected for deflection} = \frac{d_{FP} + (6 \times d_M) + d_{AP}}{8}$$

Correction of midships draft to true mean draft when CF not midships

$$= \frac{\text{Distance of CF from midships} \times \text{Trim (true trim at perp's)}}{\text{LBP}}$$

Second trim correction for position of CF if trimmed hydrostatics are not supplied

$$\text{(form correction)} = \frac{\text{True trim} \times (\text{MCTC}_2 - \text{MCTC}_1)}{2 \times \text{TPC} \times \text{LBP}}$$

$$\text{Alternative form correction} = \frac{50 \times \text{True trim}^2 \times (\text{MCTC}_2 - \text{MCTC}_1)}{\text{LBP}}$$

Summary

Always write your formula first in letters. If you then make a mathematical error you will at least obtain some marks for a correct formula.

APPENDIX II

SQA/MCA 2004 Syllabuses for Masters and Mates

(a) Chief Mate Unlimited Stability and Structure Syllabus

This MCA-approved syllabus was prepared by the IAMI Deck Subgroup and subsequently amended, following consultation with all IAMI colleges, in November 2002 through to June 2004.

Notes

1. The syllabus is based on the HND in Nautical Science. It covers Outcomes 1 and 2 of Unit 24 (Ship Stability), and part of Outcomes 1 and 3 of Unit 25 (Structures and Maintenance).
2. Calculations are to be based wherever possible on the use of real stability information.
3. Longitudinal stability calculations are to be based on taking moments about the after perpendicular and using the formula:

$$\text{Trim} = \text{Displacement} \times (\text{LCB} \sim \text{LCG})/\text{MCTC}$$

4. Formula sheets will be provided to candidates for the examination.

1. Stability information carried on board ship. The inclining experiment

(a) Explains the use of stability information to be carried on board ship.
(b) Explains the purpose of the inclining experiment.
(c) Identifies the occasions when the inclining experiment must be undertaken.
(d) Describes the procedure and precautions to be taken before and during the inclining experiment.
(e) Calculates the lightship KG and determines the lightship displacement for specified experiment conditions.
(f) Explains why a vessel's lightship displacement and KG will change over a period of time.

2. Application of 'free surface effect'

(a) Describes free-surface effect (FSE) as a virtual loss of GM and relates it to the free-surface correction (FSC).

(b) Calculates FSC given rectangular area tank dimensions and tank liquid density.
(c) Describes the effect on FSC of longitudinal subdivisions in tanks.
(d) Calculates FSC given free-surface moments (FSM).
(e) Applies FSC or FSM to all calculations as necessary.

3. **The effect on vessel's center of gravity of loading, discharging, or shifting weights. Final list. Requirements to bring vessel upright**
   (a) Calculates the final position of vessel's center of gravity relative to the keel and centerline taking into account loaded, discharged, and shifted weights.
   (b) Calculates the resultant list.
   (c) Calculates the minimum GM required prior to loading/discharging/ shifting weights to limit the maximum list.
4. **Stability during drydocking. Using real ship stability information**
   (a) Explains the virtual loss of metacentric height during drydocking and the requirements to ensure adequate stability.
   (b) Calculates the virtual loss of metacentric height and hence effective GM during drydocking.
   (c) Determines the maximum trim at which a vessel can enter dry dock to maintain a specified GM.
   (d) Calculates the draft at which the vessel takes the blocks fore and aft.
   (e) Describes the practical procedures that can be taken to improve stability prior to drydocking if it is found to be inadequate.
   (f) Explains why it is beneficial to have a small stern trim when entering dry dock.
5. **Increase in draft due to list/heel. Angle of heel when turning**
   (a) Explains increase in draft due to list/heel.
   (b) Calculates increase in draft due to list/heel.
   (c) Explains angle of heel due to turning and the effect on stability.
   (d) Calculates angle of heel due to turning.
6. **The effect of loading/discharging/shifting weights on trim, draft, and stability. Using real ship stability information**
   (a) Defines 'center of flotation' with respect to waterplane area.
   (b) Defines 'longitudinal center of flotation' (LCF) with respect to the aft perpendicular and explains change in LCF with change of draft.
   (c) Defines 'true mean draft' (TMD).
   (d) Calculates TMD.
   (e) Calculates final drafts and effective GM for various conditions of loading.
   (f) Calculates where to load/discharge a weight to produce a required trim or draft aft.
   (g) Calculates the weight to load/discharge at a given position to produce a required trim or draft aft.
   (h) Calculates final drafts when vessel moves from one water density to a different water density.

(i) Calculates the maximum cargo to discharge to pass safely under a bridge.
(j) Calculates the minimum ballast to load to safely pass under a bridge.
(k) Calculates the final drafts in (i) and (j).

7. Draft survey

(a) Calculates the correction to the observed forward and after drafts to forward perpendicular and after perpendicular respectively.
(b) Calculates the correction to the observed midship draft to amidship.
(c) Calculates the correction of the amidship draft for hull deflection.
(d) Calculates the correction of the amidship draft to true mean draft (TMD) when CF is not amidship.
(e) Calculates the correction for the position of the CF if trimmed hydrostatics are not supplied.

8. Curves of righting levers (GZ), using real ship stability information. Determine compliance with 'intact stability' requirements for the current load line regulations

(a) Constructs a curve of righting levers (GZ), for a given condition.
(b) Defines 'righting moment' (moment of statical stability) and 'dynamical stability'.
(c) Extracts stability information from a curve of righting levers (GZ).
(d) Calculates appropriate areas under a curve of righting levers (GZ), using Simpson's Rules.
(e) Assesses whether vessel complies with the 'intact stability' requirements of the current load line regulations.

9. Simplified stability. Using real ship stability information

(a) Describes the appropriate use of 'simplified stability' information.
(b) Assesses whether a vessel complies with 'maximum permissible KG' requirements for a given condition.

10. Angle of loll and effective GM at angle of loll

(a) Describes the stability at an angle of loll and shows the existence of an effective GM.
(b) Calculates the angle of loll for vessel with a negative initial GM.
(c) Calculates the effective GM at an angle of loll.
(d) Describes the dangers to a vessel with an angle of loll.
(e) Distinguishes between an angle of loll and an angle of list.
(f) Describes the correct procedure for correcting an angle of loll.

11. Factors affecting a curve of righting levers (GZ)

(a) Describes the effects of variations in beam and freeboard on the curve of righting levers (GZ).
(b) Describes the effect of trim on KN values and resultant curve of righting levers (GZ).
(c) Describes the terms 'fixed trim' and 'free trim' with respect to KN values and resultant curve of righting levers (GZ).
(d) Explains the effects of being in a seaway on the curve of righting levers (GZ).

(e) Outlines the conditions for a vessel to be in the stiff or tender condition and describes the effects on the curve of righting levers (GZ).
(f) Describes the use of ballast/bunkers to ensure adequate stability throughout the voyage.
(g) Describes icing allowances.
(h) Describes the changes in stability that may take place on a voyage.
(i) Explains the effects on the curve of righting levers (GZ) of the changes described in (h).
(j) Explains the effects of an angle of list on the curve of righting levers (GZ).
(k) Explains the effects of an angle of loll on the curve of righting levers (GZ).
(l) Explains the effects of a zero initial GM on the curve of righting levers (GZ).

12. The effect on the curve of righting levers (GZ) of shift of cargo and wind heeling moments

(a) Constructs a curve of righting levers (GZ) taking into account shift of cargo/solid ballast and describes the effects on the vessel's stability.
(b) Explains the precautions to be observed when attempting to correct a large angle of list.
(c) Explains how wind heeling moments are calculated.
(d) Constructs a curve of righting moments taking into account wind heeling moments and describes the effects on the vessel's stability.
(e) Describes the minimum stability requirements taking into account wind heeling moments as specified in current *'Load Line – Instructions for the Guidance of Surveyors'*.
(f) Determines that a ship's loaded condition complies with the minimum stability requirements specified in (e).

13. Use of the current IMO Grain Rules to determine if the vessel complies with the specified stability criteria. Real ship stability information to be used

(a) Calculates the 'grain heeling moments' for a specified loading condition.
(b) Determines the 'grain heeling moment' calculated in (a) and whether the vessel complies with the stability requirements by comparison with the 'maximum permissible heeling moments'.
(c) Calculates the approximate angle of heel in (b).
(d) Constructs graphs of a righting arm curve and heeling arm curve.
(e) Assesses whether a grain-laden vessel complies with the 'minimum stability requirements' specified in the 'IMO Grain Rules'.
(f) Discusses factors to be taken into account to minimize grain heeling moments.

14. Rolling, pitching, synchronous, and parametric rolling

(a) Describes rolling and pitching.
(b) Defines rolling period.
(c) Explains factors affecting rolling period.

(d) Describes synchronous rolling and the associated dangers.
(e) Describes parametric rolling and the associated dangers.
(f) Describes actions to be taken by Ship's Officer in the event of synchronous rolling or parametric rolling.

15. The effect of damage and flooding on stability

(a) Calculates, for box-shaped vessel, the effect on draft, trim, list, freeboard, and metacentric height if the following compartments are bilged:
  (i) Symmetrical amidships compartment with permeability.
  (ii) Symmetrical amidships compartment with watertight flat below initial waterline with permeability.
  (iii) Symmetrical amidships compartment with watertight flat above initial waterline with permeability.
  (iv) Extreme end compartment with 100% permeability.
  (v) Extreme end compartment with watertight flat below the initial waterline with 100% permeability.
  (vi) Amidships compartment off the centerline with 100% permeability.
(b) Describes countermeasures which may be taken in the event of flooding.

16. Damage stability requirements for passenger vessels and Type A and Type B vessels

(a) Defines for passenger vessels … 'bulkhead deck', 'margin line', 'floodable length', 'permissible length', 'factor of subdivision'.
(b) Describes subdivision load lines for passenger vessels.
(c) Identifies 'assumed damage' for passenger vessels.
(d) Identifies 'assumed flooding' for passenger vessels.
(e) Identifies 'minimum damage stability requirements' for passenger vessels.
(f) Describes the 'Stockholm Agreement 1996' with respect to stability requirements for passenger vessels.
(g) Identifies damage stability flooding criteria for Type A, B-60, B-100 vessels.
(h) Identifies minimum equilibrium stability condition after flooding for vessels specified in (g).

17. Load line terminology and definitions for new builds

(a) Defines Type A, B, B-60, and B-100 vessels.

18. Conditions of assignment of load lines

(a) Describes 'conditions of assignment' for vessels specified in 17(a).
(b) Describes 'tabular freeboard' with respect to vessels specified in 17(a).
(c) Explains the corrections to be applied to tabular freeboard to obtain 'statutory assigned freeboard'.

19. Assignment of special load lines, for example 'timber load lines'

(a) Describes the special factors affecting the assignment of timber load lines.

(b) Describes the intact stability requirements for vessels assigned timber load lines.

20. Requirements and codes relating to the stability of specialized vessels

(a) Identifies the stability problems associated with ro-ro vessels, offshore supply vessels, and vessels when towing.

21. The preparations required for surveys

(a) Lists surveys required by the load line rules for a vessel to maintain a valid load line certificate.

(b) Lists the items surveyed at a load line survey and describes the nature of the survey for each item.

(b) OOW Unlimited Stability and Operations Syllabus

This MCA approved syllabus was prepared by the IAMI Deck Subgroup and subsequently amended, following consultation with all IAMI colleges, in November 2002 through to June 2004.

Notes

1. The syllabus is based on the HND in Nautical Science. It covers Outcomes 3 and 4 of Unit 7 (Cargo Work), all Outcomes of Unit 8 (Ship Stability) and part of Outcomes 2 of Unit 10 (Maritime Law and Management).
2. Formula sheets will be provided to candidates for the examination.

1. Hydrostatics

(a) Defines mass, volume, density, relative density, Archimedes' principle, FWA, DWA, and TPC.

(b) Determines TPC and displacement at varying drafts using hydrostatic tables.

(c) Calculates small and large changes in displacement making appropriate use of either TPC or displacement tables.

(d) Defines waterline length, LBP, freeboard, waterplane area, C_W, and C_b.

(e) Calculates the weight to load or discharge to obtain small changes in draft or freeboard.

(f) Explains the reasons for load lines and load line zones.

(g) Calculates the weight to load or discharge in relation to load line dimensions, appropriate marks, TPC, FWA, and DWA.

2. Statical stability at small angles

(a) Defines center of gravity, center of buoyancy, initial transverse metacenter, and initial metacentric height (GM).

(b) Calculates righting moments given GM and displacement.

(c) Explains stable, neutral, and unstable equilibrium.

(d) Explains the relationship between equilibrium and the angle of loll.
(e) Identifies, from a given GZ curve, range of stability, initial GM, maximum GZ, angle of vanishing stability, angle of deck edge immersion, angle of loll, and angle of list.
(f) Explains the difference between typical GZ curves for stiff and tender vessels.
(g) Sketches typical GZ curves for vessels at an angle of loll.

3. Transverse stability

(a) Calculates the shift of ship's G, vertically and horizontally, after loading/discharging/shifting a weight.
(b) Calculates the final KG or GM by moments about the keel after loading/discharging/shifting weights including appropriate free-surface correction.
(c) Calculates distance of G horizontally from the centerline after loading/discharging/shifting weights.
(d) Calculates the effect on stability of loading or discharging a weight using ship's gear.
(e) Calculates the angle of list resulting from (a)–(d).
(f) Explains the difference between list and loll, and methods of correction.
(g) Explains the consequences and dangers of a free surface.
(h) Explains that the free-surface effect can be expressed as virtual rise of G or as a free-surface moment.
(i) Describes the effects on free surface of longitudinal subdivision of a tank.

4. Longitudinal stability

(a) Defines LCF, LCG, LCB, AP, trim, trimming moment, and MCTC.
(b) Calculates the effect on drafts of loading, discharging, and shifting weights longitudinally by taking moments about the AP.

5. Maintaining a deck watch (alongside or at anchor).

6. Pollution prevention.

7. Legislation.

Note. Parts 5, 6, and 7 are outside the remit of this book on Ship Stability.

APPENDIX III

Specimen Exam Questions with Marking Scheme

Recent Examination Questions for Ship Stability: STCW 95 Chief Mate/Master Reg. II/2 (Unlimited)

032-74 – Stability and Structure, July 2010

Marks for each part are shown in parentheses.

1. A vessel is upright, starboard side alongside, at a draught of 6.00 m in salt water. KG = 8.30 m. There is a 27 t boiler on deck, KG = 7.78 m, which is to be discharged using the vessel's crane, the head of which is 26.0 m above the keel. The boiler is to be lifted from a position on the vessel's centerline and landed on a railway truck ashore. The distance of the railway truck from the vessel's centerline is 23.30 m.

   Using the *Hydrostatic Particulars* included in the *Stability Data Booklet*, calculate each of the following:
   (a) the maximum angle of heel during discharge; (23)
   (b) the maximum angle of heel during discharge if the vessel was first listed 5° to port prior to the discharge of the boiler. (12)
2. A vessel, initially upright, is to be subjected to an inclining test. Present displacement = 4800 t, KM = 10.58 m.

   Total weights on board during the experiment:

Ballast	400 t	KG = 3.52 m	Tank full
Bunkers	175 t	KG = 3.86 m	Free surface moment 997 t m
Fresh water	85 t	KG = 4.46 m	Free surface moment 810 t m
Inclining weights	52 t	KG = 8.42 m	

At the time of the experiment the boilers are empty. They would usually contain a total of 24 t of water, KG = 4.18 m, with a free-surface moment of 124 t m.

A deck crane, weight 18 t and still ashore, will be fitted on the vessel at a KG of 9.85 m before the next cargo operation.

The plumbline has an effective vertical length of 7.88 m. The inclining weights are shifted transversely 7.20 m on each occasion and the mean horizontal deflection of the plumbline is 0.66 m.

(a) Calculate the vessel's lightship KG. (30)

(b) Identify five possible causes of a change in the vessel's lightship KG over a period of time. (5)

3. A vessel is to load a cargo of grain (stowage factor 1.62 m^3/t). Initial displacement = 5900 tonnes, initial KG = 6.65 m. All five holds are to be loaded full of grain.

The tween decks are to be loaded as follows:

No. 1 TD	Part full – ullage 1.75 m
No. 2 TD	Full
No. 3 TD	Part full – ullage 2.50 m
No. 4 TD	Full

The *Stability Data Booklet* provides the necessary cargo compartment data for the vessel.

(a) Using the *Maximum Permissible Grain Heeling Moment Table* included in the *Stability Data Booklet*, determine whether the vessel complies with the minimum criteria specified in the *International Grain Code (IMO).* (30)

(b) Calculate the ship's approximate angle of heel in the event of the grain shifting as assumed by the *International Grain Code (IMO).* (5)

4. (a) A vessel loads a packaged timber cargo on deck such that there is an increase in the vessel's KG and an effective increase in freeboard. Using a sketch, show the effect of loading this cargo on the vessel's GZ curve. (14)

(b) Sketch how the GZ curve for a vessel with a zero GM is affected by each of the following:

(i) a rise in the vessel's KG; (8)

(ii) a reduction in the vessel's KG. (8)

5. (a) Discuss the factors affecting the virtual loss of GM due to a free surface within an undivided rectangular tank. (8)

(b) Explain the effect on the virtual loss of GM due to the free surface when the slack tank is equally divided in each of the following situations:

(i) by a longitudinal bulkhead; (5)

(ii) by a transverse bulkhead. (5)

(c) Explain why stability information relating to free surface for a particular tank is usually expressed as a *Free-Surface Moment (FSM).* (5)

(d) A double-bottom tank, initially empty, is to be ballasted full of salt water. Sketch a graph to show the way in which the effective KG of the ship will change from the instant of starting to fill the tank until it is full. (12)

6. (a) State the surveys required in order that an International Load Line Certificate remains valid. (5)

(b) List the items to be inspected during the surveys stated in Q6(a), stating the nature of the examination required for each. (25)

032-74 – Stability and Structure, March 2011

1. A vessel's present particulars are: Forward draught 7.982 m, aft draught 9.218 m at an upriver berth in fresh water. The vessel is to proceed downriver to cross a sand bar at the river entrance where the relative density of the water is 1.018.

   During the river passage the following items of deadweight will be consumed:

56 t of heavy fuel oil	from No. 3 D.B. center tank
24 t of diesel oil	from No. 4 D.B. port tank
24 t of diesel oil	from No. 4 D.B. starboard tank
15 t of fresh water	from forward fresh water tank

   (a) Using the *Stability Data Booklet*, calculate the clearance of the vessel if the depth of water over the sand bar is 9.440 m. (30)

   (b) State the maximum clearance over the sand bar if the vessel is brought to an even keel condition by internal transfer of ballast. (5)

2. A box-shaped vessel floating at an even keel in salt water has the following particulars: Length 100.00 m, breadth 15.00 m, depth 9.00 m, draught 5.000 m, KG = 5.400 m. A midship watertight compartment 20.00 m long and extending the full breadth and depth of the vessel is bilged. Permeability of the compartment is 0.85. Calculate each of the following:

   (a) the new draught; (8)

   (b) the change in GM; (12)

   (c) the righting moment in the flooded condition for an angle of heel of 20°. (15)

3. A vessel completes underdeck loading in salt water with the following particulars: Displacement 15,040 tonnes, KG = 8.00 m. The *Stability Data Booklet* provides the necessary data for the vessel. During the passage 200 tonnes of heavy fuel oil (R.D. 0.80) are consumed from No. 3 D.B. tank center, which was full on departure (use VCG of tank for consumption purposes).

   Calculate the maximum weight of timber to load on deck KG = 12.60 m assuming 15% water absorption during the passage in order to arrive at the destination with the minimum GM (0.05 m) allowed under the Load Line Regulations. (35)

   Note: Assume KM constant.

4. A vessel operating in severe winter conditions may suffer from nonsymmetrical ice accretion on decks and superstructure. Describe, with the aid of a sketch of the vessel's curve of statical stability, the effects on the overall stability of the vessel. (35)

5. List 10 items of the stability and stress data required to be supplied to ships under the current Load Line Regulations, stating for each how such information might be used. (30)
6. (a) Explain why the values of trim and metacentric height in the free-floating condition are important when considering the suitability of a vessel for drydocking. (10)
   (b) Describe two methods of determining the upthrust (P force) *during* the critical period. (10)
   (c) Describe how the metacentric height and trim can be adjusted prior to drydocking so as to improve the vessel's stability at the critical instant. (10)

032-74 – Stability and Structure, July 2011

1. A vessel is to transit a canal with a minimum clearance of 0.40 m under a bridge, the underside of which is 21.26 m above the waterline.

   Present draughts in fresh water (R.D. 1.000): Forward 5.380 m, aft 6.560 m.

   The fore mast is 110 m foap and extends 26.00 m above the keel.

   The aft mast is 36 m foap and extends 27.20 m above the keel.

   Using the *Stability Data Booklet*, calculate each of the following:
   (a) the final draughts forward and aft in order to pass under the bridge with minimum clearance; (17)
   (b) the maximum weight of cargo that can be discharged in order to pass under the bridge with minimum clearance. (18)

   Note: Assume masts are perpendicular to the waterline throughout.
2. A vessel entered dry dock for emergency repairs while carrying cargo. Particulars on entry (salt water):

   Displacement 15,716 t, Trim 0.64 m by stern, KG = 8.10 m

   While in drydock 360 tonnes of cargo, KG = 3.80 m, LCG = 82.40 m, were discharged and are no longer on board. The vessel is now planning to leave dry dock.
   (a) Using the *Stability Data Booklet*, calculate the vessel's effective GM at the critical instant on departure. (35)
   (b) State the stability measures that should be considered prior to flooding the dock in order to ensure the safe undocking of this vessel. (5)
3. (a) Explain why a vessel will heel when turning. (7)
   (b) A vessel has the following particulars: Even keel draught 8.720 m, maximum breadth 22.60 m, KG = 7.96 m, KM = 8.38 m, KB = 4.18 m.
      (i) Calculate the angle and direction of heel when turning to *port* in a circle of diameter 500 m at a speed of 18.2 knots. (20)
      (ii) Calculate the new maximum draught during the turn in Q3(b)(i), assuming the midships cross-section can be considered rectangular. (8)

   Note: assume 1 nautical mile = 1852 m and g = 9.81 m/s^2.

4. Sketch a vessel's curve of statical stability, showing the effects of each of the following:
   (a) a transverse shift of cargo (e.g. ro-ro unit); (10)
   (b) an increase in beam; (10)
   (c) a reduction in GM. (10)
5. Explain the corrections to be applied to the *Tabular Freeboard* for a 'Type A' ship in order to obtain the *Assigned Freeboard*, and the reason why the freeboard could be increased or decreased in each case. (30)
6. (a) State the minimum stability requirements for a vessel under the current Load Line Regulations. (6)
   (b) At ballast draught a vessel complies in every respect with the stability requirements of the Load Line Regulations. At load draught, with the same GM, the vessel does not comply. Explain, with the aid of a diagram, why the vessel no longer complies. (14)
   (c) A vessel has the following particulars: Initial displacement 10,250 t, KG = 8.620 m. A double-bottom tank is full of fuel oil. During the passage 250 t of oil, KG = 0.500 m, will be consumed from the tank resulting in a free-surface moment (FSM) of 1040 t m. Using the *Maximum KG* table contained in the *Stability Data Booklet*, determine whether the vessel will still comply with the minimum intact stability criteria specified in the Load Line Regulations. (10)

STCW 95 Officer in Charge of Navigational Watch Reg. II/1 (Unlimited)

034-84 – Stability and Operations, March 2011

Marks for each part are shown in parentheses.

Section A

1. (a) Sketch a curve of statical stability for a vessel listed 7° with a range of positive stability of 66°. (6)
   (b) Using Datasheet Q1 – GZ Curve (see over), determine each of the following:
      (i) the condition of stability of the vessel; (1)
      (ii) range of positive stability; (1)
      (iii) angle of vanishing stability; (1)
      (iv) approximate angle of deck edge immersion; (1)
      (v) maximum GZ; (1)
      (vi) angle at which maximum GZ occurs; (1)
      (vii) approximate initial GM. (1)
   (c) (i) A ship has a displacement of 7263 t, KM = 10.25 m, KG = 9.3 m. Calculate the righting moment at an angle of heel of 7°. (3)
      (ii) Explain the difference between a *righting lever* and a *righting moment*. (4)

2. A vessel is initially upright in salt water at an even keel draught of 5.9 m. Two weights are to be loaded as follows:
   70 t at a KG of 2.5 m, 9 m to port of the centerline
   25 t at a KG of 10.5 m, 5.5 m to starboard of the centerline
   The initial KG (solid) prior to loading is 7.3 m. A total Free-Surface Moment of 2523 t m exists on the vessel at this time. Using Datasheet Q2 – *Hydrostatic Particulars 'A'* (see p. 517), determine each of the following:
   (a) the final angle and direction of list after the two weights have been loaded; (14)
   (b) the amount of ballast to be transferred, and in which direction, between No 3 port and starboard ballast tanks so that the vessel will completely upright. (6)

   Note: Each ballast tank is rectangular, has a breadth of 10 m, and is already slack.

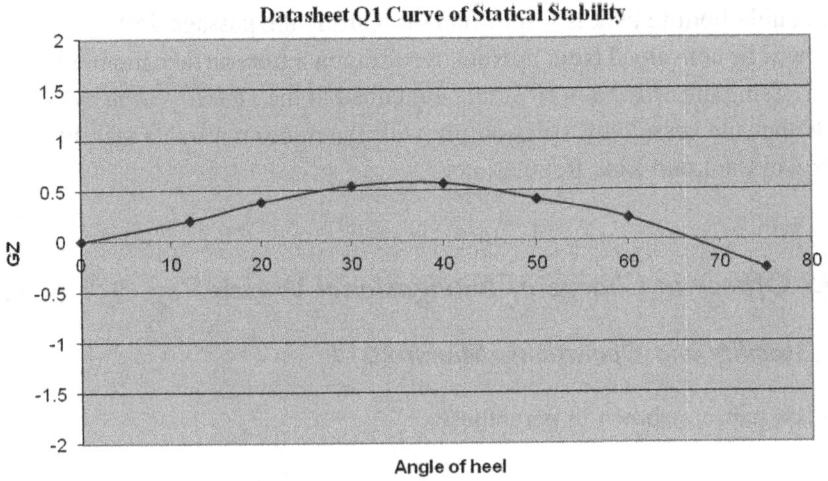

034-84 – *Stability and Operations, June 2011*

Section A

1. (a) Explain each of the following terms:
   (i) TPC; (3)
   (ii) FWA. (4)

   (b) A vessel is loading in port in dock water, RD 1.015. Initial draught (even keel) 5.62 m, TPC_{SW} 20. The ship has a summer displacement of 22,400 t, which corresponds to a summer load draught of 6.10 m. Calculate the maximum weight of cargo that can be loaded if the ship is to sail at her summer draught in salt water, given that 50 t of bunkers are still to be taken before the vessel sails. (13)
2. (a) Explain why the LCF of a vessel might change with draught. (3)

(b) A vessel of LBP 135 m is floating in salt water at an even keel draught of 5.20 m. The following cargo operations are then carried out:
Discharge 1560 t from LCG 81 m foap
Load 1700 t at LCG 90 m foap
Load 2100 t at LCG 35 m foap
Using Datasheet Q2 – *Hydrostatic Particulars 'A'* (see p. 517), calculate the final draughts fore and aft. (12)

(c) A ship has completed loading with the following draughts: Forward 7.35 m, aft 8.90 m, MCTC 170 t m. Calculate the weight of ballast to transfer between the fore peak tank (LCG 129 m foap) and the aft peak tank (LCG at the aft perpendicular) so that the ship sails with a trim of 1.00 m by the stern. (5)

Section B

3. (a) A general cargo vessel is in port and is to load a general cargo that includes palletized units, drums, timber, and bagged cargo. With reference to cargo operations, list EACH of the following:
   (i) the duties of the Officer of the Watch prior to loading; (7)
   (ii) the duties of the Officer of the Watch during loading operations. (7)
   (b) List the basic information that should be detailed on a *Cargo Damage Report.* (6)
4. A container vessel is in port for loading operations. The cargo to be loaded includes a number of containers with packaged dangerous goods.
   (a) State which publications and documents must be consulted when deciding on the stowage location and securing of dangerous goods. (5)
   (b) Explain what the Officer of the Watch must ensure when the containers with dangerous goods are being loaded. (12)
   (c) Describe what the Officer of the Watch must ensure with respect to the dangerous goods paperwork, after loading operations are completed. (3)
5. A ship of more than 400 GT is to discharge *machinery space bilge water* inside a special area. With reference to MARPOL 73/78 Annex I:
   (a) list the discharge criteria that must be complied with; (7)
   (b) state the document in which this discharge should be recorded; (2)
   (c) list the operations that must be recorded in this particular document; (9)
   (d) state who should sign the entries and who should sign the pages of this document. (2)

034-84 – Stability and Operations, July 2011

Section A

1. (a) Explain each of the following:
   (i) DWA; (4)
   (ii) C_b. (3)

(b) With reference to Datasheet Q2 – *Hydrostatic Particulars 'A'*:
   (i) A ship arrives in a salt water port with an even keel draught of 5.0 m. Calculate how much cargo must be discharged so that the vessel will sail with a draught of 4.5m. (3)
   (ii) A ship displaces 13,657 t in salt water. A total of 1500 t of cargo is then discharged. Calculate the final draught in salt water, using only the displacement and draught values. (3)
   (iii) A ship initially has a fresh water displacement of 10,260 t and then loads 864 t of cargo. Calculate the final draught in fresh water, using the appropriate TPC values. (7)

2. (a) A ship is initially upright in salt water at an even keel draught of 6.4 m. The solid KG is 7.7 m. The following cargo operations are then carried out:
   Discharge 358 t from KG = 4.1 m
   Discharge 320 t from KG = 3.5 m
   Shift 200 t from KG = 5.3 m to KG = 8.5 m
   At the time the ship has a total Free-Surface Moment of 4258 t m. Using Datasheet Q2 – *Hydrostatic Particulars 'A'* (see p. 517), calculate the effective metacentric height after completion of cargo operations. (10)
   (b) (i) Define the term *free-surface effect*, explaining how it affects a ship's stability. (6)
   (ii) State the possible consequences of free-surface effect. (4)

Section B

3. (a) Outline the procedures and precautions that must be taken when a ship is to load a heavy lift. (18)
   (b) State the document that must be checked to ensure that the lifting gear has been adequately maintained and inspected. (2)
4. (a) With reference to the International Maritime Dangerous Goods (IMDG) Code, define each of the following, explaining what they are used for:
   (i) Mfag; (3)
   (ii) EmS Guide. (3)
   (b) With reference to the Code of Safe Working Practices for Merchant Seamen:
   (i) describe what a '*dangerous space*' is; (3)
   (ii) list the precautions that should be taken before a potentially dangerous space is entered. (7)
   (c) Explain what a risk assessment is. (4)
5. (a) With reference to MARPOL 73/78, Annex 1:
   (i) list the conditions that must be complied with for the discharge of oil or oily mixtures from the cargo area of an oil tanker at sea; (7)

(ii) state the document in which this discharge should be recorded; (2)

(iii) list any other FIVE operations that would be recorded in this document. (5)

(b) State the criteria that must be complied with for discharge under Annex IV of MARPOL 73/78. (6)

Datasheet for use in Q2 of each 034-84 – Stability and Operations exam, Hydrostatic Particulars 'A'

Draught (m)	Displacement (t) SW RD 1.025	Displacement (t) FW RD 1.000	TPC (t) SW RD 1.025	TPC (t) FW RD 1.000	MCTC (t m) SW RD 1.025	MCTC (t m) FW RD 1.000	KM_T (m)	KB (m)	LCB foap (m)	LCF foap (m)
7.00	14,576	14,220	23.13	22.57	184.6	180.1	8.34	3.64	70.03	67.35
6.90	14,345	13,996	23.06	22.50	183.0	178.5	8.35	3.58	70.08	67.46
6.80	14,115	13,771	22.99	22.43	181.4	177.0	8.36	3.53	70.12	67.57
6.70	13,886	13,548	22.92	22.36	179.9	175.5	8.37	3.48	70.16	67.68
6.60	13,657	13,324	22.85	22.29	178.3	174.0	8.38	3.43	70.20	67.79
6.50	13,429	13,102	22.78	22.23	176.8	172.5	8.39	3.38	70.24	67.90
6.40	13,201	12,879	22.72	22.17	175.3	171.0	8.41	3.33	70.28	68.00
6.30	12,975	12,658	22.66	22.11	173.9	169.6	8.43	3.28	70.32	68.10
6.20	12,748	12,437	22.60	22.05	172.5	168.3	8.46	3.22	70.35	68.20
6.10	12,523	12,217	22.54	21.99	171.1	167.0	8.49	3.17	70.38	68.30
6.00	12,297	11,997	22.48	21.93	169.8	165.7	8.52	3.11	70.42	68.39
5.90	12,073	11,778	22.43	21.87	168.5	164.4	8.55	3.06	70.46	68.43
5.80	11,848	11,559	22.37	21.82	167.3	163.2	8.59	3.01	70.50	68.57
5.70	11,625	11,342	22.32	21.77	166.1	162.1	8.63	2.95	70.53	68.65
5.60	11,402	11,124	22.26	21.72	165.0	161.0	8.67	2.90	70.57	68.73
5.50	11,180	10,908	22.21	21.66	163.9	160.0	8.71	2.85	70.60	68.80
5.40	10,958	10,691	22.15	21.61	162.9	158.9	8.76	2.80	70.64	68.88
5.30	10,737	10,476	22.10	21.56	161.8	157.9	8.81	2.74	70.68	68.95
5.20	10,516	10,260	22.05	21.51	160.8	156.9	8.86	2.69	70.72	69.02
5.10	10,296	10,045	22.00	21.46	159.8	155.9	8.92	2.63	70.75	69.09
5.00	10,076	9830	21.95	21.41	158.8	154.9	8.98	2.58	70.79	69.16
4.90	9857	9616	21.90	21.36	157.9	154.0	9.06	2.53	70.82	69.23
4.80	9638	9403	21.85	21.32	156.9	153.1	9.13	2.48	70.86	69.29
4.70	9420	9190	21.80	21.27	156.0	152.2	9.22	2.43	70.90	69.35
4.60	9202	8978	21.75	21.22	155.1	151.3	9.30	2.38	70.93	69.42
4.50	8985	8766	21.70	21.17	154.2	150.5	9.40	2.32	70.96	69.48
4.40	8768	8554	21.65	21.12	153.3	149.6	9.49	2.27	71.00	69.55
4.30	8552	8344	21.60	21.07	152.4	148.7	9.60	2.22	71.04	69.62
4.20	8336	8133	21.55	21.02	151.5	147.8	9.71	2.17	71.08	69.68
4.10	8121	7923	21.50	20.97	150.6	146.9	9.83	2.12	71.12	69.74
4.00	7906	7713	21.45	20.93	149.7	146.0	9.96	2.07	71.15	69.81
3.90	7692	7505	21.40	20.88	148.7	145.1	10.11	2.01	71.18	69.88
3.80	7478	7296	21.35	20.83	147.8	144.2	10.25	1.96	71.22	69.94
3.70	7265	7088	21.30	20.78	146.8	143.3	10.41	1.91	71.25	70.00
3.60	7052	6880	21.24	20.72	145.9	142.3	10.57	1.86	71.29	70.07
3.50	6840	6673	21.19	20.67	144.9	141.3	10.76	1.81	71.33	70.14

These hydrostatic particulars have been developed with the vessel floating on even keel.

(ii) note the document in which this discrepancy should be recorded;

(iii) list any other CIVE operations that would be needed to [illegible]

(b) State the criteria that must be complied with for discharge under Annex IV of MARPOL 73/78.

Data sheet for use in Q2 of each [illegible] Stability and [illegible]

Particulars [illegible]

[illegible]

APPENDIX IV

100 Revision One-Liners

The following are 100 one-line questions acting as an aid to examination preparation. They are similar in effect to using mental arithmetic when preparing for a mathematics exam. Elements of questions may well appear in the written papers or in the oral exams. Good luck.

1. What is another name for the KG?
2. Define the term ‘storage factor’ for a grain compartment.
3. If the angle of heel is less than 10°, what is the equation for GZ?
4. What are the formulae for TPC and MCTC for a ship in salt water?
5. Give two formulae for the metacenter, KM.
6. How may free-surface effects be reduced on a ship?
7. What is another name for KB?
8. List four requirements before an inclining experiment can take place.
9. With the aid of a sketch, define LOA and LBP.
10. What are cross curves of stability used for?
11. What is the longitudinal center of a waterplane called?
12. Adding a weight to a ship usually causes two changes. What are these changes?
13. What is Simpson’s First Rule for a parabolic shape with seven equally spaced ordinates?
14. What is KB for (a) a box-shaped vessel and (b) a triangular-shaped vessel?
15. What are hydrostatic curves used for on board a ship?
16. Using sketches, define the block, the waterplane, and midship form coefficients.
17. Sketch a statical stability curve and label six important points on it.
18. What are the minimum values allowed by the DfT for GZ and for transverse GM?
19. List three ways in which a ship’s end drafts may be changed.
20. GM is 0.45 m. Radius of gyration is 7 m. Estimate the natural rolling period in seconds.
21. What is a deadweight scale used for?
22. What exactly is ‘ahm’ in the storage of grain?
23. Sketch a set of hydrostatic curves.
24. List three characteristics of an angle of loll.
25. Define (a) a moment and (b) a moment of inertia.
26. For a merchant ship, what is the TCD in terms of a ship’s LBP in deep waters?
27. What is the ‘theory of parallel axis’ formula?
28. What are the effects on a statical stability curve for increased breadth and increased freeboard?

29. Sketch a metacentric diagram for a box-shaped vessel and a triangular-shaped vessel.
30. Block coefficient is 0.715. Midship coefficient is 0.988. Calculate the prismatic coefficient.
31. Describe the use of Simpson's Third Rule.
32. What is the wall-sided formula for GZ?
33. Define 'permeability'. Give two examples relating to contents in a hold or tank.
34. Give the equations for BM, box-shaped vessels, and triangular-shaped vessels.
35. List three characteristics of an angle of list.
36. Sketch a typical navigational channel trench.
37. For a curve of seven equally spaced ordinates, give Simpson's Second Rule.
38. What does 'foap' signify?
39. When a weight is lifted from a jetty by a ship's derrick, whereabouts does its CG act?
40. Sketch a set of freeboard marks and label dimensions as specified by the DfT.
41. Sketch a displacement curve.
42. What is Morrish's formula for VCB?
43. For an inclining experiment how is the tangent of the angle of list obtained?
44. What do 'a moment of statical stability' and 'dynamical stability' mean?
45. Show the range of stability on an S/S curve having a very small initial negative GM.
46. Breadth is 45 m. Draft is 15 m. What is the increase in draft at a list of 2°?
47. What is the formula for loss of GM due to free-surface effects in a slack tank?
48. For what purpose is the inclining experiment made on ships?
49. What is the 'true mean draft' on a ship?
50. When drydocking a ship there is a virtual loss in GM. Give two formulae for this loss.
51. With Simpson's Rules, give formulae for moment of inertia about (a) amidships and (b) centerline.
52. Discuss the components involved for estimating an angle of heel whilst turning a ship.
53. What is a 'stiff ship' and a 'tender ship'. Give typical GM values.
54. With the lost buoyancy method, how does VCG change, after bilging has occurred?
55. Sketch a deadweight moment curve and label the important parts.
56. What type of vessel is a Qflex?
57. What are 'Bonjean curves' and for what purpose are they used?
58. Define 'ship squat' and 'blockage factor'.
59. Why are bulbous bows fitted on merchant ships?
60. Sketch a typical Statical Stability curve and place on it the point where the deck edge just becomes immersed.
61. What happens to cause a vessel to be in unstable equilibrium?
62. What causes hogging in a vessel?
63. Which letters signify the metacentric height?
64. Give typical C_b values for fully loaded VLCC, general cargo ships, and passenger liners.
65. What is the air draft and why is it important to know its value?

66. What are Type 'A' ships, Type 'B' ships, and Type 'B-60' ships.
67. What is synchronous rolling?
68. What is DfT tabular freeboard?
69. When dealing with subdivision of passenger vessels, what is the 'margin line'?
70. Sketch a set of freeboard marks for a timber ship.
71. Describe how icing can affect the stability of a vessel.
72. List six corrections made to a tabular freeboard to obtain an assigned freeboard.
73. Give the formula for the amidships draft for a vessel with longitudinal deflection.
74. Give the formula for the heaving period T_H in seconds.
75. In grain loading, what is the angle of repose?
76. Sketch a GZ curve together with a wind lever curve (up to 40° angle of heel).
77. Show on the sketch for the previous answer the angle of heel due to shift of grain.
78. Give a typical range for stowage rates for grain stowed in bulk.
79. Define exactly the term 'mean draft' and state whereabouts it is measured.
80. What is parametric rolling?
81. What is the difference between an ullage and a sounding in a partially filled tank?
82. What are sounding pads and why are they fitted in tanks?
83. What is a calibration book used for?
84. Sketch a gunwhale plate and indicate the placement of the freeboard Dk mark.
85. Seasonal allowances depend on three criteria. What are these three criteria?
86. What is the factor of subdivision F_S?
87. What is the criterion of service numeral C_S?
88. Discuss the term 'quasi-static stability'.
89. Do transverse bulkheads reduce free-surface effects?
90. For a DfT standard ship, what is the depth D in terms of floodable length F_L?
91. For timber ships, LW to LS is LS/n. What is the value of n?
92. The WNA mark is not fitted on ships above a particular length. What is that length?
93. According to the DfT, what is standard sheer in terms of LBP?
94. What is a Whessoe gage used for on ships?
95. List four reasons why a ship's overall G can be raised leading to loss of stability.
96. What exactly defines a bulkhead deck?
97. List four methods for removing ice that has formed on upper structures of ships.
98. Sketch a set of FL and PL curves.
99. Sketch the curves for Type 'A' ships, Type 'B' ships, and Type 'B-60' ships.
100. In the loading of grain, what is the approximate formula for angle of heel relative to 12°, due to shift of grain?

APPENDIX V

How to Pass Exams in Maritime Studies

To pass exams you have to be like a successful football or hockey team. You will need:

Ability, Tenacity, Consistency, Good Preparation, and Luck!!

The following tips should help you to obtain extra marks that could turn that 36% into a 42%+ pass or an 81% into an Honours 85%+ award. Good luck!!

Before Your Examination

1. Select 'bankers' for each subject. Certain topics come up very often. Certain topics you will have fully understood. Bank on these appearing on the exam paper.
2. Don't swat 100% of your course notes. Omit about 10% and concentrate on the 90%. In that 10% will be some topics you will never be able to understand fully.
3. Do all your coursework/homework assignments to the best of your ability. Many a student has passed on a weighted marking scheme, because of a good coursework mark compensating for a bad performance on examination day.
4. In the last days prior to examination, practice final preparation with your revision one-liners.
5. Work through passed exam papers in order to gage the standard and the time factor to complete the required solution.
6. Write all formulae discussed in each subject on pages at the rear of your notes.
7. In your notes circle each formula in a red outline or use a highlight pen. In this way formulae will stand out from the rest of your notes. Remember, formulae are like spanners. Some you will use more than others but all can be used to solve a problem. After January 2005, a formulae sheet will be provided in examinations for masters and mates.
8. Underline in red important key phrases or words. Examiners will be looking for these in your answers. Oblige them and obtain the marks.
9. Revise each subject in carefully planned sequence so as not to be rusty on a set of notes that you have not read for some time whilst you have been sitting other exams.
10. Be aggressive in your mental approach to do your best. If you have prepared well there will be less nervous approach and, like the football team, you will gain your goal. Success will come with proper training.

In Your Examination

1. If the examination is a mixture of descriptive and mathematical questions, select as your first question a descriptive or 'talkie-talkie' question to answer. It will settle you down for when you attempt the later questions to answer.
2. Select mathematical questions to answer if you wish to obtain higher marks for good answers. If they are good answers you are more likely to obtain full marks. For a 'talkie-talkie' answer a lot may be subjective to the feelings and opinion of the examiner. They may not agree with your feelings or opinions.
3. Use big sketches. Small sketches tend to irritate examiners.
4. Use colored pencils. Drawings look better with a bit of color.
5. Use a 150 mm rule to make a better sketch and a more professional drawing.
6. Have big writing to make it easier to read. Make it neat. Use a pen rather than a biro. Reading a piece of work written in biro is harder to read, especially if the quality of the biro is not very good.
7. Use plenty of paragraphs. It makes the text easier to read.
8. Write down any data you wish to remember. To write it makes it easier and longer to retain in your memory.
9. Be careful in your answers that you do not suggest things or situations that would endanger the ship or the onboard personnel.
10. Reread your answers near the end of the exam. Omitting the word 'not' does make such a difference.
11. Reread your question as you finish each answer. Don't miss, for example, part (c) of an answer and throw away marks you could have obtained.
12. Treat the exam as an advertisement of your ability rather than an obstacle to be overcome. If you think you will fail, then you probably will fail.

After the Examination

1. Switch off. Don't worry. Nothing you can do now will alter your exam mark be it good or bad. Students anyway tend to underestimate their performances in exams.
2. Don't discuss your exam answers with your student colleagues. It can falsely fill you with hope or with despair.
3. Turn your attention to preparing for the next subject in which you are to be examined.

APPENDIX VI

Ship Stability Data Sheets

MARITIME AND COASTGUARD AGENCY

in conjunction with

SCOTTISH QUALIFICATIONS AUTHORITY

EXAMINATION FOR CHIEF MATE CERTIFICATE OF COMPETENCY

SHIP STABILITY DATA SHEETS

The data sheets are intended for exercise and examination purposes only and are limited extracts from the original ship data.

General Particulars

Ship's Name	Ship A
Molded dimensions	LBP = 137.50 meters Breadth = 20.42 meters Depth = 11.7 5 meters
Summer load draught	8.867 meters
Block coefficient	0.7437
Summer load displacement	19,006 tonnes
Light displacement	3831 tonnes
Light KG	8.211 meters
Deadweight	15,175 tonnes
Gross tonnage	8996 tonnes
Net tonnage	6238 tonnes

Plan Showing Cargo Spaces, Storerooms, and Tanks

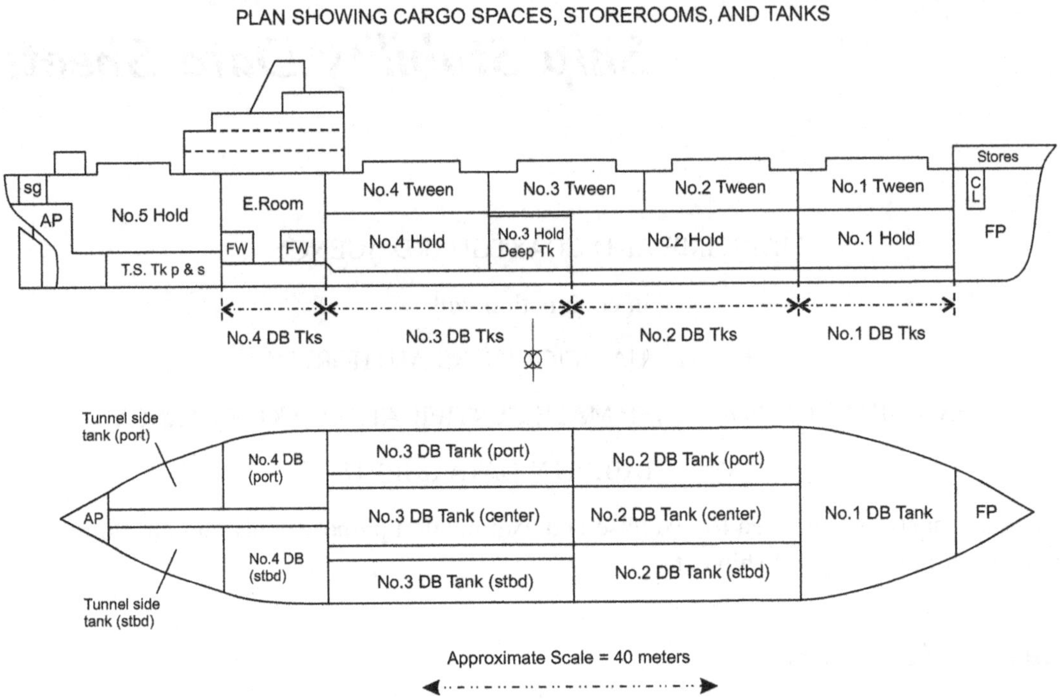

Cargo and Tank Capacities

Cargo Capacities

Compartment	Grain Capacity (m^3)	Bale Capacity (m^3)	LCG (foap) (m)	VCG Above Keel (m)
No. 1 hold	2215	1966	114.481	5.089
No. 2 hold	4672	4254	89.971	4.947
No. 3 hold	1742	1536	68.907	4.940
No. 4 hold	3474	3161	51.774	4.950
No. 5 hold	2605	2371	17.256	8.764
Total holds	14,708	13,288		
No. 1 tween deck	1695	1581	115.515	11.256
No. 2 tween deck	1676	1583	95.590	10.784
No. 3 tween deck	1626	1523	74.052	10.585
No. 4 tween deck	1674	1561	51.668	10.567
Total tween decks	6671	6248		
Total holds and tween decks	21,379	19,536		

Tank Capacities (FSMs for even keel with no list)

	Tonnes	LCG (foap) (m)	VCG Above Keel (m)	FSM for FW (m^4)
		Water Ballast		
Fore peak	554	130.556	8.434	519
After peak	108	3.067	7.726	325
No. 3 hold	1786	68.905	4.939	8113
No. 1 D.B. tank across	255	113.951	0.614	3722
No. 2 D.B. tank center	271	89.996	0.590	1021
No. 2 D.B. tank port	223	89.467	0.603	680
No. 2 D.B. tank starboard	223	89.467	0.603	680
		Heavy Fuel Oil		
No. 3 D.B. tank center	271	57.015	0.598	1142
No. 3 D.B. tank port	153	57.871	0.626	275
No 3 D.B. tank starboard	153	57.871	0.626	275
Tunnel side tank port	198	21.076	2.285	246
Tunnel side tank starboard	198	21.076	2.285	246
		Diesel Oil		
No. 4 D.B. tank port	55	35.661	0.829	168
No. 4 D.B. tank starboard	52	35.502	0.833	150
		Fresh Water		
Forward tank	51	32.468	7.350	29
After tank	44	28.669	7.383	46

Hydrostatic Particulars

Draught (m)	Displacement (t) SW RD 1.025	Displacement (t) FW RD 1.000	TPC (t) SW RD 1.025	TPC (t) FW RD 1.000	MCTC (tm) SW RD 1.025	MCTC (tm) FW RD 1.000	KM_T (m)	KB (m)	LCB foap (m)	LCF foap (m)
10.00	21,789	21,258	24.85	24.24	224.8	219.3	8.69	5.25	68.71	65.11
9.90	21,541	21,016	24.80	24.20	223.6	218.1	8.67	5.20	68.75	65.16
9.80	21,293	20,774	24.75	24.15	222.4	217.0	8.64	5.15	68.79	65.20
9.70	21,046	20,533	24.70	24.10	221.2	215.8	8.62	5.10	68.83	65.25
9.60	20,799	20,292	24.65	24.05	220.0	214.6	8.60	5.04	68.87	65.29
9.50	20,553	20,052	24.60	24.00	218.8	213.5	8.58	4.99	68.92	65.34
9.40	20,307	19,812	24.55	23.95	217.6	212.3	8.56	4.93	68.96	65.39
9.30	20,062	19,573	24.50	23.90	216.4	211.1	8.54	4.88	69.00	65.45

(Continued)

: —cont'd

Draught (m)	Displacement (t) SW RD 1.025	Displacement (t) FW RD 1.000	TPC (t) SW RD 1.025	TPC (t) FW RD 1.000	MCTC (tm) SW RD 1.025	MCTC (tm) FW RD 1.000	KM_T (m)	KB (m)	LCB foap (m)	LCF foap (m)
9.20	19,817	19,334	24.45	23.85	215.2	210.0	8.52	4.82	69.04	65.50
9.10	19,573	19,096	24.40	23.80	213.0	207.8	8.50	4.77	69.09	65.56
9.00	19,329	18,858	24.35	23.76	212.7	207.5	8.48	4.72	69.13	65.62
8.90	19,086	18,620	24.30	23.71	211.5	206.3	8.47	4.67	69.18	65.68
8.80	18,843	18,383	24.24	23.65	210.2	205.1	8.45	4.61	69.22	65.74
8.70	18,601	18,147	24.18	23.59	208.0	202.9	8.43	4.56	69.27	65.81
8.60	18,359	17,911	24.13	23.54	207.7	202.6	8.42	4.50	69.31	65.87
8.50	18,119	17,677	24.08	23.49	206.4	201.4	8.41	4.45	69.36	65.95
8.40	17,878	17,442	24.02	23.43	205.1	200.1	8.39	4.39	69.40	66.02
8.30	17,639	17,208	23.96	23.38	203.8	198.8	8.38	4.34	69.45	66.10
8.20	17,399	16,975	23.90	23.32	202.4	197.5	8.37	4.28	69.49	66.17
8.10	17,161	16,742	23.84	23.26	201.0	196.1	8.36	4.23	69.54	66.25
8.00	16,922	16,509	23.78	23.20	199.6	194.7	8.35	4.17	69.58	66.33
7.90	16,685	16,278	23.71	23.13	198.2	193.4	8.35	4.12	69.63	66.42
7.80	16,448	16,047	23.65	23.07	196.8	192.0	8.34	4.07	69.67	66.51
7.70	16,212	15,817	23.59	23.01	195.4	190.6	8.34	4.02	69.72	66.61
7.60	15,976	15,586	23.52	22.95	193.9	189.2	8.33	3.96	69.76	66.71
7.50	15,742	15,358	23.45	22.88	192.4	187.7	8.33	3.91	69.81	66.82
7.40	15,507	15,129	23.39	22.82	190.9	186.2	8.33	3.85	69.85	66.92
7.30	15,274	14,901	23.33	22.76	189.4	184.8	8.33	3.80	69.90	67.03
7.20	15,040	14,673	23.26	22.69	187.8	183.2	8.33	3.75	69.94	67.13
7.10	14,808	14,447	23.19	23.32	186.2	181.7	8.34	3.70	69.99	67.24
7.00	14,576	14,220	23.13	22.57	184.6	180.1	8.34	3.64	70.03	67.35
6.90	14,345	13,996	23.06	22.50	183.0	178.5	8.35	3.58	70.08	67.46
6.80	14,115	13,771	22.99	22.43	181.4	177.0	8.36	3.53	70.12	67.57
6.70	13,886	13,548	22.92	22.36	179.9	175.5	8.37	3.48	70.16	67.68
6.60	13,657	13,324	22.85	22.29	178.3	174.0	8.38	3.43	70.20	67.79
6.50	13,429	13,102	22.78	22.23	176.8	172.5	8.39	3.38	70.24	67.90
6.40	13,201	12,879	22.72	22.17	175.3	171.0	8.41	3.33	70.28	68.00
6.30	12,975	12,658	22.66	22.11	173.9	169.6	8.43	3.28	70.32	68.10
6.20	12,748	12,437	22.60	22.05	172.5	168.3	8.46	3.22	70.35	68.20
6.10	12,523	12,217	22.54	21.99	171.1	167.0	8.49	3.17	70.38	68.30
6.00	12,297	11,997	22.48	21.93	169.8	165.7	8.52	3.11	70.42	68.39
5.90	12,073	11,778	22.43	21.87	168.5	164.4	8.55	3.06	70.46	68.43
5.80	11,848	11,559	22.37	21.82	167.3	163.2	8.59	3.01	70.50	68.57
5.70	11,625	11,342	22.32	21.77	166.1	162.1	8.63	2.95	70.53	68.65
5.60	11,402	11,124	22.26	21.72	165.0	161.0	8.67	2.90	70.57	68.73
5.50	11,180	10,908	22.21	21.66	163.9	160.0	8.71	2.85	70.60	68.80
5.40	10,958	10,691	22.15	21.61	162.9	158.9	8.76	2.80	70.64	68.88

: —cont'd

Draught (m)	Displacement (t) SW RD 1.025	Displacement (t) FW RD 1.000	TPC (t) SW RD 1.025	TPC (t) FW RD 1.000	MCTC (tm) SW RD 1.025	MCTC (tm) FW RD 1.000	KM_T (m)	KB (m)	LCB foap (m)	LCF foap (m)
5.30	10,737	10,476	22.10	21.56	161.8	157.9	8.81	2.74	70.68	68.95
5.20	10,516	10,260	22.05	21.51	160.8	156.9	8.86	2.69	70.72	69.02
5.10	10,296	10,045	22.00	21.46	159.8	155.9	8.92	2.63	70.75	69.09
5.00	10,076	9830	21.95	21.41	158.8	154.9	8.98	2.58	70.79	69.16
4.90	9857	9616	21.90	21.36	157.9	154.0	9.06	2.53	70.82	69.23
4.80	9638	9403	21.85	21.32	156.9	153.1	9.13	2.48	70.86	69.29
4.70	9420	9190	21.80	21.27	156.0	152.2	9.22	2.43	70.90	69.35
4.60	9202	8978	21.75	21.22	155.1	151.3	9.30	2.38	70.93	69.42
4.50	8985	8766	21.70	21.17	154.2	150.5	9.40	2.32	70.96	69.48
4.40	8768	8554	21.65	21.12	153.3	149.6	9.49	2.27	71.00	69.55
4.30	8552	8344	21.60	21.07	152.4	148.7	9.60	2.22	71.04	69.62
4.20	8336	8133	21.55	21.02	151.5	147.8	9.71	2.17	71.08	69.68
4.10	8121	7923	21.50	20.97	150.6	146.9	9.83	2.12	71.12	69.74
4.00	7906	7713	21.45	20.93	149.7	146.0	9.96	2.07	71.15	69.81
3.90	7692	7505	21.40	20.88	148.7	145.1	10.11	2.01	71.18	69.88
3.80	7478	7296	21.35	20.83	147.8	144.2	10.25	1.96	71.22	69.94
3.70	7265	7088	21.30	20.78	146.8	143.3	10.41	1.91	71.25	70.00
3.60	7052	6880	21.24	20.72	145.9	142.3	10.57	1.86	71.29	70.07
3.50	6840	6673	21.19	20.67	144.9	141.3	10.76	1.81	71.33	70.14
3.40	6628	6466	21.13	20.61	143.9	140.4	10.95	1.75	71.37	70.20
3.30	6418	6261	21.08	20.56	142.9	139.4	11.18	1.70	71.41	70.27
3.20	6207	6056	21.02	20.51	141.9	138.4	11.40	1.65	71.44	70.33
3.10	5998	5852	20.96	20.45	140.9	137.5	11.66	1.60	71.48	70.40
3.00	5788	5647	20.90	20.39	139.9	136.5	11.92	1.55	71.52	70.46
2.90	5580	5444	20.84	20.33	138.9	135.5	12.22	1.50	71.56	70.53
2.80	5371	5240	20.78	20.27	137.9	134.5	12.52	1.44	71.60	70.59
2.70	5164	5038	20.72	20.21	136.9	133.6	12.87	1.39	71.64	70.66
2.60	4957	4836	20.65	20.15	135.9	132.6	13.21	1.34	71.67	70.73
2.50	4752	4636	20.58	20.08	134.9	131.6	13.63	1.29	71.71	70.80
2.40	4546	4435	20.51	20.01	133.9	130.6	14.04	1.23	71.75	70.87
2.30	4342	4236	20.44	19.94	132.9	129.6	14.56	1.18	71.79	70.94
2.20	4138	4037	20.36	19.86	131.8	128.6	15.07	1.13	71.83	71.01
2.10	3936	3840	20.28	19.79	130.7	127.5	15.72	1.08	71.87	71.08
2.00	3733	3642	20.20	19.71	129.5	126.3	16.36	1.02	71.91	71.15
1.90	3532	3446	20.12	19.63	128.3	125.2	17.19	0.97	71.96	71.22
1.80	3331	3250	20.03	19.54	127.0	123.9	18.01	0.92	72.00	71.29
1.70	3132	3055	19.93	19.45	125.6	122.5	19.08	0.87	72.05	71.37
1.60	2932	2860	19.83	19.35	124.1	121.1	20.15	0.82	72.09	71.44

These hydrostatic particulars have been developed with the vessel floating on even keel.

Tabulated KN Values

KN values in meters

KN values calculated for vessel on even keel and fixed angle of heel

Displacement (tonnes)	Angle of Heel (°)						
	12	20	30	40	50	60	75
20,000	1.80	2.90	4.14	5.14	5.92	6.51	6.84
19,500	1.79	2.90	4.17	5.19	5.97	6.55	6.86
19,000	1.78	2.91	4.20	5.24	6.02	6.59	6.88
18,500	1.77	2.92	4.23	5.29	6.07	6.63	6.90
18,000	1.75	2.93	4.27	5.36	6.12	6.67	6.92
17,500	1.74	2.94	4.30	5.43	6.18	6.71	6.94
17,000	1.73	2.95	4.34	5.48	6.23	6.75	6.96
16,500	1.73	2.97	4.37	5.54	6.29	6.79	6.98
16,000	1.72	2.98	4.40	5.60	6.35	6.83	7.00
15,500	1.72	2.98	4.44	5.66	6.44	6.87	7.02
15,000	1.72	2.98	4.48	5.72	6.48	6.91	7.04
14,500	1.73	2.98	4.51	5.79	6.58	6.95	7.07
14,000	1.74	2.98	4.53	5.81	6.68	7.00	7.10
13,500	1.75	2.99	4.56	5.86	6.73	7.05	7.13
13,000	1.76	3.00	4.59	5.90	6.78	7.10	7.16
12,500	1.77	3.03	4.64	5.96	6.83	7.15	7.19
12,000	1.78	3.06	4.68	6.02	6.88	7.20	7.21
11,500	1.80	3.10	4.73	6.07	6.93	7.25	7.24
11,000	1.82	3.14	4.78	6.12	6.98	7.30	7.26
10,500	1.83	3.19	4.81	6.17	7.02	7.35	7.29
10,000	1.86	3.24	4.85	6.21	7.08	7.40	7.31
9500	1.93	3.28	4.91	6.25	7.11	7.45	7.34
9000	2.00	3.36	4.96	6.28	7.18	7.50	7.36
8500	2.05	3.43	5.03	6.32	7.20	7.55	7.39
8000	2.10	3.52	5.09	6.35	7.22	7.60	7.41
7500	2.17	3.62	5.17	6.38	7.24	7.65	7.43
7000	2.22	3.70	5.25	6.41	7.26	7.70	7.45
6500	2.32	3.85	5.35	6.44	7.27	7.70	7.47
6000	2.42	4.00	5.45	6.48	7.28	7.70	7.49
5500	2.57	4.15	5.55	6.53	7.29	7.68	7.47
5000	2.72	4.32	5.67	6.58	7.30	7.66	7.45
4500	2.92	4.55	5.79	6.64	7.25	7.60	7.42
4000	3.15	4.75	5.91	6.71	7.22	7.52	7.40
3500	3.45	5.00	6.08	6.78	7.20	7.42	7.38

KN values are for hull and forecastle only.

Maximum KG to Comply with Minimum Intact Stability Criteria Specified in the Current Loadline Rules

Displacement (t)	KG (m)
19,500	7.849
19,000	7.932
18,500	8.015
18,000	8.100
17,500	8.169
17,000	8.203
16,500	8.189
16,000	8.181
15,500	8.180
15,000	8.185
14,500	8.196
14,000	8.215
13,500	8.243
13,000	8.283
12,500	8.341
12,000	8.415
11,500	8.502
11,000	8.599
10,500	8.712
10,000	8.854
9500	9.030
9000	9.238
8500	9.476
8000	9.729
7500	9.817
7000	9.619
6500	9.424
6000	9.180
5500	8.838
5000	8.398

Angle at Which Flooding Could Occur

Displacement (tonnes)	Angle to Immerse Deck Edge (°)	Angle of Flooding (°)	Wind Moment (meter tonnes)
4000	54.90	76.75	897
5000	50.30	73.25	869
6000	46.70	69.80	841
7000	43.60	66.65	814

(*Continued*)

: —cont'd

Displacement (tonnes)	Angle to Immerse Deck Edge (°)	Angle of Flooding (°)	Wind Moment (meter tonnes)
8000	40.70	63.55	786
9000	38.00	60.60	760
10,000	35.30	57.80	733
11,000	32.90	55.20	707
12,000	30.80	52.80	681
13,000	28.40	50.55	655
14,000	26.30	48.45	630
15,000	24.30	46.45	604
16,000	22.30	44.50	580
17,000	20.30	42.55	555
18,000	18.20	40.60	531
19,000	16.00	38.65	507
20,000	13.75	36.65	483

Grain Heeling Moments

Full Holds

Hold	Grain Capacity (m^3)	Horizontal Volumetric Heeling Moment (m^4)	VCG of Hold (m)
No. 1	2215	409.5	5.089
No. 2	4672	1284.9	4.947
No. 3	1742	475.4	4.940
No. 4	3474	910.4	4.950
No. 5	2605	454.9	8.764

Full Tween Decks

Tween Deck	Grain Capacity (m^3)	Horizontal Volumetric Heeling Moment (m^4)	VCG of Tween Deck (m)
No. 1	1695	352.4	11.256
No. 2	1676	539.0	10.784
No. 3	1626	578.2	10.585
No. 4	1674	604.4	10.567

No. 1 Tween Deck: Volumetric Heeling Moments when Partly Filled

Ullage datum: Top of hatch side coaming at its mid length

No centerline division

Ullage (m)	Volume of Grain (m^3)	Horizontal Heeling Moment (m^4)	KG of Grain (m)
0.25	1686	598	11.238
0.50	1668	659	11.187
0.75	1649	746	11.134
1.00	1628	864	11.070
1.25	1607	1016	11.006
1.50	1510	1176	10.942
1.75	1416	1372	10.880
2.00	1324	1577	10.815
2.25	1232	1700	10.747
2.50	1144	2017	10.689
2.75	1059	2218	10.633
3.00	970	2388	10.592
3.25	883	2512	10.547
3.50	800	2579	10.504
3.75	714	2575	10.450
4.00	633	2500	10.386
4.25	550	2362	10.308
4.50	467	2155	10.206
4.75	384	1908	10.197
5.00	302	1592	9.978
5.25	222	1239	9.806
5.50	143	848	9.561
5.75	64	380	9.267
5.95	0	0	8.700
	Ullage for Maximum Horizontal Heeling Moment		
3.60	764	2578	10.484

The horizontal heeling moment for the slack tween deck has been obtained by calculating the maximum horizontal heeling moment and increasing it by 12%. The KG of the level grain is to be used when calculating the KG of the ship.

So, the maximum horizontal heeling moment $\times$ 1.12 is equal to the horizontal heeling moment (see Chapter 37):

$$\text{Volumetric heeling moments} \times 1.12 = \text{ahm}$$

No. 2 Tween Deck: Volumetric Heeling Moments when Partly Filled

Ullage datum: Top of hatch side coaming at its mid length

No centerline division

Ullage (m)	Volume of Grain (m^3)	Horizontal Heeling Moment (m^4)	KG of Grain (m)
0.25	1659	975	10.762
0.50	1634	1019	10.735
0.75	1609	1122	10.699
1.00	1584	1344	10.689
1.25	1560	1642	10.653
1.50	1472	1948	10.623
1.75	1369	2257	10.589
2.00	1264	2566	10.555
2.25	1165	2874	10.495
2.50	1062	3160	10.440
2.75	959	3350	10.379
3.00	858	3460	10.323
3.25	757	3514	10.278
3.50	658	3432	10.239
3.75	550	3218	10.183
4.00	452	2925	10.112
4.25	349	2473	10.001
4.50	248	1868	9.853
4.75	143	1160	9.621
5.00	47	368	9.228
5.12	0	0	8.700
	Ullage for Maximum Horizontal Heeling Moment		
3.25	757	3514	10.278

The horizontal heeling moment for the slack tween deck has been obtained by calculating the maximum horizontal heeling moment and increasing it by 12%. The KG of the level grain is to be used when calculating the KG of the ship.

So, the maximum horizontal heeling moment × 1.12 is equal to the horizontal heeling moment (see Chapter 37):

$$\text{Vhm} \times 1.12 = \text{ahm}$$

No. 3 Tween Deck: Volumetric Heeling Moments when Partly Filled

Ullage datum: Top of hatch side coaming at its mid length

No centerline division

Ullage (m)	Volume of Grain (m^3)	Horizontal Heeling Moment (m^4)	KG of Grain (m)
0.25	1598	1028	10.550
0.50	1569	1051	10.526
0.75	1540	1122	10.501
1.00	1512	1268	10.481
1.25	1476	1547	10.469
1.50	1399	1900	10.452
1.75	1308	2236	10.415
2.00	1203	2558	10.356
2.25	1086	2863	10.315
2.50	979	3142	10.254
2.75	866	3304	10.190
3.00	758	3339	10.134
3.25	649	3291	10.080
3.50	538	3109	10.042
3.75	427	2750	9.981
4.00	317	2262	9.861
4.25	206	1610	9.617
4.50	97	796	9.367
4.72	0	0	8.700
	Ullage for Maximum Horizontal Heeling Moment		
3.00	7.58	3339	10.134

The horizontal heeling moment for the slack tween deck has been obtained by calculating the maximum horizontal heeling moment and increasing it by 12%. The KG of the level grain is to be used when calculating the KG of the ship.

So, the maximum horizontal heeling moment × 1.12 is equal to the horizontal heeling moment (see Chapter 37):

$$\text{Vhm} \times 1.12 = \text{ahm}$$

No. 4 Tween Deck: Volumetric Heeling Moments when Partly Filled

Ullage datum: Top of hatch side coaming at its mid length

No centerline division

Ullage (m)	Volume of Grain (m^3)	Horizontal Heeling Moment (m^4)	KG of Grain (m)
0.25	1646	1072	10.581
0.50	1618	1147	10.559
0.75	1589	1248	10.541
1.00	1561	1398	10.517
1.25	1530	1592	10.495
1.50	1452	1847	10.470
1.75	1360	2150	10.425
2.00	1243	2512	10.371
2.25	1132	2857	10.324
2.50	1014	3116	10.268
2.75	900	3278	10.203
3.00	786	3323	10.140
3.25	672	3283	10.085
3.50	555	3094	10.038
3.75	444	2794	9.977
4.00	329	2250	9.847
4.25	213	1526	9.671
4.50	98	813	9.371
4.72	0	0	8.700
Ullage for Maximum Horizontal Heeling Moment			
3.00	786	3323	10.140

The horizontal heeling moment for the slack tween deck has been obtained by calculating the maximum horizontal heeling moment and increasing it by 12%. The KG of the level grain is to be used when calculating the KG of the ship.

So, the maximum horizontal heeling moment × 1.12 is equal to the horizontal heeling moment (see Chapter 37):

$$\text{Vhm} \times 1.12 = \text{ahm}$$

Table of Maximum Permissible Grain Heeling Moments

D	Fluid KG (m)														
(tonnes)	5.500	5.600	5.700	5.800	5.900	6.000	6.100	6.200	6.300	6.400	6.500	6.600	6.700	6.800	6.900	7.000
5000	8425	8315	8204	8094	7983	7873	7762	7651	7541	7430	7320	7209	7099	6988	6877	6767
5500	8336	8214	8093	7971	7850	7728	7606	7485	7363	7241	7120	6998	6877	6755	6633	6512
6000	8158	8026	7893	7760	7628	7495	7362	7230	7097	6964	6832	6699	6566	6434	6301	6168
6500	7987	7843	7699	7556	7412	7268	7124	6981	6837	6693	6550	6406	6262	6118	5975	5831
7000	7891	7737	7582	7427	7272	7117	6963	6808	6653	6498	6343	6189	6034	5879	5724	5569
7500	7911	7745	7579	7413	7248	7082	6916	6750	6584	6418	6252	6087	5921	5755	5589	5423
8000	7979	7802	7625	7448	7271	7094	6917	6741	6564	6387	6210	6033	5856	5679	5502	5325
8500	8006	7818	7630	7442	7254	7066	6879	6691	6503	6315	6127	5939	5751	5563	5375	5187
9000	8017	7818	7619	7420	7221	7022	6823	6624	6425	6226	6027	5828	5629	5430	5231	5032
9500	8062	7852	7642	7432	7222	7012	6802	6592	6382	6171	5961	5751	5541	5331	5121	4911
10,000	8151	7930	7709	7488	7267	7046	6824	6603	6382	6161	5940	5719	5498	5276	5055	4834
10,500	8270	8038	7800	7573	7341	7109	6877	6645	6412	6180	5948	5716	5484	5251	5019	4787
11,000	8376	8133	7890	7647	7403	7160	6917	6674	6430	6187	5944	5701	5457	5214	4971	4728
11,500	8437	8182	7928	7674	7419	7165	6911	6656	6402	6148	5893	5639	5385	5130	4876	4622
12,000	8511	8245	7980	7714	7449	7184	6918	6653	6388	6122	5857	5591	5326	5061	4795	4630
12,500	8655	8379	8102	7826	7549	7273	6997	6702	6444	6167	5891	5614	5338	5062	4785	4509
13,000	8809	8522	8234	7947	7659	7372	7084	6797	6509	6222	5934	5647	5359	5072	4784	4497
13,500	8909	8611	8312	8013	7715	7416	7118	6819	6521	6222	5924	5625	5327	5028	4730	4431
14,000	9053	8743	8434	8124	7815	7505	7195	6886	7576	6267	5957	5647	5338	5028	4719	4409
14,500	9347	9026	8706	8385	8065	7744	7423	7103	6782	6461	6141	5820	5499	5179	4858	4537
15,000	9702	9371	9039	8707	8376	8044	7712	7380	7049	6717	6385	6053	5722	5390	5058	4727
15,500	10,010	9667	9325	8982	8639	8296	7954	7611	7268	6925	6583	6240	5897	5554	5211	4869
16,000	10,352	9998	9644	9290	8937	8583	8229	7875	7521	7167	6814	6460	6106	5752	5398	5045
16,500	10,823	10,458	10,093	9728	9363	8998	8634	8269	7904	7539	7174	6809	6444	6079	5715	5350
17,000	11,329	10,953	10,577	10,201	9826	9450	9074	8698	8322	7946	7570	7194	6818	6442	6066	5690
17,500	11,762	11,375	10,988	10,601	10,214	9827	9440	9053	8666	8279	7892	7505	7118	6731	6344	5957
18,000	12,173	11,775	11,377	10,979	10,581	10,183	9785	9387	8989	8590	8192	7794	7396	6998	6600	6202
18,500	12,626	12,217	11,808	11,398	10,989	10,580	10,171	9762	9353	8944	8535	8125	7716	7307	6898	6489
19,000	13,040	12,619	12,199	11,779	11,539	10,939	10,519	10,098	9678	9258	8838	8418	7998	7577	7157	6737
19,500	13,376	12,945	12,514	12,082	11,651	11,220	10,789	10,357	9926	9495	9064	8633	8201	7770	7339	6908

(Continued)

: —cont'd

D	Fluid KG (m)													
(tonnes)	7.100	7.200	7.300	7.400	7.500	7.600	7.700	7.800	7.900	8.000	8.100	8.200	8.300	8.400
5000	6656	6546	6435	6325	6214	6103	5993	5882	5772	5661	5551	5440	5329	
5500	6390	6268	6147	6025	5903	5782	5660	5539	5417	5295	5174	5052	4930	4809
6000	6035	5903	5770	5637	5505	5372	5239	5107	4974	4841	4709	4576	4443	4311
6500	5687	5543	5400	5256	5112	4968	4825	4681	4537	4393	4250	4106	3962	3818
7000	5415	5260	5105	4950	4795	4641	4486	4331	4176	4021	3867	3712	3557	3402
7500	5257	5091	4926	4760	4594	4428	4262	4096	3930	3765	3599	3433	3267	3101
8000	5148	4971	4795	4618	4441	4264	4087	3910	3733	3556	3379	3202	3025	2848
8500	4999	4811	4623	4435	4247	4059	3871	3683	3495	3307	3119	2931	2743	2555
9000	4833	4634	4435	4236	4037	3838	3639	3440	3241	3041	2842	2643	2444	2245
9500	4701	4491	4281	4071	3860	3650	3440	3230	3020	2810	2600	2390	2180	1970
10,000	4613	4392	4171	3950	3728	3507	3286	3065	2844	2623	2402	2180	1959	1738
10,500	4555	4323	4090	3858	3626	3394	3162	2929	2697	2465	2233	2001	1768	1536
11,000	4484	4241	3998	3755	3511	3268	3025	2782	2538	2295	2052	1808	1565	1322
11,500	4368	4113	3859	3605	3350	3096	2842	2587	2333	2079	1824	1570	1316	
12,000	4265	3999	3734	3468	3203	2938	2672	2407	2142	1876	1611	1346		
12,500	4232	3956	3679	3403	3127	2850	2574	2297	2021	1744	1468			
13,000	4209	3922	3634	3347	3059	2772	2484	2197	1909	1622	1335			
13,500	4132	3834	3535	3237	2938	2640	2341	2043	1744	1446				
14,000	4099	3790	3480	3171	2861	2551	2242	1932	1623	1313				
14,500	4217	3896	3575	3255	2934	2613	2293	1972	1651	1331				
15,000	4395	4063	3731	3400	3068	2736	2405	2073	1741	1409				
15,500	4526	4183	3840	3498	3155	2812	2469	2127	1784	1441				
16,000	4691	4337	3983	3629	3275	2922	2568	2214	1860	1506				
16,500	4985	4620	4255	3890	3525	3160	2795	2431	2066	1701				
17,000	5314	4938	4562	4186	3810	3434	3059	2683	2079	1064				
17,500	5570	5183	4796	4409	4022	3635	3248	2411	1324	383				
18,000	5804	5406	5008	4610	4212	3814	2563	1493	542					
18,500	6080	5671	5262	4853	3721	2662	1681	793	14					
19,000	6317	5897	5076	3995	2974	2029	1156	349						
19,500	6476	5616	4490	3427	2436	1514	644							

Worksheet: Trim and Stability

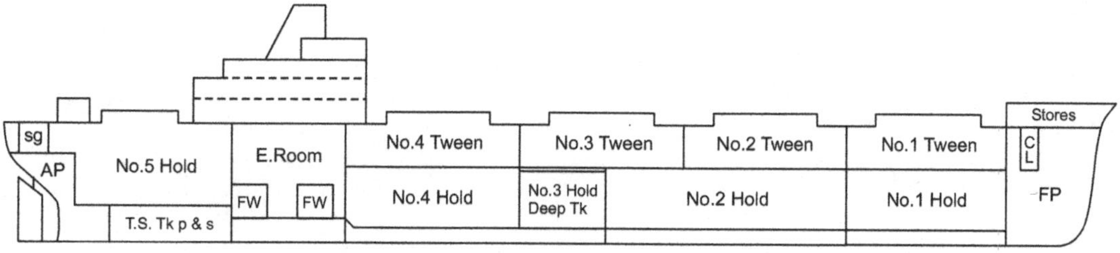

CONDITION:								
Compartment	**Capacity (m^3)**	**Stowage Factor (m^3/t)**	**Weight (t)**	**KG (m)**	**Vertical Moment (tm)**	**Free Surface Moment (tm)**	**LCG foap (m)**	**Longitudinal Moment (tm)**
Deadweight								
Lightship								
DISPLACEMENT								
HYDROSTATICS				True Mean Draught =		LCB = m foap		LCF = m foap
				MCTC =				
						KM_T = m		
LCB from LCG =								
Trim between perpendiculars								
DRAUGHTS:				F.			A.	

APPENDIX VII

Capsize of the Herald of Free Enterprise — A Journalistic Review

Introduction

The *Herald of Free Enterprise*, a ro-ro vessel, was delivered to her owners in 1980. The general particulars of the ship are listed in Table 1. She was one of three sister ships. The other two were the *Pride of Free Enterprise* and the *Spirit of Free Enterprise*.

On 6 March 1987 the *Herald of Free Enterprise* capsized shortly after leaving Zeebrugge in Belgium. She was a passenger and freight ferry and was bound for Dover. This Zeebrugge to Dover transit was scheduled to be four crossings per day, with each crossing taking about 4 hours.

The *Herald of Free Enterprise* left Zeebrugge at 6.05 p.m., about 78 minutes after high water. When static and when at forward speed she was trimming by the bow. She left the port of Zeebrugge with both bow doors open. She was carrying 459 passengers, 80 crew, 81 cars, three buses/coaches, and 47 lorries.

Table 1: General Particulars of the *Herald of Free Enterprise*.

WA = 131.9 m	Breadth molded = 22.70 m
LBP = 126.1 m	Service speed = 22 knots
Ship's displacement	8874–9250 tonnes
Mean draft (static)	5.68–5.85 m
Draft forward (static)	6.06–6.26 m
Draft aft (static)	5.31–5.43 m
Trim by the bow (static)	0.75–0.83 m
VCG (fluid) with slack tanks	9.73–9.75 m
GM_T (fluid) with slack tanks	2.04–2.09 m
Ship speed just prior to grounding	18.00–18.50 knots
C_b at draft molded of 5.30 m	0.525
Three props aft, one prop forward	Brake power = 20,142 kW
Sulzer diesel engines fitted	CP propellers fitted
Hinged stabilizer fins fitted	Deadweight = 2000 t
Cost	Of the order of £30 million
Number of lives lost	193

Figure 1:
'Herald of Free Enterprise'. Grounded and capsized on 6th March 1987.

She was turned to starboard. It was this action that saved a greater loss of life. The vessel ended up on a sandbank. In doing so, it turned a sinkage into a capsize incident. When capsized, her starboard side was above the water. Approximately half her breadth molded remained above the waterline (see Figure 1).

She eventually ended up about 930 m off the centerline of the navigation channel. The disaster unfolded in just one and a half minutes in calm water conditions. At the time that she went aground it was suggested by Maritime Press that her forward speed was of the order of 18–20 knots (see Figure 2); 193 people lost their lives.

What were the possible reasons for this ferry to capsize in this manner? Can lessons be learned? This chapter is written to consider and suggest at least 30 possible reasons. Furthermore, it then considers improvements arising from those reasons for retrofits on existing ferries and future design policy for new ro-ro ferry designs.

Reasons Why the Herald of Free Enterprise *Capsized*

It can take only one and a half minutes for a ro-ro ferry to sink after she has lost her seaworthiness.

1. A larger ship than normally used was selected for the run.
2. H of FE was usually on the Dover Calais run.
3. The ship had to be trimmed by the bow to bring cars on board.
4. Larger ship meant smaller static underkeel clearances than normal. (See Figure 2)

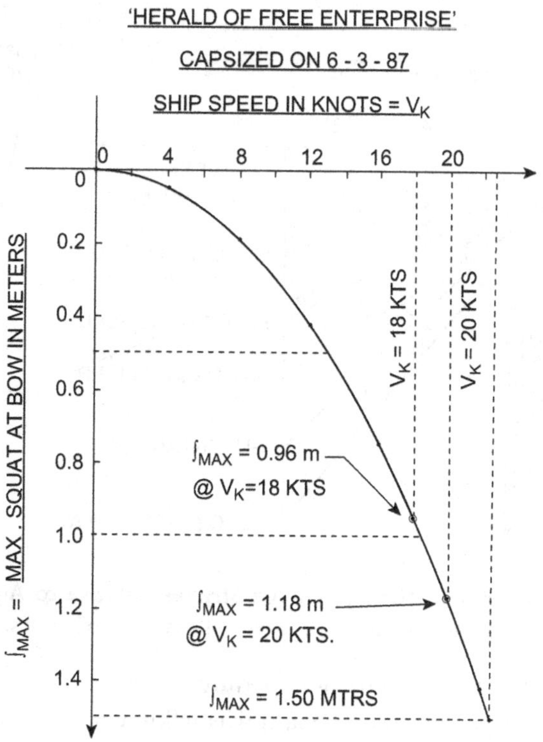

Figure 2:
Squat Curve for the *Herald of Free Enterprise*.

5. The ship developed too great a speed just prior to the capsize. This caused build-up of a pressure wave at the bow and also excessive squat.
6. The bow doors were left open.
7. Water entered through these open bow doors.
8. Ingress of water caused rise of 'G' due to loading.
9. Ingress of water caused a rise of 'G' due to free-surface effects (FSE).
10. The ship's 'G' rose above ship's 'M' because of loading and free-surface effects. See Figure 3.
11. At forward speed, squat produced trim by the bow. This was added to static trim by the bow.
12. Water ballast was put in No. 1 double-bottom tank to produce trim by the bow.
13. At speed, an order to turn the ship to starboard was given. This would cause the ship to heel and produce loss of UKC at bilge strakes.
14. To bring the ship nearer to even keel, a pump was used to transfer water from No. 1 double-bottom (forward) to No. 14 double-bottom tank (aft).
15. A more powerful pump was requested. This was refused.
16. The ship's lightweight was inaccurate.

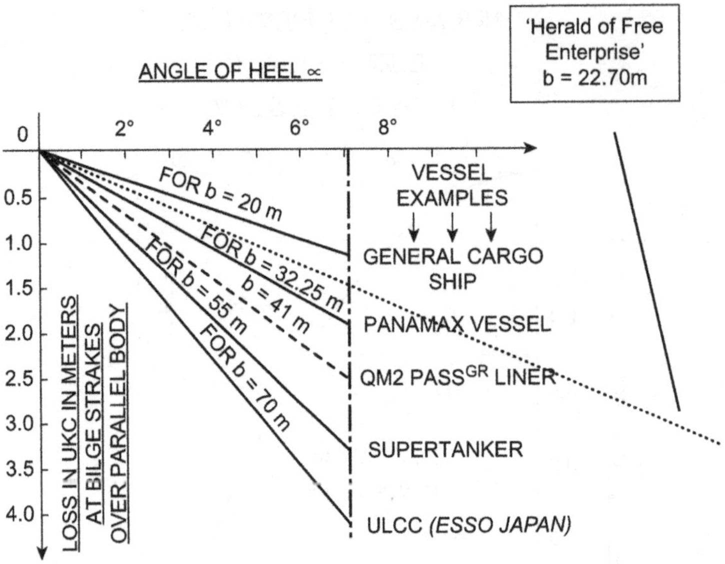

Figure 3: Loss of UKC [UKC] at Bilge Strakes Relative to Angle of Heel.
Ship is stationary at zero forward speed. Loss of UKC = ½sin α meters.

17. The vessel was overloaded at time of departure.
18. The exact number of passengers on board was unknown.
19. Aft draft was unable to be read.
20. No end draft indicators linked to the bridge were fitted.
21. The vessel departed 5 minutes late.
22. There were no onboard stability data regarding condition of trimming the bow.
23. Estimates of individual car weights were incorrect.
24. There was no audio, telephone, or television linkup between the bridge and the bow doors.
25. Other occasions had occurred when the bow doors had been left open.
26. There were fatigue problems with officers and crew.
27. There were too many changes in sailing schedules.
28. There were very wide open transverse spaces at car deck level.
29. There were commercial pressures for a quick turnaround in port.
30. Ship handling is less efficient in shallow water than in deep water.
31. Communication problems existed, producing human error, or as the Marine Inquiry Judge quoted, 'sloppy seamanship'.

Suggested Changes in Ship Construction and Operation after the Herald of Free Enterprise *Capsize*

1. Close bow and stern doors before leaving port.
2. Fit or retrofit longitudinal and transverse bulkheads.

3. Fit polythene drums or spheres to reduce the permeability in compartments that are most likely to be flooded.
4. Increase breadth molded for new ro-ro ferries.
5. Eliminate the entry of cars through the bow. Only have entrance and departure of cars through stern doors.
6. Install side-loading ramps for cargo and cars.
7. Improve escape facilities, for example using chutes or capsules.
8. Have all cars/lorries firmly fastened to deck and encased in an inflatable padding.
9. Whenever possible, ships leaving port must be upright and on even keel.
10. Fit television cameras to ensure stern and bow doors are closed before ship gets underway.
11. Decide upon maximum ship speeds within a certain distance of a port during departure and arrival maneuvers. This will counteract ship squat effects.
12. Fit a cut-out mechanism within the engine system, if stern or bow doors remain open.
13. Have larger GM values, say greater than 2.00 meters.
14. Have end draft indicators linking readings to the bridge.
15. Use hand-held instruments to count people and cars on and off ship for every voyage.
16. Fit high-pressure pump(s), to be used for quick transfer of water ballast or oil.
17. Improve onboard stability data to include conditions having trim by the stern and by the bow.
18. Reconsider all the working conditions of personnel to decrease fatigue and complacency.
19. Make officers and crew, via retraining courses, aware of the consequences of 'sloppy seamanship'.
20. Promote 'responsibility and accountability' attitude on board ship.
21. Pay utmost attention to ship speed, especially when the ship is in shallow waters. If the speed is halved then the squat is quartered.
22. Do not attempt to turn the ship at high speed when she is in shallow waters. Transverse squat could occur at the bilge plating.
23. If there is any danger of FSE with slack water tanks, then fill them up completely or totally empty them.
24. Do not request rudder helm if the ship is speeding in shallow waters.

Conclusions

To try to ascertain further the reasons why the *Herald of Free Enterprise* capsize took place, a sister ship, the *Pride of Free Enterprise*, had full-scale tests carried out. This was on 10 May 1987, about two months after the Zeebrugge incident. Ship-model simulation studies and mathematical modeling programs were also carried out.

The *Herald of Free Enterprise* was repaired and in fact using ship surgery methods was lengthened. She was renamed *Herald of Free Spirit*. Later still, she was again renamed, this time the *Flushing Range*. Her final voyage was to Alang in India. She was broken up for scrap in Alang in 1988.

Since the *Herald of Free Enterprise* capsize in 1987, following accidents/incidents involving ferries, over 3000 people have lost their lives.

In 1990, new SOLAS regulations stated that the freeboard for new ro-ro vessels should be raised from 76 cm (30 inches) to 125 cm (49 inches). This freeboard is the depth from the vehicle deck to the waterline at amidships.

References

No.	Prepared by:	Date	Title	Publisher	ISBN
1	Barrass C.B.	1999, 2006	Two editions – *Ship Stability for Masters and Mates*	Elsevier Ltd	978-0-7506-67845
2	Barrass C.B.	2001	*Ship Stability – Notes and Examples*	Elsevier Ltd	0 7506 4850 3
3	Barrass C.B.	1977	*A Unified Approach to Ship Squat*	Institute of Nautical Studies	NA
4	Barrass C.B.	2004	*Ship Design and Performance*	Elsevier Ltd	0 7506 6000 7
5	Cahill R.A.	2002	*Collisions and their Causes*	Institute of Nautical Studies	1 870077 60 1
6	Cahill R.A.	1985	*Strandings and their Causes*, 1st edition	Fairplay Publications	0 9050 4560 2
7	Cahill R.A.	2002	*Strandings and their Causes*, 2nd edition	Institute of Nautical Studies	1 870077 61 X
8	Carver A.	2002	*Simple Ship Stability – A Guide for Seafarers*	Fairplay Publications	1 870093 65 8
9	Clark I.C.	2005	*Ship Dynamics for Mariners*	Institute of Nautical Studies	1 870077 68 7
10	Dand I.W.	1988	*Hydrodynamic Aspects of the Sinking of HFE*	Royal Institution of Naval Architects	NA
11	Dibble W.J. and Mitchell P.	1994	*Draught Surveys – A Guide to Good Practice*	Mid-C Consultancy	0 9521164 1 3
12	Durham C.F.	1982	*Marine Surveys – An Introduction*	Fairplay Publications	0 905045 33 5
13	Gilardoni E.O.	2006	*Manejo del buque en aguas restringidas*	Instituto de Derecho Maritimo	978-987-20610-1-3
14	Gilardoni E.O.	2006	*Handling of a Ship in Restricted Waters*	Instituto de Derecho Maritimo	978-987-20610-1-3
15	House D.	2000	*Cargo Work*	Elsevier Ltd	0 7506 3988 1
16	IMO	2002	*Load Lines – 2002*	International Maritime Organization	92-801-5113-4
17	IMO	2004	*Solas Consolidated Edition – 2004*	International Maritime Organization	92-801-4183-X

(*Continued*)

: —cont'd

No.	Prepared by:	Date	Title	Publisher	ISBN
18	IMO	2002	*Code of Intact Stability for all Types of Ships*	International Maritime Organization	92-801-5117-7
19	IMO	1991	*International Grain Code*	International Maritime Organization	92-801-1275-9
20	Merchant Shipping Notice	1998	*M. S. (Load Line) Regulations*	M' Notice MSN 1752 (M)	NA
21	Molland A.F.	2008	*The Maritime Enmgineering Reference Book*	Elsevier Ltd	978-0-7506-8987-8
22	Munro-Smith R.	1975	*Elements of Ship Design*	Marine Media Management Ltd	0 900976 30 X
23	Rawson K.J. and Tupper E.C.	2001	*Basic Ship Theory*	Elsevier Ltd	0 7506 5397 3
24	RINA	2006–2011	*Significant Ships – Annual Publications*	Royal Institution of Naval Architects	NA
25	*Shipping Telegraph*	2009–2011	Technical articles re Maritime developments	Nautilus International	NA
26	Stokoe E.A.	1991	*Naval Architecture for Marine Engineeers*	Thomas Reed Ltd	0 94 7637 85 0
27	*The Naval Architect*	2006–2011	Monthly articles re Maritime developments	Royal Institution of Naval Architects	NA
28	Tupper E.C.	2004	*Introduction to Naval Architecture*	Elsevier Ltd	0 7506 6554 8
29	Van Dockum K. et al.	2010	*Ship Stability*	Dokmar, The Netherlands	978-90-71500-15-2

Answers to Exercises

Exercise 1

1. 7612.5 tonnes **2.** 4352 tonnes **3.** 1.016 m **4.** 7.361 m **5.** 0.4875 m **6.** 187.5 tonnes **7.** 13,721.3 tonnes **8.** 6406.25 tonnes **9.** 6.733 m F, 6.883 m A **10.** 228 tonnes **11.** 285 tonnes **12.** 8515 tonnes, 11,965 tonnes **13.** 27 mm **14.** 83.2 mm

Exercise 2

5. 2604 tonnes meters

Exercise 3

1-3 Review chapter notes

Exercise 4

1. (b) 3.78 tonnes, 4.42 tonnes, (c) 2.1 m **2.** (b) 6.46 tonnes, 7.8 tonnes, (c) 3.967 m **3.** 4.53 m **4.** (b) 920 tonnes, (c) 3.3 m, (d) 6.16 tonnes **5.** (a) 2.375 m, (b) 3092 tonnes, (c) 1125 tonnes **6.** (b) 3230 tonnes, (c) 1.625 m **7.** (b) 725 tonnes, (c) 4.48 m, (d) 5 tonnes **8.** (b) 5150 tonnes, 4.06 m, (c) 5.17 m

Exercise 5

1. (b) 12.3 tonnes **2.** (b) 8302.5 tonnes **3.** 12,681.3 tonnes **4.** 221 tonnes **5.** 180 tonnes

Exercise 7

1. 508 m^2, 5.2 tonnes, 0.8 m aft of amidships **2.** 488 m^2, 5 tonnes, 0.865, 0.86 m aft of amidships **3.** 122 mm, 6.563 m forward of amidships **4.** 30,476.7 tonnes, 371.4 mm, 15.6 m

5. 5062.5 tonnes **6.** 978.3 m^2, 15.25 cm, 2.03 m aft of amidships **7.** 9993.75 tonnes, 97.44 mm, 4.33 m **8.** 671.83 m^2, 1.57 m aft of amidships **9.** 12.125 m^2 **10.** 101 m^2 **11.** 1513 m^3, 8.01 tonnes **12.** 2893.33 m^3 or 2965.6 tonnes, 3 m

Exercise 8

1. 0.707 **2.** $8a^4/3$ **3.** 9:16 **4.** 63,281 cm^4 **5.** 3154 m^4, 28,283 m^4 **6.** 18,086 m^4, 871,106 m^4 **7.** BM_L 206.9 m, BM_T 8.45 m **8.** I_{CL} 35,028 m^4, I_{CF} 1,101,540 m^4 **9.** I_{CL} 20,267 m^4, I_{CF} 795,417 m^4 **10.** I_{CL} 13,227 m^4, I_{CF} 396,187 m^4

Exercise 9

1. 658.8 tonnes, 4.74 m **2.** 17.17 m **3.** 2.88 m, 313.9 tonnes **4.** 309.1 tonnes, 1.74 m **5.** 5.22 m

Exercise 10

1. 5 m **2.** 1.28 m, 4.56 m **3.** 1.78 m, 3 m **4.** No, unstable when upright **5.** (a) 6.2 m, 13.78 m, (b) 4.9 m **6.** (a) 10.6 m, 5.13 m, (b) 4.9 m at 4.9 m draft **7.** (a) 6.31 m, 4.11 m, (b) 4.08 m **8.** (b) $GM = +1.8$ m, so this vessel is in stable equilibrium, (c) GM is zero, $KG = KM$, so ship is in neutral equilibrium

Exercise 11

1. 2.84 m **2.** 3.03 m **3.** 3.85 m **4.** 5.44 m **5.** 0.063 m **6.** 1466.67 tonnes **7.** 1525 tonnes **8.** 7031.3 tonnes in L.H. and 2568.7 tonnes in T.D. **9.** 1.2 m **10.** 1.3 m **11.** 55 tonnes **12.** 286.3 tonnes **13.** 1929.67 tonnes

Exercise 12

1. 6° 03′ to starboard **2.** 4.2 m **3.** 216.5 tonnes to port and 183.5 tonnes to starboard **4.** 9° 30′ **5.** 5.458 m **6.** 12° 57′ **7.** 91.9 tonnes **8.** 282.75 tonnes to port, 217.25 tonnes to starboard **9.** 8.52 m to port, $GM = 0.864$ m **10.** 14° 04′ to port **11.** 13° 24′ **12.** 50 tonnes

Exercise 13

1. 4.62 m **2.** 6.08 m **3.** 0.25 m **4.** 3° 48′

Exercise 14

1. 3.6° **2.** 11.57° **3.** 5.7° **4.** 4.9° **5.** 8°

Exercise 15

1. 4° 34′ **2.** 6° 04′ **3.** 5° 44′ **4.** 6° 48′ to starboard

Exercise 16

1. No, 35° 49′ **2.** 39° 14′ **3.** Probable cause of the list is a negative GM **4.** Discharge timber from the high side first

Exercise 17

1. 674.5 tonnes meters **2.** 7.773 m **3.** 546.2 tonnes.m **4.** 6.027 m, 2000 tonnes meters **5.** (a) 83.43 tonnes meters, (b) 404.4 tonnes meters **6.** (a) 261.6 tonnes meters, (b) 2647 tonnes meters **7.** 139.5 tonnes meters, 1366 tonnes meters **8.** 0.522 m **9.** (b) Angle of loll is 14.96°, KM is 2.62 m, GM is −0.05 m

Exercise 18

1. 218.4 tonnes in No. 1 and 131.6 tonnes in No. 4 **2.** 176.92 tonnes **3.** 5.342 m A, 5.152 m F **4.** 6.726 m A, 6.162 m F **5.** 668.4 tonnes from No. 1 and 1331.6 tonnes from No. 4 **6.** 266.7 tonnes **7.** 24.4 cm **8.** 380 tonnes, 6.56 m F

Exercise 19

1. 42.9 tonnes in No. 1 and 457.1 tonnes in No. 4, GM is 0.79 m **2.** 402.1 tonnes from No. 1 and 47.9 tonnes from No. 4 **3.** 4.340 m A, 3.118 m F **4.** 5.56 m A, 5.50 m F **5.** 5.901 m A, 5.679 m F **6.** 4 meters aft **7.** 3.78 meters aft **8.** 4.44 meters aft

Exercise 20

1. 55.556 meters forward **2.** 276.75 tonnes, 13.6 meters forward **3.** 300 tonnes, 6.3 mF **4.** 200 tonnes, 7.6 mF **5.** 405 tonnes in No. 1 and 195 tonnes in No. 4 **6.** 214.3 tonnes **7.** 215.4 tonnes, 5.96 m F **8.** 200 meters **9.** 240 meters **10.** 8.23 m A, 7.79 m F, trim by the stern is 0.44 m, Dwt is 9195 tonnes

Exercise 21

1. GM is 2 m, range is 0−84.5°, max GZ is 2.5 m at 43.5° heel **2.** GM is 4.8 m, max moment is 67,860 tonnes meters at 42.25° heel, range is 0−81.5°, **3.** GM is 3.07 m, max GZ is 2.43 m at

41° heel, range is 0–76°, moment at 10° is 16,055 tonnes meters, moment at 40° is 59,774 tonnes meters **4.** (b) Moment at 10° is 16,625 tonnes meters, GM is 2 m, max GZ is 2.3 m at 42° heel, range is 0–82° **5.** GM is 3.4 m, range is 0–89.5°, max GZ is 1.93 m at 42° heel **6.** (a) 0–95°, (b) 95°, (c) 3.18 m at 47.5°, **7.** (a) 0–75°, (b) 75°, (c) 2.15 m at 40°, **8.** (a) 1.60 m, (b) 40°, (c) 12°, (d) 72.5° **9.** (a) 1.43 m, (b) 39.5°, (c) 44,000 tonnes meters, (d) 70°

Exercise 23

1. 1344 m tonnes **2.** 2038.4 m tonnes **3.** 1424 m tonnes **4.** 13.67 m tonnes **5.** 107.2 m tonnes

Exercise 25

1. 6.61 m **2.** 8.37m **3.** 25.86 m forward **4.** 39,600 tonnes, 513 tonnes m, 40.5 tonnes, 9.15 m **5.** 4.53 m aft, 3.15 m forward

Exercise 27

1. 1.19 m **2.** 1.06 m **3.** 0.64 m **4.** 7 m **5.** 4.62 m

Exercise 28

1. Transfer 41.94 tonnes from starboard to port and 135 tonnes from forward to aft **2.** Transfer 125 tonnes from forward to aft and 61.25 tonnes from port to starboard. Final distribution: No. 1 port 106.9 tonnes, No. 1 starboard 168.1 tonnes, No. 4 port 31.90 tonnes, No. 4. starboard 93.1 tonnes **3.** 13° 52′, 3.88 m F, 4.30 m A **4.** Transfer 133.33 tonnes from each side of No. 5. Put 141.6 tonnes into No. 2 port and 125.1 tonnes into No. 2 starboard

Exercise 29

1. 0.148 m **2.** 0.431 m **3.** 1.522 m **4.** 7° 02′ **5.** Dep. GM is 0.842 m, arr. GM is 0.587 m **6.** 3° exactly **7.** 112.4 tonnes.m **8.** 6.15 m, 0.27 m

Exercise 30

1. (b) 5.12 m **2.** 0.222 m **3.** 0.225 m **4.** 0.558 m **5.** 0.105 m **6.** 0.109 m **7.** 0.129 m **8.** 3.55 m F, 2.01 m A **9.** 5.267 m A, 7.529 m F **10.** 2.96 m A, 6.25 m F **11.** 3.251 m A, 5.598 m F **12.** 4.859 m A, 5.305 m F

Exercise 31

1 and 2. Review chapter notes.

Exercise 32

1 and 2. Review chapter notes.

Exercise 34

1. Review chapter notes. **2.** 0.734 **3.** Review chapter notes. **4.** 53.82

Exercise 35

1–4. Review chapter notes **5.** 556 mm

Exercise 36

1 and 2. Review chapter notes **3.** 551 mm **4.** Review chapter notes

Exercise 37

1. (b) 11.5° **2.** Review chapter notes **3.** 0.240 m, 0.192 m, 0.216 m **4.** Review chapter notes

Exercise 38

1. 33 tonnes **2.** 514 tonnes **3.** 639 tonnes **4.** 146.3 tonnes, sag 0.11 m, 0.843 m

Exercise 39

1. 15.6 cm **2.** 1962 tonnes, 4.087 m **3.** 4.576 m **4.** 5000 tonnes **5.** 39.1 cm **6.** 10.67 m **7.** 2.92 m **8.** 4.24 m **9.** (a) 0.48 m, 1.13°, (b) 8050 tonnes, 8.51 m

Exercise 40

1. 1.19 m **2.** 69.12 tonnes, 1.6 m

Exercise 41

1. −0.2 m or −0.25 m **2.** +0.367 m or +0.385 m **3.** +0.541 m or +0.573 m, safe to drydock vessel **4.** +0.550 m or +0.564 m, safe to drydock vessel, **5.** Maximum trim 0.896 m or 0.938 m by the stern

Exercise 42

1. (a) 366.5 m, (b) 0.100, (c) 0.65 m at the stern, (d) 0.85 m **2.** (a) W of I is 418 m as it is >350 m, (b) Review your notes, (c) 9.05 knots **3.** 10.56 knots **4.** 0.50 m at the stern using detailed formula and 0.53 m using 'K' format

Exercise 43

1. (a) 1.41, (b) 1.60, (c) 2.00 **2.** 1° 38′

Exercise 44

1. Passenger liner (0.78 m), general cargo ship (0.66 m) **2.** Review chapter notes **3.** Review chapter notes

Exercise 45

1. 20.05 seconds **2.** 15.6 seconds **3.** 15.87 seconds **4.** T_R is 26.35 seconds – a 'tender ship' **5.** 24, 7.37, and 6.70 seconds

Exercise 46

1 and **2.** Review chapter notes

Exercise 47

1. 8.20 m A, 4.10 m F **2.** 4.39 m A, 2.44 m F **3.** 6.91 m A, 6.87 m F

Exercise 50

1. 420 tonnes

Exercise 51

1. (a) 10,745 t, (b) 14,576 t, (c) 8.03 m, (d) 6.04 m, (e) 1.03 m

Exercise 55

1. Review chapter notes

General Particulars of Selected Merchant Ships, Delivered 2007–2011

Delivery Date	Name of Vessel	Cargo Gas (m^3)	LBP (L) (m)	Br. Mld (B) (m)	L/B Value	Depth (D) (m)	Draft (H) (m)	H/D Value	Speed (knots)
			Gas Carriers						
Feb 2010	*Barcelona Knutsen*	173,650	279.0	45.80	6.09	26.50	11.95	0.45	19.00
May 2009	*Express*	151,000	280.00	43.40	6.45	26.00	11.60	0.45	19.20
Aug 2008	*Navigation Aries*	20,664	152.2	25.60	5.95	16.40	8.30	0.51	16.00
July 2007	*British Emerald*	155,000	275.0	44.20	6.22	26.00	11.47	0.44	20.00
		Deadweight (tonnes)							
			ULCCs						
March 2002	*TI Europe* Largest	441,893	366.0	68.00	5.38	34.00	24.50	0.72	16.40
1980–2007	*Knock Nevis* Breaker's yard	564,769	440.0	68.80	6.40	29.80	24.61	0.83	13.00
			Passenger Liners/Cruise Ships						
1 Oct 2010	*Allure of the Seas* Largest	25,000	330.0	47.00	7.02	22.5	9.30	0.41	22.60
Sept 2010	*Queen Elizabeth*	8920	265.2	32.25	8.22	10.80	8.00	0.74	22.30
June 2010	*Niew Amsterdam*	7631	254.0	32.25	7.88	10.80	7.85	0.73	22.70
Sept 2009	*Carnival Dream*	10,250	269.2	37.20	7.24	11.20	8.20	0.73	20.00
June 2011	*Celebrity Silhouette*	12,074	294.0	36.80	7.99	11.30	8.60	0.76	24.00
			VLCCs						
Oct 2010	*Dar Salwa*	293,046	320.0	60.00	5.33	30.50	21.00	0.69	16.20
Feb 2009	*C Galaxy*	289,100	324.0	60.00	5.40	29.60	20.50	0.69	16.00
April 2008	*Sepid*	151,890	264.0	50.00	5.28	24.40	16.00	0.66	15.40
Aug 2006	*Promitheas*	117,050	239.0	44.00	5.43	22.70	15.40	0.68	15.10
Aug 2011	*Samco Amazon*		333.0	60.00		30.00			16.00
			Bulk Carriers						
Mar 2010	*E.R. Borneo*	197,842	283.5	45.00	6.30	24.70	16.50	0.67	15.40
May 2011	*Vale Brasil* Largest	400,000	347.5	65.00	5.35	30.40	23.00	0.76	15.00
Jan 2010	*Ocean Garnet*	73,795	222.0	38.00	5.84	20.70	12.50	0.60	14.35
Nov 2008	*Mariloula*	180,000	283.0	45.00	6.29	24.70	16.50	0.67	15.40
			Cargo Ships						
June 2009	*Star Kirkenes*	47,914	197.4	32.20	6.13	19.50	12.34	0.63	16.00
May 2009	*Corella Arrow* Largest	72,863	216.0	32.25	6.70	20.56	14.42	0.70	15.50
Dec 2008	*STX Rose 1*	16,715	165.0	40.00	4.13	8.50	5.00	0.59	11.70

(*Continued*)

: —cont'd

Delivery Date	Name of Vessel	Cargo Gas (m^3)	LBP (L) (m)	Br. Mld (B) (m)	L/B Value	Depth (D) (m)	Draft (H) (m)	H/D Value	Speed (knots)
Mar 2007	*Eileen C*	5000	84.7	14.50	5.84	7.35	6.30	0.86	13.00
XXX	*S020 (3rd generation)*	19,684	152.5	22.80	6.69	12.70	9.20	0.72	15.00
			Chemical Carriers						
Mar 2010	*Bunga Bakawali*	39,589	174.0	32.00	5.44	18.90	11.00	0.58	15.00
Mar 2009	*Emmy Schulte*	16,500	135.6	23.00	5.90	12.52	8.80	0.70	13.50
Jan 2007	*Overseas Houston*	40,600	174.0	32.20	5.40	18.80	11.00	0.59	14.60
Oct 2006	*Marida Boreas*	15,000	139.3	22.00	6.33	12.95	8.50	0.66	15.50
Aug 2011	*Samco Amazon*		333.0	60.00		30.00			16.00
			Container Ships						
Mar 2010	*Frisia Bonn*	22,860	178.0	16.50	10.79	27.60	10.00	0.36	21.00
Mar 2009	*M S C Beatrice*	13,840	350.0	50.20	6.97	29.90	14.50	0.48	24.30
Aug 2006	*Emma Maersk* Large	156,907	381.0	56.00	6.80	30.00	14.00	0.47	25.00
For 2013 −2015	*Triple 'E'* Largest	165,000	384.0	59.00	6.51	NG	14.50	NK	23.00
			Roll-on/Roll-off Vessels						
April 2010	*Grande Marocco*	23,820	196.8	32.26	6.10	13.51	9.40	0.70	20.90
May 2010	*Stena Hollandica*	11,600	230.0	32.00	7.19	7.95	6.40	0.81	22.00
Dec 2008	*City of Hamburg*	3500	117.0	20.60	5.68	6.65	5.50	0.83	15.00
April 2007	*Star*	4700	170.0	27.70	6.14	9.50	6.50	0.68	27.00
Jan 2011	*Spirit of Britain*		210.0	31.40			6.50		22.0
Mar 2011	*Towsberg* Largest		265.0	32.00			8.14		17.9

Reference sources:
Significant Ships of 2006 to 2010, published annually by RINA, London.
Data published in *Nautilus Shipping Telegraph* up to September 2011.
XXX as per *Standard General Cargo Ships* (Fairplay Publications).

Nomenclature of Ship Terms

aft terminal	The aftmost point on a ship's floodable length.
ahm	Actual volumetric heeling moments of grain in a compartment.
air draft	The vertical distance from the waterline to the highest point on the ship.
amidships	A position midway between the aft perpendicular (AP) and the forward perpendicular (FP).
angle of list	An angle of heel where G is transversely offset from the ship's centerline and transverse GM is positive.
angle of loll	An angle of heel where G is on the ship's centerline and transverse GM is zero.
angle of repose	An angle created by a shift of grain.
apparent trim	The difference between the drafts observed at the forward and aft draft marks.
appendage	A small attachment to a main area or body.
assigned freeboard	The ship's freeboard, after corrections have been made to the DfT Tabular Freeboard value.
blockage factor	Area of ship's midship section divided by the cross-sectional area of a river or canal.
boot-topping	The vertical distance between the lightdraft and the SLWL.
bow height	A vertical distance measured at the FP, from the waterline to the uppermost deck exposed to the weather.
bulbous bow	A type of bow to give a ship extra forward speed, or for the same speed less fuel consumption.
bulk carriers	Workhorse vessels, built to carry such cargoes as ore, coal, grain, and sugar in large quantities.
calibrated tank	A tank giving volumes or weights at vertical increments of say 1 cm in the tank's depth.
cargo-passenger ship	A vessel that carries cargo and up to 12 paying passengers.
C_b	Block coefficient: linking the volume of displacement with LBP, breadth Mld, and draft.
channel trench	A trench at the base of a navigational channel to allow transit of deeper draft vessels.
computer packages	Packages for estimating stability, trim, end drafts, shear forces, and bending moments for a condition of loading.
confined channel	A river or canal where there is a nearby presence of banks.
C_w	Waterplane area coefficient: linking the waterplane area with the LBP and the ship's breadth Mld.
deadweight	The weight that a ship carries.
deck camber	Transverse curvature of a deck, measured from deck height at centerline to deck height at side, at amidships.

(*Continued*)

: —cont'd

deck sheer	Longitudinal curvature of a deck, measured vertically from amidships to the deck, at aft perp or forward perp.
depth molded	Measured from top of keel to underside of uppermost continuous deck, at amidships.
DfT	Department for Transport.
displacement	For all conditions of loading, it is the lightweight plus any deadweight.
domain of ship	Mainly the area in which the pressure bulbs exist on a moving vessel.
draft molded	Distance from the waterline to the top of keel, measured at amidships.
dwt	Abbreviation for deadweight, the weight that a ship carries.
dynamical stability	The area under a statical stability curve multiplied by the ship's displacement.
even keel	A vessel with no trim: where the aft draft has the same value as the forward draft.
F and A drafts	Forward and aft drafts.
factor of subdivision F_S	The floodable length ordinates times F_S gives the permissible length ordinates.
F_B	Breadth of influence in open water conditions.
FLNG	Floating liquefied natural gas platform for storing LNG (acting like a warehouse).
foap	Forward of the aft perpendicular.
forward terminal	The foremost point on a ship's floodable length.
freeboard ratio	Freeboard to the margin line at amidships/ship's draft molded.
free-surface effects	Loss of stability caused by liquids in partially filled tanks.
general particulars	LBP, breadth mld, depth mld, draft Mld, lightweight, deadweight, displacement, C_b, service speed, etc.
grain cargo	This covers wheat, maize, corn, oats, rye, barley, seeds, rice, etc.
grain heeling moments	Transverse moments caused by the transverse movement of grain in a compartment.
heaving motion	The vertical movement of a ship's VCG.
hydrostatic curves	Used for calculating the trim and stability values for various conditions of loading.
hydrostatic table	A tabular statement of values on hydrostatic curves.
icing allowances	Must be considered when dealing with loss of stability and change of trim.
IMO	International Maritime Organization.
inertia coefficient	Used for obtaining moments of inertias of waterplanes.
interaction	Action and reaction of ships when they get too close to one another or too close to a river bank.
knot	1852 m per hour.
lightship draft	Draft of ship when ship is empty, with a deadweight of zero.
lightweight	Weight of empty ship, with boilers topped up to working level.
LNG ships	Liquefied natural gas carrier with cargo at −161 °C.
LOA	Length overall, from the foremost part of the bow to the aftermost part of the stern.
longitudinal	Running from bow to stern in a fore and aft direction.
LPG ships	Liquefied petroleum gas carriers with cargo at −42 °C.
margin line	A line that is 75 mm below the bulkhead deck at side.

: —cont'd

MCA	Marine Coastguard Agency.
mean draft	Draft measured from waterline to keel, at a position immediately under the LCF.
moment	An area times a lever or a weight times a lever.
mpm	Maximum permissible moments of grain in a compartment.
NA	Data not available or not applicable.
NK	Data not known.
open water	A stretch of water where there are no adjacent river or canal banks.
P and S	Port and starboard.
Panamax vessel	A vessel having a breadth mld of 32.26 m.
parametric rolling	Additional rolling of a ship, caused by having a bluff stern interacting with a sharp streamlined bow form.
passenger-cargo ship	A vessel that carries cargo and more than 12 paying passengers.
passenger liners	Vessels traveling between definite ports, with timetabled departure and arrival dates.
permeability	The amount of water that can enter a bilged compartment, usually expressed as a percentage.
pitching motion	Vertical see-saw movement of a ship at the bow and at the stern.
Plimsoll disk	Center of this disk is in line with the SLWL and its center is spot on amidships.
point of contraflexure	Angle of heel at which the deck edge just becomes immersed.
port	Left side of a ship when looking forward.
pressure bulbs	Bulbs of pressure that build up around a moving vessel and disappear when vessel stops.
Qflex	Type of LNG vessel, about 210,000 cubic meters of gas capacity.
Qmax	Type of LNG vessel, about 260,000 cubic meters of gas capacity.
quasi statical stability	Almost a loss in statical stability.
retrofit	A structure later added or deducted after the delivery of the ship to the shipowner.
righting lever	A lever that will bring a stable ship back to the upright after being displaced by temporary external forces.
righting moment	A moment that will bring a stable ship back to the upright. Equals W $\times$ GZ tonnes m.
rolling motion	A ship roll from extreme port to extreme starboard and back again to extreme port.
ro-ro ships	Roll-roll vessels that carry cars/lorries and passengers.
SF	The stowage factor in a grain compartment.
shallow water	Where the depth of water reduces the ship speed and prop revs, increases squat, reduces rolling motions, etc.
sheer ratio	Deck sheer forward or aft/ship's draft molded.
shifting boards	Used to prevent a transverse shift of grain.
ship surgery	Lengthening, deepening, or widening a ship after cutting her transversely, along or longitudinally.
Simpson's rules	Used for calculating areas, moments, volumes, and inertias, without needing to use calculus.
SLWL	Summer load waterline similar to draft molded, in a density of water of 1.025 t/cubic meter.

(Continued)

: —cont'd

sounding	A vertical distance measured from the bottom of a tank to the surface of a liquid in the tank.
sounding pad	A steel pad at the base of a sounding pipe.
sounding pipe	A pipe of about 37.5 mm that can be used for taking soundings with a steel sounding tape.
sounding tube/glass	A vertical tube used for reading calibrated values of a liquid in a tank.
SQA	Scottish Qualifications Authority.
squat	Loss of underkeel clearance as a ship moves forward from rest.
St Lawrence Seaway max.	Where the ship's breadth mld is 23.8 m.
starboard	Right side of a ship when looking forward.
stiff ship	A vessel with a quick rapid rolling period.
stowage factor	Volume per unit weight for a grain cargo.
stringer plate	The line of deck plates nearest to the shearstrake or gunwhale plating.
summer freeboard	The minimum freeboard for a ship floating in water with a density of salt water of 1.025 t/cubic meter.
supertanker	Similar to a VLCC: having a dwt of 100,000–300,000 t.
synchronous rolling	The roll of the ship being in phase with the action of the waves.
tabular freeboard	Minimum summer freeboard for a DfT standard ship.
TCD	Turning circle diameter, generally about three to four times ship's LBP in deep water conditions.
tender ship	A vessel with a long slow lazy rolling period.
TEU	Tonnes equivalent unit, equal to a container of about 6.10 m length × 2.44 m breadth × 2.59 m depth.
timber loadlines	Marked on ships that carry timber on the main deck.
tonne	Equivalent to 1000 kg.
transverse	Running from port to starboard across the ship.
transverse squat	Squat at the bilge plating, caused by ships overtaking or passing in a river.
trim	The difference between the forward draft and the aft draft.
Triple 'E' ship	Latest designs of container vessels for Maerske Line, about 165,000 t dwt.
true trim	The difference between the drafts at forward perpendicular and after perpendicular.
Type 'A' ship	A ship that carries liquid in bulk.
Type 'B' ship	A ship other than a Type 'A' ship.
ukc	Underkeel clearance, being depth of water minus a ship's draft.
ULCC	Ultra large crude carrier, say over 300,000 t dwt.
ullage	Vertical distance from surface of a liquid in the tank to top of the sounding pipe or top of ullage plug.
VLCC	Very large crude carrier, say 100,000–300,000 t dwt.
wind heel moments	Moments caused by the wind on side of the ship causing ship to have angle of heel.

Index

Note: Page numbers with "f" denote figures; "t" tables

S

T

U

V

W

Z

Printed in the United States
By Bookmasters